The Finite Element Method

The Finite Element Method

From Theory to Practice

Patrick Ciarlet
Eric Lunéville

iSTE

WILEY

First published 2023 in Great Britain and the United States by ISTE Ltd and John Wiley & Sons, Inc.

ISTE Ltd
27-37 St George's Road
London SW19 4EU
UK

www.iste.co.uk

John Wiley & Sons, Inc.
111 River Street
Hoboken, NJ 07030
USA

www.wiley.com

Library of Congress Control Number: 2023933918

British Library Cataloguing-in-Publication Data
A CIP record for this book is available from the British Library
ISBN 978-1-78630-768-2

Contents

Preface

Numerical simulation is becoming a powerful means of investigation that plays a very important role in science and technology, second only to the classical experimental approaches. In physics, chemistry and mechanics, models based upon systems of partial differential equations are used frequently (*elastodynamics, acoustics, electromagnetism, aerodynamics, hydrodynamics, ab initio chemistry*, etc.). More recently, this type of modeling has started to appear in other fields such as biology and economics (finance, in particular). Apart from a few special cases, it is not possible to solve these systems of equations analytically, and therefore it is necessary to use approximation techniques instead. Semi-analytical techniques, probabilistic methods, finite difference methods, spectral methods and finite element methods are just a few principle ones. Since being introduced in the 1950s, the finite element method has undergone several developments and today is used in numerical simulation software. Any engineer who works in an environment where numerical simulation plays an important role will have encountered, and moreover should understand, its fundamental principles, up to the most recent refinements.

Similar to other approximation techniques, the finite element method must be employed while obeying certain rules, which include usage constraints in addition to various kinds of quality factors in particular. The aim of this book is to provide the necessary components, from the most theoretical to the most concrete, so that one can master these constraints and factors. With this in mind, we want to present the different aspects of the method: the mathematical framework, numerical analysis and efficient implementation, highlighting each's importance, all while providing the key aspects in addition to applying them to situations typically encountered in practice. This work is devoted to presenting the fundamental concepts of the finite element method applied to stationary elliptical problems. The body of the text presents the basic notions that one should understand completely, and some additional results are also provided. Exercises along with their answers can be found at the end of each chapter and in the appendix, to allow the reader to further their understanding and also introduce them to more advanced subjects.

Chapters 1 and 2 are devoted to studying *elliptical* problems and the *variational theory* of elliptical equations (in the context of the finite element method), which are covered in a rigorous functional framework and only require a basic knowledge of analysis. Most of the operational tools of functional analysis (Hilbert analysis and distribution theory) will be recalled to allow for an easier reading for those unfamiliar with these areas. The finite element method is then the subject of Chapters 3 and 4, for which we start by providing a concrete presentation of the method by applying it to a simple yet important example, before presenting the formal framework that allows for a wide range of finite elements to be constructed, with the main results of *error estimation* being discussed in detail. In Chapter 5, we will study the practical and algorithmic aspects of the method that are used in its *computer implementation*, for which *digital illustrations* made with MATLAB[1] are provided as examples. In Appendix 1, we will present some elementary – albeit essential – applications to the *resolution of linear systems* and, in particular, *sparse* linear systems that follow from the finite element approximation. Detailed answers to the exercises are provided in Appendix 2, and the definitions for the usual differential operators are in Appendix 3.

March 2023

1 MATLAB is a trademark of MathWorks, Inc.

1

Theoretical Aspects of Elliptic Equations

1.1. Some vocabulary

There are several steady-state physical systems (i.e. the state reached when $t \to \infty$), which are modeled using elliptic partial differential equations[1]. For example:

– **heat diffusion** can be modeled with the help of a second-order equation (derivatives of order 2):

$$\text{div }(k\nabla T) = f$$

where T denotes the temperature, k is the thermal conductivity and f is the superposition of heat sources and wells;

– the **elastic deformation** of an isotropic homogeneous linear body gives rise to the model:

$$\sum_{j=1}^{\mathrm{d}} \frac{\partial}{\partial x_j} \sigma_{ij}(\mathbf{u}) = 0,\ 1 \leq i \leq \mathrm{d}$$

where $\mathbf{u}$ denotes the displacement vector in $\mathbb{R}^{\mathrm{d}}$ (usually $\mathbb{R}^3$) and:

$$\sigma_{ij}(\mathbf{u}) = \lambda \left(\sum_{k=1}^{\mathrm{d}} \varepsilon_{kk}(\mathbf{u}) \right) \delta_{ij} + 2\mu\varepsilon_{ij}(\mathbf{u}), \quad \varepsilon_{ij}(\mathbf{u}) = \frac{1}{2}\left(\frac{\partial u_i}{\partial x_j} + \frac{\partial u_j}{\partial x_i} \right)$$

1 Many detailed examples can be found in Dautray and Lions (1990).

are the stress and deformation tensors, respectively, λ and μ are the Lamé parameters, and δ_{ij} is the Kronecker symbol;

– the **electrostatic equilibrium** in a dielectric medium is governed by the equation:

$$-\operatorname{div}(\varepsilon\nabla\varphi) = \rho$$

where φ denotes the electrostatic potential, ε is the electric permittivity of the material and ρ is the charge density;

– the **bending of beams** or **of elastic plates** is modeled with the help of a fourth-order differential operator:

$$-\Delta^2 u = f$$

where u denotes the displacement that is transverse to the beam or to the plate, $\Delta^2 u = \Delta(\Delta u)$ (Δ is the Laplacian of order 2) and f is the volume loading (from the effect of gravity, for example);

– the **propagation of acoustic waves** in a compressible and homogeneous medium, in the harmonic regime (i.e. there is a dependence on $\cos(\omega t)$ or on $\sin(\omega t)$), leads to the Helmholtz equation:

$$-\Delta p - \omega^2 p = f$$

where p denotes the acoustic pressure.

None of these equations have a characteristic curve. Consequently, any property and technique used to study and approximate these equations will be specific to each one, and will be very different from those used in the hyperbolic case (see, for example, Després (2010)). In order to understand the techniques specific to elliptic equations, we will focus on the standard and academic example of the **Laplace equation** (or the Laplace–Poisson equation):

$$-\Delta u = f \ \text{ in } \Omega \tag{1.1}$$

where Ω is an open and bounded subset[2] of $\mathbb{R}^d$ ($d \geq 1$), whose boundary we will denote by $\partial\Omega$ and which we will assume to be Lipschitz, or "sufficiently regular"[3], and where f is a function defined on Ω.

2 The case of open but unbounded subsets gives rise to further difficulties that come from boundary conditions located at infinity, which will not be discussed in this textbook.

3 These notions will be made precise in what will follow (see definitions 1.8 and 1.10).

After considering the boundary conditions that can be imposed on Laplace's equation, we will introduce the notion of a **classical solution** (i.e. twice differentiable solutions in the classical sense). We can easily prove some uniqueness results for these solutions, but it is very difficult to prove their existence. This is why we will introduce the **variational theory** of elliptic equations in detail, which will grant us access to **weak solutions** (functions whose derivatives are square-integrable in the distribution sense). Finally, the fundamental properties of weak solutions (positivity, the maximum principle and regularity) are covered in section 2.5.

1.1.1. *Boundary conditions for the Laplace equation*

It is clear that if v is a regular harmonic function (i.e. $\Delta v = 0,\ v \in \mathcal{C}^2(\Omega)$), then the function $u + v$ is again a solution of [1.1]. This is why we must consider boundary conditions on the boundary $\partial\Omega$ for equation [1.1] to have any chance of finding a unique solution.

Classically, there are three types of boundary conditions:

$$u = g \text{ on } \partial\Omega \quad \text{Dirichlet conditions} \qquad [1.2]$$

$$\frac{\partial u}{\partial n} = g \text{ on } \partial\Omega \quad \text{Neumann conditions} \qquad [1.3]$$

$$\frac{\partial u}{\partial n} + \lambda u = g \text{ on } \partial\Omega \quad \text{Fourier conditions} \qquad [1.4]$$

where $\mathbf{n}$ denotes the unit normal vector that points outwards from $\partial\Omega$, $\lambda \in \mathbb{R} \setminus \{0\}$, $\frac{\partial u}{\partial n} = \nabla u \cdot \mathbf{n}$, and g is a given function on $\partial\Omega$.

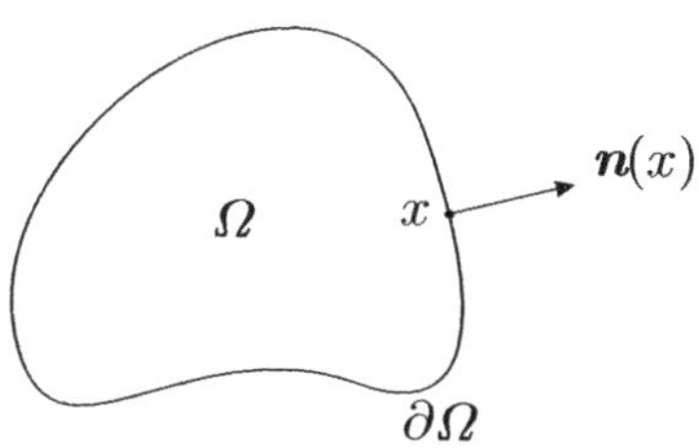

Figure 1.1. *The domain Ω along with a choice of outward normal*

One may interpret these conditions as thermal properties in the following way:

– Dirichlet conditions correspond to a fixed temperature along the boundary, for example $T = T_{ext}$;

– Neumann conditions can be interpreted as a thermal flux condition, with heat passing through $\partial\Omega$ ($g = 0$ corresponds to an isolated wall, $g > 0$ represents heat energy gained from passing inwards through the wall and $g < 0$ represents heat energy lost outwards through the wall);

– Fourier conditions correspond to energy losses that are proportional to the temperature difference if $\lambda > 0$, from:

$$\frac{\partial(T - T_{ext})}{\partial n} = -\lambda(T - T_{ext}) \quad \text{(Fourier's law)}.$$

After considering the boundary conditions [1.2]–[1.4], respectively, along with equation [1.1], we obtain a boundary problem that we denote by P_D (Dirichlet problem), P_N (Neumann problem) and P_F (Fourier problem), from now on. When the given g function vanishes (in either Dirichlet or Neumann problems), we say that the problem is a **homogeneous problem** in this case.

Furthermore, there are problems where the boundary conditions are of a different nature on different regions of the boundary, in which case we say that these are problems with **mixed-boundary conditions**. For example, the mixed Dirichlet–Neumann ($\lambda = 0$) or the mixed Dirichlet–Fourier ($\lambda \neq 0$) problems take on the following form:

$$\begin{cases} -\Delta u = f & \text{in } \Omega \\ u = g_1 & \text{on } \Gamma_1 \\ \dfrac{\partial u}{\partial n} + \lambda\, u = g_2 & \text{on } \Gamma_2 \end{cases} \qquad [1.5]$$

where Γ_1 and Γ_2 are two open subsets of $\partial\Omega$ with a non-zero measure, such that $\overline{\Gamma_1} \cup \overline{\Gamma_2} = \partial\Omega$ and $\Gamma_1 \cap \Gamma_2 = \emptyset$.

1.1.2. *Other boundary conditions*

Just as we have seen in the case of *second-order* elliptic equations (Δ, for example), *a boundary condition* on $\partial\Omega$ is necessary for the problem to be well-posed (i.e. its solution is unique and exists) whereas, for *fourth-order* equations (Δ^2 for example), *two boundary conditions* must be imposed on $\partial\Omega$ for the problem to be well-posed, for example:

$$u = g_1 \text{ and } \frac{\partial u}{\partial n} = g_2 \qquad \text{on } \partial\Omega.$$

However, other conditions are equally possible ($\dfrac{\partial}{\partial n}(\Delta u) = g$, for example). Their treatment is similar to the case that we are going to study.

Note that we often encounter boundary conditions which are said to be "oblique", for example in $\mathbb{R}^2$:

$$\alpha\frac{\partial u}{\partial n} + \beta\frac{\partial u}{\partial \tau} = g$$

where τ denotes the unit tangent vector to the boundary. The analysis of equation [1.1] along with this boundary condition is delicate, and will not be considered here (see Grisvard 1985, 1992).

Finally, boundary conditions with an order that is greater than 1 for equation [1.1] (e.g. $\frac{\partial^2 u}{\partial \tau^2} = g$ on $\partial\Omega$) leads to ill-posed problems (i.e. non-existence), apart from in a few particular cases.

In section 1.2, we shall present some elementary results regarding classical solutions. The variational theory of elliptic equations makes up the topic of sections 1.3 and 2.3.

1.2. Classical solutions to Laplace's equation

Let us begin by stating concisely just what is meant by a classical solution (i.e. differentiable in the classical sense) for the Dirichlet, Neumann, and Fourier problems. So that the boundary conditions make sense, the solution has to satisfy some regularity properties along the boundary $\partial\Omega$.

DEFINITION 1.1.– *We say that a classical solution to the Dirichlet problem* (P_D) *is any function* $u \in \mathcal{C}^2(\Omega) \cap \mathcal{C}^0\left(\overline{\Omega}\right)$ *satisfying* (P_D).

DEFINITION 1.2.– *We say that a classical solution to the Neumann problem* (P_N) *(respectively, Fourier* (P_F)*) is any function* $u \in \mathcal{C}^2(\Omega) \cap \mathcal{C}^1\left(\overline{\Omega}\right)$ *satisfying* (P_N) *(respectively,* (P_F)*).*

Proofs for the existence and uniqueness for classical solutions depend on maximum principles, which are linked to properties of harmonic functions.

1.2.1. *Green's formulae*

The main properties of harmonic functions are derived from Green's formulae, which are presented next[4] and are valid on some open subset Ω of $\mathbb{R}^d$, which may or may not be bounded, and are subject to some regularity assumptions on the boundary $\partial\Omega$:

4 See the formula in Appendix 3 for a reminder of how differential operators are defined.

PROPOSITION 1.1 (Green's formulae).–

– Let $\mathbf{w} \in \left(\mathcal{C}^1(\Omega) \cap \mathcal{C}^0\left(\overline{\Omega}\right)\right)^{\mathrm{d}}$ *and* $v \in \mathcal{C}^1(\Omega) \cap \mathcal{C}^0\left(\overline{\Omega}\right)$ *with bounded support, then we have that:*

$$\int_\Omega (\operatorname{div} \mathbf{w})\, v\, d\Omega = -\int_\Omega \mathbf{w} \cdot \nabla v\, d\Omega + \int_{\partial\Omega} \mathbf{w} \cdot \mathbf{n}\, v\, d\Gamma. \qquad [1.6]$$

– Let $u, v \in \mathcal{C}^1(\Omega) \cap \mathcal{C}^0\left(\overline{\Omega}\right)$ *with bounded support, then we with for each* $i = 1$ *to* d *(*n_i *: ith component of* $\mathbf{n}$*):*

$$\int_\Omega \frac{\partial u}{\partial x_i}\, v\, d\Omega = -\int_\Omega u\, \frac{\partial v}{\partial x_i}\, d\Omega + \int_{\partial\Omega} u\, v\, n_i\, d\Gamma. \qquad [1.7]$$

– Let $u \in \mathcal{C}^2(\Omega) \cap \mathcal{C}^1\left(\overline{\Omega}\right)$ *and* $v \in \mathcal{C}^1(\Omega) \cap \mathcal{C}^0\left(\overline{\Omega}\right)$ *with bounded support, then we have that:*

$$\int_\Omega \Delta u\, v\, d\Omega = -\int_\Omega \nabla u \cdot \nabla v\, d\Omega + \int_{\partial\Omega} \frac{\partial u}{\partial n} v\, d\Gamma. \qquad [1.8]$$

– Let $u, v \in \mathcal{C}^2(\Omega) \cap \mathcal{C}^1\left(\overline{\Omega}\right)$ *with bounded support, then we have that:*

$$\int_\Omega \Delta u\, v\, d\Omega = \int_\Omega u\, \Delta v\, d\Omega + \int_{\partial\Omega} \left(\frac{\partial u}{\partial n} v - \frac{\partial v}{\partial n} u\right) d\Gamma. \qquad [1.9]$$

Formula [1.6] is Stokes' fundamental formula, the proof of which is outside the scope of this textbook (see, for example, Abraham et al. 1988). Formula [1.7] is an immediate consequence of [1.6] (take $\mathbf{w} = u\mathbf{e}_i$; see also exercise 1). Formula [1.8], which is said to be Green's first formula, can be deduced from [1.6] by taking $\mathbf{w} = \nabla u$, and formula [1.9], said to be Green's second formula, trivially follows from [1.8]. These result are valid subject to the regularity assumptions related to the boundary.

Consider now the harmonic function $u \in \mathcal{C}^2(\Omega) \cap \mathcal{C}^0\left(\overline{\Omega}\right)$, that is, it satisfies:

$$\Delta u = 0 \quad \text{in } \Omega. \qquad [1.10]$$

We then have the following mean-value formulae

PROPOSITION 1.2 (Spherical mean formula).– *Let* $u \in \mathcal{C}^2(\Omega) \cap \mathcal{C}^0\left(\overline{\Omega}\right)$ *be a harmonic function on an open subset* Ω *of* $\mathbb{R}^{\mathrm{d}}$*. For each ball* B *centered at* y *and of radius* R *such that* $B \subset \Omega$*, we have that:*

$$u(y) = \frac{1}{\omega_{\mathrm{d}} R^{\mathrm{d}-1}} \int_{\partial B} u d\Gamma. \qquad [1.11]$$

$$u(y) = \frac{\mathrm{d}}{\omega_{\mathrm{d}} R^{\mathrm{d}}} \int_B u d\Omega \qquad [1.12]$$

where ω_{d} *denotes the measure (that is to say the surface) of the unit sphere in* $\mathbb{R}^{\mathrm{d}}$.

Proof. Let $0 \leq \rho \leq R$, then after applying Green's formula [1.8] to the ball B_ρ that is centered at y of radius ρ with $v = 1$:

$$\int_{\partial B_\rho} \frac{\partial u}{\partial n} d\Gamma = \int_{B_\rho} \Delta u \, d\Omega = 0.$$

Denote $r = |x - y|$ and $\omega = \dfrac{x - y}{r}$. For every x, we have: $u(x) = u(y + r\omega)$ from which we deduce that:

$$\begin{aligned}
\int_{\partial B_\rho} \frac{\partial u}{\partial n} d\Gamma &= \int_{\partial B_\rho} \frac{\partial u}{\partial r}(y + \rho\omega) \, d\Gamma \\
&= \int_{|\omega|=1} \frac{\partial u}{\partial r}(y + \rho\omega)\rho^{\mathrm{d}-1} d\omega \quad \text{change of variable } d\Gamma = \rho^{\mathrm{d}-1} d\omega) \\
&= \rho^{\mathrm{d}-1} \frac{\partial}{\partial \rho} \left(\int_{|\omega|=1} u(y + \rho\omega) \, d\omega \right) \qquad (\text{since } \frac{\partial}{\partial r} = \frac{\partial}{\partial \rho}) \\
0 &= \rho^{\mathrm{d}-1} \frac{\partial}{\partial \rho} \left(\rho^{1-\mathrm{d}} \int_{\partial B_\rho} u \, d\Gamma \right) \quad (\text{change of variable } d\omega = \rho^{1-\mathrm{d}} d\Gamma).
\end{aligned}$$

from which we derive that $\rho^{1-\mathrm{d}} \int_{\partial B_\rho} u \, d\Gamma = R^{1-\mathrm{d}} \int_{\partial B_R} u \, d\Gamma$. However, $\lim_{\rho \to 0} \rho^{1-\mathrm{d}} \int_{\partial B_\rho} u \, d\Gamma = \omega_{\mathrm{d}} u(y)$ since:

$$\begin{aligned}
\rho^{1-\mathrm{d}} \int_{\partial B_\rho} u \, d\Gamma &= \rho^{1-\mathrm{d}} \int_{\partial B_\rho} (u(y) + O(\rho)) \, d\Gamma \quad (\text{uniform continuity of } u \text{ on } B) \\
&= \rho^{1-\mathrm{d}} \left(\rho^{\mathrm{d}-1} u(y) \int_{|\omega|=1} d\omega + \rho^{\mathrm{d}-1} \rho \int_{|\omega|=1} O(1) \, d\omega \right) \\
&= \omega_{\mathrm{d}} u(y) + \rho \int_{|\omega|=1} O(1) \, d\omega \quad (\text{second term} \longrightarrow 0 \text{ when } \rho \longrightarrow 0)
\end{aligned}$$

which obtains the stated formula in [1.11].

To obtain [1.12], we integrate formula [1.11] for $0 \leq \rho \leq R$: $\omega_{\mathrm{d}} \rho^{\mathrm{d}-1} u(y) = \int_{\partial B_\rho} u \, d\Gamma$, which gives:

$$\int_0^R \omega_{\mathrm{d}} \rho^{\mathrm{d}-1} u(y) \, d\rho = \int_{\rho=0}^R \int_{\partial B_\rho} u \, d\Gamma d\rho.$$

Let: $\omega_d u(y) \int_0^R \rho^{d-1} d\rho = \int_{B_R} u \, d\Omega$, so that finally: $\omega_d u(y) \frac{R^d}{d} = \int_{B_R} u \, d\Omega$. ■

1.2.2. *The maximum principle*

A consequence of the spherical mean-value formula is the **maximum principle**. More precisely, we have:

PROPOSITION 1.3 (Maximum principle).– *Let $u \in C^2(\Omega) \cap C^0\left(\overline{\Omega}\right)$ be a harmonic function on a connected, bounded and open subset Ω. If the function u is non-constant on $\overline{\Omega}$, then we have that:*

$$\inf_{\partial\Omega} u < u(x) < \sup_{\partial\Omega} u \qquad \forall x \in \Omega. \tag{1.13}$$

In other words, either the maximum or the minimum of a non-constant harmonic function on a bounded and open subset Ω is attained on its boundary $\partial\Omega$. In one dimension, the only regular harmonic functions are those of the form $u(x) = ax + b$ and it is indeed clear that both the maximum and the minimum are attained at the ends of an interval.

Proof. Let us prove that $u < \sup_{\partial\Omega} u$, with the proof for the opposite inequality being similar. Suppose that there exists $y \in \Omega$ such that $u(y) = \sup_{\overline{\Omega}} u = M$, and consider the set:

$$\Omega_M = \{x \in \Omega \text{ such that } u(x) = M\}.$$

Note that $\Omega_M \neq \emptyset$ since $y \in \Omega_M$. Since u is continuous, the inverse image of $\{M\}$ under u, Ω_M, is a closed subset of Ω. Let $z \in \Omega_M$, and then by applying the spherical mean-value formula [1.11] to the function $u - M$ (indeed, we have that $\Delta(u - M) = 0$) in the ball B_R centered at z such that $B_R \subset \Omega$, we have that:

$$0 = u(z) - M = \frac{d}{\omega_d R^d} \int_{B_R} (u - M) d\Gamma.$$

However, $u - M \leq 0$ (by the definition of M), from which we see that $u = M$ on B_R for any R such that $B_R \subset \Omega$, and hence Ω_M is an open subset of Ω. Finally, since Ω is connected, we have that $\Omega_M = \Omega$, showing that u is constant on $\overline{\Omega}$ and thus contradicts our hypothesis. ■

The maximum principle allows us to prove a uniqueness result for Dirichlet problems in a straightforward way, which can be stated as:

THEOREM 1.1 (Uniqueness of the Dirichlet problem).– *The Dirichlet problem* (P_D) *admits at most one classical solution in* $\mathcal{C}^2(\Omega) \cap \mathcal{C}^0\left(\overline{\Omega}\right)$.

Proof. Suppose that (u_1, u_2) are two classical solutions to the problem (P_D), and $v = u_1 - u_2$. Then $v \in \mathcal{C}^2(\Omega) \cap \mathcal{C}^0\left(\overline{\Omega}\right)$ as well, and observe that:

$$\begin{cases} \Delta v = 0 \text{ in } \Omega \\ v = 0 \quad \text{on } \partial\Omega \end{cases}.$$

Since $\inf_{\partial\Omega} v = \sup_{\partial\Omega} v = 0$, we deduce from the maximum principle (proposition 1.3) that $v \equiv 0$ on $\overline{\Omega}$, and consequently $u_1 \equiv u_2$ on $\overline{\Omega}$. ■

The maximum principle [1.13] does not prove the uniqueness of classical solutions for either Fourier or Neumann problems, or even for mixed-problems involving the Dirichlet condition. Indeed, there is no simple method to estimate the minimum or the maximum to a solution along the boundary $\partial\Omega$. However, maximum principles that allow us to address these problems do exist (see Gilbarg and Trudinger 1977).

Let us point out an obvious fact of the Neumann problem: the solution, should it exist, is not unique. Indeed, if u is a solution of (P_N), then $\tilde{u} = u + \beta$, where β is an arbitrary real number; it is again a solution to the Neumann problem. In fact, one can show that the solution is unique up to a constant (see exercise 2).

The question regarding the existence of solutions to the Dirichlet, Neumann and Fourier problems is a difficult one. Indeed, regularity of the open subset Ω is a crucial factor and any "direct proofs" (Perron's method for example; see Gilbarg and Trudinger 1977) are all rather complex.

Furthermore, we will see that variational methods easily allow one to obtain results related to the existence of weak solutions, from which the regularity results for weak solutions then ensure the existence of classical ones.

1.3. Weak solutions

Consider a classical solution $u \in \mathcal{C}^2(\Omega) \cap \mathcal{C}^1\left(\overline{\Omega}\right)$, which is a solution to the homogeneous Dirichlet problem:

$$\begin{cases} -\Delta u = f \text{ in } \Omega \quad \text{with } f \in \mathcal{C}^0(\Omega) \\ u = 0 \qquad \text{on } \partial\Omega \end{cases}. \qquad [1.14]$$

and v is a **test function** belonging to $\mathcal{C}^1(\Omega) \cap \mathcal{C}^0\left(\overline{\Omega}\right)$.

From Green's formula [1.8], we get the following relation:

$$\int_{\Omega} \nabla u \cdot \nabla v \, d\Omega = \int_{\Omega} f v \, d\Omega + \int_{\partial\Omega} \frac{\partial u}{\partial n} v \, d\Gamma. \qquad [1.15]$$

On the other hand, if we choose v such that $v = 0$ on $\partial\Omega$, then [1.15] reduces to:

$$\int_{\Omega} \nabla u \cdot \nabla v d\Omega = \int_{\Omega} f v d\Omega. \qquad [1.16]$$

Finding such a function u, which vanishes along the boundary $\partial\Omega$ and satisfies [1.16] for *every test function* v *vanishing on* $\partial\Omega$, is called a **variational formulation** of the Dirichlet problem [1.14].

What are the advantages of such a formulation?

– On the one hand, we have *lowered the derivation order applied* to the function u. Thus, there may exist once-differentiable solutions u that satisfy [1.16].

– On the other hand, we can consider functions such that each one of their derivatives are square-integrable in the distribution sense, that is:

$$u \text{ measurable such that } \int_{\Omega} u^2 d\Omega + \int_{\Omega} |\nabla u|^2 d\Omega < +\infty, \qquad [1.17]$$

for which the formulation [1.16] is always well defined.

– If the variational formulation [1.16] admits such a solution, are they also classical solutions to the problem written in the form of a PDE [1.14]? Nothing guarantees this. Formally, in the sense of distributions, these solutions satisfy:

$$-\Delta u = f \quad \text{in } \Omega.$$

This is why we call these solutions **weak solutions** to the Dirichlet problem and the variational formulation [1.16] of the **weak formulation** the Dirichlet problem.

– The set of functions satisfying [1.17] defines a Hilbert space, while the set of functions C^1 only defines a Banach space. The structure of a Hilbert space is richer than that of Banach spaces (inner products, Hilbert projection theorem, Riesz's representation theorem, in particular all apply; see section 1.4). This will allow us to prove very generally the existence of weak solutions.

– In fact, this Hilbert space can be thought of as the space of energies for these solutions. Indeed, if we consider the energy functional associated with the function u which solves [1.16]:

$$E(u) = \frac{1}{2}\int_{\Omega} |\nabla u|^2 d\Omega - \int_{\Omega} fu\, d\Omega$$

then an elementary calculation shows that for every v satisfying [1.17] and such that $v = 0$ on $\partial\Omega$:

$$E(u+v) - E(u) = \int_{\Omega} \nabla u \cdot \nabla v\, d\Omega - \int_{\Omega} fv\, d\Omega + \frac{1}{2}\int_{\Omega} |\nabla v|^2 d\Omega$$

which therefore shows (Ciarlet 1989) that the functional E is differentiable with respect to u, with derivative:

$$dE(u).v = \int_{\Omega} \nabla u \cdot \nabla v\, d\Omega - \int_{\Omega} fv\, d\Omega.$$

In other words, solutions to the variational formulation are the functions that minimize the energy E within the space of energies. This physical interpretation agrees with the principle of virtual work, which is dear to many mechanicians.

– Finally, the variational formulation can be interpreted as a **pseudo-matrix system**. Indeed, if $(w_i)_{i\geq 0}$ is an "infinite basis" for the space in which we are searching for u, then, by taking the functions w_i to be test functions, [1.16] can be rewritten "formally" as:

$$\mathbb{A}U = F$$

with:

$$\mathbb{A}_{ij} = \int_{\Omega} \nabla w_j \cdot \nabla w_i d\Omega, \; u = \sum_j U_j w_j \text{ and } F_i = \int_{\Omega} f w_i d\Omega.$$

It is upon this idea that the finite element method is based.

1.4. Reminders on Banach and Hilbert spaces

We are going to emphasize these notions by recalling three fundamental theorems in the analysis of Hilbert spaces, which will serve us to define the spaces of energies and to justify the integration by parts formulae, in particular. For the proof of these results, we refer to Brezis (1983, 2011) and McLean (2000).

1.4.1. *Banach spaces*

We denote by X a **Banach space**, that is, a vector space endowed with a norm $\|\cdot\|_X$ and which is complete with respect to this norm: every Cauchy sequence in $X^{\mathbb{N}}$ is convergent.

By definition, two norms $\|\cdot\|_1$ and $\|\cdot\|_2$ of X are **equivalent** if there exist two constants[5] $C, C' > 0$ such that, for every $v \in X$, $C\,\|v\|_1 \leq \|v\|_2 \leq C'\,\|v\|_1$. In this case, the topologies induced by each of the two norms are identical. Concretely, the measures are comparable in the sense that, for each sequence $(v_k)_k \in X^{\mathbb{N}}$ and every element $v \in X$:

$$\lim_{k\to\infty} \|v_k - v\|_1 = 0 \quad \Longleftrightarrow \quad \lim_{k\to\infty} \|v_k - v\|_2 = 0.$$

We denote by X' the **dual** space of the space X, which is to say the space of continuous linear functionals on X. By definition, a linear functional $\ell(\cdot)$ on X is **continuous** if:

$$\sup_{v\in X\setminus\{0\}} \left(|\ell(v)|/\|v\|_X\right) < \infty.$$

The dual space is endowed with the canonical norm:

$$\|\ell\|_{X'} = \sup_{v\in X\setminus\{0\}} \frac{|\ell(v)|}{\|v\|_X}.$$

For $\ell(v)$, we generally use the notation provided by the **duality brackets** $\langle \ell, v\rangle_{X',X}$. Endowed with $\|\cdot\|_{X'}$, X' becomes a Banach space. When using this definition, we can check that if V and X are two Banach spaces such that V is dense in X with a continuous injection[6], then their respective dual spaces are such that $X' \subset V'$ with a continuous injection too.

---

5 Here and in what follows, we will denote the "generic" positive constants by C, C' (to lighten the notation). Hence, two instances of C may refer to two different constants.

6 That is to say that:

(density) $\forall v \in X$, $\exists (v_k)_k \in V^{\mathbb{N}}$ such that $\lim_{k\to\infty} \|v - v_k\|_X = 0$

(continuous injection) $\exists C > 0$ such that $\forall v \in V$, $\|v\|_X \leq C\,\|v\|_V$

The injective nature of the inclusion of X' into V' comes from the density of V in X (see exercise 4). The notation $X \subset Y$ for two topological vector spaces X and Y signifies that the injection from X into Y is *continuous*.

Let Y be another Banach space. A bilinear form $a(\cdot,\cdot)$ on $X \times Y$ is **continuous** if:

$$\text{(modulus of continuity)} \quad M_a = \sup_{v\in X\setminus\{0\},\, w\in Y\setminus\{0\}} \frac{|a(v,w)|}{\|v\|_X \|w\|_Y} < \infty.$$

A linear mapping A from X into Y is **continuous** if:

$$\sup_{v\in X\setminus\{0\}} \frac{\|Av\|_Y}{\|v\|_X} < \infty.$$

Its norm is:

$$|||A||| = \sup_{v\in X\setminus\{0\}} \frac{\|Av\|_Y}{\|v\|_X}.$$

We denote by $\mathcal{L}(X,Y)$ the set of continuous linear mappings from X into Y and by $\mathcal{L}(X)$ the set of continuous linear mappings from X into itself. A linear mapping A in $\mathcal{L}(X,Y)$ is **compact** if, for every bounded sequence in X, we can extract a subsequence whose image under A converges in Y. We can check that the composition of a continuous linear mapping with a compact linear mapping (or *vice versa*) is compact.

Let $A \in \mathcal{L}(X)$, then λ is an **eigenvalue** of A if there exists $v \in X \setminus \{0\}$ such that $Av = \lambda v$. We say that v is an **eigenvector**, and that $\dim(\{v \in X \text{ such that } Av = \lambda v\})$ is the **multiplicity** corresponding to λ.

Finally, the space X is **separable** if it contains a countable and dense subset. In this case, there exists a countable family of elements which makes up a basis of it.

1.4.2. *Hilbert spaces*

We denote by H a **Hilbert space**, that is, a vector space endowed with an inner product, and which is complete with respect to the norm induced by this inner product. In particular, every Hilbert space is a Banach space. We denote by $(\cdot,\cdot)_H$ the inner product, and $\|\cdot\|_H$ the associated norm.

We say that a **Hilbert basis** is a sequence of elements $(e_n)_{n\geq 1}$ of H such that:

i) $(e_n, e_m)_H = \delta_{nm}$ for each $n, m \geq 1$;

ii) the vector space generated by the finite linear combinations of the $(e_n)_{n\geq 1}$ is *dense* in H.

If $(e_n)_{n\geq 1}$ is a Hilbert basis, then, for every $f \in H$, we have:

$$f = \sum_{n\geq 1}(f, e_n)_H e_n, \qquad \text{with } \|f\|_H = \Big(\sum_{n\geq 1}((f, e_n)_H)^2\Big)^{1/2}.$$

If we introduce the sequence $(\tilde{f}_N)_{N\geq 1}$ with $\tilde{f}_N = \sum_{1\leq n\leq N}(f, e_n)_H e_n$ for $N \geq 1$, we have that $\lim_{N\to\infty} \|f - \tilde{f}_N\|_H = 0$. Conversely, to any sequence $(a_n)_{n\geq 1}$ in ℓ^2, we can associate $f = \sum_{n\geq 1} a_n e_n$ to it, which belongs to H with $(f, e_n)_H = a_n$ for each $n \geq 1$, and $\|f\|_H = (\sum_{n\geq 1} a_n^2)^{1/2}$.

Finally, when H is *separable*, we have seen that there exists a countable family of elements constituting a basis for H. We may then orthogonalize this basis to obtain a Hilbert basis for H.

Let $A \in \mathcal{L}(H)$: we say that A is **positive** if $(Af, f)_H \geq 0$ for every $f \in H$; we say that A is **self-adjoint** if $(Af, g)_H = (f, Ag)_H$ for every $f, g \in H$.

Two essential properties that Hilbert spaces possess are as follows: the existence of a projection onto every non-empty convex and closed subset, on the one hand, and the representation of a continuous linear functional by the inner product, on the other hand.

THEOREM 1.2 (Projection).– *Let K be a convex and closed non-empty subset of H. Then, for every $f \in H$, there exists a unique element of K, which we denote by $P_K f$, such that:*

$$\|f - P_K f\|_H = \min_{v\in K} \|f - v\|_H \qquad [1.18]$$

and is characterized by:

$$(f - P_K f, v - P_K f)_H \leq 0 \quad \forall v \in K. \qquad [1.19]$$

Furthermore, the mapping P_K is a contraction:

$$\|P_K f_1 - P_K f_2\|_H \leq \|f_1 - f_2\|_H \quad \forall f_1, f_2 \in H. \qquad [1.20]$$

REMARK 1.1.– *While K is a closed vector subspace of H, and if we set $J(u) = \|u - f\|_H^2$, condition [1.19] becomes $dJ(P_K f).v = 0\ \forall v \in K$, which connects the energy interpretation with the variational formulation. Moreover, $P_K f$ is characterized by: $P_K f \in K$ and $(f - P_K f, v)_H = 0$, $\forall v \in K$. Then we say that P_K is the* ***orthogonal projection*** *from H onto K.*

COROLLARY 1.1.– *Let K be a closed vector subspace of H, then there exists a decomposition $H = K \oplus K^{\perp}$, where $K^{\perp}$ is the closed vector subspace called the* **orthogonal** *subspace of K, and is defined by:*

$$K^{\perp} = \{f \in H \text{ such that } (f, v)_H = 0, \ \forall v \in K\}$$

and additionally $(K^{\perp})^{\perp} = K$.

Let us recall Riesz's representation theorem (or the Riesz–Fréchet theorem).

THEOREM 1.3 (Riesz's representation theorem).– *Let $\ell \in H'$, then there exists a unique element $f \in H$ such that:*

$$\ell(v) = (f, v)_H \qquad \forall v \in H$$

and that:

$$\|f\|_H = \|\ell\|_{H'}.$$

REMARK 1.2.– *Consequently, H and its dual H' are isometrically isomorphic. Indeed, we know that the linear form $\ell \in H'$ defined by $\ell(v) = (f, v)_H$, $\forall v \in H$ is such that $\|\ell\|_{H'} = \|f\|_H$, for every $f \in H$. Conversely, from Riesz's representation theorem, for every $\ell \in H'$, there exists a (unique) $f \in H$ such that $(f, v)_H = \ell(v)$, $\forall v \in H$ and $\|f\|_H = \|\ell\|_{H'}$. In finite dimension, this theorem is used naturally when we identify the differential operator with the gradient vector.*

Finally, the last result that is essential is a method used to obtain a Hilbert basis with the help of compact linear mappings.

THEOREM 1.4 (Spectral-I).– *Suppose that H is separable. Let $A \in \mathcal{L}(H)$ be self-adjoint and compact. Then:*

i) the non-zero *eigenvalues of A each have a finite multiplicity. Moreover, if the number of distinct eigenvalues is infinite, then they make up a sequence of real numbers that tend to 0;*

ii) there exists a Hilbert basis of H made up of the eigenvectors of A.

When the linear map A satisfies further properties, we have the following results.

THEOREM 1.5 (Spectral-II).– *Suppose that $dim(H) = +\infty$. Let $A \in \mathcal{L}(H)$ be self-adjoint, compact, injective and positive. Then:*

i) the eigenvalues of A all have finite multiplicity, and constitute a sequence of strictly positive real numbers tending toward 0;

ii) there exists a Hilbert basis of H made up of eigenvectors of A.

1.5. Sobolev spaces

In all of what follows, Ω denotes an open subset of $\mathbb{R}^{\mathrm{d}}$ with boundary $\partial\Omega$. We denote by $\mathcal{D}(\Omega)$ the subspace of functions $\mathcal{C}^{\infty}$ on Ω with *compact support* in Ω.

An element $\alpha = (\alpha_1, \cdots, \alpha_{\mathrm{d}})$ of $\mathbb{N}^{\mathrm{d}}$ is called a multi-index of length $|\alpha| = \sum_{j=1}^{\mathrm{d}} \alpha_j$, and the partial derivatives of order α are denoted by:

$$\partial_\alpha f = \frac{\partial^{|\alpha|} f}{\partial x_1^{\alpha_1} \cdots \partial x_{\mathrm{d}}^{\alpha_{\mathrm{d}}}}.$$

Let us begin with Lebesgue spaces of type L^p, where $1 \leq p \leq \infty$. We will only consider real-valued functions, but in what follows each definition can be extended to the case of complex-valued functions with $|.|$ as the modulus. Let $d\Omega = dx_1 dx_2 \cdots dx_{\mathrm{d}}$ be the Lebesgue measure on Ω, *considered as a subspace of* $\mathbb{R}^{\mathrm{d}}$.

DEFINITION 1.3.– *The space $L^p(\Omega)$ consists of functions f that are defined almost everywhere, measurable in the Lebesgue sense on Ω, and satisfy:*

$$\left| \begin{array}{lll} \textit{for } 1 \leq p < \infty & \|f\|_{L^p(\Omega)} = \left(\displaystyle\int_\Omega |f|^p \, d\Omega \right)^{1/p} < \infty \\ \textit{for } p = \infty & \|f\|_{L^\infty(\Omega)} = \textit{ess} \sup\limits_{x \in \Omega} |f(x)| < \infty. \end{array} \right. \qquad [1.21]$$

Endowed with the norm $\|\cdot\|_{L^p(\Omega)}$, $L^p(\Omega)$ becomes a Banach space. The space $L^p_{loc}(\Omega)$ consists of the functions f such that, for every measurable and compact subset K of Ω, f belongs to $L^p(K)$. We say that f locally belongs to $L^p(\Omega)$.

Let $p \in [1, \infty]$, then $f_1 = f_2$ in $L^p(\Omega)$ indicates that $f_1(x) = f_2(x)$ almost everywhere on Ω. Also recall that, if $f_1 \in L^p(\Omega)$ and $f_2 \in L^\infty(\Omega)$, then the product $f_1 f_2$ also belongs to $L^p(\Omega)$. We have the following density result.

PROPOSITION 1.4.– *For $1 \leq p < \infty$, $\mathcal{D}(\Omega)$ is a dense subspace in $L^p(\Omega)$. In other words:*

$$\forall f \in L^p(\Omega), \ \exists (f_k)_k \in \big(\mathcal{D}(\Omega)\big)^{\mathbb{N}}, \ \lim_{k \to \infty} \|f - f_k\|_{L^p(\Omega)} = 0.$$

Therefore, $L^2(\Omega)$ is the space of functions (that are defined almost everywhere on Ω), measurable in the Lebesgue sense and square-integrable on Ω. Endowed with the inner product[7]:

$$(u, v)_{L^2(\Omega)} = \int_\Omega uv \, d\Omega.$$

induced from the associated norm:

$$\|u\|_{L^2(\Omega)} = \Big(\int_\Omega u^2 d\Omega \Big)^{1/2}$$

and so $L^2(\Omega)$ is a Hilbert space. On the other hand, $\mathcal{D}(\Omega)$ is dense in $L^2(\Omega)$. Once again note that the equality $v = w$ in $L^2(\Omega)$ indicates that $v(x) = w(x)$ almost everywhere on Ω. In what follows, we *will identify* $L^2(\Omega)$ with its dual. To be precise, every continuous linear functional ℓ on $L^2(\Omega)$ is also an element of $L^2(\Omega)$, and conversely we will write:

$$\ell(v) = (\ell, v)_{L^2(\Omega)} = \int_\Omega \ell \, v \, d\Omega \quad \forall v \in L^2(\Omega).$$

1.5.1. *Differentiation in the distribution sense*

Denote by $\mathcal{D}'(\Omega)$ the space of continuous linear functionals on $\mathcal{D}(\Omega)$ that is endowed with the inductive limit topology, that is, the space of distributions on Ω (see, for example, Yosida 1980; Adams and Fournier 2003). In practice, we define the topology by considering sequences that *converge in the limit*; let $(f_k)_k$ be a sequence of elements in $\mathcal{D}(\Omega)$, then we say that it converges in $\mathcal{D}(\Omega)$ to f if, and only if:

i) there exists a compact subset K of Ω such that $\text{supp}(f_k) \subset K$, for each k; $\text{supp}(f_k - f) \subset K$, for each k;

ii) for each multi-index α, $(\partial_\alpha f_k)_k$ converges uniformly in K to $\partial_\alpha f$.

DEFINITION 1.4.– *A continuous linear functional T defined on $\mathcal{D}(\Omega)$ is called a distribution. The space of distributions is denoted by $\mathcal{D}'(\Omega)$. Let $T \in \mathcal{D}'(\Omega)$ and $f \in \mathcal{D}(\Omega)$: the pairing of T with f is written via the duality brackets, that is:*

$$\langle T, f \rangle.$$

7 As remarked above, the same results can be extended to functions with values in $\mathbb{C}$. Then we would use the Hermitian product $(u, v)_{L^2(\Omega)} = \int_\Omega u\bar{v} \, d\Omega$ induced from the associated norm $\|u\|_{L^2(\Omega)} = \left(\int_\Omega |u|^2 d\Omega\right)^{1/2}$, and with which $L^2(\Omega)$ is then a Hilbert space on $\mathbb{C}$.

From the definition of the topology on $\mathcal{D}(\Omega)$ in the sense of limits, T is continuous provided that:

$$\forall (f_k)_k, f \in \mathcal{D}(\Omega) \text{ such that } f_k \to f \text{ in } \mathcal{D}(\Omega), \quad \langle T, f_k \rangle \to \langle T, f \rangle.$$

Some examples are proposed in what will follow (see [1.23] and [1.27]). When we wish to consider whether a sequence of elements in $\mathcal{D}'(\Omega)$ converges, we will use the following definition.

DEFINITION 1.5.– *Let $(T_k)_k$ be a sequence of elements in $\mathcal{D}'(\Omega)$: it converges in $\mathcal{D}'(\Omega)$ to T if, and only if, for every f in $\mathcal{D}(\Omega)$, $\langle T_k, f \rangle \to \langle T, f \rangle$.*

With this in hand, we can easily prove the following inclusion $L^1_{loc}(\Omega) \subset \mathcal{D}'(\Omega)$, by *identifying* every element f in $L^1_{loc}(\Omega)$ with a distribution that is again denoted by f, via:

$$\langle f, g \rangle = \int_\Omega f\, g\, d\Omega \quad \forall g \in \mathcal{D}(\Omega). \qquad [1.22]$$

In particular, the identification [1.22] holds for every element f of $L^2(\Omega)$, and we have the following sequence of inclusions:

$$\mathcal{D}(\Omega) \subset L^2(\Omega) \subset \mathcal{D}'(\Omega). \qquad [1.23]$$

Let us recall a very useful property for later.

PROPOSITION 1.5.– *Let f_1 and f_2 be two elements in $L^1_{loc}(\Omega)$. The correspondence $\langle f_1, g \rangle = \langle f_2, g \rangle$ for every $g \in \mathcal{D}(\Omega)$ implies that $f_1(x) = f_2(x)$ almost everywhere in Ω.*

Now, let us come to differentiation in the sense of distributions.

DEFINITION 1.6.– *Let $T \in \mathcal{D}'(\Omega)$. Its jth partial derivative ($j = 1, \cdots, \mathrm{d}$) is defined by:*

$$\langle \frac{\partial T}{\partial x_j}, f \rangle = -\langle T, \frac{\partial f}{\partial x_j} \rangle \quad \forall f \in \mathcal{D}(\Omega).$$

Of course, the derivative of a distribution is once again a distribution.

PROPOSITION 1.6.– *The mapping $T \mapsto \partial_j T$ is continuous and linear from $\mathcal{D}'(\Omega)$ into $\mathcal{D}'(\Omega)$.*

Example of a derivative in the sense of distributions. Let Ω be a bounded and open set, partitioned into two disjoint open subsets Ω_1 and Ω_2 such that $\overline{\Omega} = \overline{\Omega_1} \cup \overline{\Omega_2}$ and with $\overline{\Omega_1} \cap \overline{\Omega_2} = F$ their interface. Consider a function $v \in L^2(\Omega)$ such that $v_j = v_{|\Omega_j} \in \mathcal{C}^1\left(\overline{\Omega_j}\right)$ for $j = 1, 2$.

By the definition of a derivative in the distribution sense, for $i = 1, \cdots, \mathrm{d}$ we have that:

$$\langle \frac{\partial v}{\partial x_i}, \varphi \rangle = -\langle v, \frac{\partial \varphi}{\partial x_i} \rangle = -\int_\Omega v \frac{\partial \varphi}{\partial x_i} d\Omega \quad \forall \varphi \in \mathcal{D}(\Omega).$$

Let:

$$\begin{aligned} \langle \frac{\partial v}{\partial x_i}, \varphi \rangle &= -\int_{\Omega_1} v_1 \frac{\partial \varphi}{\partial x_i} d\Omega - \int_{\Omega_2} v_2 \frac{\partial \varphi}{\partial x_i} d\Omega \\ (\text{[1.7] on } \Omega_j) &= \int_{\Omega_1} \frac{\partial v_1}{\partial x_i} \varphi \, d\Omega + \int_{\Omega_2} \frac{\partial v_2}{\partial x_i} \varphi \, d\Omega - \int_F \left(v_{1|F} \, (n_1)_i + v_{2|F} \, (n_2)_i \right) \varphi \, d\Gamma \\ (\text{since } \mathbf{n}_1 = -\mathbf{n}_2) &= \int_{\Omega_1} \frac{\partial v_1}{\partial x_i} \varphi \, d\Omega + \int_{\Omega_2} \frac{\partial v_2}{\partial x_i} \varphi \, d\Omega - \int_F \left(v_{1|F} - v_{2|F} \right) (n_1)_i \, \varphi \, d\Gamma. \end{aligned}$$

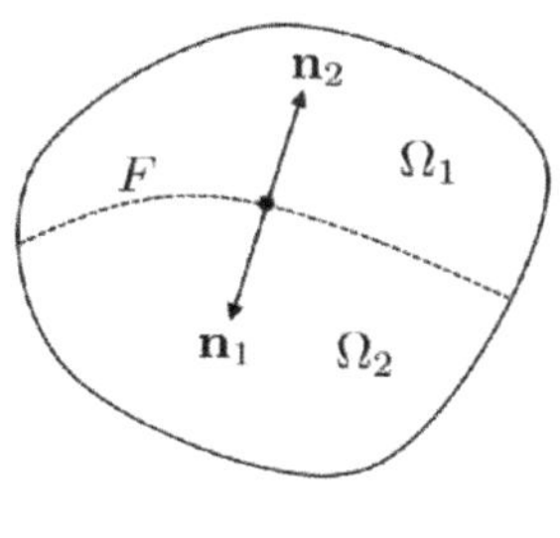

Figure 1.2. *Example of an interface*

Moreover, if $v \in \mathcal{C}^0\left(\overline{\Omega}\right)$, then the integral over the interface F vanishes, and we get:

$$\langle \frac{\partial v}{\partial x_i}, \varphi \rangle = \int_{\Omega_1} \frac{\partial v_1}{\partial x_i} \varphi \, d\Omega + \int_{\Omega_2} \frac{\partial v_2}{\partial x_i} \varphi \, d\Omega.$$

Therefore $(\partial_i v)_{|\Omega_1} = \partial_i v_1$ and $(\partial_i v)_{|\Omega_2} = \partial_i v_2$, and $\nabla v \in \left(L^2(\Omega)\right)^{\mathrm{d}}$ since $v_j \in \mathcal{C}^1(\overline{\Omega_j})$ for $j = 1, 2$.

However, if $v \notin \mathcal{C}^0\left(\overline{\Omega}\right)$, then v is not continuous when traversing over the interface F. In this case, the reasoning above shows that at least one of the derivatives $\partial_i v$ does not belong to $L^2(\Omega)$, due to the appearance of the term:

$$-\int_F \left(v_{1|F} - v_{2|F}\right)(n_1)_i \varphi \, d\Gamma$$

which is comparable to the Dirac measure along F where v is discontinuous and where $(n_1)_i \neq 0$ (see exercise 3).

Since $L^2(\Omega)$ is a subspace of $\mathcal{D}'(\Omega)$ (by identification, see [1.23]), we may differentiate its elements in the distribution sense (see also exercise 10). Consequently, this gives rise to notion of a Sobolev space:

$$H^1(\Omega) = \{v \in L^2(\Omega) \text{ such that } \nabla v \in \left(L^2(\Omega)\right)^{\mathrm{d}}\}$$

where the derivative $(\nabla\cdot)$ is taken in the distribution sense.

According to the example we have just seen, we can deduce the following result, which is a fundamental tool that is used in the finite element method.

PROPERTY 1.1.– *Let Ω be a bounded and open set, partitioned into two disjoint open subsets Ω_1 and Ω_2 ($\overline{\Omega} = \overline{\Omega_1} \cup \overline{\Omega_2}$). Let $v \in L^2(\Omega)$ be such that $v_j = v_{|\Omega_j} \in \mathcal{C}^1\left(\overline{\Omega_j}\right)$ for $j = 1, 2$: $v \in \mathcal{C}^0\left(\overline{\Omega}\right)$ if, and only if, $v \in H^1(\Omega)$.*

The space $H^1(\Omega)$ is endowed with the inner product:

$$(u, v)_{H^1(\Omega)} = \int_\Omega uv \, d\Omega + \int_\Omega \nabla u \cdot \nabla v \, d\Omega \qquad [1.24]$$

from which one derives the norm:

$$\|u\|_{H^1(\Omega)} = \left((u, u)_{H^1(\Omega)}\right)^{1/2}$$

making $H^1(\Omega)$ into a Hilbert space. We introduce the following **semi-norm** of $H^1(\Omega)$:

$$|u|_{1,\Omega} = \|\nabla u\|_{(L^2(\Omega))^{\mathrm{d}}}. \qquad [1.25]$$

PROPOSITION 1.7.– *Let $f \in L^2(\Omega)$. Then f belongs to $H^1(\Omega)$ if, and only if, there exist $h_1, \cdots, h_{\mathrm{d}} \in L^2(\Omega)$ such that, for $j = 1, \cdots, \mathrm{d}$:*

$$\int_\Omega h_j g \, d\Omega = -\int_\Omega f \partial_j g \, d\Omega \quad \forall g \in \mathcal{D}(\Omega).$$

Proof. By definition, a function f belongs to $H^1(\Omega)$ if, and only if, $f \in L^2(\Omega)$ and $\partial_j f \in L^2(\Omega)$, $j = 1, \cdots, \mathrm{d}$, where the partial derivatives are taken in the distribution sense. From definition 1.6, $\partial_1 f \in \mathcal{D}'(\Omega)$ and, for every g in $\mathcal{D}(\Omega)$, we have the equality:

$$\langle \partial_1 f, g \rangle = -\langle f, \partial_1 g \rangle = -\int_\Omega f \frac{\partial g}{\partial x_1} \, d\Omega$$

since $L^2(\Omega)$ can be identified (see [1.23]) with a subspace of $\mathcal{D}'(\Omega)$. Again from this identification, we deduce that $\partial_1 f$ belongs to $L^2(\Omega)$ *if, and only if*, there exist $h_1 \in L^2(\Omega)$ such that:

$$\int_\Omega h_1 g \, d\Omega = -\int_\Omega f \frac{\partial g}{\partial x_1} \, d\Omega \quad \forall g \in \mathcal{D}(\Omega).$$

This is true for each partial derivative ($j = 1, \cdots, \mathrm{d}$). In other words, each component of the gradient ∇f is equal to an element h_j of $L^2(\Omega)$, for $j = 1, \cdots, \mathrm{d}$. ■

We have the following result regarding differentiation in the weak sense, or weak derivatives.

PROPOSITION 1.8.– *Let $f \in L^2(\Omega)$. Then f belongs to $H^1(\Omega)$ if, and only if, there exists a constant $C \geq 0$ such that, for $j = 1, \cdots, \mathrm{d}$:*

$$\left| \int_\Omega f \partial_j g \, d\Omega \right| \leq C \, \|g\|_{L^2(\Omega)} \quad \forall g \in \mathcal{D}(\Omega).$$

Proof. From proposition 1.4, we know that $\mathcal{D}(\Omega)$ is a vector subspace that is dense in $L^2(\Omega)$. For each index j, from the hypothesis we can continuously extend the linear functional $g \mapsto -\int_\Omega f \partial_j g \, d\Omega$ (see exercise 4) to a continuous linear functional on $L^2(\Omega)$: we denote it by $\ell_j(\cdot)$. From Riesz's representation theorem, we can deduce from this that there exists a (unique) element $h_j \in L^2(\Omega)$ such that:

$$\ell_j(g) = \int_\Omega h_j g \, d\Omega, \quad \forall g \in L^2(\Omega).$$

Then proposition 1.7 lets us conclude the result. ■

Let us continue: using definition 1.6 along with a multi-index α, we can recursively define the partial derivative of a distribution of order α.

DEFINITION 1.7.– *Let $T \in \mathcal{D}'(\Omega)$, its partial derivative of order α is defined by:*

$$\langle \partial_\alpha T, f \rangle = (-1)^{|\alpha|} \langle T, \partial_\alpha f \rangle, \quad \forall f \in \mathcal{D}(\Omega).$$

When $\alpha = (0, 0, 0)$, no differentiation takes place.

This definition lets us construct Sobolev spaces of any order. Given $m \in \mathbb{N}$, we define:

$$H^m(\Omega) = \left\{ v \in L^2(\Omega) \text{ such that } \partial_\alpha v \in L^2(\Omega),\ \forall \alpha \in \mathbb{N}^{\mathrm{d}},\ |\alpha| \leq m \right\}.$$

It is also a Hilbert space, endowed with the inner product:

$$(u, v)_{H^m(\Omega)} = \int_\Omega \Big(\sum_{\alpha \in \mathbb{N}^{\mathrm{d}},\ |\alpha| \leq m} \partial_\alpha u\, \partial_\alpha v \Big) d\Omega$$

which induces the norm:

$$\|u\|_{H^m(\Omega)} = \Big(\sum_{\alpha \in \mathbb{N}^{\mathrm{d}},\ |\alpha| \leq m} \|\partial_\alpha u\|^2_{L^2\Omega)} \Big)^{1/2}.$$

We introduce the following **semi-norm** for $H^m(\Omega)$:

$$|u|_{m,\Omega} = \Big(\sum_{\alpha \in \mathbb{N}^{\mathrm{d}},\ |\alpha| = m} \|\partial_\alpha u\|^2_{L^2(\Omega)} \Big)^{1/2}. \qquad [1.26]$$

In particular, we have the following sequence of inclusions (for $m \geq 2$):

$$\mathcal{D}(\Omega) \subset H^m(\Omega) \subset H^1(\Omega) \subset L^2(\Omega). \qquad [1.27]$$

One can also define Sobolev spaces with fractional orders $s \in \mathbb{R}^+ \backslash \mathbb{N}$, again denoted by $H^s(\Omega)$: either by using interpolation theory (see Lions and Magenes (1973) and exercise 6), or by giving a direct definition (see Grisvard 1985). For $\sigma \in (0, 1)$, first we define:

$$H^\sigma(\Omega) = \{ v \in L^2(\Omega) \text{ such that } \int_\Omega \int_\Omega \frac{|v(x) - v(y)|^2}{|x - y|^{\mathrm{d}+2\sigma}} d\Omega_x\, d\Omega_y < \infty \}$$

endowed with the norm:

$$\|u\|_{H^\sigma(\Omega)} = \Big(\|u\|^2_{L^2(\Omega)} + \int_\Omega \int_\Omega \frac{|u(x) - u(y)|^2}{|x - y|^{\mathrm{d}+2\sigma}} d\Omega_x\, d\Omega_y \Big)^{1/2}.$$

Then, for $s = m + \sigma$, with $m \in \mathbb{N}$ and $\sigma \in (0, 1)$, we define:

$$H^s(\Omega) = \{ v \in H^m(\Omega) \text{ such that } \partial_\alpha v \in H^\sigma(\Omega),\ \forall \alpha \in \mathbb{N}^{\mathrm{d}} \text{ with } |\alpha| = m \}$$

which we endow with the norm:

$$\|u\|_{H^s(\Omega)} = \Big(\|u\|^2_{H^m(\Omega)} + \sum_{\alpha\in\mathbb{N}^{\mathrm{d}},\ |\alpha|=m} \int_\Omega\int_\Omega \frac{|\partial_\alpha u(x) - \partial_\alpha u(y)|^2}{|x-y|^{\mathrm{d}+2\sigma}} d\Omega_x\, d\Omega_y\Big)^{1/2}.$$

Finally, the **semi-norm** for $H^s(\Omega)$ is:

$$|u|_{s,\Omega} = \Big(\sum_{\alpha\in\mathbb{N}^{\mathrm{d}},\ |\alpha|=m} \int_\Omega\int_\Omega \frac{|\partial_\alpha u(x) - \partial_\alpha u(y)|^2}{|x-y|^{\mathrm{d}+2\sigma}} d\Omega_x\, d\Omega_y\Big)^{1/2}. \quad [1.28]$$

In the definitions above, the exponents of the double integrals depend on the dimension d and on the non-integer part of the order, s.

The two definitions of $H^s(\Omega)$ are equivalent when $\Omega = \mathbb{R}^{\mathrm{d}}$, or when the boundary $\partial\Omega$ is Lipschitz in the sense of definition 1.8 (see Adams and Fournier 2003). Moreover, for each $s > s' \geq 0$, we have that $H^s(\Omega) \subset H^{s'}(\Omega)$.

In addition to $L^2(\Omega)$, we will focus mostly on the Sobolev spaces $H^1(\Omega)$ and $H^2(\Omega)$, and therefore we will specify some of their properties[8] as well as some of their vector subspaces.

To begin with, we can prove that the subset $\mathcal{D}(\Omega)$ is always dense in $L^2(\Omega)$, but that it is not dense in $H^1(\Omega)$ in general. This is why we defined the space $H_0^1(\Omega)$, the closure of $\mathcal{D}(\Omega)$ in $H^1(\Omega)$:

$$H_0^1(\Omega) = \overline{\mathcal{D}(\Omega)}^{H^1(\Omega)} \quad [1.29]$$

that is, the set of limits of sequences (for the norm of $H^1(\Omega)$) of elements in $\mathcal{D}(\Omega)$. It is a Hilbert space since it is a closed subspace of another Hilbert space.

1.5.2. *"Sufficiently regular" boundaries*

So far, the results are valid in any open subset Ω of $\mathbb{R}^{\mathrm{d}}$, that is, without any assumption on the boundary $\partial\Omega$. To continue further, we will have to make the notion of a "sufficiently regular" boundary precise, which we do in two stages.

DEFINITION 1.8.– *Let Ω be an open subset of $\mathbb{R}^{\mathrm{d}}$. We say that its boundary $\partial\Omega$ is **Lipschitz** if, and only if:*

8 See, for example, Raviart and Thomas (1983) for elementary "proofs". For a much more substantial exposition of Sobolev space properties, see Grisvard (1985) and Adams and Fournier (2003).

– *at any point x of $\partial\Omega$, there exists a Lipschitz mapping (defined on the hypercube in $\mathbb{R}^{d-1}$ with values in $\mathbb{R}$), whose graph locally represents $\partial\Omega$ on an (open) neighborhood of x;*

– *at any point x of $\partial\Omega$, Ω is locally on one side of $\partial\Omega$.*

In this definition, both statements are necessary. Figure 1.3 presents two examples of Lipschitz mappings (we also present two **local charts**).

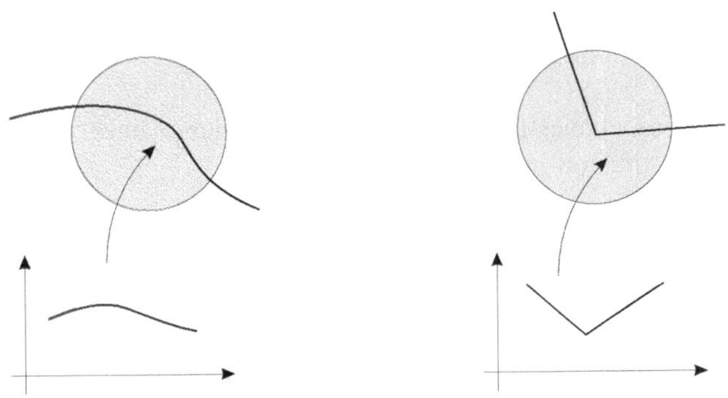

Figure 1.3. *Two examples of a local chart*

In Figure 1.4, we present two examples of open sets whose boundaries are not Lipschitz. In the left-hand figure (the open "crack"), the open set is not locally on one side of the boundary due to the extremity of the "crack". In the right-hand figure (a cusp), the map that represents it locally is not Lipschitz because of the cusp point on its border.

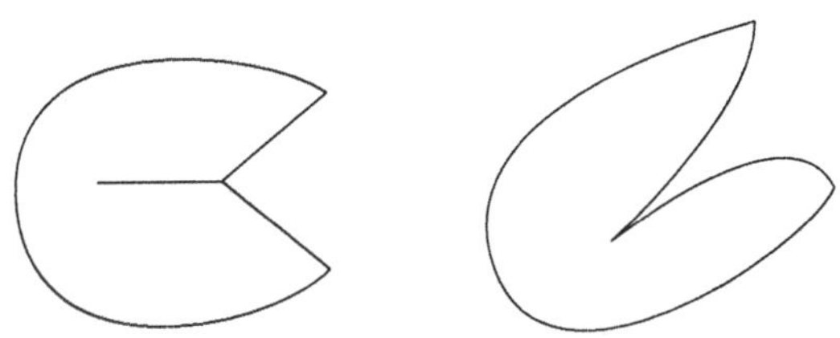

Figure 1.4. *Two examples of a non-Lipschitz boundary*

We have the following result concerning the outwards unit normal vector to $\partial\Omega$.

PROPOSITION 1.9.– *Let Ω be an open set with Lipschitz boundary. Then the outwards unit normal vector to $\partial\Omega$ belongs to $(L^\infty(\partial\Omega))^d$.*

On an open subset of $\mathbb{R}^d$ whose boundary is Lipschitz, one can prove some very useful density results. For this, let us define the set of ad hoc regular functions.

DEFINITION 1.9.– *We say that $\mathcal{C}_c^\infty(\overline{\Omega})$ is the space consisting of restricted functions from $\mathcal{D}(\mathbb{R}^d)$ to $\overline{\Omega}$.*

By construction, $\mathcal{C}_c^\infty(\overline{\Omega}) \subset \mathcal{C}^k(\overline{\Omega})$ for each $k \geq 0$.

PROPOSITION 1.10.– *Let $m \in \mathbb{N}$. For any open subset Ω with a Lipschitz boundary, $\mathcal{C}_c^\infty(\overline{\Omega})$ is dense in $H^m(\Omega)$:*

$$\forall f \in H^m(\Omega),\ \exists (f_k)_k \in \left(\mathcal{C}_c^\infty(\overline{\Omega})\right)^{\mathbb{N}},\ \lim_{k\to\infty} \|f - f_k\|_{H^m(\Omega)} = 0.$$

In what follows, we will consider subclasses of open sets whose boundaries are Lipschitz, which are encountered quite frequently in practice. These subsets are open and also have "sufficiently regular" boundaries.

DEFINITION 1.10.– *Let Ω be an open subset of $\mathbb{R}^d$ with Lipschitz boundary $\partial\Omega$. Its boundary is said to be **"sufficiently regular"** if one of the following cases holds:*

– for $\Omega \subset \mathbb{R}^d$, the boundary of Ω is regular, that is, at that every point x of $\partial\Omega$, the mapping whose graph locally represents $\partial\Omega$ is of class $\mathcal{C}^\infty$;

– for $\Omega \subset \mathbb{R}^3$, the boundary of Ω is polyhedral (with straight faces), or curvilinear polyhedral (with curved faces);

– for $\Omega \subset \mathbb{R}^2$, the boundary of Ω is polygonal (with straight edges), or curvilinear polygonal (with curved edges).

Figure 1.5 presents some examples of open sets with "sufficiently regular" boundaries.

1.5.3. *Differentiation in the classical or distribution sense?*

Let us continue this overview of Sobolev spaces by highlighting their connection with the Banach spaces, $\mathcal{C}^k(\overline{\Omega})$, $k \geq 0$[9].

PROPOSITION 1.11 (Regularity of Sobolev spaces).– *Let Ω be a bounded and open subset of $\mathbb{R}^d$ with Lipschitz boundary, and let $m > \dfrac{d}{2} + k$ ($k \geq 0$). Then $H^m(\Omega)$ is continuously embedded into $\mathcal{C}^k(\overline{\Omega})$, that is, there exists a constant C such that:*

$$\|u\|_{\mathcal{C}^k(\overline{\Omega})} \leq C\,\|u\|_{H^m(\Omega)}, \qquad \forall u \in H^m(\Omega). \qquad [1.30]$$

9 We recall that $\|u\|_{\mathcal{C}^k(\overline{\Omega})} = \sum_{\alpha\in\mathbb{N}^d,\ |\alpha|\leq k} \sup_{x\in\overline{\Omega}} |\partial_\alpha u(x)|$.

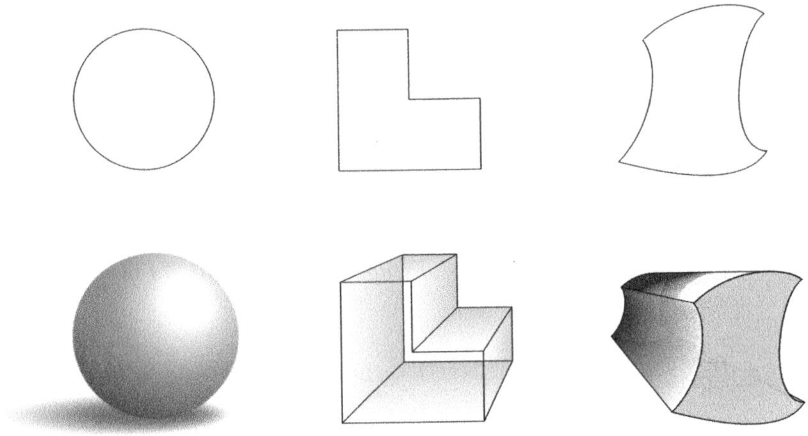

Figure 1.5. *Examples of open sets with "sufficiently regular" boundaries*

In the strict sense, an element u of a Sobolev space $H^m(\Omega)$ is not defined everywhere on Ω but only almost everywhere, since $H^m(\Omega)$ is a vector subspace of $L^2(\Omega)$. Thus, when we consider u as a function u^* belonging to $\mathcal{C}^k(\overline{\Omega})$, we must redefine it on a subset of Ω of measure zero, so that $u^* = u$ almost everywhere with u^* a continuous function.

In particular, we have the following general result:

– *in dimension* $\mathrm{d} = 1$*, functions in* $H^1(\Omega)$ *are continuous but not necessarily differentiable;*

– *in dimensions* $\mathrm{d} \in \{2, 3\}$*, functions in* $H^1(\Omega)$ *are not necessarily continuous, but functions in* $H^2(\Omega)$ *are continuous.*

1.5.4. *A compactness result*

Finally, we recall a result (Brezis 1983, 2011) that is valid only when the open set Ω is *bounded*. In particular, this result allows us to establish the equivalence between the norms of each vector subspaces of $H^1(\Omega)$ (see exercises 7 and 8), and therefore lets us specify how the solutions to our problems will be measured.

THEOREM 1.6 (Rellich).– *Let* Ω *be a bounded and open subset of* $\mathbb{R}^{\mathrm{d}}$ *with Lipschitz boundary. Then, from any bounded sequence in* $H^1(\Omega)$*, we can extract a subsequence that converges in* $L^2(\Omega)$*.*

In other terms, the canonical injection of $H^1(\Omega)$ into $L^2(\Omega)$ is *compact*.

1.6. Existence of the trace and integration by parts formulae

Let us consider $v \in H^1(\Omega)$ and ask ourselves the question: Can we define its "value on the boundary", its trace on $\partial\Omega$? This is not a naïve question!

Indeed, first of all let us note that functions in $L^2(\Omega)$ do not necessarily have a trace (e.g. take the function $v(x) = x^{-1/4}$ which belongs to $L^2((0,1))$). Moreover, for $\mathrm{d} \geq 2$, functions in $H^1(\Omega)$ are not continuous in general (take $\mathrm{d} = 2$, $\Omega = B(0,R)$ and $v(r) = |\log r|^p$ with $p \in (0, \frac{1}{2})$; see exercise 3 to be convinced). The definition of the trace of such functions is therefore not straightforward.

However, for any function $v \in C_c^\infty\left(\overline{\Omega}\right)$, one can define its value on the boundary $\partial\Omega$, $v_{|\partial\Omega}$, which satisfies:

$$v_{|\partial\Omega} \in L^2(\partial\Omega) = \left\{ w \text{ is measurable on } \partial\Omega \text{ with } \int_{\partial\Omega} w^2 d\Gamma < +\infty \right\}$$

where $d\Gamma$ denotes the surface measure of $\partial\Omega$. In particular, when $\partial\Omega$ is Lipschitz, $L^2(\partial\Omega)$ is well defined. Endowed with the inner product:

$$(w, z)_{L^2(\partial\Omega)} = \int_{\partial\Omega} wz \, d\Gamma$$

from which we derive the norm

$$\|w\|_{L^2(\partial\Omega)} = \left(\int_{\partial\Omega} w^2 \, d\Gamma \right)^{1/2}$$

we see that $L^2(\partial\Omega)$ becomes a Hilbert space.

We say that $\partial\Omega$ is **bounded** if $\int_{\partial\Omega} d\Gamma < \infty$. In the case where Ω is bounded and open with a Lipschitz boundary $\partial\Omega$, then it is always bounded.

The implicitly defined mapping $\gamma_0 : v \mapsto \gamma_0 v = v_{|\partial\Omega}$ extends to functions in $H^1(\Omega)$, so we can state in a more precise way that:

THEOREM 1.7 (Existence of the trace).– *Let Ω be an open subset of $\mathbb{R}^\mathrm{d}$ with a bounded and Lipschitz boundary. Then the trace mapping:*

$$\gamma_0 : \left| \begin{array}{lcl} C_c^\infty\left(\overline{\Omega}\right) & \to & L^2(\partial\Omega) \\ v & \mapsto & \gamma_0 v = v_{|\partial\Omega} \end{array} \right.$$

can be continuously extended to a continuous linear mapping, again denoted by γ_0, from $H^1(\Omega)$ into $L^2(\partial\Omega)$. Hence there exists a constant C such that:

$$\|\gamma_0 v\|_{L^2(\partial\Omega)} \leq C \|v\|_{H^1(\Omega)}, \qquad \forall v \in H^1(\Omega). \qquad [1.31]$$

The existence of the trace mapping makes it possible to characterize the space $H_0^1(\Omega)$:

THEOREM 1.8 (Characterization of $H_0^1(\Omega)$).– *If Ω is an open subset of $\mathbb{R}^d$ with a bounded and Lipschitz boundary, then:*

$$H_0^1(\Omega) = \left\{v \in H^1(\Omega) \text{ such that } \gamma_0 v = 0\right\}.$$

For an elementary proof of the existence of traces, see exercise 5. However, the mapping γ_0 *is not surjective* from $H^1(\Omega)$ onto $L^2(\partial\Omega)$. In other words, there exist functions in $L^2(\partial\Omega)$ that are not the traces of functions in $H^1(\Omega)$. On the other hand, the trace mapping γ_0 is *surjective* onto $H^{1/2}(\partial\Omega)$, the fractional-index Sobolev space satisfying $H^1(\partial\Omega) \subset H^{1/2}(\partial\Omega) \subset L^2(\partial\Omega)$, and whose study is rather detailed and beyond the scope of this book; see exercise 6 for a simple case, and for the general theory we refer the reader to Grisvard (1985) and Adams and Fournier (2003). In particular, the following properties can be established for the space:

$$H^{1/2}(\partial\Omega) = \left\{w \in L^2(\partial\Omega) \text{ such that } \exists v \in H^1(\Omega),\ w = \gamma_0 v\right\}$$

when endowed with the norm:

$$\|w\|_{H^{1/2}(\partial\Omega)} = \inf_{v \in H^1(\Omega)\ st\ \gamma_0 v = w} \|v\|_{H^1(\Omega)}. \qquad [1.32]$$

PROPOSITION 1.12.– *Let Ω be an open subset of $\mathbb{R}^d$ with a bounded and Lipschitz boundary. Endowed with the norm [1.32], the space of traces $H^{1/2}(\partial\Omega)$ becomes a Banach space, and moreover $H^{1/2}(\partial\Omega)$ is dense in $L^2(\partial\Omega)$.*

Let $w \in H^{1/2}(\partial\Omega)$: for every $v \in H^1(\Omega)$ such that $\gamma_0 v = w$, we have (theorem 1.7) $\|w\|_{L^2(\partial\Omega)} \leq C\|v\|_{H^1(\Omega)}$, with C being the continuity constant from [1.31]. With the help of definition [1.32], we deduce from this that:

$$\|w\|_{L^2(\partial\Omega)} \leq C \inf_{v \in H^1(\Omega)\ st\ \gamma_0 v = w} \|v\|_{H^1(\Omega)} = C\,\|w\|_{H^{1/2}(\partial\Omega)}$$

which expresses the fact that the injection from $H^{1/2}(\partial\Omega)$ into $L^2(\partial\Omega)$ is continuous.

In using theorem 1.7 and proposition 1.10, we may generalize Green's integration by parts formula [1.7].

PROPOSITION 1.13 (Green's formula).– *Let Ω be an open subset of $\mathbb{R}^d$ with a bounded and Lipschitz boundary, and let $u, v \in H^1(\Omega)$. Then we have, from $i = 1$ to d:*

$$\int_\Omega \frac{\partial u}{\partial x_i}\, v\, d\Omega = -\int_\Omega u\, \frac{\partial v}{\partial x_i}\, d\Omega + \int_{\partial\Omega} u_{|\partial\Omega}\, v_{|\partial\Omega}\, n_i\, d\Gamma. \qquad [1.33]$$

Proof. From proposition 1.10, there exist two sequences $(u_k)_k$ and $(v_m)_m$ in $\mathcal{C}_c^\infty\left(\overline{\Omega}\right)$ such that:

$$\lim_{k\to\infty} u_k = u \text{ and } \lim_{m\to\infty} v_m = v \text{ in } H^1(\Omega).$$

We note that we can apply Green's integration by parts formula [1.7] to each pair (u_k, v_m).

Furthermore, $u_k \to u$ in $L^2(\Omega)$ and $\partial_i u_k \to \partial_i u$ in $L^2(\Omega)$. Moreover, $v_m \to v$ in $L^2(\Omega)$ and $\partial_i v_m \to \partial_i v$ in $L^2(\Omega)$. Therefore, in letting k and m tend toward infinity, we see that:

$$\int_\Omega \frac{\partial u_k}{\partial x_i}\, v_m \, d\Omega \to \int_\Omega \frac{\partial u}{\partial x_i}\, v \, d\Omega \text{ and } \int_\Omega u_k \,\frac{\partial v_m}{\partial x_i}\, d\Omega \to \int_\Omega u\, \frac{\partial v}{\partial x_i}\, d\Omega.$$

As for the boundary term, by the continuity of the trace mapping γ_0 (see theorem 1.7), we know that $\lim_{k\to\infty} \gamma_0 u_k = \gamma_0 u$ and $\lim_{m\to\infty} \gamma_0 v_m = \gamma_0 v$ in $L^2(\partial\Omega)$.

Finally, since the boundary $\partial\Omega$ is Lipschitz, we have the following result $n_i \in L^\infty(\partial\Omega)$ (see proposition 1.9). In this case, when k and m tend to infinity we can conclude that:

$$\int_{\partial\Omega} \gamma_0 u_k \, \gamma_0 v_m \, n_i \, d\Gamma \to \int_{\partial\Omega} \gamma_0 u \, \gamma_0 v \, n_i \, d\Gamma = \int_{\partial\Omega} u_{|\partial\Omega}\, v_{|\partial\Omega}\, n_i \, d\Gamma.$$

If we now use formula [1.7] for each pair (u_k, v_m), we can take the limits (in both k and m) to obtain the result [1.33]. ■

We have seen that functions belonging to the space $H^1(\Omega)$ are not necessarily continuous, provided that $\mathrm{d} \geq 2$. However, we can demonstrate that these functions do not jump when going across any interface. Also, if they are continuous on either side of the interface, then they are in fact continuous globally. This last result should be compared to the converse of property 1.1. So we have:

PROPERTY 1.2.– *Let Ω, Ω_1 and Ω_2, be two open subsets with a bounded and Lipschitz boundary in $\mathbb{R}^{\mathrm{d}}$, such that $\Omega_1 \cap \Omega_2 = \emptyset$ and $\overline{\Omega} = \overline{\Omega_1} \cup \overline{\Omega_2}$. Let $F = \overline{\Omega_1} \cap \overline{\Omega_2}$ be the interface. Given $v \in H^1(\Omega)$, denote $v_j = v_{|\Omega_j}$ for $j = 1, 2$. Then we have:*

$$\nabla v_j = (\nabla v)_{|\Omega_j} \text{ a. e. on } \Omega_j, \text{ and } v_j \in H^1(\Omega_j), \text{ for } j = 1, 2$$
$$v_{1|F} = v_{2|F} \text{ a. e. on } F.$$

In particular, when Ω is bounded and if $v_j \in \mathcal{C}^0(\overline{\Omega_j})$ for $j = 1, 2$, then $v \in \mathcal{C}^0\left(\overline{\Omega}\right)$.

Proof. Consider $j = 1, 2$. For every function $z \in L^2(\Omega_j)$, denote by $\tilde{z}$ its extension by 0 to the entirety of Ω. For $i = 1, \cdots, \mathrm{d}$, we have that $\forall \varphi \in \mathcal{D}(\Omega_j)$:

$$\begin{aligned} \langle \frac{\partial v_j}{\partial x_i}, \varphi \rangle &= -\langle v_j, \frac{\partial \varphi}{\partial x_i} \rangle = -\int_{\Omega_j} v_j \frac{\partial \varphi}{\partial x_i} \, d\Omega = -\int_{\Omega} v \widetilde{\frac{\partial \varphi}{\partial x_i}} \, d\Omega = -\int_{\Omega} v \frac{\partial \tilde{\varphi}}{\partial x_i} \, d\Omega \\ [1.33] \quad &= \int_{\Omega} \frac{\partial v}{\partial x_i} \tilde{\varphi} \, d\Omega = \int_{\Omega_j} \left(\frac{\partial v}{\partial x_i} \right)_{|\Omega_j} \varphi \, d\Omega = \langle \left(\frac{\partial v}{\partial x_i} \right)_{|\Omega_j}, \varphi \rangle. \end{aligned}$$

From this, we deduce that $\partial_i v_j = (\partial_i v)_{|\Omega_j} \in L^2(\Omega_j)$ for $i = 1, \cdots, \mathrm{d}$, and therefore that $v_j \in H^1(\Omega_j)$.

Furthermore, we have $\forall \varphi \in \mathcal{D}(\Omega)$:

$$\begin{aligned} \int_{\Omega} \frac{\partial v}{\partial x_i} \varphi \, d\Omega &= -\int_{\Omega} v \frac{\partial \varphi}{\partial x_i} \, d\Omega = -\int_{\Omega_1} v_1 \frac{\partial \varphi}{\partial x_i} \, d\Omega - \int_{\Omega_2} v_2 \frac{\partial \varphi}{\partial x_i} \, d\Omega \\ ([1.33] \text{ on } \Omega_j) \quad &= \int_{\Omega_1} \frac{\partial v_1}{\partial x_i} \varphi \, d\Omega + \int_{\Omega_2} \frac{\partial v_2}{\partial x_i} \varphi \, d\Omega - \int_F \left(v_{1|F} \, (n_1)_i + v_{2|F} \, (n_2)_i \right) \varphi_{|F} dF \\ (\mathbf{n}_1 = -\mathbf{n}_2) \quad &= \int_{\Omega_1} \frac{\partial v_1}{\partial x_i} \varphi \, d\Omega + \int_{\Omega_2} \frac{\partial v_2}{\partial x_i} \varphi \, d\Omega - \int_F \left(v_{1|F} - v_{2|F} \right) (n_1)_i \, \varphi_{|F} dF \\ &= \int_{\Omega} \frac{\partial v}{\partial x_i} \varphi \, d\Omega - \int_F \left(v_{1|F} - v_{2|F} \right) (n_1)_i \, \varphi_{|F} dF. \end{aligned}$$

Since the magnitude of the normal vector equals one, there exists at least one non-zero component of it, from which we can conclude by a density argument[10] that $v_{1|F} = v_{2|F}$ in $L^2(F)$, and consequently:

$$v_{1|F} = v_{2|F} \quad \text{a. e. on } F.$$

When $v_j \in \mathcal{C}^0\left(\overline{\Omega_j}\right)$, the equality above is in fact true everywhere, proving that $v \in \mathcal{C}^0\left(\overline{\Omega}\right)$. ■

How can we generalize Stokes' fundamental formula [1.6]? Intuitively, both integrals over Ω need to be well defined. Note that, if $v \in H^1(\Omega)$ and if each component of $\mathbf{w}$, $(w_i)_{i=1,\mathrm{d}}$, *as well as* its divergence, all belong to $L^2(\Omega)$, then each of these integrals can be written in the form of a inner product:

$$\int_{\Omega} (\operatorname{div} \mathbf{w}) \, v \, d\Omega = (\operatorname{div} \mathbf{w}, v)_{L^2(\Omega)}$$

and:

$$\int_{\Omega} \mathbf{w} \cdot \nabla v \, d\Omega = \sum_{i=1}^{\mathrm{d}} \int_{\Omega} w_i \frac{\partial v}{\partial x_i} \, d\Omega = \sum_{i=1}^{\mathrm{d}} \left(w_i, \frac{\partial v}{\partial x_i} \right)_{L^2(\Omega)} = (\mathbf{w}, \nabla v)_{(L^2(\Omega))^{\mathrm{d}}}.$$

10 More precisely, we can apply the following result to $\lambda = (v_{1|F} - v_{2|F})\,(n_1)_i$: let $\lambda \in L^2(F)$ be such that $\int_F \lambda \, \varphi_{|F} \, d\Gamma = 0$, $\forall \varphi \in \mathcal{D}(\Omega)$, then $\lambda = 0$.

Consider now the Sobolev space:

$$H(\mathrm{div};\Omega) = \left\{\mathbf{v} \in \left(L^2(\Omega)\right)^{\mathrm{d}} \text{ such that } \mathrm{div}\,\mathbf{v} \in L^2(\Omega)\right\}$$

where the divergence is considered in the distribution sense, and then endow $H(\mathrm{div};\Omega)$ with the norm:

$$\|\mathbf{u}\|_{H(\mathrm{div};\Omega)} = \left(\|\mathbf{u}\|^2_{(L^2(\Omega))^{\mathrm{d}}} + \|\,\mathrm{div}\,\mathbf{u}\|^2_{L^2(\Omega)}\right)^{1/2}.$$

If $(\mathbf{w}, v) \in H(\mathrm{div};\Omega) \times H^1(\Omega)$, then the integrals over Ω are well defined in formula [1.6]. What about the integral over $\partial\Omega$? From the trace theorem 1.7, we know that $v_{|\partial\Omega} \in L^2(\partial\Omega)$. Unfortunately, if $\mathbf{w}$ is an element of $H(\mathrm{div};\Omega)$, there is no guarantee that $(\mathbf{w}\cdot\mathbf{n})_{|\partial\Omega} \in L^2(\partial\Omega)$. That being said, we have seen that the trace of an element of $H^1(\Omega)$ always belongs to $H^{1/2}(\partial\Omega)$, where $H^{1/2}(\partial\Omega)$ is a strict subspace of $L^2(\partial\Omega)$. Thus, for the integral over $\partial\Omega$ to be well defined, a sufficient condition would be if:

$$v \quad \mapsto \quad \int_{\partial\Omega} (\mathbf{w}\cdot\mathbf{n})_{|\partial\Omega} v_{|\partial\Omega}\, d\Gamma$$

defined a continuous linear functional on $H^{1/2}(\partial\Omega)$: then we may say that $(\mathbf{w}\cdot\mathbf{n})_{|\partial\Omega} \in \left(H^{1/2}(\partial\Omega)\right)'$ and use the duality brackets:

$$\left\langle (\mathbf{w}\cdot\mathbf{n})_{|\partial\Omega}, v_{|\partial\Omega}\right\rangle_{\left(H^{1/2}(\partial\Omega)\right)', H^{1/2}(\partial\Omega)}$$

to represent the action of the continuous linear functional on an element of $H^{1/2}(\partial\Omega)$. This is the subject of the result discussed, taken from Girault and Raviart (1986), and proven in exercise 12.

PROPOSITION 1.14 (Normal trace).– *Let Ω be an open subset of $\mathbb{R}^{\mathrm{d}}$, with a bounded and Lipschitz boundary. Then:*

– $\left(C_c^\infty\left(\overline{\Omega}\right)\right)^{\mathrm{d}}$ *is dense in* $H(\mathrm{div};\Omega)$;

– *the normal trace mapping:*

$$\gamma_n : \left| \begin{array}{ccc} \left(C_c^\infty\left(\overline{\Omega}\right)\right)^{\mathrm{d}} & \to & \left(H^{1/2}(\partial\Omega)\right)' \\ \mathbf{w} & \mapsto & \gamma_n\mathbf{w} = (\mathbf{w}\cdot\mathbf{n})_{|\partial\Omega} \end{array}\right.$$

can be continuously extended to a continuous and surjective linear mapping, again denoted by γ_n, from $H(\mathrm{div};\Omega)$ onto $\left(H^{1/2}(\partial\Omega)\right)'$, and satisfies the inequality:

$$\|\gamma_n\mathbf{w}\|_{(H^{1/2}(\partial\Omega))'} \le \|\mathbf{w}\|_{H(\mathrm{div};\Omega)}, \qquad \forall \mathbf{w} \in H(\mathrm{div};\Omega).$$

– *let* $\mathbf{w} \in H(\text{div};\Omega)$ *and* $v \in H^1(\Omega)$, *then we have Stokes' integration by parts formula:*

$$\int_\Omega (\text{div}\,\mathbf{w})\, v \, d\Omega = -\int_\Omega \mathbf{w} \cdot \nabla v \, d\Omega + \langle \gamma_n \mathbf{w}, \gamma_0 v \rangle_{(H^{1/2}(\partial\Omega))', H^{1/2}(\partial\Omega)}. \quad [1.34]$$

In what follows, we will omit the explicit references to the trace mappings γ_0, γ_n, $_{|\partial\Omega}$, etc., in the duality brackets and in the integrals along the boundary when there is no ambiguity.

First, a few details: along the boundary, $L^2(\partial\Omega)$ can be identified with its dual. So we have the strict inclusions:

$$H^{1/2}(\partial\Omega) \subset L^2(\partial\Omega) \subset \left(H^{1/2}(\partial\Omega)\right)' \quad [1.35]$$

and, for a pair $(z_0, z_1) \in L^2(\partial\Omega) \times H^{1/2}(\partial\Omega)$, we can write:

$$\langle z_0, z_1 \rangle_{(H^{1/2}(\partial\Omega))', H^{1/2}(\partial\Omega)} = (z_0, z_1)_{L^2(\partial\Omega)} = \int_{\partial\Omega} z_0 \, z_1 \, d\Gamma. \quad [1.36]$$

Again according to Girault and Raviart (1986), note that the dual of $H^{1/2}(\partial\Omega)$ can be identified with the Sobolev space defined on $\partial\Omega$ whose index is $-1/2$: $H^{-1/2}(\partial\Omega) = \left(H^{1/2}(\partial\Omega)\right)'$.

Now generalize Green's formula [1.8]. For that, notice that if $u \in H^1(\Omega)$ with $\Delta u \in L^2(\Omega)$, then $\nabla u \in H(\text{div};\Omega)$: indeed, on the one hand, $u \in H^1(\Omega)$ leads to $\nabla u \in \left(L^2(\Omega)\right)^{\mathrm{d}}$ and, on the other hand, $\text{div}(\nabla u) = \Delta u$. We can therefore apply the results of proposition 1.14. So, let:

$$H^1(\Delta;\Omega) = \left\{ v \in H^1(\Omega) \text{ such that } \Delta v \in L^2(\Omega) \right\} \quad [1.37]$$

where the Laplacian is to be understood in the distribution sense. We endow $H^1(\Delta;\Omega)$ with the norm:

$$\|u\|_{H^1(\Delta;\Omega)} = \left(\|u\|^2_{H^1(\Omega)} + \|\Delta u\|^2_{L^2(\Omega)} \right)^{1/2}.$$

Recall that the trace of the normal derivative, if it exists, is defined by $(\nabla u \cdot \mathbf{n})_{|\partial\Omega} = \left(\dfrac{\partial u}{\partial n}\right)_{|\partial\Omega}$. We have the following results (for surjectivity, see exercise 12):

THEOREM 1.9 (Green's formulae).– *Let* Ω *be an open subset of* $\mathbb{R}^{\mathrm{d}}$ *with a bounded and Lipschitz boundary. Then:*

– the trace mapping of the normal derivative:

$$\gamma_1 : \left| \begin{array}{lcl} H^1(\Delta;\Omega) & \to & \left(H^{1/2}(\partial\Omega)\right)' \\ v & \mapsto & \gamma_1 v = \left(\dfrac{\partial v}{\partial n}\right)_{|\partial\Omega} \end{array} \right.$$

is linear, continuous and surjective. Moreover, we have the inequality:

$$\|\gamma_1 v\|_{(H^{1/2}(\partial\Omega))'} \leq \|v\|_{H^1(\Delta;\Omega)}, \qquad \forall v \in H^1(\Delta;\Omega) \qquad [1.38]$$

– let $u \in H^1(\Delta;\Omega)$ and $v \in H^1(\Omega)$, then we have Green's integration by parts formula:

$$\int_\Omega \Delta u\, v\, d\Omega = -\int_\Omega \nabla u \cdot \nabla v\, d\Omega + \langle \frac{\partial u}{\partial n}_{|\partial\Omega}, v_{|\partial\Omega}\rangle_{(H^{1/2}(\partial\Omega))', H^{1/2}(\partial\Omega)}. \quad [1.39]$$

REMARK 1.3.– *The reader unfamiliar with function spaces may be a little dissuaded by the delicate notions presented just now, in particular the use of duality brackets for the boundary terms. In the following, we will* abuse notation *by replacing these brackets with an integral along the boundary as often as possible. In doing so, we will preserve the spirit of the "generalized" integration by parts formulae ([1.34] and [1.39]).*

Moreover, we are going to see that if the test functions are just a bit more regular, then the "generalized" integration by parts formulae can be written – this time without any abuse of notation – as an integral along the boundary. Let us start with Stokes's fundamental formula [1.6], and assume this time that $v \in H^1(\Omega)$ and $\mathbf{w} \in \left(H^1(\Omega)\right)^{\mathrm{d}}$. From the trace theorem 1.7, we know that $\mathbf{w}_{|\partial\Omega} \in \left(L^2(\partial\Omega)\right)^{\mathrm{d}}$. Moreover, since the outwards unit normal vector to $\partial\Omega$, $\mathbf{n}$, belongs to $\left(L^\infty(\partial\Omega)\right)^{\mathrm{d}}$ (see proposition 1.9), from this we can deduce that $(\mathbf{w}\cdot\mathbf{n})_{|\partial\Omega} \in L^2(\partial\Omega)$. In formula [1.6], the integral along the boundary can therefore be expressed as:

$$\int_{\partial\Omega} (\mathbf{w}\cdot\mathbf{n})\, v\, d\Gamma = ((\mathbf{w}\cdot\mathbf{n})_{|\partial\Omega}, v_{|\partial\Omega})_{L^2(\partial\Omega)}.$$

Hence for $\mathbf{w} \in \left(H^1(\Omega)\right)^{\mathrm{d}}$ and $v \in H^1(\Omega)$, we obtain Stokes' formula:

$$\int_\Omega (\operatorname{div} \mathbf{w})\, v\, d\Omega = -\int_\Omega \mathbf{w} \cdot \nabla v\, d\Omega + \int_{\partial\Omega} (\mathbf{w}\cdot\mathbf{n})\, v\, d\Gamma. \qquad [1.40]$$

Moreover, if $u \in H^2(\Omega)$, then $\nabla u \in \left(H^1(\Omega)\right)^{\mathrm{d}}$ and $\left(\dfrac{\partial u}{\partial n}\right)_{|\partial\Omega} \in L^2(\partial\Omega)$. In the "more regular" framework, we have the following results:

PROPOSITION 1.15 (Trace of the normal derivative and Green's formula).– *Let Ω be an open subset of $\mathbb{R}^{\mathrm{d}}$ with a bounded and Lipschitz boundary. Then:*

– the trace mapping of the normal derivative:

$$\gamma_1 : \left| \begin{array}{ccl} C_c^\infty\left(\overline{\Omega}\right) & \to & L^2(\partial\Omega) \\ v & \mapsto & \gamma_1 v = \left(\dfrac{\partial v}{\partial n}\right)_{|\partial\Omega} \end{array} \right.$$

can be continuously extended to a continuous linear mapping, denoted again by γ_1, from $H^2(\Omega)$ into $L^2(\partial\Omega)$: there is a constant C such that:

$$\|\gamma_1 v\|_{L^2(\partial\Omega)} \leq C\|v\|_{H^2(\Omega)}, \qquad \forall v \in H^2(\Omega). \qquad [1.41]$$

– let $u \in H^2(\Omega)$ and $v \in H^1(\Omega)$. Then we have the Green's integration by parts formula:

$$\int_\Omega \Delta u\, v\, d\Omega = -\int_\Omega \nabla u \cdot \nabla v\, d\Omega + \int_{\partial\Omega} \left(\frac{\partial u}{\partial n}\right)_{|\partial\Omega} v_{|\partial\Omega}\, d\Gamma. \qquad [1.42]$$

Hence, it will be possible here and in the subsequent chapters to write some of the boundary terms as integrals in the integration by parts formulae. However, note that as soon as the dimension d is strictly greater than 1, we have the *strict inclusions* $\left(H^1(\Omega)\right)^{\mathrm{d}} \subset H(\mathrm{div};\Omega)$ and $H^2(\Omega) \subset H^1(\Delta;\Omega)$, thus making it necessary to use the duality brackets in certain situations.

Finally, let us come back to the regularity results for Sobolev spaces and, more precisely, for the sets of solutions which we are looking for. We introduce to following:

$$\left| \begin{array}{l} \Psi_D(\Omega) = \left\{v \in H^1(\Delta;\Omega) \text{ such that } \gamma_0 v = 0\right\} \\ \Psi_N(\Omega) = \left\{v \in H^1(\Delta;\Omega) \text{ such that } \gamma_1 v = 0\right\} \end{array} \right. \qquad [1.43]$$

which are endowed with the norm $\|\cdot\|_{H^1(\Delta;\Omega)}$. According to the continuity properties for the trace mappings γ_0 and γ_1, we note that $\Psi_D(\Omega)$ and $\Psi_N(\Omega)$ are closed subspaces of $H^1(\Delta;\Omega)$.

PROPOSITION 1.16 (Regularity of the space of solutions).– *Let Ω be an open and bounded subset of $\mathbb{R}^{\mathrm{d}}$, $\mathrm{d} = 2, 3$, with a "sufficiently regular" boundary. Then $\Psi_D(\Omega)$ and $\Psi_N(\Omega)$ are continuously embedded into $C^0(\overline{\Omega})$.*

On the other hand, this result is false for $H^1(\Delta;\Omega)$, which is not continuously embedded into $C^0(\overline{\Omega})$ except when $\mathrm{d} = 1$. It is the extra information of the homogeneous boundary condition that makes this possible.

1.7. Exercises – statements

Exercise 1. Integration by parts formulae

Let Ω be an open and bounded subset of $\mathbb{R}^3$, whose boundary $\partial\Omega$ is Lipschitz. Denote by $\mathbf{n}$ the outwards unit normal vector to the boundary.

a) By a calculation, establish the equivalence between the following formulae:

$$\int_\Omega (u\frac{\partial v}{\partial x_i} + \frac{\partial u}{\partial x_i} v)\, d\Omega = \int_{\partial\Omega} uvn_i\, d\Gamma \ (1 \le i \le 3), \forall u, v \in \mathcal{C}^1(\Omega) \cap \mathcal{C}^0(\overline{\Omega})$$

$$\int_\Omega (u \operatorname{div} \mathbf{v} + \nabla u \cdot \mathbf{v})\, d\Omega = \int_{\partial\Omega} u(\mathbf{v}\cdot\mathbf{n})\, d\Gamma, \quad \begin{array}{l}\forall u \in \mathcal{C}^1(\Omega)\cap\mathcal{C}^0(\overline{\Omega})\\ \forall \mathbf{v} \in \left(\mathcal{C}^1(\Omega)\cap\mathcal{C}^0(\overline{\Omega})\right)^3.\end{array}$$

b) Let $\mathbf{u}$ and $\mathbf{v}$ be two elements of $\left(\mathcal{C}^1(\Omega)\cap\mathcal{C}^0(\overline{\Omega})\right)^3$. By using the integration by parts formula, show that:

$$\int_\Omega (\mathbf{u}\cdot\mathbf{curl}\,\mathbf{v} - \mathbf{curl}\,\mathbf{u}\cdot\mathbf{v})\, d\Omega = \int_{\partial\Omega} (\mathbf{u}\times\mathbf{n})\cdot\mathbf{v}\, d\Gamma.$$

Exercise 2. Classical solutions to the Laplacian

Let Ω be a bounded and connected open subset of $\mathbb{R}^3$, whose boundary $\partial\Omega$ is Lipschitz. Consider the following boundary value problem:

$$\left\{\begin{array}{ll}\text{Find } u \text{ such that} & \\ \beta\, u - \Delta u = f & \text{in } \Omega \\ \dfrac{\partial u}{\partial n} = g & \text{on } \partial\Omega.\end{array}\right. \qquad [P2]$$

Above, f and g are continuous on $\overline{\Omega}$ and $\partial\Omega$ respectively, and β is a non-negative constant.

a) First choose $\beta = 0$, i.e., the case of the Neumann problem (P_N). We assume that the solution u belongs to $\mathcal{C}^2(\Omega)\cap\mathcal{C}^1(\overline{\Omega})$. State the variational formulation [F2$_a$] associated to [P2], set *in the space* $\mathcal{C}^1(\overline{\Omega})$. By a direct calculation, establish the uniqueness of the solution up to a constant to the variational formulation [F2$_a$].

b) Now choose $\beta = 1$. We still assume that the solution u belongs to $\mathcal{C}^2(\Omega)\cap\mathcal{C}^1(\overline{\Omega})$. State the variational formulation [F2$_b$] associated to [P2], set *in the space* $\mathcal{C}^1(\overline{\Omega})$. Establish the uniqueness of the solution to the variational formulation [F2$_b$].

c) Still in the case $\beta = 1$, what can you conclude about the existence of a solution to [F2$_b$]? By identifying the bilinear form of [F2$_b$] with an inner product, in which space should we place ourselves to obtain a well-posed problem, i.e., to guarantee the existence and uniqueness of the solution?

Exercise 3. Examples of elements of H^1

a) *In one dimension*: Do the following functions defined, respectively, by:

$$v(x) = \begin{cases} 1 & \text{on } (0,1) \\ 2-x & \text{on } (1,2) \end{cases}, w(x) = \begin{cases} a & \text{on } (0,1) \\ b & \text{on } (1,2) \end{cases}, z(x) = x^p, \ p \in \mathbb{R}$$

belong to $H^1((0,2))$?

b) *In two dimensions*: Let $\Omega_2 = B(0,1/2)$ be an open ball in $\mathbb{R}^2$. Show that the function defined by:

$$u(x,y) = |\log(x^2+y^2)|^p \text{ for } (x,y) \neq (0,0)$$

belongs to $H^1(\Omega_2)$ if $p < 1/2$, and that it does not belong to $C^0(\overline{\Omega_2})$ as soon as $p > 0$.

Exercise 4. Continuous extensions

a) Let X be a normed real vector space and let V be a vector subspace of X, such that V is dense in X. Let ℓ be a linear functional defined on V, for which there exists a constant $C_\ell > 0$ such that, for all $v \in V$, we have that $|\ell(v)| \leq C_\ell \|v\|_X$. Show that we can continuously extend ℓ to a continuous linear functional $\tilde{\ell}$ from X into $\mathbb{R}$. Show that $\tilde{\ell}$ is unique.

b) Using the same hypotheses as in the previous question, additionally let Y be a Banach space. Let ℓ be a linear map from V into Y, for which there exists a constant $C_\ell > 0$ such that, for all $v \in V$, we have $\|\ell(v)\|_Y \leq C_\ell \|v\|_X$. Show that we can continuously extend ℓ to a continuous linear mapping $\tilde{\ell}$ from X into Y uniquely.

Exercise 5. Trace theorem

1) *In one dimension.*

a) Let $\gamma_\star : \left| \begin{array}{l} C^1([0,1]) \to \mathbb{R} \\ v \mapsto v(0) \end{array} \right.$ be the trace mapping. Show that there exists a constant $C > 0$ such that:

$$|v(0)| \leq C \|v\|_{H^1((0,1))}, \quad \forall v \in C^1([0,1]).$$

b) By applying the results of exercise 4, deduce that we can continuously extend the trace mapping $\gamma_\star$ from $H^1((0,1))$ into $\mathbb{R}$.

c) Is $\gamma_\star$ continuous from $L^2((0,1))$ into $\mathbb{R}$?

2) *Hint*: for Question (a), one can write $v(0) = v(x) - \displaystyle\int_0^x v'(t)\,dt$.

3) *In two dimensions.*

d) Let $\Omega_2 = (0,1) \times \mathbb{R}$. Show that there exists a constant C' such that:

$$\|v(0,\cdot)\|_{L^2(\mathbb{R})} + \|v(1,\cdot)\|_{L^2(\mathbb{R})} \leq C'\|v\|_{H^1(\Omega_2)}, \qquad \forall v \in C_c^\infty(\overline{\Omega_2}).$$

e) Using the method just explained, establish that the trace mapping:

$$\gamma_0 \ : \left| \begin{array}{l} H^1(\Omega_2) \to L^2(\partial\Omega_2) \\ v \mapsto \gamma_0 v \end{array} \right.$$

is linear and continuous.

Exercise 6. Sobolev spaces of fractional orders and the trace theorem

Let Ω be a bounded and open subset of $\mathbb{R}^{\mathrm{d}}$ with a Lipschitz boundary. According to the spectral corollary (exercise 18), by solving the eigenvalue problem:

$$\left| \begin{array}{l} \text{Find } (u,\mu) \in \big(H^1(\Omega) \setminus \{0\}\big) \times \mathbb{R} \text{ such that} \\ (u,v)_{H^1(\Omega)} = (\mu+1)\,(u,v)_{L^2(\Omega)}, \ \forall v \in H^1(\Omega) \end{array} \right.$$

we can construct a Hilbert basis $(u_k)_{k\geq 0}$ of $L^2(\Omega)$ consisting of eigenfunctions, along with the corresponding Hilbert basis $(u_k/\sqrt{\mu_k+1})_{k\geq 0}$ of $H^1(\Omega)$.

We say therefore that an *interpolated space* of order s ($s \in [0,1]$) is the space:

$$H^s(\Omega) = \Big\{ v \in L^2(\Omega) \text{ such that } v = \sum_{k\geq 0} \alpha_k u_k \text{ with } \sum_{k\geq 0} \mu_k^s |\alpha_k|^2 < \infty \Big\}$$

endowed with $\|v\|_{H^s(\Omega)} = \sum_{k\geq 0}(1+\mu_k^s)\,|\alpha_k|^2$. We can identify $L^2(\Omega) = H^0(\Omega)$.

a) *Staircase function*: Consider the closed interval $[0,L]$, for $L > 0$.

- Characterize the space $H^s([0,L])$.
- Let $v = \mathbf{1}_{[0,L/2]}$: for which values of s do we have that $v \in H^s([0,L])$?

b) *Trace mapping*: Consider the product of closed intervals $\Omega_2 = [0,L_1] \times [0,L_2]$ for $L_1, L_2 > 0$, and also consider the subset of the boundary $\Gamma = \{0\} \times [0,L_2]$. Denote by γ_Γ the trace mapping on Γ. Establish that the trace mapping γ_Γ is continuous from $H^1(\Omega_2)$ into $L^2(\Gamma)$.

c) *Trace mapping*: Generalize the previous result and establish that the trace mapping γ_Γ is continuous from $H^1(\Omega_2)$ into $H^s(\Gamma)$, if $s < 1/2$.

Exercise 7. Equivalence of norms: Poincaré inequalities in one dimension

a) Show that there exists a constant $C_1 > 0$ such that:

$$\|v\|_{L^2((0,1))} \leq C_1 \left(\|v'\|_{L^2((0,1))} + |v(0)|\right), \quad \forall v \in H^1((0,1)).$$

b) What happens to this inequality in $H_0^1((0,1))$?

c) Show that there exists a constant $C_2 > 0$ such that:

$$\|v\|_{L^2((0,1))} \leq C_2 \left(\|v'\|_{L^2((0,1))} + |m(v)|\right), \quad \forall v \in H^1((0,1))$$

where $m(v) = \displaystyle\int_0^1 v(y)\,dy$ the mean-value of v.

d) Give some examples of subspaces of $H^1((0,1))$ for which the semi-norm of $H^1((0,1))$:

$$|\cdot|_{1,(0,1)} \ : \ v \mapsto \|v'\|_{L^2((0,1))}$$

is a norm that is equivalent to $\|\cdot\|_{H^1((0,1))}$.

Exercise 8. Equivalence of norms: Poincaré inequalities in two dimensions or higher

Let Ω be a bounded and connected open subset of $\mathbb{R}^d$ (d $\geq$ 2), whose boundary $\partial\Omega$ is Lipschitz.

a) Denote by Γ a measurable part of $\partial\Omega$ with a non-zero measure. Show that there is a constant $C_1 > 0$ such that:

$$\|v\|_{L^2(\Omega)} \leq C_1 \left(\|\nabla v\|_{(L^2(\Omega))^d} + \|\gamma_\Gamma v\|_{L^2(\Gamma)}\right), \quad \forall v \in H^1(\Omega)$$

where γ_Γ is the trace mapping $\gamma_\Gamma \ : \ H^1(\Omega) \to L^2(\Gamma)$.

b) Show that there exists a constant $C_2 > 0$ such that:

$$\|v\|_{L^2(\Omega)} \leq C_2 \left(\|\nabla v\|_{(L^2(\Omega))^d} + |m(v)|\right), \quad \forall v \in H^1(\Omega)$$

with $m(v) = \dfrac{1}{\text{meas}\,\Omega}\displaystyle\int_\Omega v\,d\Omega$ the mean-value of v.

c) Let $L \ : \ H^1(\Omega) \to \mathbb{R}$ be a continuous linear functional. Under what condition on L can we prove that there exists a constant $C_3 > 0$, such that:

$$\|v\|_{L^2(\Omega)} \leq C_3 \left(\|\nabla v\|_{(L^2(\Omega))^d} + |L(v)|\right), \quad \forall v \in H^1(\Omega).$$

d) Give some example subspaces of $H^1(\Omega)$ for which the semi-norm of $H^1(\Omega)$:

$$|\cdot|_{1,\Omega} \;:\; v \mapsto \|\nabla v\|_{(L^2(\Omega))^{\mathrm{d}}}$$

is a norm that is equivalent to $\|\cdot\|_{H^1(\Omega)}$.

Hint: We can reason by absurdity to answer questions (a), (b) and (c), with the help of Rellich's theorem 1.6.

Exercise 9. Equivalence of norms and semi-norms

Let Ω be a connected and bounded open subset of $\mathbb{R}^{\mathrm{d}}$, whose boundary $\partial\Omega$ is Lipschitz. Let $m \in \mathbb{N}$, and define $\mathbb{P}^m(\Omega)$ to be the space of polynomials of degree at most m on Ω. The aim of the exercise is to prove that there exists a constant $C_m > 0$ depending only on Ω and m, such that:

$$\inf_{p\in\mathbb{P}^m(\Omega)} \|v+p\|_{H^{m+1}(\Omega)} \leq C_m|v|_{m+1,\Omega},\ \forall v \in H^{m+1}(\Omega) \qquad [I9]$$

where the semi-norm $|\cdot|_{m+1,\Omega}$ is defined by [1.26]. This result is called the Deny-Lions lemma. For this, we need several preliminary results. (Questions (a), (b) and (c)).

a) Let $m \in \mathbb{N}$ and for $v \in H^{m+1}(\Omega)$ non-zero, set:

$$J_v : \left| \begin{array}{l} \mathbb{P}^m(\Omega) \to \mathbb{R} \\ p \mapsto \|v+p\|^2_{H^{m+1}(\Omega)}. \end{array} \right.$$

Show that there exists a unique $p_v \in \mathbb{P}^m(\Omega)$ at which J_v attains its minimum.

b) *Generalization of Rellich's theorem*: By induction on m, show that for any m and for any bounded sequence in $H^{m+1}(\Omega)$, one can extract a subsequence which converges in $H^m(\Omega)$.

c) *Characterization of the elements with zero semi-norm*: By induction on m, show that for all m, if $w \in H^{m+1}(\Omega)$ is such that $|w|_{m1,\Omega} = 0$, then $w \in \mathbb{P}^m(\Omega)$.

d) Let $m \in \mathbb{N}$ and, by reasoning by absurdity, prove [I9].

Exercise 10. Differentiation in the distribution sense

Let Ω be an open subset of $\mathbb{R}^{\mathrm{d}}$: recall that $\left(H_0^1(\Omega)\right)'$ is the space of continuous linear functionals on $H_0^1(\Omega)$.

a) Let I be an open interval in $\mathbb{R}$. Denote by v any element in $L^2(I)$: establish that its derivative *in the distribution sense* belongs to $\left(H_0^1(I)\right)'$. Check that the map which associates v' to v is continuous from $L^2(I)$ into $\left(H_0^1(I)\right)'$.

b) Let Ω be an open subset of $\mathbb{R}^d$. Denote by v any element of $L^2(\Omega)$: establish that its ith partial derivative *in the distribution sense* belongs to $\left(H_0^1(\Omega)\right)'$. Check that the map associating ∂_i to v is continuous from $L^2(\Omega)$ into $\left(H_0^1(\Omega)\right)'$.

c) Let Ω be an open subset of $\mathbb{R}^3$. Denote by $\mathbf{v}$ any element of $\left(L^2(\Omega)\right)^3$: to which space does its divergence and its curl belong to, respectively, *in the distribution sense*?

Exercise 11. Extension and restriction

Let Ω be an open subset of $\mathbb{R}^d$ with a bounded and Lipschitz boundary $\partial\Omega$. Denote by $\mathbf{n}$ the outwards unit normal vector to the boundary.

a) Decompose Ω as $\overline{\Omega} = \overline{\Omega_1} \cup \overline{\Omega_2}$, where Ω_1 and Ω_2 are two disjoint open subsets with Lipschitz boundaries, and denote by F the (open) interface between Ω_1 and Ω_2. Let $v \in L^2(\Omega)$: denote by v_j the restriction of v to Ω_j, for $j = 1, 2$. Suppose that $v_j \in H^1(\Omega_j)$ for $j = 1, 2$ and $\gamma_1 v_1 = \gamma_2 v_2$ on F. Show that $v \in H^1(\Omega)$, and also establish the converse.

b) *Extension*: Let $v \in H_0^1(\Omega)$, which we extend by 0 to $\mathbb{R}^d \setminus \overline{\Omega}$. Denote by $\tilde{v}$ its extension. Show that $\tilde{v} \in H^1(\mathbb{R}^d)$.

c) *Restriction*: Let $v \in H^1(\Omega)$, and denote by $\tilde{v}$ its extension by 0 to $\mathbb{R}^d \setminus \overline{\Omega}$. Show that if $\tilde{v} \in H^1(\mathbb{R}^d)$, then $v \in H_0^1(\Omega)$.

Hint: We can use the following result, which is valid for any measurable and Lipschitz subset Γ of $\partial\Omega$:

$$\text{let } \lambda \in L^2(\Gamma) \text{ be such that } \int_\Gamma \lambda (\mathbf{z}\cdot\mathbf{n})_{|\Gamma} d\Gamma = 0,\ \forall \mathbf{z} \in \left(\mathcal{D}(\mathbb{R}^d)\right)^d, \text{ then } \lambda = 0.$$

Exercise 12. The function space $H(\mathrm{div}, \Omega)$

Let Ω be an open subset of $\mathbb{R}^d$, with a bounded and Lipschitz boundary $\partial\Omega$. The aim is to prove proposition 1.14.

a) *Density*: Show that $\left(\mathcal{C}_c^\infty(\overline{\Omega})\right)^d$ is dense in $H(\mathrm{div}, \Omega)$.

b) *Normal trace*: Let $\mathbf{w} \in \left(\mathcal{C}_c^\infty(\overline{\Omega})\right)^d$. With the help of the definition of $H^{1/2}(\partial\Omega)$, show that $\lambda \mapsto \int_{\partial\Omega} (\mathbf{w}\cdot\mathbf{n})_{|\partial\Omega} \lambda\, d\Gamma$ defines a continuous linear functional on $H^{1/2}(\partial\Omega)$, denoted by $(\mathbf{w}\cdot\mathbf{n})_{|\partial\Omega}$: $(\mathbf{w}\cdot\mathbf{n})_{|\partial\Omega} \in \left(H^{1/2}(\partial\Omega)\right)'$, and show that its action can be written as $\langle (\mathbf{w}\cdot\mathbf{n})_{|\partial\Omega}, \lambda \rangle_{(H^{1/2}(\partial\Omega))', H^{1/2}(\partial\Omega)}$.

c) *Normal trace mapping* γ_n: Show that the normal trace mapping:

$$\gamma_n : \left| \begin{array}{lcl} \left(\mathcal{C}_c^\infty(\overline{\Omega})\right)^{\mathrm{d}} & \to & \left(H^{1/2}(\partial\Omega)\right)' \\ \mathbf{w} & \mapsto & \gamma_n \mathbf{w} = (\mathbf{w}\cdot\mathbf{n})_{|\partial\Omega} \end{array} \right.$$

is continuous with respect to the norm of $H(\mathrm{div};\Omega)$, we can extend it to a continuous linear mapping from $H(\mathrm{div};\Omega)$ into $\left(H^{1/2}(\partial\Omega)\right)'$, again denoted by γ_n, which furthermore satisfies:

$$\|\gamma_n \mathbf{w}\|_{(H^{1/2}(\partial\Omega))'} \le \|\mathbf{w}\|_{H(\mathrm{div};\Omega)}, \qquad \forall \mathbf{w} \in H(\mathrm{div};\Omega).$$

d) *Integration by parts*: Let $\mathbf{w} \in H(\mathrm{div};\Omega)$ and $v \in H^1(\Omega)$. Establish the integration by parts formula:

$$\int_\Omega (\mathrm{div}\,\mathbf{w})\, v\, d\Omega = -\int_\Omega \mathbf{w}\cdot\nabla v\, d\Omega + \langle \gamma_n \mathbf{w}, \gamma_0 v\rangle_{(H^{1/2}(\partial\Omega))', H^{1/2}(\partial\Omega)}.$$

e) *Surjectivity*: Let $g \in \left(H^{1/2}(\partial\Omega)\right)'$ and consider the variational formulation:

$$\left| \begin{array}{l} \text{Find } u \in H^1(\Omega) \text{ such that} \\ (u,v)_{H^1(\Omega)} = \langle g, \gamma_0 v\rangle_{(H^{1/2}(\partial\Omega))', H^{1/2}(\partial\Omega)} \quad \forall v \in H^1(\Omega). \end{array} \right. \qquad \text{[F12]}$$

- Show that [F12] has a unique solution, and interpret it in terms of partial differential equations and boundary conditions.

- Conclude that the trace mapping of the normal derivative γ_1 is surjective.

- Construct $\mathbf{w} \in H(\mathrm{div};\Omega)$ such that $\gamma_n \mathbf{w} = g$, and deduce that the normal trace mapping γ_n is also surjective.

f) *Closure* $H_0(\mathrm{div},\Omega)$: The Sobolev space $H_0(\mathrm{div},\Omega)$ is defined to be the closure of $\left(\mathcal{D}(\Omega)\right)^{\mathrm{d}}$ in $H(\mathrm{div},\Omega)$. Let $\mathbf{w} \in H(\mathrm{div},\Omega)$: How can one (intuitively) characterize the fact that it belongs to $H_0(\mathrm{div},\Omega)$?

Hint: For part (a), one can use the equivalence below which is valid for any Hilbert space H and any vector subspace $W \subset H$: W is dense in H if, and only if:

$$\{\ell \in H' \text{ such that } \ell(w) = 0,\ \forall w \in W\} = \{0\}.$$

2

Variational Formulations and Their Solutions

For classical solutions, we saw in the previous chapter that we can go from a problem with second-order derivatives to a variational formulation with just first-order derivatives. By using the generalized integration-by-parts formulae, now we will justify this transition to the lower order derivatives in the general case. Once that is done, we will solve variational formulations that are defined in Sobolev spaces, allowing us to establish the existence and uniqueness of solutions to the Dirichlet, Neumann and Fourier problems. We place ourselves in Ω, an open set of $\mathbb{R}^{\mathrm{d}}$ whose boundary is Lipschitz and bounded.

2.1. Variational formulations of the Dirichlet, Neumann and Fourier problems

In this section, we will see which spaces naturally appear within the variational formulations of the Dirichlet, Neumann and Fourier problems, which were first introduced in Chapter 1. Given [1.17], a priori we seek solutions that belong to $H^1(\Omega)$.

2.1.1. *Homogeneous Dirichlet problem*

$$\begin{cases} -\Delta u = f & \text{in } \Omega \\ u = 0 & \text{on } \partial\Omega \end{cases} \qquad [2.1]$$

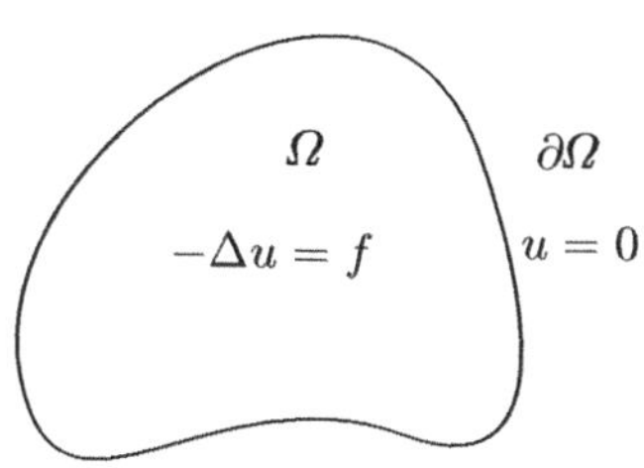

In the following, we assume that $f \in L^2(\Omega)$.

By the virtue of Green's formula [1.39] and since $u \in H^1(\Delta;\Omega)$, for any test function v in $H^1(\Omega)$ observe that:

$$\int_\Omega \nabla u \cdot \nabla v \, d\Omega = \int_\Omega f v \, d\Omega + \langle \frac{\partial u}{\partial n}, v \rangle_{(H^{1/2}(\partial\Omega))', H^{1/2}(\partial\Omega)}.$$

As a general rule, we will avoid specifying that we are actually considering a trace (in the above, we wrote v instead of $\gamma_0 v$ or $v_{|\partial\Omega}$ in the last term), when there is no risk of confusion.

If we consider $v \in H_0^1(\Omega)$, then, given the characterization of $H_0^1(\Omega)$ (theorem 1.8), we have that:

$$\int_\Omega \nabla u \cdot \nabla v \, d\Omega = \int_\Omega f v \, d\Omega \quad \forall v \in H_0^1(\Omega)$$

and hence obtain the **variational formulation** of the **homogeneous Dirichlet problem** [2.1] in the Hilbert space $H_0^1(\Omega)$:

$$\left| \begin{array}{l} \text{Find } u \in H_0^1(\Omega) \text{ such that} \\ \displaystyle\int_\Omega \nabla u \cdot \nabla v \, d\Omega = \int_\Omega f v \, d\Omega \quad \forall v \in H_0^1(\Omega). \end{array} \right. \qquad [2.2]$$

REMARK 2.1.– *The test functions v and the solution u belong to the same space, which in this case is $H_0^1(\Omega)$. Moreover, in this space the Dirichlet boundary condition ($u = 0$) appears explicitly. The Dirichlet boundary condition is said to be an* ***essential condition****, as opposed to a* ***natural boundary condition****, which does not appear in space (see note 2.4).*

We have established that any solution $u \in H^1(\Omega)$ to equation [2.1] also satisfies the variational formulation [2.2]. In fact, the converse also holds. Let us choose a test function $v \in \mathcal{D}(\Omega)$ for [2.2], which is possible since $\mathcal{D}(\Omega) \subset H_0^1(\Omega)$, and then identifying $\partial_i u \in L^2(\Omega)$ with an element of $\mathcal{D}'(\Omega)$. Then, by differentiating in the distribution sense:

$$\begin{aligned} \int_\Omega \nabla u \cdot \nabla v \, d\Omega &= \sum_{i=1}^{d} \int_\Omega \frac{\partial u}{\partial x_i} \frac{\partial v}{\partial x_i} \, d\Omega \\ &= \sum_{i=1}^{d} \langle \frac{\partial u}{\partial x_i}, \frac{\partial v}{\partial x_i} \rangle = -\sum_{i=1}^{d} \langle \frac{\partial^2 u}{\partial x_i^2}, v \rangle = -\langle \Delta u, v \rangle. \end{aligned}$$

Furthermore, if we replace the integral on the right-hand side of [2.2] with the duality bracket, by considering Ω in the distribution sense, we find that:

$$-\Delta u = f \quad \text{in} \quad \mathcal{D}'(\Omega).$$

But since $f \in L^2(\Omega)$ and $L^2(\Omega) \subset \mathcal{D}'(\Omega)$, we have shown furthermore that:

$$-\Delta u = f \quad \text{in } L^2(\Omega)$$

that is, in the function sense (almost everywhere).

There is an alternative reasoning available that does not resort to an integration-by-parts formula, however. Indeed, since $u \in H_0^1(\Omega)$, by working in the distribution sense we can still go from [2.1] to [2.2]. This time, however, not only do we use the fact that $\mathcal{D}(\Omega) \subset H_0^1(\Omega)$, but rather the stronger result that states that $\mathcal{D}(\Omega)$ is dense in $H_0^1(\Omega)$, by definition. Let $v \in H_0^1(\Omega)$, and let $(v_k)_k$ be a sequence in $(\mathcal{D}(\Omega))^{\mathbb{N}}$ converging to v in $H^1(\Omega)$. According to [2.1], we see that:

$$\langle f, v_k\rangle = -\langle \Delta u, v_k\rangle = -\sum_{i=1}^{\mathrm{d}} \langle \frac{\partial^2 u}{\partial x_i^2}, v_k\rangle = \sum_{i=1}^{\mathrm{d}} \langle \frac{\partial u}{\partial x_i}, \frac{\partial v_k}{\partial x_i}\rangle.$$

However, on the one hand, we have that $f \in L^2(\Omega)$, whereas, on the other hand, $\partial_i u \in L^2(\Omega)$ for $i = 1, \mathrm{d}$. Thus, we can replace the first and last duality brackets with their respective integrals:

$$\int_\Omega f v_k \, d\Omega = \sum_{i=1}^{\mathrm{d}} \int_\Omega \frac{\partial u}{\partial x_i} \frac{\partial v_k}{\partial x_i} \, d\Omega.$$

Finally, as $(v_k)_k$ converges to v in $H^1(\Omega)$, we see that $(v_k)_k$ converges to v in $L^2(\Omega)$, but we see also that $(\partial_i v_k)_k$ converges to $\partial_i v$ in $L^2(\Omega)$ for $i = 1, \mathrm{d}$. We can therefore pass to the limit on both the left-hand and the right-hand sides, arriving at:

$$\int_\Omega f v \, d\Omega = \sum_{i=1}^{\mathrm{d}} \int_\Omega \frac{\partial u}{\partial x_i} \frac{\partial v}{\partial x_i} \, d\Omega = \int_\Omega \nabla u \cdot \nabla v \, d\Omega$$

which indeed corresponds to the variational formulation [2.2].

PROPOSITION 2.1.– *Let $f \in L^2(\Omega)$. Then $u \in H_0^1(\Omega)$ satisfies equation [2.1] if and only if u satisfies the variational formulation [2.2].*

By construction, observe that $u \in \Psi_D(\Omega)$, which was the closed subspace of $H^1(\Delta;\Omega)$ introduced at the end of Chapter 1.

REMARK 2.2.– *In the statement of proposition 2.1, we can replace both $L^2(\Omega)$ and $H_0^1(\Omega)$, respectively, with $\mathcal{C}^0(\Omega)$ and $\mathcal{C}^2(\Omega) \cap \mathcal{C}^1(\overline{\Omega})$ (see exercise 2). This relates weak solutions to their classical ones.*

2.1.2. *Non-homogeneous Dirichlet problem*

$$\begin{cases} -\Delta u = f & \text{in } \Omega \\ u = g & \text{on } \partial\Omega \end{cases} \qquad [2.3]$$

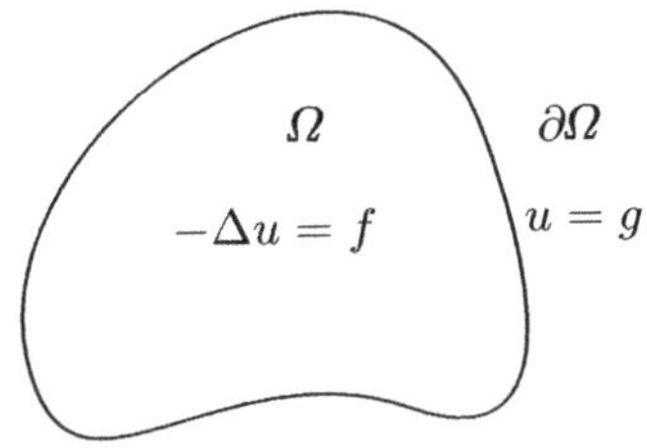

Let us continue to assume that $f \in L^2(\Omega)$. The regularity of the data g is more intricate. Indeed, recall that the trace mapping γ_0 is not surjective from $H^1(\Omega)$ into $L^2(\partial\Omega)$, but it is however surjective from $H^1(\Omega)$ into $H^{1/2}(\partial\Omega)$. Therefore, we will make the following assumption on g:

$$\left| \begin{array}{l} g \in H^{1/2}(\partial\Omega) \textit{ or, equivalently} \\ \text{there exists } \tilde{g} \in H^1(\Omega) \text{ such that } \gamma_0\tilde{g} = g. \end{array} \right. \qquad [2.4]$$

The function $\tilde{g}$ is called an **extension** of the function g, and moreover we can use $\|\tilde{g}\|_{H^1(\Omega)}$ to measure g. Indeed, from [1.32], we see that $\|g\|_{H^{1/2}(\partial\Omega)} \leq \|\tilde{g}\|_{H^1(\Omega)}$.

We are now able to establish the variational formulation for the non-homogeneous Dirichlet problem. First, let us consider the product of the equation on Ω with a test function $v \in H_0^1(\Omega)$, then by using Green's formula [1.39], we get:

$$\int_\Omega \nabla u \cdot \nabla v \, d\Omega = \int_\Omega f\, v \, d\Omega.$$

Indeed, given the fact that the trace of v vanishes on $\partial\Omega$, the boundary term disappears and we end up with the variational formulation:

$$\left| \begin{array}{l} \text{Find } u \in H^1(\Omega) \text{ such that } u = g \text{ on } \partial\Omega \text{ and} \\ \displaystyle\int_\Omega \nabla u \cdot \nabla v \, d\Omega = \int_\Omega f\, v d\Omega \quad \forall v \in H_0^1(\Omega). \end{array} \right. \qquad [2.5]$$

Note that the test function space – which here is $H_0^1(\Omega)$ – is not the same as the solution space – which is instead $H^1(\Omega)$ here. This is why one makes the change in variable:

$$\tilde{u} = u - \tilde{g}$$

which leads to the variational formulation for just one of the two spaces:

$$\left|\begin{array}{l} \text{Find } \tilde{u} \in H_0^1(\Omega) \text{ such that} \\ \displaystyle\int_\Omega \nabla\tilde{u}\cdot\nabla v d\Omega = \int_\Omega f v d\Omega - \int_\Omega \nabla\tilde{g}\cdot\nabla v d\Omega \qquad \forall v \in H_0^1(\Omega) \end{array}\right. \tag{2.6}$$

with the solution u given by $u = \tilde{u} + \tilde{g}$.

REMARK 2.3.– *In fact, the problem has reduced to the homogeneous one.*

We could have also reasoned in the distribution sense, in order to pass from [2.3] to the variational formulation [2.6]. Conversely, from [2.6] and with the change in variable $\tilde{u} = u - \tilde{g}$, we arrive back at the equation $-\Delta u = f$, just like in the homogeneous case. And since $\tilde{u} = 0$ on $\partial\Omega$, we recover the second equation $u = g$ on $\partial\Omega$. To summarize, we can go from [2.6] to [2.3]. This time, therefore, we have the following result:

PROPOSITION 2.2.– *Consider $f \in L^2(\Omega)$, $g \in H^{1/2}(\partial\Omega)$ and let $\tilde{g}$ be an extension of g. Then $u \in H^1(\Omega)$ satisfies equation [2.3] if and only if $\tilde{u} = u - \tilde{g}$ satisfies the variational formulation [2.6].*

2.1.3. *Neumann and Fourier problems*

We are going to deal with the Neumann and the Fourier cases simultaneously by introducing the general problem:

$$\left\{\begin{array}{ll} -\Delta u + \beta u = f & \text{in } \Omega \\ \dfrac{\partial u}{\partial n} + \lambda u = g & \text{on } \partial\Omega \end{array}\right. \tag{2.7}$$

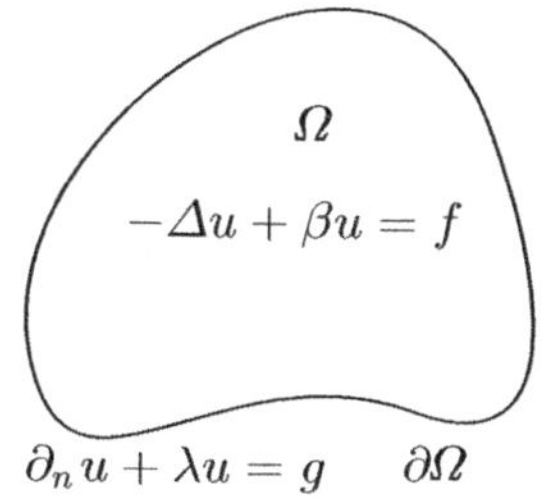

where β and λ are real constants[1], $f \in L^2(\Omega)$ and $g \in L^2(\partial\Omega)$.

1 The case where $\beta < 0$ corresponds to the Helmholtz equation, whereas the $\lambda = 0$ case corresponds to Neumann conditions.

Multiply the first equation of [2.7] with a test function $v \in H^1(\Omega)$ and then by integrating on Ω, from [1.39] we obtain:

$$\int_\Omega \nabla u \cdot \nabla v \, d\Omega + \beta \int_\Omega uv \, d\Omega = \int_\Omega fv \, d\Omega + \langle \frac{\partial u}{\partial n}, v \rangle_{(H^{1/2}(\partial\Omega))', H^{1/2}(\partial\Omega)}.$$

Given the boundary conditions, we may rewrite this as:

$$\langle \frac{\partial u}{\partial n}, v \rangle_{(H^{1/2}(\partial\Omega))', H^{1/2}(\partial\Omega)} = \int_{\partial\Omega} gv \, d\Gamma - \lambda \int_{\partial\Omega} uv \, d\Gamma.$$

Due to the regularity of g and of the traces, $u_{|\partial\Omega}$ and $v_{|\partial\Omega}$, note that the terms on $\partial\Omega$ are integrals. This leads us to the variational formulation for the problem [2.7] in $H^1(\Omega)$:

$$\left| \begin{aligned} &\text{Find } u \in H^1(\Omega) \text{ such that} \\ &\int_\Omega \nabla u \cdot \nabla v \, d\Omega + \beta \int_\Omega uv d\Omega + \lambda \int_{\partial\Omega} uv \, d\Gamma \\ &\qquad = \int_\Omega fv \, d\Omega + \int_{\partial\Omega} gv \, d\Gamma \quad \forall v \in H^1(\Omega). \end{aligned} \right. \qquad [2.8]$$

REMARK 2.4.– *The main difference when dealing with the Dirichlet problem case resides in the fact that the boundary conditions appear explicitly in the variational formulation [2.8], and no longer from the space. Here, we are talking about* ***natural boundary conditions****. When $g = 0$ and $\lambda = 0$, they no longer appear even explicitly.*

There exists an equivalence still between the variational formulation and the strong formulation.

PROPOSITION 2.3.– *Let $f \in L^2(\Omega)$ and $g \in L^2(\partial\Omega)$. Then $u \in H^1(\Omega)$ satisfies equation [2.7] if and only if u satisfies the variational formulation [2.8].*

Proof. The proof here is a little more instructive than those for previous results, since the boundary conditions have to reappear. Suppose that $u \in H^1(\Omega)$ satisfies [2.8], and consider $v \in \mathcal{D}(\Omega) \subset H^1(\Omega)$. Then we have that:

$$\int_\Omega \nabla u \cdot \nabla v \, d\Omega + \beta \int_\Omega uv \, d\Omega = \int_\Omega fv \, d\Omega \qquad \text{since } v = 0 \text{ on } \partial\Omega.$$

By replacing the integrals with duality brackets, in the distribution sense, it becomes:

$$\langle f, v \rangle = \sum_{i=1}^{d} \langle \frac{\partial u}{\partial x_i}, \frac{\partial v}{\partial x_i} \rangle + \beta \langle u, v \rangle = \langle -\Delta u + \beta u, v \rangle$$

which means that: $\langle -\Delta u + \beta u, v\rangle = \langle f, v\rangle \quad \forall v \in \mathcal{D}(\Omega)$. Let $-\Delta u + \beta u = f$ in the distribution sense, and additionally in the function sense too, because $\beta u - f \in L^2(\Omega)$. In particular $u \in H^1(\Delta;\Omega)$, and furthermore:

$$\int_\Omega (-\Delta u + \beta u) v \, d\Omega = \int_\Omega f\, v \, d\Omega, \quad \forall v \in H^1(\Omega).$$

Now let us consider $v \in H^1(\Omega)$ in [2.8] and then apply Green's formula [1.39]:

$$\int_\Omega (-\Delta u + \beta u) v \, d\Omega + \langle \frac{\partial u}{\partial n}, v\rangle_{(H^{1/2}(\partial\Omega))', H^{1/2}(\partial\Omega)} + \lambda \int_{\partial\Omega} uv \, d\Gamma$$
$$= \int_\Omega fv \, d\Omega + \int_{\partial\Omega} gv \, d\Gamma$$

which becomes:

$$\langle \frac{\partial u}{\partial n}, v\rangle_{(H^{1/2}(\partial\Omega))', H^{1/2}(\partial\Omega)} + \lambda \int_{\partial\Omega} uv \, d\Gamma = \int_{\partial\Omega} gv \, d\Gamma.$$

Provided that we know $u_{|\partial\Omega}$ and g both belong to $L^2(\partial\Omega)$, then we know that $\frac{\partial u}{\partial n}_{|\partial\Omega}$ belongs to $(H^{1/2}(\partial\Omega))'$. But according to [1.35], we also have that $\lambda u_{|\partial\Omega} - g \in (H^{1/2}(\partial\Omega))'$. Given what we have just established and since the trace mapping γ_0 is surjective from $H^1(\Omega)$ into $H^{1/2}(\partial\Omega)$, we arrive at:

$$\langle \frac{\partial u}{\partial n} + \lambda u - g, w\rangle_{(H^{1/2}(\partial\Omega))', H^{1/2}(\partial\Omega)} = 0 \qquad \forall w \in H^{1/2}(\partial\Omega).$$

This means that the linear form $\frac{\partial u}{\partial n}_{|\partial\Omega} + \lambda u_{|\partial\Omega} - g = 0$ in $(H^{1/2}(\partial\Omega))'$. On the other hand, as $\lambda u_{|\partial\Omega} - g \in L^2(\partial\Omega)$, the equality $\frac{\partial u}{\partial n}_{|\partial\Omega} = g - \lambda u_{|\partial\Omega}$ holds in $L^2(\partial\Omega)$ (see [1.35] again) proving that $\frac{\partial u}{\partial n} + \lambda u = g$ almost everywhere on $\partial\Omega$. ∎

Lastly, we can deduce a more general version of proposition 2.3, so that we can solve the Neumann and Fourier problems when $f \in L^2(\Omega)$ and $g \in \left(H^{1/2}(\partial\Omega)\right)'$. In this case, the variational formulation becomes:

$$\left|\begin{array}{l} \text{Find } u \in H^1(\Omega) \text{ such that} \\ \displaystyle\int_\Omega \nabla u \cdot \nabla v \, d\Omega + \beta \int_\Omega uv d\Omega + \lambda \int_{\partial\Omega} uv \, d\Gamma \\ \displaystyle\qquad = \int_\Omega fv \, d\Omega + \langle g, v\rangle_{(H^{1/2}(\partial\Omega))', H^{1/2}(\partial\Omega)}, \ \forall v \in H^1(\Omega). \end{array}\right. \qquad [2.9]$$

PROPOSITION 2.4.– *Let $f \in L^2(\Omega)$ and $g \in \left(H^{1/2}(\partial\Omega)\right)'$. Then $u \in H^1(\Omega)$ satisfies equation [2.7] if and only if u satisfies the variational formulation [2.9].*

Proof. Left as an exercise. ■

By construction, note that the solution to the homogeneous Neumann problem (i.e. with the parameter $\lambda = 0$ and the data $g = 0$) is such that $u \in \Psi_N(\Omega)$. The treatment of the problem with mixed boundary conditions [1.5] is put forth in exercise 20.

2.2. Other variational formulations examples

The analysis made extends to even more general situations, such as with operators with variable coefficients, or with higher order equations. Moreover, one can propose even weaker formulations.

2.2.1. *Variable coefficient problems*

Consider a possibly discontinuous function $k(x)$ along with the variable coefficient problem:

$$\begin{cases} u - \operatorname{div}(k(x)\nabla u) = f & \text{in } \Omega \\ \dfrac{\partial u}{\partial n} = 0 & \text{on } \partial\Omega. \end{cases} \qquad [2.10]$$

Assume that $k \in L^\infty(\Omega)$, then it is easy to check that $k\nabla u \in H(\mathrm{div};\Omega)$. By multiplying it with a test function $v \in H^1(\Omega)$ and integrating-by-parts (see [1.34]), we arrive at the following variational formulation:

$$\left| \begin{array}{l} \text{Find } u \in H^1(\Omega) \text{ such that} \\ \displaystyle\int_\Omega uv\, d\Omega + \int_\Omega k(x)\nabla u \cdot \nabla v\, d\Omega = \int_\Omega fv\, d\Omega, \quad \forall v \in H^1(\Omega). \end{array} \right. \qquad [2.11]$$

This variational formulation makes sense even when k is a discontinuous function. As before, it is sufficient for the function k to belong to $L^\infty(\Omega)$. An example of a similar problem (heat diffusion) is provided in exercise 14.

REMARK 2.5.– *Following the same idea, one could also consider the case when a coefficient k depends on the solution u, which leads to nonlinear problems that are more difficult to study, since the solutions strongly depend on the nature of the function associating $k(u)$ to u.*

2.2.2. *Second-order elliptic problems*

The Laplace equation is just a special case of an elliptic one. Indeed, consider the partial derivative operator P, defined by:

$$Pu = -\sum_{i,j=1}^{\mathrm{d}} \frac{\partial}{\partial x_i}\left(a_{ij}\frac{\partial u}{\partial x_j}\right) + \sum_{i=1}^{\mathrm{d}} b_i \frac{\partial u}{\partial x_i} + cu.$$

and when:

$$\sum_{i,j=1}^{\mathrm{d}} a_{ij} X_i X_j \geq \alpha\,|\mathbf{X}|^2 \quad \forall \mathbf{X} \in \mathbb{R}^{\mathrm{d}}$$

where $\alpha > 0$ is independent of $\mathbf{X}$, then it is said to be uniformly strongly elliptical (the Laplacian case corresponds to $a_{ij} = \delta_{ij}$ for $i, j = 1, \cdots, \mathrm{d}$). Provided that the coefficients a_{ij}, b_i and c, which can depend on x, belong to $L^\infty(\Omega)$, then the homogeneous Dirichlet problem:

$$\begin{cases} Pu = f & \text{in } \Omega \\ u = 0 & \text{on } \partial\Omega \end{cases} \qquad [2.12]$$

has the following variational formulation:

$$\left|\begin{array}{l} \text{Find } u \in H_0^1(\Omega) \text{ such that} \\ \displaystyle\sum_{i,j=1}^{\mathrm{d}} \int_\Omega a_{ij} \frac{\partial u}{\partial x_i}\frac{\partial v}{\partial x_j}\, d\Omega + \sum_{i=1}^{\mathrm{d}} \int_\Omega b_i \frac{\partial u}{\partial x_i}\, v\, d\Omega \\ \qquad\qquad + \displaystyle\int_\Omega c\, u\, v\, d\Omega = \int_\Omega f\, v\, d\Omega, \ \forall v \in H_0^1(\Omega). \end{array}\right. \qquad [2.13]$$

2.2.3. *Higher order problems*

As an example, consider the bilaplacian along with the Dirichlet and tangential boundary conditions:

$$\begin{cases} \Delta^2 u = f & \text{in } \Omega \\ u = 0 & \text{on } \partial\Omega \\ \Delta u = 0 & \text{on } \partial\Omega. \end{cases} \qquad [2.14]$$

Multiplying it with a test function $v \in H^2(\Omega)$ such that $v = 0$ on $\partial\Omega$ and then by applying Green's formula [1.42] twice, formally we arrive at:

$$\int_\Omega \Delta^2 u\, v\, d\Omega = \int_\Omega \Delta u \Delta v\, d\Omega + \int_{\partial\Omega} \left(\frac{\partial}{\partial n}(\Delta u)\, v - \Delta u \frac{\partial v}{\partial n}\right) d\Gamma.$$

Given the boundary conditions, it then becomes:

$$\int_\Omega \Delta u \Delta v \, d\Omega = \int_\Omega f v \, d\Omega \quad \forall v \in H^2(\Omega), \ v = 0 \text{ on } \partial\Omega. \qquad [2.15]$$

which makes up the variational formulation of [2.14] for the space:

$$H^2(\Omega) \cap H_0^1(\Omega) = \left\{ v \in H^2(\Omega), v = 0 \text{ sur } \partial\Omega \right\}.$$

The condition $u = 0$ on $\partial\Omega$ appears to be an essential condition for the problem, whereas $\Delta u = 0$ on $\partial\Omega$ is a natural boundary condition. Note that the order of the differentials has been halved in the variational formulation.

2.2.4. *Less-regular data*

In previous examples, we always considered $f \in L^2(\Omega)$. In fact, we can sometimes settle for data that is less regular in Ω. For example, given the Dirichlet problem, consider[2] $f \in \left(H_0^1(\Omega)\right)'$, the dual space of $H_0^1(\Omega)$, which contains $L^2(\Omega)$ and is included in $\mathcal{D}'(\Omega)$. In this case, for all $v \in \mathcal{D}(\Omega)$, we have that $\langle f, v \rangle_{(H_0^1(\Omega))', H_0^1(\Omega)} = \langle f, v \rangle$, where $\langle \cdot, \cdot \rangle_{(H_0^1(\Omega))', H_0^1(\Omega)}$ denotes the duality bracket between $H_0^1(\Omega)$ and its dual. Thus let us introduce the following weak formulation for the homogeneous Dirichlet problem:

$$\left| \begin{array}{l} \text{Find } u \in H_0^1(\Omega) \text{ such that} \\ \displaystyle\int_\Omega \nabla u \cdot \nabla v \, d\Omega = \langle f, v \rangle_{(H_0^1(\Omega))', H_0^1(\Omega)} \quad \forall v \in H_0^1(\Omega) \end{array} \right. \qquad [2.16]$$

whose solution u belongs to $H^1(\Omega)$ and satisfies:

$$\begin{cases} -\Delta u = f & \text{in the distribution sense in } \Omega \\ u = 0 & \text{on } \partial\Omega. \end{cases}$$

For this, we may use the fact that $\mathcal{D}(\Omega)$ is a subspace of $H_0^1(\Omega)$. Let $v \in \mathcal{D}(\Omega)$, then beginning with the weak formulation and then differentiating successively in the distribution sense, we arrive at:

$$\begin{aligned} \langle f, v \rangle_{(H_0^1(\Omega))', H_0^1(\Omega)} &= \int_\Omega \nabla u \cdot \nabla v \, d\Omega \\ &= \sum_{i=1}^{d} \langle \frac{\partial u}{\partial x_i}, \frac{\partial v}{\partial x_i} \rangle = -\sum_{i=1}^{d} \langle \frac{\partial^2 u}{\partial x_i^2}, v \rangle = -\langle \Delta u, v \rangle. \end{aligned}$$

2 According to Adams and Fournier (2003), we may identify the dual of $H_0^1(\Omega)$ with the Sobolev space of index -1: $H^{-1}(\Omega) = (H_0^1(\Omega))'$.

In other words, $-\Delta u = f$ in $\mathcal{D}'(\Omega)$. Of course, the boundary condition comes from $u \in H_0^1(\Omega)$. Conversely, we can pass from the Dirichlet problem to the weak formulation this time by using the fact that $\mathcal{D}(\Omega)$ is a dense subspace of $H_0^1(\Omega)$.

It is quite clear that we cannot consider just any non-regular data (any distribution), without possibly losing the existence and uniqueness properties.

The simplest "example" of this is the following. For the Neumann and Fourier problems, if we consider $f \in (H_0^1(\Omega))'$, then a solution no longer exists. Indeed, the variational formulation [2.9] was established for any v in $H^1(\Omega)$. However, f belongs to the dual of $H_0^1(\Omega)$: therefore, a priori, we cannot always define the action of f on an element v in $H^1(\Omega) \setminus H_0^1(\Omega)$. In particular, this example illustrates the difference between essential boundary conditions (for the Dirichlet problem) and natural boundary conditions (for the Neumann and Fourier problems).

2.3. Existence and uniqueness of weak solutions

All of the variational formulations presented in the previous section fit into the following general framework:

$$\left| \begin{array}{l} \text{Find } u \in H \text{ such that} \\ a(u,v) = \ell(v) \qquad \forall v \in H \end{array} \right. \qquad [2.17]$$

where:

– H is the *solution* space, whose elements are denoted by u, and the *test function* space, whose elements are denoted by v;

– $a(\cdot,\cdot)$ is a *bilinear form* on H and contains the information regarding the equations and boundary conditions;

– $\ell(\cdot)$ is a *linear* form on H and contains the data.

Thus, for the Dirichlet [2.1] and Neumann–Fourier problems [2.7], we have Table 2.1.

Here, we have chosen the data $f \in L^2(\Omega)$ and $g \in L^2(\partial\Omega)$. The space H is a *Hilbert* space endowed with an inner product $(\cdot,\cdot)_H$, whose associated norm is denoted by $\|\cdot\|_H$.

Before covering the general existence and uniqueness theory for the solutions to the problem [2.17] in section 2.4, we will highlight the principles that underlay this theory with the help of examples. In particular, we will emphasize the *continuity* of the form ℓ, as well as the definition of a norm on H which is induced by a form a, and is *equivalent* to standard $\|\cdot\|_H$.

	Dirichlet	Neumann-Fourier
$\boldsymbol{H}$	$H_0^1(\Omega)$	$H^1(\Omega)$
$\boldsymbol{a(u,v)}$	$\int_\Omega \nabla u \cdot \nabla v d\Omega$	$\int_\Omega (\nabla u \cdot \nabla v + \beta uv) d\Omega + \int_{\partial\Omega} \lambda uv d\Gamma$
$\boldsymbol{\ell(v)}$	$\int_\Omega fv d\Omega$	$\int_\Omega fv d\Omega + \int_{\partial\Omega} gv d\Gamma$

Table 2.1. *Spaces and their respective forms*

2.3.1. *Neumann problem*

Fix $\lambda = 0$. The Fourier problem ($\lambda \neq 0$) will be dealt with in exercise 13. Let us first consider the case when $\beta = 1$. Then we have:

$$a(u,v) = \int_\Omega (\nabla u \cdot \nabla v + uv) d\Omega = (u,v)_{H^1(\Omega)} \text{ (inner product on } H^1(\Omega))$$

and:

$$\ell(v) = \int_\Omega fv d\Omega + \int_{\partial\Omega} gv d\Omega.$$

The mapping $\ell : \left| \begin{array}{ll} H^1(\Omega) & \rightarrow \mathbb{R} \\ v & \mapsto \ell(v) \end{array} \right.$ is clearly continuous and linear on $H^1(\Omega)$ thanks to the Cauchy–Schwarz inequalities on $L^2(\Omega)$ and $L^2(\partial\Omega)$:

$$\left| \int_\Omega fv \, d\Omega \right| \leq \|f\|_{L^2(\Omega)} \|v\|_{L^2(\Omega)} \leq \|f\|_{L^2(\Omega)} \|v\|_{H^1(\Omega)} \qquad [2.18]$$

$$\left| \int_{\partial\Omega} gv \, d\Gamma \right| \leq \|g\|_{L^2(\partial\Omega)} \|\gamma_0 v\|_{L^2(\partial\Omega)} \underset{\text{(continuous trace [1.31])}}{\leq} C \|g\|_{L^2(\partial\Omega)} \|v\|_{H^1(\Omega)}. \qquad [2.19]$$

Applying Riesz's representation theorem asserts the existence and uniqueness of a solution $u \in H^1(\Omega)$ for the variational formulation [2.17] of the Neumann problem, along with the following continuity estimate:

$$\|u\|_{H^1(\Omega)} = \|\ell\|_{(H^1(\Omega))'} = \sup_{v \neq 0} \frac{|\ell(v)|}{\|v\|_{H^1(\Omega)}}.$$

Now, by virtue of the estimates [2.18] and [2.19], we have:

$$\sup_{v\neq 0} \frac{|\ell(v)|}{\|v\|_{H^1(\Omega)}} \leq \max(1,C)\, \left(\|f\|_{L^2(\Omega)} + \|g\|_{L^2(\partial\Omega)}\right).$$

Now consider the case with $\beta > 0$ along with the bilinear form:

$$a(u,v) = \int_\Omega (\nabla u \cdot \nabla v + \beta\, uv) d\Omega$$

which defines an equivalent norm on $H^1(\Omega)$, and moreover is equal to the norm $\|\cdot\|_{H^1(\Omega)}$ if $\beta = 1$. Indeed, we have that:

$$\min(1,\beta)\, \|v\|^2_{H^1(\Omega)} \leq a(v,v) \leq \max(1,\beta)\, \|v\|^2_{H^1(\Omega)} \qquad \forall v \in H^1(\Omega). \quad [2.20]$$

Hence, we can endow $H^1(\Omega)$ with the inner product that is associated with the norm $v \mapsto \sqrt{a(v,v)}$: equipped with this inner product, $H^1(\Omega)$ becomes a Hilbert space. In addition, as before we check that the mapping ℓ is a linear and continuous form on $H^1(\Omega)$ with respect to this norm: we just have to go back to the previous calculations and use [2.20]. Afterwards, we just apply Riesz's representation theorem again.

Hence, we obtain the following result:

THEOREM 2.1 (Uniqueness and existence for the Neumann problem).– *Let $\beta > 0$ be given, and let $f \in L^2(\Omega)$ and $g \in L^2(\partial\Omega)$. Then there exists a unique weak solution to the Neumann problem:*

$$\begin{cases} \text{Find } u \in H^1(\Omega) \text{ such that} \\ -\Delta u + \beta\, u = f \textit{ in } \Omega \\ \dfrac{\partial u}{\partial n} = g \textit{ on } \partial\Omega. \end{cases}$$

Moreover, the solution u depends continuously on the data f and g, that is, that there exists a constant C, independent of u, f and g, such that:

$$\|u\|_{H^1(\Omega)} \leq C\left(\|f\|_{L^2(\Omega)} + \|g\|_{L^2(\partial\Omega)}\right).$$

REMARK 2.6.– *This notion of continuous dependence is sometimes also called stability with respect to data (see section 2.4.4).*

The case $\beta = 0$ is considered in exercise 19, and the case $\beta < 0$ is covered in section 2.4.3.

2.3.2. *Dirichlet problem – Poincaré's inequality*

In the case of the homogeneous Dirichlet problem, a priori the bilinear form:

$$a(u, v) = \int_\Omega \nabla u \cdot \nabla v d\Omega$$

does not define a norm which is equivalent to the one induced by $H^1(\Omega)$ on $H_0^1(\Omega)$:

$$\|v\|_{H^1(\Omega)}^2 = \int_\Omega v^2 d\Omega + \int_\Omega |\nabla v|^2 d\Omega.$$

We have only one of the two inequalities required for us to characterize the equivalence of norms, namely:

$$a(v, v) = \int_\Omega |\nabla v|^2 d\Omega \leq \|v\|_{H^1(\Omega)}^2 \quad \forall v \in H^1(\Omega). \qquad [2.21]$$

Consequently, Riesz's representation theorem does not apply to the usual inner product of $H^1(\Omega)$.

However, we can prove a converse inequality, which is satisfied only for functions belonging to $H_0^1(\Omega)$. This is called **Poincaré's inequality**.

PROPOSITION 2.5 (Poincaré's inequality).– *Assume that Ω is bounded. Then there exists a constant C_p, depending only on Ω, such that:*

$$\int_\Omega v^2 d\Omega \leq C_p \int_\Omega |\nabla v|^2 d\Omega \quad \forall v \in H_0^1(\Omega). \qquad [2.22]$$

REMARK 2.7.– *This inequality does not hold on $H^1(\Omega)$ (take $v = 1$).*

Proof. By the density of $\mathcal{D}(\Omega)$ in $H_0^1(\Omega)$, then it suffices to consider a function v of class $\mathcal{C}^\infty$ in Ω that is compactly supported on Ω. Denote by $\tilde{v}$ the extension of v by 0 outside of Ω ($\tilde{v}$ is defined on $\mathbb{R}^{\mathrm{d}}$).

As Ω is assumed to be bounded in $\mathbb{R}^{\mathrm{d}}$, it is necessarily contained in a strip $\{(x_1, \cdots, x_{\mathrm{d}}) \in \mathbb{R}^{\mathrm{d}}$ such that $a \leq x_{\mathrm{d}} \leq b\}$ (we set $x = (x', x_{\mathrm{d}})$ with $x' = (x_1, \cdots, x_{\mathrm{d}-1})$).

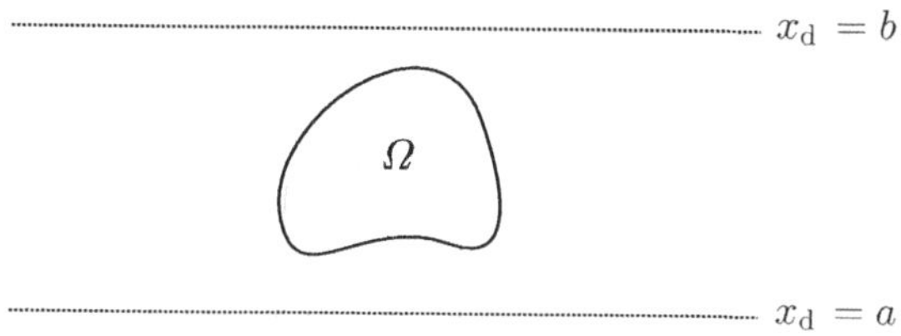

Then as v is regular, we can write: $\tilde{v}(x', x_d) = \displaystyle\int_a^{x_d} \frac{\partial \tilde{v}}{\partial x_d}(x', t)\, dt$, which by the Cauchy–Schwarz inequality leads us to the estimate:

$$|\tilde{v}(x', x_d)|^2 \leq (x_d - a) \int_a^{x_d} \left| \frac{\partial \tilde{v}}{\partial x_d}(x', t) \right|^2 dt \leq (x_d - a) \int_{-\infty}^{+\infty} \left| \frac{\partial \tilde{v}}{\partial x_d}(x', t) \right|^2 dt$$

and by using Fubini's theorem, this shows that:

$$\int_{\mathbb{R}^{d-1}} |\tilde{v}(x', x_d)|^2 \, dx' \leq (x_d - a) \int_{\mathbb{R}^d} \left| \frac{\partial \tilde{v}}{\partial x_d}(x', t) \right|^2 dx' dt.$$

By integrating from a to b with respect to x_d, finally we obtain:

$$\int_{\mathbb{R}^d} \tilde{v}^2 dx \leq \int_a^b (x_d - a)\, dx_d \int_{\mathbb{R}^d} \left| \frac{\partial \tilde{v}}{\partial x_d} \right|^2 dx = \frac{(b-a)^2}{2} \int_{\mathbb{R}^d} \left| \frac{\partial \tilde{v}}{\partial x_d} \right|^2 dx$$

which provides the desired inequality, since $\tilde{v} = 0$ outside of Ω:

$$\|v\|^2_{L^2(\Omega)} \leq \frac{1}{2}(b-a)^2 \int_\Omega |\nabla v|^2 d\Omega \quad \forall v \in \mathcal{D}(\Omega).$$

Lastly, by a density argument, this last inequality holds true for all $v \in H_0^1(\Omega)$. ■

REMARK 2.8.– *According to the proof provided, it suffices for Ω to be an open set that is* bounded in one direction *for Poincaré's inequality to hold. Indeed, after a possible change of basis in $\mathbb{R}^d$, we can always assume that we are in the case where $\Omega \subset \{(x_1, \cdots, x_d) \in \mathbb{R}^d$ such that $a \leq x_d \leq b\}$. Furthermore, the constant C_p can be chosen so that it is proportional to the square of the distance in this direction, similar to the term in $\frac{1}{2}(b-a)^2$ in the proof.*

There is also an ad absurdum proof of Poincaré's inequality based on the compact injection from $H_0^1(\Omega)$ into $L^2(\Omega)$ (see exercises 7 and 8). In the case where Ω is also connected, this proof allows us to establish very useful generalizations of Poincaré's inequality: there exists $C_{PW}, C_{PF} > 0$ such that:

– Poincaré–Wirtinger inequality:

$$\int_\Omega v^2 d\Omega \leq C_{PW}\Big(\int_\Omega |\nabla v|^2 d\Omega + \frac{1}{\text{meas}(\Omega)}\Big(\int_\Omega v\, d\Omega\Big)^2\Big) \quad \forall v \in H^1(\Omega). \qquad [2.23]$$

– Poincaré–Friedrichs inequality ($\Gamma \subset \partial\Omega$, $\text{meas}(\Gamma) > 0$):

$$\int_\Omega v^2 d\Omega \leq C_{PF}\Big(\int_\Omega |\nabla v|^2 d\Omega + \Big(\int_\Gamma v\, d\Gamma\Big)^2\Big) \quad \forall v \in H^1(\Omega). \qquad [2.24]$$

From Poincaré's inequality [2.22], immediately we see that:

$$\int_\Omega |\nabla v|^2 d\Omega \geq \frac{1}{1+C_p}\|v\|^2_{H^1(\Omega)}, \qquad \forall v \in H^1_0(\Omega). \qquad [2.25]$$

And given [2.21], this shows that:

$$a(v,v) = \int_\Omega |\nabla v|^2 d\Omega$$

defines an equivalent norm to the one on $H^1(\Omega)$, albeit *on the space* $H^1_0(\Omega)$. Thus, the semi-norm H^1 that was defined in [1.25] is therefore also a norm on $H^1_0(\Omega)$ and is associated with the inner product:

$$(u,v)_{H^1_0(\Omega)} = \int_\Omega \nabla u \cdot \nabla v\, d\Omega. \qquad [2.26]$$

This result now allows us to apply Riesz's representation theorem by choosing an inner product [2.26] on $H^1_0(\Omega)$. Hence, we obtain the existence and uniqueness theorem for the homogeneous Dirichlet problem, since $\ell(v) = \int_\Omega fv\, d\Omega$ remains a continuous linear form on $H^1_0(\Omega)$, because of the estimate:

$$\Big|\int_\Omega f\, v\, d\Omega\Big| \leq \|f\|_{L^2(\Omega)}\|v\|_{L^2(\Omega)} \underset{(Poincaré)}{\leq} C_p^{1/2}\,\|f\|_{L^2(\Omega)}|v|_1.$$

Thus we can now state the following result:

THEOREM 2.2 (Existence and uniqueness for the Dirichlet problem).– *Assume that Ω is bounded, and let $f \in L^2(\Omega)$. Then there exists a unique weak solution to the homogeneous Dirichlet problem:*

$$\begin{cases} \text{Find } u \in H^1_0(\Omega) \text{ such that} \\ -\Delta u = f \text{ in } \Omega \end{cases}$$

which depends continuously on the data f: there exists a $C > 0$, independent of u and f, such that:

$$\|u\|_{H^1(\Omega)} \leq (1+C_p)^{1/2}|u|_1 \leq C\|f\|_{L^2(\Omega)}.$$

2.3.3. *The non-homogeneous Dirichlet problem*

When Ω is bounded, the existence and uniqueness of a solution to the non-homogeneous Dirichlet problem follows from the previous analysis.

Indeed, let us recall that with the hypothesis $g \in H^{1/2}(\partial\Omega)$ (see [2.4]), with the change of variable $\tilde{u} = u - \tilde{g}$ the variational formulation reduces to one set in $H_0^1(\Omega)$, with:

$$a(\tilde{u}, v) = \int_\Omega \nabla\tilde{u} \cdot \nabla v \, d\Omega \quad \text{and} \quad \ell(v) = \int_\Omega f v \, d\Omega - \int_\Omega \nabla\tilde{g} \cdot \nabla v \, d\Omega.$$

So it suffices to check that ℓ is a continuous linear form on $H_0^1(\Omega)$ with respect to the norm [1.25]. Now, using the Cauchy–Schwarz and Poincaré inequalities, we have that:

$$\begin{aligned}|\ell(v)| &\leq \|f\|_{L^2(\Omega)}\|v\|_{L^2(\Omega)} + \|\nabla\tilde{g}\|_{(L^2(\Omega))^d}\|\nabla v\|_{(L^2(\Omega))^d} \\ &\leq {C_p}^{1/2}\|f\|_{L^2(\Omega)}|v|_1 + |\tilde{g}|_1|v|_1\end{aligned}$$

proving that a unique solution $\tilde{u} \in H_0^1(\Omega)$ exists to the variational formulation [2.6]. More specifically, we may state that:

THEOREM 2.3 (Existence and uniqueness for the non-homogeneous Dirichlet problem).– *Assume that Ω is bounded, and let $f \in L^2(\Omega)$ and $g \in H^{1/2}(\partial\Omega)$. Then there exists a unique weak solution to the non-homogeneous Dirichlet problem:*

$$\begin{cases} \text{Find } u \in H^1(\Omega) \text{ such that} \\ -\Delta u = f \textit{ in } \Omega \\ u = g \textit{ on } \partial\Omega \end{cases}$$

which depends continuously on the data f and g: there exists a $C > 0$, independent of u, f, and g, such that:

$$\|u\|_{H^1(\Omega)} \leq C(\|f\|_{L^2(\Omega)} + \|g\|_{H^{1/2}(\partial\Omega)}).$$

Proof. We have already established the existence and uniqueness for the solution $\tilde{u} \in H_0^1(\Omega)$ to the variational formulation [2.6]. Moreover, if we choose $v = \tilde{u}$ in [2.6], we find that:

$$|\tilde{u}|_1^2 \leq \|f\|_{L^2(\Omega)}\|\tilde{u}\|_{L^2(\Omega)} + |\tilde{g}|_1 \, |\tilde{u}|_1 \leq \left(C_p^{1/2}\|f\|_{L^2(\Omega)} + |\tilde{g}|_1\right) |\tilde{u}|_1$$

from which:

$$|\tilde{u}|_1 \leq C_p^{1/2}\|f\|_{L^2(\Omega)} + |\tilde{g}|_1. \quad [2.27]$$

– *Existence of* u: As $\tilde{u}$ is solution to [2.6], $u = \tilde{u} + \tilde{g}$ belongs to $H^1(\Omega)$ because $\tilde{u} \in H_0^1(\Omega)$ and $\tilde{g} \in H^1(\Omega)$, by assumption. Therefore, u is a solution to the non-homogeneous Dirichlet problem.

– *Uniqueness of* u: Let u_1 and u_2 be two solutions to the non-homogeneous Dirichlet problem. Let $w = u_1 - u_2 \in H^1(\Omega)$, and it is clear that w satisfies the homogeneous Dirichlet problem:

$$\begin{cases} -\Delta w = 0 & \text{in } \Omega \\ w = 0 & \text{on } \partial\Omega \end{cases}$$

whose solution is $w = 0$ by the virtue of theorem 2.2, thus proving that u is unique.

– *Continuous dependence of u on the data*: According to [1.32], we may choose an extension $\tilde{g}_0$ of g such that $\|\tilde{g}_0\|_{H^1(\Omega)} \leq 2\|g\|_{H^{1/2}(\partial\Omega)}$. If not, then this would contradict the fact that $\|g\|_{H^{1/2}(\partial\Omega)} = \inf_{v \in H^1(\Omega) \text{ s.t. } \gamma_0 v = g} \|v\|_{H^1(\Omega)}$. For this extension, observe that $\tilde{u}_0$ solves the variational formulation [2.6]. By the triangle inequality and [2.25], we see that:

$$\begin{aligned} \|u\|_{H^1(\Omega)} = \|\tilde{u}_0 + \tilde{g}_0\|_{H^1(\Omega)} &\leq \|\tilde{u}_0\|_{H^1(\Omega)} + \|\tilde{g}_0\|_{H^1(\Omega)} \\ &\leq (1 + C_p)^{1/2} |\tilde{u}_0|_1 + \|\tilde{g}_0\|_{H^1(\Omega)} \end{aligned}$$

which, given the estimate [2.27] and the choice of $\tilde{g}_0$, allows us to establish the desired continuity result. Indeed, let us write:

$$\begin{aligned} \|u\|_{H^1(\Omega)} &\leq (1 + C_p)^{1/2} (C_p^{1/2} \|f\|_{L^2(\Omega)} + |\tilde{g}_0|_1) + \|\tilde{g}_0\|_{H^1(\Omega)} \\ &\leq (1 + C_p)^{1/2} C_p^{1/2} \|f\|_{L^2(\Omega)} + 2(1 + (1 + C_p)^{1/2}) \|g\|_{H^{1/2}(\partial\Omega)} \\ &\leq \max\left((C_p + C_p^2)^{1/2}, 2(1 + (1 + C_p)^{1/2})\right)\left(\|f\|_{L^2(\Omega)} + \|g\|_{H^{1/2}(\partial\Omega)}\right) \end{aligned}$$

that is, that it depends continuously on the data g and f. ■

REMARK 2.9.– *From remark 2.8, the results of theorems 2.2 and 2.3 remain true even when* Ω *is bounded in one direction only.*

2.4. Existence and uniqueness – general framework

As a reminder, the abstract variational formulation that we want to consider is:

$$\left| \begin{array}{l} \text{Find } u \in H \text{ such that} \\ a(u, v) = \ell(v) \qquad \forall v \in H. \end{array} \right. \qquad [2.28]$$

where:

– H is a Hilbert space endowed with the inner product $(\cdot, \cdot)_H$;

– $a(\cdot, \cdot)$ is a continuous bilinear form on H;

– $\ell(\cdot)$ is a continuous linear form on H.

Now we will propose several general results to solve it.

2.4.1. *Lax–Milgram theorem*

The existence and uniqueness proofs for the particular cases that we have just seen highlight two essential points:

– the fact that the bilinear form a defines an equivalent norm on the Hilbert space H;

– the use of Riesz's representation theorem.

The first point arises from a very important property, which is called the **coercivity** of the bilinear form a.

DEFINITION 2.1 (Coercivity).– *A bilinear form a is said to be coercive on H if and only if there exists a constant $\alpha > 0$ such that:*

$$|a(v,v)| \geq \alpha \|v\|_H^2 \quad \forall v \in H. \tag{2.29}$$

REMARK 2.10.– *In finite dimensions, this property expresses the definiteness of a bilinear mapping (see exercise 16).*

LEMMA 2.1.– *Let a be a continuous bilinear form on H, then a is coercive with respect to the constant $\alpha > 0$ in [2.29] if and only if:*

$$a(v,v) \geq \alpha \|v\|_H^2 \quad \forall v \in H \quad \textit{or} \quad -a(v,v) \geq \alpha \|v\|_H^2 \quad \forall v \in H. \tag{2.30}$$

Proof. It is clear that if we have the alternative condition [2.30] for some constant $\alpha > 0$, then as $|a(v,v)| = \max\big(a(v,v), -a(v,v)\big)$, we see that [2.29] follows and the form a is coercive.

Conversely, suppose that [2.29] holds for some $\alpha > 0$. We will show by contradiction that $f_a : v \mapsto a(v,v)$ does not change sign on H. So suppose that it does change sign. Then there exist v_+ and v_- such that $f_a(v_+) > 0$ and $f_a(v_-) < 0$. As the form a is continuous, so is f_a, and hence there exists a $v_0 \in [v_+, v_-]$ such that $f(v_0) = 0$, that is, $a(v_0, v_0) = 0$. By the hypothesis, this is only possible if $v_0 = 0$. However, since $v_0 \in (v_+, v_-)$ (the extremities are excluded since $f_a(v_+) > 0$ and $f_a(v_-) < 0$), we may write $0 = v_0 = \lambda v_+ + (1-\lambda)v_-$ for $\lambda \in (0,1)$ and therefore $v_+ = \mu v_-$ for $\mu = 1 - \lambda^{-1} \neq 0$. From the bilinearity of the form a, we see that $f_a(v_+) = \mu^2 f_a(v_-)$, contradicting the fact that $f_a(v_+)$ and $f_a(v_-)$ have opposite signs.

Since $v \mapsto a(v,v)$ does not change sign on H, we conclude that it must be either be positive on H so that $|a(v,v)| = a(v,v)$ for all H, or be negative on H so that $|a(v,v)| = -a(v,v)$ for all H. In either case, [2.29] implies the possibility of either one of the two conditions in [2.30] for the same constant $\alpha > 0$. ■

REMARK 2.11.– *In what follows, to establish the coercivity condition for a continuous bilinear form a we will try to prove that there exists a constant $\alpha > 0$ such that $a(v, v) > \alpha\|v\|_H^2$ for all $v \in H$. Should this not be the case, then we can replace the form a with its negative $-a$.*

This notion allows us to formulate the first general existence and uniqueness result for the abstract variational formulation [2.28] which, moreover, includes the case when the bilinear form is not necessarily symmetric:

THEOREM 2.4 (Lax–Milgram).– *Let H be a Hilbert space, $a(\cdot, \cdot)$ be a coercive and continuous bilinear form on H, and $\ell(\cdot)$ be a continuous linear form on H. Then there exists a unique solution $u \in H$ to the problem:*

$$a(u, v) = \ell(v) \quad \forall v \in H.$$

Furthermore, the solution u depends continuously on the linear form ℓ: there exists a $C > 0$, independent of u and ℓ, such that:

$$\|u\|_H \leq C\,\|\ell\|_{H'}.$$

REMARK 2.12.– *The principle of continuous dependence can be interpreted in this way. If we denote a sequence in $(H')^{\mathbb{N}}$ by $(\ell_k)_k$ converging to ℓ in H' ($\lim_{k\to\infty}\|\ell - \ell_k\|_{H'} = 0$), then the sequence of the corresponding solutions $(u_k)_k$ in $H^{\mathbb{N}}$ converges to the solution u in H, that is, $\lim_{k\to\infty}\|u - u_k\|_H = 0$.*

Proof. On the one hand, by Riesz's representation theorem, there exists a unique element $L \in H$ such that:

$$\ell(v) = (L, v)_H, \quad \forall v \in H.$$

However, the mapping $v \mapsto a(u, v)$ defines a continuous linear form on H. Again thanks to Riesz's representation theorem, there exists a unique element in H, which we denote by Au, such that:

$$(Au, v)_H = a(u, v), \quad \forall v \in H$$

along with the estimate:

$$\|Au\|_H = \sup_{v\neq 0}\frac{|a(u, v)|}{\|v\|_H} \leq C_a\,\|u\|_H \ \ (C_a \text{ modulus of continuity for } a(\cdot,\cdot)). \quad [2.31]$$

This shows that the linear mapping:

$$A : \left|\begin{array}{lcl} H & \longrightarrow & H \\ u & \mapsto & Au \end{array}\right.$$

is continuous on H. Indeed, the norm of A is bounded from above:

$$|||A||| = \sup_{u \neq 0} \frac{\|Au\|_H}{\|u\|_H} = \sup_{u \neq 0} \frac{1}{\|u\|_H} \left(\sup_{v \neq 0} \frac{|a(u,v)|}{\|v\|_H} \right) \leq C_a.$$

And given the coercivity, it also satisfies:

$$\epsilon\,(Au, u)_H \geq \alpha \|u\|_H^2 \qquad [2.32]$$

with $\epsilon \in \{-1, +1\}$ and $\alpha > 0$ is independent of u, thanks to lemma 2.1.

The problem $a(u,v) = \ell(v) \;\; \forall v \in H$ is then equivalent to the problem:

$$(Au - L, v)_H = 0 \quad \forall v \in H$$

and therefore equivalent to the functional equation:

$$Au = L \quad \text{in } H. \qquad [2.33]$$

Define a linear mapping S on H as follows:

$$Sv = -\rho\epsilon(Av - L) + v \quad \text{with } \rho > 0 \text{ still to be defined}$$

we have then that:

$$\|Sv_1 - Sv_2\|_H^2 = \|-\rho\epsilon A\,(v_1 - v_2) + v_1 - v_2\|_H^2$$

by combining both of the estimates [2.31] and [2.32]:

$$\begin{aligned}\|Sv_1 - Sv_2\|_H^2 &= \|v_1 - v_2\|_H^2 + \rho^2 \|Av_1 - Av_2\|_H^2 - 2\rho\epsilon\,(Av_1 - Av_2, v_1 - v_2)_H \\ &\leq (1 - 2\rho\alpha + \rho^2 C_a^2)\, \|v_1 - v_2\|_H^2 .\end{aligned}$$

In taking $0 < \rho < \dfrac{2\alpha}{C_a^2}$, we obtain:

$$\|Sv_1 - Sv_2\|_H \leq k_\rho \|v_1 - v_2\|_H \quad \text{with } k_\rho < 1 \qquad [2.34]$$

proving that S is a contraction mapping on H. Using Banach's fixed-point theorem, we may deduce that S has a unique fixed point $u = Su$ in H. So, according to the definition of S, u satisfies [2.33].

Finally, by the virtue of [2.32], we have that:

$$\alpha \|u\|_H^2 \leq |(Au, u)_H| = |(L, u)_H| \leq \|L\|_H \|u\|_H = \|\ell\|_{H'} \|u\|_H$$

establishing that the solution u continuously depends on ℓ: $\|u\|_H \leq \alpha^{-1} \|\ell\|_{H'}$. ■

REMARK 2.13.– *When the bilinear form* $a(\cdot,\cdot)$ *is furthermore* symmetric *and, to fix ideas,* positive, *the proof becomes immediate since* $a(\cdot,\cdot)$ *defines an equivalent norm on the Hilbert space* H*, and Riesz's representation theorem then answers the question (this property has already been used previously).*

Moreover, if we introduce the functional:

$$J(u) = \frac{1}{2}a(u,u) - \ell(u)$$

we have then, according to Ciarlet (1989) and also exercise 17:

$$dJ(u).v = (Au - L, v), \text{ or } \nabla J(u) = Au - L$$

which shows that the solution u *minimizes the functional* J*, when the bilinear form is symmetric. In the context of quadratic optimization, coercivity is interpreted as a strict convexity condition on* J*.*

2.4.2. An example application of the Lax–Milgram theorem with the Laplacian

Here, we assume that the open set Ω is bounded and connected. Consider the Fourier problem with a homogeneous boundary condition ($\beta = 0$ and $\lambda > 0$):

$$\begin{cases} -\Delta u = f & \text{in } \Omega \quad \text{with } f \in L^2(\Omega) \\ \dfrac{\partial u}{\partial n} + \lambda u = 0 & \text{on } \partial\Omega \end{cases} \qquad [2.35]$$

for which a variational formulation in $H^1(\Omega)$ is:

$$\left| \begin{array}{l} \text{Find } u \in H^1(\Omega) \text{ such that} \\ \displaystyle\int_\Omega \nabla u \cdot \nabla v \, d\Omega + \lambda \int_{\partial\Omega} uv \, d\Gamma = \int_\Omega fv \, d\Omega \quad \forall v \in H^1(\Omega). \end{array} \right. \qquad [2.36]$$

Applying the Lax–Milgram theorem with $H = H^1(\Omega)$:

$$a(u,v) = \int_\Omega \nabla u \cdot \nabla v \, d\Omega + \lambda \int_{\partial\Omega} uv \, d\Gamma \text{ and } \ell(v) = \int_\Omega fv \, d\Omega.$$

As $f \in L^2(\Omega)$, ℓ is a continuous linear form on $H^1(\Omega)$ (see [2.18]).

On the other hand, a is clearly a bilinear form on $H^1(\Omega)$. It is continuous on $H^1(\Omega)$, since:

$$\left| \int_\Omega \nabla u \cdot \nabla v \, d\Omega \right| \leq \|\nabla u\|_{(L^2(\Omega))^d} \|\nabla v\|_{(L^2(\Omega))^d} \leq \|u\|_{H^1(\Omega)} \|v\|_{H^1(\Omega)}$$

and:

$$\left| \int_{\partial\Omega} uv \, d\Gamma \right| \leq \|u\|_{L^2(\partial\Omega)} \|v\|_{L^2(\partial\Omega)} \leq C^2 \|u\|_{H^1(\Omega)} \|v\|_{H^1(\Omega)} \quad \text{(Theorem 1.7).}$$

Moreover, since we have:

$$\left(\int_{\partial\Omega} v \, d\Gamma \right)^2 \leq \text{vol}(\partial\Omega) \int_{\partial\Omega} v^2 d\Gamma \quad \text{(dueto Cauchy-Schwarz),}$$

we deduce from the Poincaré–Friedrichs inequality [2.24] that there exists a $C' > 0$ such that:

$$\int_{\partial\Omega} v^2 d\Gamma + \int_{\Omega} |\nabla v|^2 d\Omega \geq C' \int_{\Omega} \left(v^2 + |\nabla v|^2 \right) d\Omega \quad \forall v \in H^1(\Omega),$$

which leads to the estimation:

$$a(v, v) \geq \min(1, \lambda) \left(\int_{\Omega} |\nabla v|^2 d\Omega + \int_{\partial\Omega} v^2 d\Gamma \right) \geq C' \min(1, \lambda) \|v\|^2_{H^1(\Omega)}$$

proving that $a(\cdot, \cdot)$ is coercive on $H^1(\Omega)$. The variational formulation [2.36] thus admits a unique solution $u \in H^1(\Omega)$. We refer to exercise 19 for the non-homogeneous Fourier boundary conditions.

When $\lambda = 0$ (and $\beta = 0$), the Fourier problem reduces to a Neumann problem, whose variational formulation follows from [2.36]:

$$\left| \begin{array}{l} \text{Find } u \in H^1(\Omega) \text{ such that} \\ \displaystyle\int_{\Omega} \nabla u \cdot \nabla v \, d\Omega = \int_{\Omega} fv \, d\Omega \quad \forall v \in H^1(\Omega). \end{array} \right.$$

By choosing $v = 1$ in this formulation, note that f must satisfy the so-called compatibility condition:

$$\int_{\Omega} f \, d\Omega = 0. \qquad [2.37]$$

This is a necessary condition on f for the problem to be well posed. Moreover, if this condition is satisfied, then it is clear that the solution is not unique, because any constant function u_{cst} on Ω satisfies:

$$\int_{\Omega} \nabla u_{cste} \cdot \nabla v \, d\Omega = 0 \quad \forall v \in H^1(\Omega).$$

From these remarks, if we consider the subspace of $H^1(\Omega)$ given by

$$\left\{v \in H^1(\Omega) \text{ such that } \int_\Omega v \, d\Omega = 0\right\}$$

then the existence, uniqueness and continuous dependence of the solution can be established (see exercise 19).

REMARK 2.14.– *We have always assumed that $a(\cdot,\cdot)$ takes its values in $\mathbb{R}$. There are some situations where $a(\cdot,\cdot)$ is complex-valued, in which case the form $a(\cdot,\cdot)$ becomes sesquilinear, and* **coercivity** *then becomes: let $|\cdot|$ represent the modulus, then there exists a constant $\alpha > 0$ such that:*

$$|a(v,v)| \geq \alpha \|v\|_H^2 \qquad \forall v \in H.$$

When the form is additionally continuous then, similarly to lemma 2.1, we can consider the following equivalent *definition (Brezis 2011): there exists a constant $\alpha > 0$ and an argument $\theta \in [0, 2\pi)$, such that:*

$$\Re(e^{\imath\theta} a(v,v)) \geq \alpha \|v\|_H^2 \qquad \forall v \in H.$$

We can still state the Lax–Milgram theorem; however, its proof is different from the one stated previously (injectivity and surjectivity of the linear mapping A associated with the sesquilinear form $a(\cdot,\cdot)$ are proven directly; see the proof of theorem 2.6 for an idea of the approach).

A useful generalization of the Lax–Milgram theorem is Stampacchia's theorem, which provides an existence and uniqueness result for the **variational inequality**[3]:

$$\left| \begin{array}{l} \text{Find } u \in K, \text{ such that} \\ a(u, v-u) \geq \ell(v-u) \quad \forall v \in K \end{array} \right. \qquad [2.38]$$

where K is a convex closed and non-empty subset of Hilbert space H.

Under the hypotheses of theorem 2.4 (and after choosing $v \mapsto a(v,v)$ with positive values), we can prove that there exists a unique solution $u \in K$ to the problem [2.38].

Indeed, using Riesz's representation theorem, the problem is reduced back down to:

$$(Au, v-u)_H \geq (L, v-u)_H \quad \forall v \in K$$

3 One often meets these types of inequalities when dealing with contact problems.

which naturally appears in the orthogonal projection on K (see the projection theorem 1.2). Then we may introduce the linear map:

$$Sv = P_K(-\rho(Av - L) + v), \quad \rho > 0$$

for which we prove, by virtue of [1.20] and for ρ small enough, that there exists a $0 < k_\rho < 1$ such that:

$$\|Sv_1 - Sv_2\|_H \leq \|P_K(v_1 - v_2 - \rho(Av_1 - Av_2))\|_H \leq k_\rho \|v_1 - v_2\|_H.$$

We conclude with a fixed point argument.

2.4.3. *The Helmholtz problem*

On the contrary, the Helmholtz problem does not fall within the scope of section 2.4.1. For a given $\omega > 0$ and a bounded Ω, recall that we wish to solve:

$$\begin{cases} -\Delta u - \omega^2 u = f & \text{in } \Omega \\ \dfrac{\partial u}{\partial n} = 0 & \text{on } \partial\Omega \end{cases}$$

with $f \in L^2(\Omega)$. In the above, we have chosen a boundary condition similar to that of the homogeneous Neumann problem. Then thanks to [2.8], we arrive at the variational formulation of the Helmholtz problem:

$$\left| \begin{array}{l} \text{Find } u \in H^1(\Omega) \text{ such that} \\ a(u, v) = \ell(v) \quad \forall v \in H^1(\Omega) \end{array} \right. \qquad [2.39]$$

with $a(u, v) = \displaystyle\int_\Omega (\nabla u \cdot \nabla v - \omega^2 uv)\, d\Omega$ and $\ell(v) = \displaystyle\int_\Omega fv\, d\Omega$.

Observe that $a(v, v) = \displaystyle\int_\Omega |\nabla v|^2\, d\Omega - \omega^2 \int_\Omega v^2\, d\Omega$ has both a positive and negative contribution; it is therefore not possible to reuse our previous arguments. To make this clear, introduce:

$$a_+(u, v) = \int_\Omega (\nabla u \cdot \nabla v + uv)\, d\Omega, \quad u, v \in H^1(\Omega).$$

Of course, we have $a(u, v) = a_+(u, v) - (\omega^2 + 1) \displaystyle\int_\Omega uv\, d\Omega$. The form a_+ is equal to the inner product on $H^1(\Omega)$: therefore, the contribution coming from $a_+(v, v)$ is now equal to $\|v\|^2_{H^1(\Omega)}$, while the negative contribution is proportional to $\|v\|^2_{L^2(\Omega)}$.

From theorem 2.1, for every $f^* \in L^2(\Omega)$, we know that there exists a unique $u^* \in H^1(\Omega)$ such that:

$$a_+(u^*, v) = \int_\Omega f^* v \, d\Omega \quad \forall v \in H^1(\Omega).$$

Furthermore, $\|u^*\|_{H^1(\Omega)} \leq C \|f^*\|_{L^2(\Omega)}$ where $C > 0$ is independent of both f^* and u^*. Therefore, we can define a continuous linear mapping $T_+ \in \mathcal{L}(L^2(\Omega), H^1(\Omega))$ by $T_+ f^* = u^*$, that is, that:

$$a_+(T_+ f^*, v) = \int_\Omega f^* v \, d\Omega \quad \forall f^* \in L^2(\Omega),\ \forall v \in H^1(\Omega). \qquad [2.40]$$

In [2.39], notice now that we can replace $\int_\Omega uv \, d\Omega$ by $a_+(T_+ u, v)$, providing the equivalent form:

$$\left| \begin{array}{l} \text{Find } u \in H^1(\Omega) \text{ such that} \\ a_+((I - (\omega^2 + 1)T_+)u, v) = \ell(v) \quad \forall v \in H^1(\Omega) \end{array} \right. \qquad [2.41]$$

where I is the identity mapping on $\mathcal{L}(H^1(\Omega))$. Hence, to solve the Helmholtz problem, we need to investigate the properties of T_+. For this, let us set $T = i_{H^1(\Omega) \to L^2(\Omega)} \circ T_+ \in \mathcal{L}(L^2(\Omega))$.

LEMMA 2.2.– *Assume that Ω is bounded. Then the linear mapping T is self-adjoint, compact, injective and positive.*

Proof. First, by Rellich's theorem 1.6 (Ω is bounded), the injection $i_{H^1(\Omega) \to L^2(\Omega)}$ is compact, and hence T is *compact*.

Then, if we have $Tf^* = 0$, by the definition of T we have that $T_+ f^* = 0$. From [2.40], this implies that $\int_\Omega f^* v \, d\Omega = 0$ for every $v \in H^1(\Omega)$, and hence by the density of $H^1(\Omega)$ in $L^2(\Omega)$ (recall that $\mathcal{D}(\Omega)$, which can be seen as a subspace of $H^1(\Omega)$, is itself dense in $L^2(\Omega)$, see proposition 1.4), we deduce that $f^* = 0$; T is *injective*.

Furthermore, for every $f^* \in L^2(\Omega)$, we have that:

$$(f^*, Tf^*)_{L^2(\Omega)} = \int_\Omega f^* (Tf^*) \, d\Omega = \int_\Omega f^* (T_+ f^*) \, d\Omega = a_+(T_+ f^*, T_+ f^*) \geq 0$$

where the last equality follows from [2.40]; T is *positive*.

Finally, for every $f^*, g^* \in L^2(\Omega)$, we have:

$$
\begin{aligned}
(g^*, Tf^*)_{L^2(\Omega)} &= \int_\Omega g^*(Tf^*)\, d\Omega = \int_\Omega g^*(T_+f^*)\, d\Omega = a_+(T_+g^*, T_+f^*) \\
&= a_+(T_+f^*, T_+g^*) = \int_\Omega f^*(T_+g^*)\, d\Omega = \int_\Omega f^*(Tg^*)\, d\Omega = (f^*, Tg^*)_{L^2(\Omega)}
\end{aligned}
$$

where we have used [2.40] twice as well as the symmetry of a_+; T is *self-adjoint*. ■

We can therefore apply the spectral theorem 1.5 to $T \in \mathcal{L}(L^2(\Omega))$. Let $(e_n)_{n\geq 1}$ be a Hilbert basis of $L^2(\Omega)$, which is formed of the eigenvectors to T. Denote the corresponding eigenvalues by $(\mu_n)_{n\geq 1}$, which are all strictly positive and have finite multiplicities, and moreover that $\lim_{n\to\infty} \mu_n = 0$. By the construction of T, we note that $e_n \in H^1(\Omega)$ for each $n \geq 1$. Indeed, we have that $T_+e_n \in H^1(\Omega)$, and:

$$
Te_n = \mu_n e_n \Longrightarrow i_{H^1(\Omega)\to L^2(\Omega)}(T_+e_n) = \mu_n e_n \Longrightarrow e_n = (\mu_n)^{-1} T_+ e_n.
$$

To obtain the last equality, we used the property that $i_{H^1(\Omega)\to L^2(\Omega)} v = v$, for every $v \in H^1(\Omega)$. Now, let us denote $\lambda_n = (\mu_n)^{-1}$. Since $e_n = \lambda_n T_+ e_n$ in $H^1(\Omega)$ for each $n \geq 1$, from [2.40] we get that:

$$
a_+(v, e_n) = \lambda_n a_+(v, T_+e_n) = \lambda_n \int_\Omega v e_n\, d\Omega \quad \forall n \geq 1,\ \forall v \in H^1(\Omega).
$$

Thus, if u is a solution to [2.39], then in particular we have that:

$$
\int_\Omega f e_n\, d\Omega = \left(\lambda_n - (\omega^2 + 1)\right) \int_\Omega u e_n\, d\Omega \quad \forall n \geq 1. \qquad [2.42]
$$

We can now prove the following results:

THEOREM 2.5 (Helmholtz problem).– *Suppose that Ω is bounded. Then, for a given value $\omega > 0$, one of two things can occur:*

1) For all $f \in L^2(\Omega)$, there exists a unique weak solution to the Helmholtz problem:

$$
\begin{cases}
\text{Find } u \in H^1(\Omega) \text{ such that} \\
-\Delta u - \omega^2 u = f \textit{ in } \Omega \\
\dfrac{\partial u}{\partial n} = 0 \textit{ on } \partial\Omega
\end{cases}
$$

which depends continuously on the data f: there exists a $C > 0$, independent of u and f, such that:

$$
\|u\|_{H^1(\Omega)} \leq C \|f\|_{L^2(\Omega)}.
$$

2) There exists an $f \in L^2(\Omega)$ for which the Helmholtz problem has no solution. Furthermore, any weak solution $f \in L^2(\Omega)$ to the Helmholtz problem is not unique.

Proof. If u is a solution to [2.39], then in particular u satisfies [2.42]. From this, it follows that there are two *distinct possibilities*.

a) Let $\omega^2 \notin \{\lambda_n - 1,\ n \geq 1\}$: in this case, the solution $u \in H^1(\Omega) \subset L^2(\Omega)$ must *necessarily* be of the form:

$$u = \sum_{m \geq 1} (u, e_m)_{L^2(\Omega)} e_m = \sum_{m \geq 1} \frac{1}{\lambda_m - (\omega^2 + 1)} (f, e_m)_{L^2(\Omega)} e_m.$$

Furthermore, it can be established that $(e_n/\sqrt{\lambda_n})_{n \geq 1}$ forms a Hilbert basis of $H^1(\Omega)$ (see exercise 18 for a proof). Let us denote:

$$a_n = \frac{\sqrt{\lambda_n}}{\lambda_n - (\omega^2 + 1)} (f, e_n)_{L^2(\Omega)} \text{ for } n \geq 1.$$

Then we may check that $(a_n)_{n \geq 1} \in \ell^2$:

$$\begin{aligned}
\sum_{m \geq 1} a_m^2 &= \sum_{m \geq 1} \Big(\frac{\sqrt{\lambda_m}}{(\lambda_m - (\omega^2 + 1))} \Big)^2 ((f, e_m)_{L^2(\Omega)})^2 \\
&\leq \Big(\sup_{n \geq 1} \frac{\sqrt{\lambda_n}}{(\lambda_n - (\omega^2 + 1))} \Big)^2 \sum_{m \geq 1} ((f, e_m)_{L^2(\Omega)})^2 \\
&= \Big(\sup_{n \geq 1} \frac{\sqrt{\lambda_n}}{(\lambda_n - (\omega^2 + 1))} \Big)^2 \|f\|^2_{L^2(\Omega)}.
\end{aligned}$$

However, since $\omega^2 \notin \{\lambda_n - 1,\ n \geq 1\}$ and $\lim_{n \to \infty} \lambda_n = +\infty$, we have that:

$$C_\omega = \sup_{n \geq 1} \frac{\sqrt{\lambda_n}}{(\lambda_n - (\omega^2 + 1))} < \infty$$

and it follows that $(a_n)_{n \geq 1} \in \ell^2$. From the identity $u = \sum_{m \geq 1} a_m (e_m/\sqrt{\lambda_m})$, we can conclude from this that $u \in H^1(\Omega)$, with $\|u\|_{H^1(\Omega)} \leq C_\omega \|f\|_{L^2(\Omega)}$.

Thus, u is indeed a solution to [2.39], since, as $(e_n/\sqrt{\lambda_n})_{n \geq 1}$ forms a Hilbert basis of $H^1(\Omega)$, it suffices to check that [2.39] is satisfied for every function $v \in \{e_n/\sqrt{\lambda_n},\ n \geq 1\}$, which follows immediately from [2.42]. We have proven the existence and uniqueness of the solution, which depends continuously with respect to the solution.

b) Let $\omega^2 \in \{\lambda_n - 1,\ n \geq 1\}$: denote by $\mathbb{E}_\omega$ the finite and non-empty set of n indices such that: $\lambda_n - 1 = \omega^2$. Then for u to exist, according to [2.42] we must *necessarily* have that:

$$\int_\Omega f e_n \, d\Omega = 0 \quad \forall n \in \mathbb{E}_\omega.$$

If at least one of these conditions is false, then *there is no solution* to the Helmholtz problem. On the contrary, if they are all true, then we can check (this is the same calculation as before) that:

$$u = \sum_{m \geq 1, m \notin \mathbb{E}_\omega} \frac{1}{\lambda_m - (\omega^2 + 1)} (f, e_m)_{L^2(\Omega)} e_m + \sum_{m \in \mathbb{E}_\omega} \alpha_m e_m$$

is a solution of the Helmholtz problem, regardless of whatever $(\alpha_n)_{n \in \mathbb{E}_\omega}$ is. Hence the solution *is no longer unique*. ■

Thus, for the Helmholtz problem[4], the existence of a solution for any given f *is equivalent to* the uniqueness of the solution for every f.

2.4.4. *Well-posed problems*

First, let us propose a general definition to establish when a problem, that is written in the variational formulation [2.28], is **well-posed**.

DEFINITION 2.2 (Hadamard).– *The problem [2.28] is well posed if, and only if, for every $\ell \in H'$, it admits a solution u which is unique with a continuous dependence: there exists a $C > 0$, independent of ℓ and u, such that:*

$$\|u\|_H \leq C \, \|\ell\|_{H'}, \qquad \forall \ell \in H'.$$

In particular, the above inequality expresses the stability, or rather the continuity, of the solution u with respect to the given data ℓ.

As in the proof of the Lax–Milgram theorem, let us introduce a continuous linear mapping A from H into H, such that:

$$(Au, v)_H = a(u, v), \qquad \forall u, v \in H.$$

4 See the exercise 34 for other results regarding the Helmholtz problem.

Then we have the following result:

PROPOSITION 2.6.– *Let H be a Hilbert space and $a(\cdot,\cdot)$ be a continuous bilinear form on H. Then the problem [2.28] is well posed if, and only if, the inverse of A from H into H, denoted by A^{-1}, is well defined and continuous.*

Proof. Suppose that the problem [2.28] is well posed and let $L \in H$. Then there is a unique solution to the problem [2.28] for the continuous linear form $v \mapsto (L,v)_H$; let us write it as u_L. Moreover, we know that $\|u_L\|_H \leq C\,\|L\|_H$, where $C > 0$, is independent of L. It is easy to check that $B : L \mapsto u_L$ is a linear mapping (by the linearity of a) and continuous from H into H.

Furthermore, for $L \in H$:

$$(ABL, v)_H = a(BL, v) = (L, v)_H, \qquad \forall v \in H$$

and so $ABL = L$. Similarly, for $u \in H$:

$$a(BAu-u, v) = a(B(Au), v) - a(u, v) = (Au, v)_H - a(u, v) = 0, \qquad \forall v \in H$$

and therefore $BAu = u$, by the uniqueness of the solution to the problem [2.28]. We conclude that $BA = AB = I_H$, that is, that B is the (continuous) inverse of A.

Conversely, if the inverse of A exists and is continuous, then consider $\ell \in H'$: from Riesz's representation theorem, there exists a unique $L \in H$ such that $(L,v)_H = \ell(v)$ $\forall v \in H$, with $\|L\|_H = \|\ell\|_{H'}$. Denoting $u_L = A^{-1}L$, then, by construction, $Au_L = A(A^{-1}L) = L$ on H. Let:

$$a(u_L, v) = (L, v)_H = \ell(v), \qquad \forall v \in H.$$

In other words, u_L solves the problem [2.28] with the data ℓ, and:

$$\|u_L\|_H = \|A^{-1}L\|_H \leq |||A^{-1}|||\;\|L\|_H = |||A^{-1}|||\;\|\ell\|_{H'}.$$

Therefore, we have determined that the solution u_L depends continuously on the data. Finally, let us check that the solution is unique. For a given ℓ, if u_1, u_2 are two solutions, then $A(u_1-u_2) = 0$. Therefore $u_1-u_2 = A^{-1}A(u_1-u_2) = 0$, concluding the proof. ■

2.4.5. *The Banach–Necas–Babuska theorem*

Sometimes, we cannot apply the Lax–Milgram theorem, in particular, when there exists a *non-zero* element v_0 for which the bilinear form a equals $a(v_0, v_0) = 0$. In this case, *the form a is not coercive* (see exercise 22, for example). However, other mathematical tools remain available, in particular **stability conditions**, which are also called **inf-sup conditions**. The terminology "Banach–Necas–Babuska theorem" is not universal, and sometimes it is also called the "**Ladyzhenskaya–Babuska–Brezzi theorem**". These tools can be used in more general frameworks than that of coercivity, namely those involving bilinear forms that are defined on the Cartesian product of two Hilbert spaces, either of which *may be different*. Therefore, we consider two Hilbert spaces, H_1 and H_2, here.

DEFINITION 2.3 (Inf-sup condition).– *We say that a bilinear form a on $H_1 \times H_2$ satisfies a* stability condition, *or an* inf-sup condition, *if, and only if, there exists a constant $\alpha' > 0$ such that:*

$$\sup_{v_2 \in H_2 \setminus \{0\}} \frac{|a(v_1, v_2)|}{\|v_2\|_{H_2}} \geq \alpha' \, \|v_1\|_{H_1} \qquad \forall v_1 \in H_1. \qquad [2.43]$$

We will also need the following **solvability condition**. Intuitively, this second condition expresses the fact that we should not have "too many" test-functions.

DEFINITION 2.4.– *A bilinear form a on $H_1 \times H_2$ is said to satisfy a* solvability condition *if, and only if:*

$$\{v_2 \in H_2 \text{ such that } a(v_1, v_2) = 0, \ \forall v_1 \in H_1\} = \{0\}. \qquad [2.44]$$

REMARK 2.15.– *For a bilinear form defined on $H_1 = H_2 = H$, we have the choice between either coercivity, on the one hand, or stability plus solvability, on the other hand. If the form is coercive (with a constant $\alpha > 0$), then it is easy to check that it satisfies both the stability condition (with $\alpha' = \alpha$) as well as the solvability condition. The converse to this is* false, *since one may find continuous bilinear forms whose sign changes, yet still satisfy the stability and solvability conditions (see, e.g., exercise 22).*

Now, we consider the abstract variational framework with respect to two spaces:

$$\left| \begin{array}{l} \text{Find } u_1 \in H_1 \text{ such that} \\ a(u_1, v_2) = \ell_2(v_2) \qquad \forall v_2 \in H_2 \end{array} \right. \qquad [2.45]$$

where:

– H_1 and H_2 are two Hilbert spaces;

– $a(\cdot, \cdot)$ is a continuous bilinear form on $H_1 \times H_2$;

– $\ell_2(\cdot)$ is a continuous linear form on H_2.

The space H_1 is called the *solution* space, and H_2 is called the *test-function* space.

DEFINITION 2.5 (Hadamard).– *The problem [2.45] is said to be well posed if, and only if, for every $\ell_2 \in (H_2)'$, there exists a unique solution u_1 to the problem with a continuous dependence: there exists a $C > 0$, independent of both ℓ_2 and u_1, such that:*

$$\|u_1\|_{H_1} \leq C \, \|\ell_2\|_{H_2'}, \qquad \forall \ell_2 \in H_2'.$$

Let us introduce a continuous linear mapping A from H_1 into H_2, such that:

$$(Au_1, v_2)_{H_2} = a(u_1, v_2), \qquad \forall (u_1, v_2) \in H_1 \times H_2.$$

Just like in the case where $H_1 = H_2$, we have that:

PROPOSITION 2.7.– *The problem [2.45] is well posed if, and only if, the inverse of A is well defined from H_2 into H_1 and continuous.*

Proof. Similar to proposition 2.6, it is left as an exercise. ■

THEOREM 2.6 (Banach–Necas–Babuska).– *Let H_1, H_2 be two Hilbert spaces, and let $a(\cdot,\cdot)$ be a continuous bilinear form on $H_1 \times H_2$. The problem [2.45] is well posed if, and only if, conditions [2.43] and [2.44] are satisfied.*

Proof. Suppose that the problem [2.45] is well posed. From the previous proposition, the inverse A^{-1} is well defined and continuous from H_2 into H_1. Let $v_1 \in H_1 \setminus \{0\}$, then:

– on the one hand, $a(v_1, Av_1) = (Av_1, Av_1)_{H_2} = \|Av_1\|^2_{H_2}$;

– on the other hand, $\|v_1\|_{H_1} = \|A^{-1}(Av_1)\|_{H_1} \leq |||A^{-1}||| \, \|Av_1\|_{H_2}$ (in particular, $Av_1 \neq 0$).

Therefore:

$$\sup_{v_2 \in H_2 \setminus \{0\}} \frac{|a(v_1, v_2)|}{\|v_2\|_{H_2}} \geq \frac{|a(v_1, Av_1)|}{\|Av_1\|_{H_2}} = \|Av_1\|_{H_2} \geq \frac{1}{|||A^{-1}|||} \|v_1\|_{H_1}.$$

The stability condition [2.43] is satisfied with $\alpha' = 1/|||A^{-1}|||$.

Let $v_2 \in H_2$ be such that $a(v_1, v_2) = 0 \; \forall v_1 \in H_1$. Then in particular:

$$0 = a(A^{-1}v_2, v_2) = (A(A^{-1}v_2), v_2)_{H_2} = \|v_2\|^2_{H_2}, \text{ so } v_2 = 0.$$

Furthermore, the solvability condition [2.44] is equally satisfied.

Conversely, suppose now that conditions [2.43] and [2.44] are satisfied. We are going to prove that the linear mapping A associated with the form a is bijective, with a continuous inverse. Proposition 2.7 will then allow us to conclude the proof.

Let $u_1 \in H_1$ be such that $Au_1 = 0$: then, $a(u_1, v_2) = (Au_1, v_2)_{H_2} = 0 \, \forall v_2 \in H_2$ and, from [2.43], we have that $\|u_1\|_{H_1} = 0$, that is, $u_1 = 0$. Hence, A is injective.

Next, we will show that the image of A, $\mathrm{Im}(A) = \{v_2 \in H_2 : \exists v_1 \in H_1,\ v_2 = Av_1\}$, is a closed vector subspace of H_2, whose orthogonal subspace is $\{0\}$. It is clear that $\mathrm{Im}(A)$ is a vector subspace of H_2. For the closedness property, let us show that any sequence of elements $(v_2^k)_k$ in $\mathrm{Im}(A)$ converging in H_2 has its limit in $\mathrm{Im}(A)$. By definition, for each k, there exists a $v_1^k \in H_1$ such that $v_2^k = Av_1^k$. Using the Cauchy–Schwarz inequality, we have that:

$$\begin{aligned} |a(v_1^k - v_1^m, v_2)| &= |(Av_1^k - Av_1^m, v_2)_{H_2}| = |(v_2^k - v_2^m, v_2)_{H_2}| \\ &\leq \|v_2^k - v_2^m\|_{H_2} \|v_2\|_{H_2}, \qquad \forall k, m,\ \forall v_2 \in H_2. \end{aligned}$$

Then using [2.43], we can deduce that:

$$\|v_1^k - v_1^m\|_{H_1} \leq \frac{1}{\alpha'} \sup_{v_2 \in H_2 \setminus \{0\}} \frac{|a(v_1^k - v_1^m, v_2)|}{\|v_2\|_{H_2}} \leq \frac{1}{\alpha'} \|v_2^k - v_2^m\|_{H_2}, \quad \forall k, m.$$

As the sequence $(v_2^k)_k$ converges in H_2, it is therefore a Cauchy sequence: $\lim_{k,m \to \infty} \|v_2^k - v_2^m\|_{H_2} = 0$; the same is therefore true of $(v_1^k)_k$ in H_1, according to the above results. Since H_1 is a complete space, there exists a $v_1 \in H_1$ such that $\lim_{k \to \infty} \|v_1^k - v_1\|_{H_1} = 0$. Therefore, we see that:

$$\|v_2^k - Av_1\|_{H_2} = \|A(v_1^k - v_1)\|_{H_2} \leq |||A|||\ \|v_1^k - v_1\|_{H_1} \to 0 \text{ when } k \to \infty.$$

Thus, $(v_2^k)_k$ converges to Av_1, which belongs to $\mathrm{Im}(A)$, and is therefore closed in H_2.

Now let $v_2 \in (\mathrm{Im}(A))^{\perp}$, the orthogonal subspace to $\mathrm{Im}(A)$ in H_2: by definition, for all $v_1 \in H_1$, $0 = (Av_1, v_2)_{H_2} = a(v_1, v_2)$. According to [2.44], we have that $v_2 = 0$. Corollary 1.1 tells us that $\mathrm{Im}(A) = H_2$, which is to say that A is surjective. According to the previous results, for every $L_2 \in H_2$, there exists a unique $u_1 \in H_1$ such that $Au_1 = L_2$. And, according to [2.43] and the Cauchy–Schwarz inequality, we find that:

$$\|u_1\|_{H_1} \leq \frac{1}{\alpha'} \sup_{v_2 \in H_2 \setminus \{0\}} \frac{|a(u_1, v_2)|}{\|v_2\|_{H_2}} \leq \frac{1}{\alpha'} \|Au_1\|_{H_2} = \frac{1}{\alpha'} \|L_2\|_{H_2}$$

hence providing the continuous dependence with constant $1/\alpha'$, that is, independent of L_2. Finally, we can conclude that the problem [2.45] is well posed in the Hadamard sense.

■

In particular, when the solution space H_1 and the test-function space H_2 coincide, that is, $H_1 = H_2 = H$, the stability and solvability properties are therefore necessary and sufficient conditions to ensure the well posedness of the problem [2.28]. If the form a is also symmetric, then we obtain the following result.

COROLLARY 2.1 (Banach–Necas–Babuska).– *Let $H = H_1 = H_2$ be a Hilbert space, and $a(\cdot,\cdot)$ a symmetric and continuous bilinear form on $H \times H$. The problem [2.28] is well posed if, and only if, condition [2.43] is satisfied.*

Proof. When the form a is symmetric, it suffices to check that condition [2.43] implies condition [2.44]. Let $v_2 \in H_2$ satisfy $a(v_1, v_2) = 0$ for all v_1 in H_1. As the form a is symmetric, we have therefore that:

$$a(v_2, v_1) = 0 \quad \forall v_1 \in H_1.$$

Then according to [2.43], we see that $v_2 = 0$, hence proving the result. ■

2.4.6. *Fredholm alternative*

Now, we consider the variational formulation [2.28] again, albeit in a specific framework. Precisely, we will now assume that H is separable, and that the form a decomposes as:

$$a(u, v) = a_+(u, v) + c(u, v)$$

along with the *assumptions* that:

– $a_+(\cdot,\cdot)$ is a symmetric, coercive, continuous bilinear form on $H \times H$;
– $c(\cdot,\cdot)$ is a continuous bilinear form, and such that $C_+ \in \mathcal{L}(H)$ defined by:

$$a_+(C_+u, v) = c(u, v) \qquad \forall u, v \in H$$

is compact, satisfying $a_+(C_+u, v) = a_+(u, C_+v)$ for every u, v in H.

Also, we will endow H with the following inner product $a_+(\cdot,\cdot)$, along with its associated norm $\|\cdot\|_{H_+}$, defined by $v \mapsto (a_+(v, v))^{1/2}$. On the one hand, C_+ is self-adjoint on $(H, a_+(\cdot,\cdot))$, whereas, on the other, the form ℓ appearing in [2.28] is continuous and linear on $(H, \|\cdot\|_{H_+})$. Let:

$$\|\ell\|_{(H_+)'} = \sup_{v\in H\setminus\{0\}} \frac{|\ell(v)|}{\|v\|_{H_+}}.$$

The problem that we wish to solve can be written therefore in the following variational form:

$$\left|\begin{array}{l}\text{Find } u \in H \text{ such that}\\ a_+(u, v) + c(u, v) = \ell(v) \quad \forall v \in H.\end{array}\right. \qquad [2.46]$$

REMARK 2.16.– *The Helmholtz problem, studied in section 2.4.3 and in exercise 34, is a particular case of [2.46].*

We have the following results:

THEOREM 2.7 (Fredholm alternative).– *Assume that the previous hypotheses on H, a_+ and c hold. Then one of two possibilities can take place:*

a) that -1 is not an eigenvalue of C_+: the problem [2.46] is well posed;

b) that -1 is an eigenvalue of C_+ of multiplicity $n-1$: for a given $\ell \in H'$, the problem [2.46] admits a solution if and only if:

$$\ell(v) = 0 \quad \forall v \in N_{-1}$$

where N_{-1} is the vector space of dimension n_{-1} that is generated by the eigenvalues of C_+ associated with -1. Under this condition, the set of solutions to [2.46] is an affine subspace of dimension n_{-1}.

Proof. According to the spectral theorem 1.4, there exists a Hilbert basis $(u_n)_{n\geq 1}$ of $(H, a_+(\cdot,\cdot))$, made up of the eigenvectors to C_+. Denote the associated eigenvalues by $(\mu_n)_{n\geq 1}$. As $(u_n)_{n\geq 1}$ is a basis of H, solving [2.46] is equivalent to:

$$\left| \begin{array}{l} \text{Find } u \in H \text{ such that} \\ a_+(u, u_n) + c(u, u_n) = \ell(u_n) \quad \forall n \geq 1. \end{array} \right.$$

First, by the definition of C_+ and by its self-adjoint property, for each $n \geq 1$ we have that:

$$c(u, u_n) = a_+(C_+u, u_n) = a_+(u, C_+u_n) = \mu_n\, a_+(u, u_n).$$

Moreover, we can write $u = \sum_{m\geq 1} \alpha_m u_m$, with $\alpha_m = a_+(u, u_m)$ for each $m \geq 1$. It follows that [2.46] is equivalent now to:

$$\left| \begin{array}{l} \text{Find } (\alpha_m)_{m\geq 1} \in \mathbb{R}^{\mathbb{N}} \text{ such that} \\ (1 + \mu_n)\alpha_n = \ell(u_n) \quad \forall n \geq 1. \end{array} \right. \qquad [2.47]$$

Then there are two (exclusive) cases which may arise:

a) That $-1 \notin \{\mu_n,\ n \geq 1\}$: in this case, the system [2.47] admits only one solution $(\alpha_m)_{m\geq 1}$, with $\alpha_n = (1+\mu_n)^{-1}\ell(u_n)$ for each $n \geq 1$. Recall furthermore that, if we denote by $f \in H$ the element such that $\ell(v) = a_+(f, v)$ for every $v \in H$, then from Riesz's representation theorem we see that:

$$\|\ell\|^2_{(H_+)'} = \|f\|^2_{H_+} = \sum_{m\geq 1}(a_+(f, u_m))^2 = \sum_{m\geq 1}(\ell(u_m))^2.$$

Hence, we get that:

$$\begin{aligned}\|u\|^2_{H_+} &= \sum_{m\geq 1}\alpha_m^2 = \sum_{m\geq 1}\left(\frac{1}{1+\mu_m}\right)^2(\ell(u_m))^2 \\ &\leq \left(\sup_{n\geq 1}\frac{1}{1+\mu_n}\right)^2\sum_{m\geq 1}(\ell(u_m))^2 = \left(\sup_{n\geq 1}\frac{1}{1+\mu_n}\right)^2\|\ell\|^2_{(H_+)'}.\end{aligned}$$

By the virtue of the spectral theorem 1.4, one of two things can now occur: either the cardinality of $\{\mu_n,\ n \geq 1\}$ is finite, or $\lim_{n\to\infty}\mu_n = 0$. In either case, we find that:

$$C_c = \sup_{n\geq 1}\frac{1}{(1+\mu_n)} < \infty$$

and it follows that $\|u\|_{H_+} \leq C_c\,\|\ell\|_{(H_+)'}$. Therefore, we have asserted the existence and uniqueness of the solution u to the problem [2.46], and it depends continuously on the data ℓ;

b) That $-1 \in \{\mu_n,\ n \geq 1\}$: in this case, for a solution to [2.47] to exist, we must have that $\ell(u_n) = 0$ for each $n \in \mathbb{E}_{-1}$, where $\mathbb{E}_{-1}$ is the finite and non-empty set of n indices for which $\mu_n = -1$. If not, then no solution u exists. On the other hand, if this is the case then one can choose:

$$\alpha_n = \frac{\ell(u_n)}{1+\mu_n} \quad \forall n \in \mathbb{N}^* \setminus \mathbb{E}_{-1};\ \alpha_n \text{ arbitrary} \quad \forall n \in \mathbb{E}_{-1}.$$

And, with the same calculations as in the above, we can check that $u = \sum_{m\geq 1}\alpha_m u_m \in H$ is a solution to [2.46]. ■

2.5. Some properties of weak solutions

Let us end this chapter by briefly highlighting some examples of results that are related to weak solutions: the positivity principle, the maximum principle and the regularity of solutions with respect to that of the data. Compare these with the properties that are related to classical solutions, which were presented in section 1.2.

2.5.1. *Maximum principle*

In the case of the Dirichlet problem, we are going to highlight an approach allowing us to obtain a positivity principle for weak solutions, thereby obtaining a maximum principle.

Consider the following Dirichlet problem (see section 2.3.2):

$$\begin{cases} \text{Find } u \in H^1(\Omega) \text{ such that} \\ -\Delta u = f \ \text{ in } \Omega \\ u = g \qquad \text{on } \partial\Omega \end{cases} \qquad [2.48]$$

with $f \in L^2(\Omega)$ and $g \in H^{1/2}(\partial\Omega)$.

We have the following positivity principle:

PROPOSITION 2.8 (Positivity).– *Let $f \geq 0$ almost everywhere on Ω and $g \geq 0$ almost everywhere on $\partial\Omega$ (respectively, $f \leq 0$ and $g \leq 0$). Then the solution $u \in H^1(\Omega)$ to the Dirichlet problem [2.48] is positive (respectively, negative) almost everywhere on Ω.*

Proof. Let us consider the case $f \geq 0$, $g \geq 0$, as the other case is similar. Observe that $u \in H^1(\Omega)$ satisfies:

$$u = g \text{ on } \partial\Omega, \qquad \int_\Omega \nabla u \cdot \nabla v \, d\Omega = \int_\Omega f v \, d\Omega \quad \forall v \in H_0^1(\Omega). \qquad [2.49]$$

Set $u_- = \text{Inf}(u, 0)$ and $u_+ = \text{Sup}(u, 0)$, as well as $\Omega_- = \text{Support}(u \leq 0)$ and $\Omega_+ = \text{Support}(u \geq 0)$.

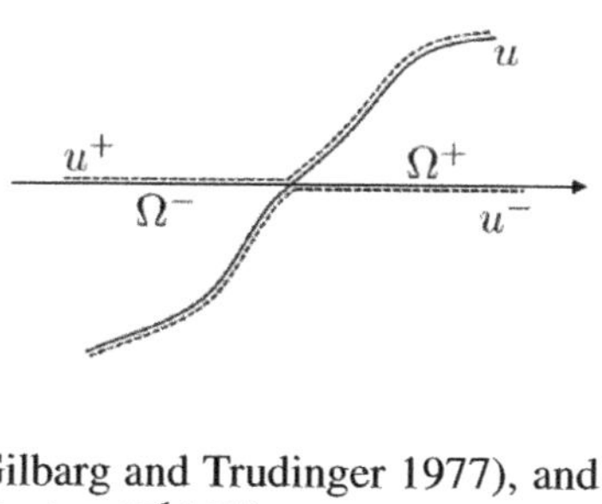

Then one can show (Gilbarg and Trudinger 1977), and that we will use here, that if $v \in H^1(\Omega)$ (respectively, $H_0^1(\Omega)$), then both v_+ and v_- belong to $H^1(\Omega)$ (respectively, $H_0^1(\Omega)$), and furthermore:

$$\nabla v_-(x) = \begin{cases} \nabla v(x) & \text{if } x \in \Omega_- \\ 0 & \text{if } x \notin \Omega_- \end{cases} \quad \text{and} \quad \nabla v_+(x) = \begin{cases} \nabla v(x) & \text{if } x \in \Omega_+ \\ 0 & \text{if } x \notin \Omega_+. \end{cases} \qquad [2.50]$$

Let us choose $v = u_-$ in [2.49], which is possible since $u_- \in H_0^1(\Omega)$:

$$(u_-)_{|\partial\Omega} = \inf(g, 0) = 0 \quad \text{if } g \geq 0.$$

Then after observing that $f \geq 0$ and $u_- \leq 0$, we see that:

$$0 \geq \int_{\Omega_-} f u_- d\Omega = \int_{\Omega_-} \nabla u \cdot \nabla u_- d\Omega = \int_{\Omega_-} \nabla(u_+ + u_-) \cdot \nabla u_- d\Omega = \int_{\Omega_-} |\nabla u_-|^2 d\Omega.$$

(Above we have used the property $u = u_+ + u_-$, as well as $\nabla u_+ \perp \nabla u_-$ almost everywhere.)

This proves that $\nabla u_- = 0$ and therefore $u_- = 0$, because $u_- \in H_0^1(\Omega)$. The proof is similar if $f \leq 0$ and $g \leq 0$ (choose $v = u_+$). ■

From this proposition, we can now deduce the following maximum principle:

THEOREM 2.8 (Maximum principle).– *Let $f \in L^2(\Omega)$ and $g \in H^{1/2}(\partial\Omega)$. Then the unique solution $u \in H^1(\Omega)$ to the Dirichlet problem [2.48] satisfies the following maximum principle:*

$$\begin{array}{lll} f \geq 0 & \Longrightarrow \quad u \geq \inf\limits_{\partial\Omega} g & \text{a. e. on } \Omega \\ f \leq 0 & \Longrightarrow \quad u \leq \sup\limits_{\partial\Omega} g & \text{a. e. on } \Omega \\ f = 0 & \Longrightarrow \quad \inf\limits_{\partial\Omega} g \leq u \leq \sup\limits_{\partial\Omega} g & \text{a. e. on } \Omega. \end{array}$$

Proof. Case $f \geq 0$. Let us denote $K = \inf\limits_{\partial\Omega} g$. If $K = -\infty$, it is clear that $u \geq -\infty$ almost everywhere. If K is finite, then $-\Delta K = 0$ on Ω. Set $v = K - u$ then, by construction, v satisfies the two equations:

$$\begin{cases} -\Delta v = -f & (\leq 0) \quad \text{in } \Omega \\ v = K - g & (\leq 0) \quad \text{on } \partial\Omega \end{cases}$$

proving that $v \leq 0$ by virtue of the preceding proposition, that is, that $u \geq K$. The other cases can be treated in the same way. ■

REMARK 2.17.– *For the modified Dirichlet problem:*

$$\begin{cases} \text{Find } u \in H^1(\Omega) \text{ such that} \\ -\Delta u + u = f \ \text{ in } \Omega \\ u = g \qquad\quad \text{on } \partial\Omega \end{cases}$$

we can obtain an even more precise result:

$$\min(\inf_{\Omega} f, \inf_{\partial\Omega} g) \leq u \leq \max(\sup_{\Omega} f, \sup_{\partial\Omega} g) \quad \text{a. e. on } \Omega$$

for which the proof is left as an exercise.

2.5.2. *Regularity of the solutions*

Variational formulations only provide weak solutions (belonging to $H^1(\Omega)$ for the Laplace problem). Can the weak solutions that we find be more regular or not? One can consult exercise 21 for a regularity estimate for the solutions to the Laplace equation.

This question is very technical, because it is then necessary to carefully study the behavior "at the boundary", the regularity of the boundary, the shape of the open subset (with a $\mathcal{C}^\infty$ boundary, polygonal or polyhedral, convex or not, etc.) and how they all crucially interplay with one another. Let Ω be a bounded open set of $\mathbb{R}^d$.

First, we will provide two regularity results in the case when the boundary is assumed to be *highly regular* (see Necas (1967), Brezis (1983, 2011) and Grisvard (1985) for the proofs of these and other more general results). Then we will continue by investigating the situation when the bounded open subset Ω is a polygon in $\mathbb{R}^2$, or a (Lipschitz) polyhedron in $\mathbb{R}^3$. Here, we will reproduce some results from Grisvard (1985, 1992). Finally, to finish, we will state a "local" regularity result, which is valid inside the open set.

2.5.2.1. *Dirichlet problem ($\mathcal{C}^\infty$ boundary)*

Let $f \in H^k(\Omega)$ and $g \in L^2(\partial\Omega)$ be such that there exists a $\tilde{g} \in H^\ell(\Omega)$ with $\ell \geq 1$, and a $\tilde{g} = g$ on $\partial\Omega$. Then the solution to the Dirichlet problem:

$$\begin{cases} \text{Find } u \in H^1(\Omega) \text{ such that} \\ -\Delta u = f \quad \text{in } \Omega \\ u = g \qquad\quad \text{on } \partial\Omega \end{cases}$$

is an element of $H^m(\Omega)$ with $m = \mathrm{Min}(k+2, \ell)$.

From the Sobolev spaces injections into the space(s) $\mathcal{C}^p(\overline{\Omega})$ (see proposition 1.11), observe that $k > \frac{d}{2}$ and $\ell > \frac{d}{2}+2$ are necessary for $H^m(\Omega) \subset \mathcal{C}^2\left(\overline{\Omega}\right)$. This provides the sufficient conditions for u to be a classical solution to the Dirichlet problem.

For example, it holds that:

$$\left| \begin{array}{l} f \in L^2(\Omega) \\ g = 0 \end{array} \right. \Longrightarrow u \in H^2(\Omega)$$

and more generally that:

$$\left| \begin{array}{l} f \in H^k(\Omega) \\ g = 0 \end{array} \right. \Longrightarrow u \in H^{k+2}(\Omega).$$

2.5.2.2. *Neumann problem ($\mathcal{C}^\infty$ boundary)*

Let $f \in H^k(\Omega)$ and $g \in L^2(\partial\Omega)$ be such that there exist a $\tilde{g} \in H^\ell(\Omega)$ with $\ell \geq 2$, and $\frac{\partial \tilde{g}}{\partial n} = g$ on $\partial\Omega$. Then the solution to the Neumann problem:

$$\left\{ \begin{array}{ll} \text{Find } u \in H^1(\Omega) \text{ such that} & \\ -\Delta u + u = f & \text{in } \Omega \\ \dfrac{\partial u}{\partial n} = g & \text{on } \partial\Omega \end{array} \right.$$

belongs to $H^m(\Omega)$ with $m = \text{Min}(k+2, \ell)$. This is a classical solution provided that $k > \dfrac{\mathrm{d}}{2}$ and $\ell > \dfrac{\mathrm{d}}{2} + 2$.

Be aware that these results do not extend naturally to the mixed Dirichlet–Neumann or Dirichlet–Fourier problems, as the regularity of the solutions is related to the interplay between the boundary conditions (see Grisvard (1985, 1992)).

2.5.2.3. *Homogeneous Dirichlet problem (polygonal or polyhedral Ω)*

Here, we consider the problem ($f \in L^2(\Omega)$):

$$\left\{ \begin{array}{l} \text{Find } u \in H_0^1(\Omega) \text{ such that} \\ -\Delta u = f \text{ in } \Omega. \end{array} \right.$$

REMARK 2.18.– *Indeed, if Ω is an open subset of $\mathbb{R}$, we have that $u'' \in L^2(\Omega)$ and u always belongs to $H^2(\Omega)$ in this case.*

– If Ω *is convex*, then automatically we have that $u \in H^2(\Omega)$.

– If Ω *is not convex*, then it is not guaranteed that u belongs to $H^2(\Omega)$.

To be precise, if we come back to fractional-order Sobolev spaces (see section 1.5), then we can establish that there exists a $\sigma_D \in (1/2, 1)$ which depends *only* on Ω, and such that:

i) $\forall f \in L^2(\Omega)$, $u \in \bigcap_{s<\sigma_D} H^{1+s}(\Omega)$;

ii) $\exists f \in L^2(\Omega)$ such that $u \notin H^{1+\sigma_D}(\Omega)$.

Finally, apart from any neighborhood ω of the reentrant corners of $\partial\Omega$ for d $= 2$, or of the reentrant corners and reentrant edges for d $= 3$, we have the following result:

iii) $u_{|\Omega\setminus\overline{\omega}} \in H^2(\Omega \setminus \overline{\omega})$.

In a "more colorful" way, for a given f in $L^2(\Omega)$, the **regularity limit exponent** for the solution to the homogeneous Dirichlet problem is 2 if Ω is convex, and is $1 + \sigma_D$ if Ω is non-convex with $\frac{3}{2} < 1 + \sigma_D < 2$. Moreover, in the second case, the "loss of regularity" of the solution is *localized* to the reentrant corners (and edges) of the boundary. Finally, returning to the solution space $\Psi_D(\Omega)$ defined in [1.43], we have the following *continuous* injections:

– $\Psi_D(\Omega) \subset H^2(\Omega)$, *if* Ω *is convex;*

– $\Psi_D(\Omega) \subset \bigcap_{s<\sigma_D} H^{1+s}(\Omega)$, *if* Ω *is non-convex.*

To illustrate this, if we were to solve the homogeneous Dirichlet problem on a convex polygon or on a convex polyhedron, we have that: $\|u\|_{H^2(\Omega)} \le C\,\|u\|_{H^1(\Delta;\Omega)}$, with $C > 0$ being the continuity modulus for the injection of $\Psi_D(\Omega)$ into $H^2(\Omega)$. Then we may write:

$$\begin{aligned} \|u\|_{H^1(\Delta;\Omega)} &= \left(\|u\|^2_{H^1(\Omega)} + \|\Delta u\|^2_{L^2(\Omega)}\right)^{1/2} \\ (-\Delta u = f \text{ in } \Omega) &= \left(\|u\|^2_{H^1(\Omega)} + \|f\|^2_{L^2(\Omega)}\right)^{1/2} \le C'\|f\|_{L^2(\Omega)} \end{aligned}$$

where the constant $C' > 0$ is independent of f (see theorem 2.2). From this, we conclude that: $\|u\|_{H^2(\Omega)} \le C''\,\|f\|_{L^2(\Omega)}$, where $C'' > 0$ is independent of f.

2.5.2.4. *Homogeneous Neumann problem (polygonal or polyhedral Ω)*

Now, let us consider the problem:

$$\begin{cases} \text{Find } u \in H^1(\Omega) \text{ such that} \\ -\Delta u + u = f \quad \text{in } \Omega \\ \dfrac{\partial u}{\partial n} = 0 \qquad\quad \text{on } \partial\Omega \end{cases}$$

with a given f that belongs to $L^2(\Omega)$.

The results here are similar to those we reported for the homogeneous Dirichlet problem.

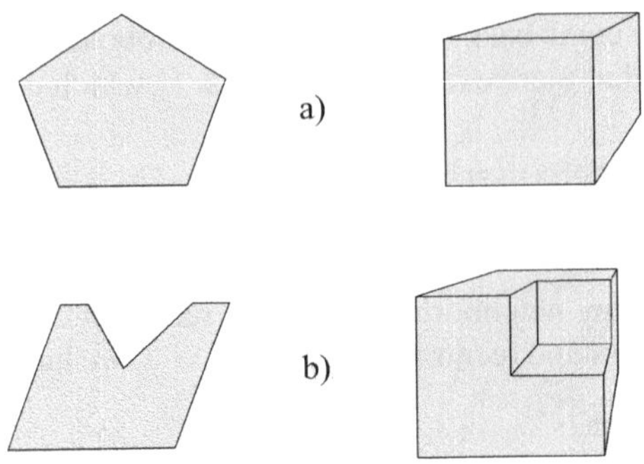

Figure 2.1. *Convex (a) and non-convex (b) domains for* $\mathrm{d} = 2$ *and* $\mathrm{d} = 3$

– *If* Ω *is convex*, then $u \in H^2(\Omega)$.

– *If* Ω *is not convex*, then there exist a $\sigma_N \in (1/2, 1)$, depending *only* on Ω, such that:

i) $\forall f \in L^2(\Omega), u \in \bigcap_{s<\sigma_N} H^{1+s}(\Omega)$;

ii) $\exists f \in L^2(\Omega)$, such that $u \notin H^{1+\sigma_N}(\Omega)$.

Finally, apart from any neighborhood ω' of the reentrant corners (and edges) of $\partial\Omega$, we get the following result:

iii) $u_{|\Omega\setminus\overline{\omega}'} \in H^2(\Omega \setminus \overline{\omega}')$.

When Ω is a polygon in $\mathbb{R}^2$, we can establish that $\sigma_D = \sigma_N$. Conversely, when Ω is a (Lipschitz) polyhedron in $\mathbb{R}^3$, we can have that $\sigma_D \neq \sigma_N$.

And returning to the solution space $\Psi_N(\Omega)$ defined in [1.43], this time we have the following *continuous* injections:

– $\Psi_N(\Omega) \subset H^2(\Omega)$, *if* Ω *is convex*;

– $\Psi_N(\Omega) \subset \bigcap_{s<\sigma_N} H^{1+s}(\Omega)$, *if* Ω *not convex*.

For the Neumann and Dirichlet problems, the same type of results can be obtained for curvilinear polygons or polyhedra. In this case, the convexity or the non-convexity of the open subset Ω should be taken into account only when in the neighborhood of the corners (and edges) of the boundary.

2.5.2.5. *Local regularity of the solution to a Laplace problem*

Let $u \in H^1(\Omega)$ be such that:

$$\begin{cases} -\Delta u = f \text{ in } \Omega \\ \text{boundary condition on } \partial\Omega. \end{cases}$$

So, using a cut-off function along with a partition of unity (Adams and Fournier 2003), we may establish the following result.

Let $B(C,R)$ be a ball of center C and radius R, such that $\overline{B(C,R)} \subset \Omega$. If $f \in H^k(B(C,R))$, then $u \in H^{k+2}(B(C,R'))$ for all $R' < R$.

In other words, for the Laplace problem (and independently of any boundary condition), if the data f are "locally" H^k-regular, then the solution u is "locally" H^{k+2}-regular too.

2.6. Exercises – statements

In the following, we let Ω be a bounded open subset of $\mathbb{R}^{\mathrm{d}}$, where $2 \leq \mathrm{d} \leq 3$, with a Lipschitz boundary, unless stated otherwise.

Exercise 13. Variational formulation for Fourier boundary condition problems

Consider the following boundary value problem ($\lambda > 0$, $f \in L^2(\Omega)$ and $g \in L^2(\partial\Omega)$):

$$\begin{cases} \text{Find } u \in H^1(\Omega) & \text{such that} \\ u - \Delta u = f & \text{in } \Omega \\ \dfrac{\partial u}{\partial n} + \lambda\, u = g & \text{on } \partial\Omega. \end{cases} \qquad \text{[P13]}$$

a) Construct the variational formulation [F13] associated with [P13].

b) Prove the existence and the uniqueness of [F13].

c) Establish the equivalence between [P13] and [F13].

d) What happens when $\lambda < 0$?

Exercise 14. Variational formulation for heat diffusion

Consider the following boundary value problem:

$$\begin{cases} \text{Find } u^\star \in H^1(\Omega) & \text{such that} \\ -\operatorname{div}(k(x)\nabla u^\star) = f & \text{in } \Omega \\ u^\star = \tau_0 & \text{on } \partial\Omega. \end{cases} \qquad \text{[P14]}$$

where $\tau_0 > 0$ is a constant, $f \in L^2(\Omega)$, $k \in L^\infty(\Omega)$, and $k(x) \geq k_{min} > 0$, for almost all x in Ω.

a) Set $u = u^\star - \tau_0$. What problem, which we denote by [P14$_b$], does u satisfy?

b) Construct the variational formulation [F14$_b$] associated with [P14$_b$].

c) Prove the uniqueness of the solution to [F14$_b$].

d) Establish the equivalence between [P14] and [F14$_b$].

Exercise 15. Problem with a discontinuous positive coefficient

Decompose Ω – with a "sufficiently regular" boundary – into $\overline{\Omega} = \overline{\Omega}_1 \cup \overline{\Omega}_2$, where Ω_1 and Ω_2 are two disjoint and open subsets whose boundaries are "sufficiently regular", and label the interface between Ω_1 and Ω_2 by Σ: assume that $\partial\Omega_1 = \Sigma$, $\Sigma \cap \partial\Omega = \emptyset$ and $\partial\Omega \subset \partial\Omega_2$, and let $\kappa_1, \kappa_2 \in \mathbb{R}$ be strictly positive. Denote by κ the function defined by $\kappa_{|\Omega_j} = \kappa_j$, for $j = 1, 2$. Lastly, let $g \in L^2(\Sigma)$. Then solve the problem that is written as the following variational formulation:

$$\left| \begin{array}{l} \text{Find } u \in H_0^1(\Omega) \text{ such that} \\ \displaystyle\int_\Omega \kappa \nabla u \cdot \nabla v \, d\Omega = \int_\Sigma g v \, d\Sigma, \quad \forall v \in H_0^1(\Omega). \end{array} \right. \qquad \text{[F15]}$$

Interpret [F15] in terms of a system of partial differential equations on Ω_1 and Ω_2, first with the boundary conditions, and then with conditions on the interface, Σ (reason "formally" for the latter).

Exercise 16. An elementary study of coercivity

Let H be a real Hilbert space, and a be a continuous bilinear form on $H \times H$. Recall that a is said to be *definite* if $\forall v \in H \setminus \{0\}, a(v, v) \neq 0$.

a) If $H = \mathbb{R}^p$ with $p \in \mathbb{N} \setminus \{0\}$, show that a being definite is equivalent to a being coercive.

b) What happens if H is infinite-dimensional instead?

Exercise 17. Variational formulation with a symmetric form

Let H be a real Hilbert space, a be a positive-definite and *symmetric* bilinear form on $H \times H$ and ℓ be a linear form on H.

Recall that a is *positive-definite* if $\forall v \in H \setminus \{0\}, a(v, v) > 0$.

Our aim is to solve:

$$\left| \begin{array}{l} \text{Find } u \in H \text{such that} \\ a(u, v) = \ell(v) \quad \forall v \in H. \end{array} \right. \qquad \text{[F17]}$$

a) Prove that u is a solution to [F17] if, and only if, u minimizes the functional:

$$J : \left| \begin{array}{l} H \rightarrow \mathbb{R} \\ v \mapsto \frac{1}{2}a(v,v) - \ell(v). \end{array} \right.$$

b) Establish that the point which minimizes J, should it exist, is unique.

c) Assume that a is continuous and coercive on $H \times H$, and that ℓ is continuous on H. Establish that the point that realizes the minimum of J exists.

Exercise 18. Spectral corollary: an application of the spectral theorem

Let:

– H be a Hilbert space;

– $a_+(\cdot,\cdot)$ be a symmetric and coercive continuous bilinear form on $H \times H$;

– L be a second Hilbert space such that $H \subset L$, with H dense in L, and suppose that the injection of H into L is compact.

Our aim is to solve the eigenvalue problem:

$$\left| \begin{array}{l} \text{Find } (u,\lambda) \in (H \setminus \{0\}) \times \mathbb{R} \text{ such that} \\ a_+(u,v) = \lambda(u,v)_L \quad \forall v \in H. \end{array} \right. \qquad \text{[F18]}$$

To do this, to each element $u^* \in L$ we associate another element $T_+u^* \in H$ that is defined by:

$$\left| \begin{array}{l} \text{Find } T_+u^* \in H \text{ such that} \\ a_+(T_+u^*,v) = (u^*,v)_L \quad \forall v \in H. \end{array} \right. \qquad \text{[F18}_b\text{]}$$

a) Show that $T_+ \in \mathcal{L}(L,H)$. Set $T = i_{H\to L} \circ T_+ \in \mathcal{L}(L)$: what are its properties? Deduce from this that the eigenvectors of T constitute a Hilbert basis $(e_n)_{n\geq 1}$ of L, and denote by $(\mu_n)_{n\geq 1}$ their respective eigenvalues.

b) Check that the set of solutions λ to the problem [F18] coincides with $\{\lambda_n,\ n \geq 1\}$, where $\lambda_n = (\mu_n)^{-1}$ for $n \geq 1$. What are the solutions u to the problem [F18]?

Now we want to show that $((\lambda_n)^{-1/2}e_n)_{n\geq 1}$ forms a Hilbert basis for H.

c) Decompose $u \in L$ into $u = \sum_{n\geq 1}(u,e_n)_L e_n$. Show that if u belongs to H, then the sequence $(\sum_{n=1,N}(u,e_n)_L e_n)_N$ converges in H (to u). In particular, show that $a_+(u,u) = \sum_{n\geq 1} \lambda_n((u,e_n)_L)^2$.

d) Conclude from this that $((\lambda_n)^{-1/2}e_n)_{n\geq 1}$ forms a Hilbert basis for H if we endow it with the inner product $(u,v) \mapsto a_+(u,v)$, whose associated norm is $u \mapsto (\sum_{n\geq 1} \lambda_n((u,e_n)_L)^2)^{1/2}$.

Exercise 19. Solving the Laplacian with non-homogeneous conditions

Assume that the open subset Ω is connected, and consider the boundary value problem:

$$\begin{cases} \text{Find } u \in H^1(\Omega) & \text{suchthat} \\ -\Delta u = f & \text{in } \Omega \\ \textit{boundary condition} & \text{on } \partial\Omega \end{cases}$$

where $f \in L^2(\Omega)$. As for the boundary condition on $\partial\Omega$, we have a choice between the following two:

$$\text{Fourier } (\lambda > 0,\, g \in L^2(\partial\Omega)) : \frac{\partial u}{\partial n} + \lambda\, u = g. \qquad \text{[Fou19]}$$

$$\text{Neumann } (g \in L^2(\partial\Omega)) : \frac{\partial u}{\partial n} = g. \qquad \text{[Neu19]}$$

For each set of boundary conditions, propose an *ad hoc* framework that shows that the problem is well posed.

Exercise 20. Dirichlet–Neumann or Dirichlet–Fourier mixed boundary condition problems

Decompose the boundary into $\partial\Omega = \overline{\Gamma_1} \cup \overline{\Gamma_2}$, with $\Gamma_1 \cap \Gamma_2 = \emptyset$ and where Γ_1, Γ_2 are of non-zero measure. Let $\lambda \geq 0$. Our aim is to solve the mixed boundary condition problem:

$$\begin{cases} \text{Find } u \in H^1(\Omega) & \text{such that} \\ -\Delta u = f & \text{in } \Omega \\ u = 0 & \text{on } \Gamma_1 \\ \dfrac{\partial u}{\partial n} + \lambda u = g & \text{on } \Gamma_2. \end{cases} \qquad \text{[P20]}$$

where $f \in L^2(\Omega)$ and $g \in L^2(\Gamma_2)$.

a) Write down the associated variational formulation ([F20]). Why do we have to introduce $H^1_{0,\Gamma_1}(\Omega) = \{v \in H^1(\Omega) \text{ such that } v_{|\Gamma_1} = 0\}$ in it?

b) Show that [F20] is well posed.

c) Check that the solution to [F20] is equivalently a solution to [P20]. For the boundary condition on Γ_2 assume that the space consisting of the traces on Γ_2 of elements in $H^1_{0,\Gamma_1}(\Omega)$ is dense in $L^2(\Gamma_2)$.

Exercise 21. Regularity of the solutions to Laplace's equation

Consider the open subset Ω of $\mathbb{R}^2$ in the left-hand figure below.

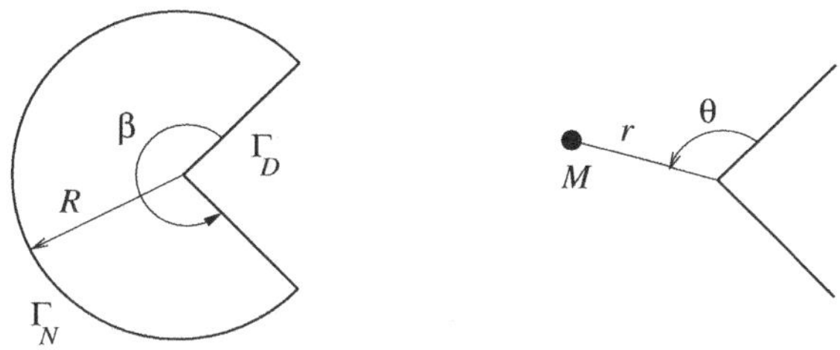

In the following, fix $\beta = \frac{3}{2}\pi$ and $R = 1$, and adopt the polar coordinates given in the right-hand figure, noting that $\Omega = \{(r, \theta)$ such that $0 < r < 1, 0 < \theta < \frac{3}{2}\pi\}$.

Decompose the boundary into $\partial\Omega = \overline{\Gamma_D} \cup \overline{\Gamma_N}$, with $\Gamma_N = \{(1, \theta)$ such that $0 < \theta < \frac{3}{2}\pi\}$, where Γ_D further decomposing into $\overline{\Gamma_D} = \overline{\Gamma_D^1} \cup \overline{\Gamma_D^2}$, where $\Gamma_D^1 = \{(r, 0)$ such that $0 < r < 1\}$ and $\Gamma_D^2 = \{(r, \frac{3}{2}\pi)$ such that $0 < r < 1\}$.

We are interested in the function u_s that is defined by:

$$u_s(r, \theta) = r^{-2/3} \sin(\frac{2}{3}\theta), \text{ for } 0 < r < 1, \ 0 \leq \theta \leq \frac{3}{2}\pi.$$

a) Check that $u_s \in L^2(\Omega)$ (and calculate $\|u_s\|_{L^2(\Omega)}$).

b) Recall the expression for the Laplacian in polar coordinates:

$$\Delta v = \frac{\partial^2 v}{\partial r^2} + \frac{1}{r}\frac{\partial v}{\partial r} + \frac{1}{r^2}\frac{\partial^2 v}{\partial \theta^2}.$$

Calculate Δu_s, and conclude that u_s is a solution to the Laplace equation.

c) What is the trace of u_s on Γ_D^1, and on Γ_D^2, equal to? On Γ_N, we have that $\dfrac{\partial v}{\partial n} = \dfrac{\partial v}{\partial r}\Big|_{r=1}$ since Γ_N is included within the circle centered at $(0, 0)$ of radius 1. When applied to u_s on Γ_N, what is the trace of the normal derivative mapping equal to?

d) Collect together the results of the previous questions. What are your thoughts?

Exercise 22. Problem with a discontinuous coefficient that changes sign

Again, we are interested in a problem where there is a transition between two subsets of the open set Ω – with a "sufficiently regular" boundary – that is to say,

$\overline{\Omega} = \overline{\Omega}_1 \cup \overline{\Omega}_2$, where Ω_1 and Ω_2 are two disjoint open subsets whose boundaries are "sufficiently regular", and call Σ the interface between Ω_1 and Ω_2. Let $\sigma_1 > 0$ and $\sigma_2 < 0$. Denote by σ the function defined by $\sigma_{|\Omega_j} = \sigma_j$, for $j = 1, 2$. Our goal is to solve the problem:

$$\begin{cases} \text{Find } u \in H_0^1(\Omega) \text{ such that} \\ -\operatorname{div} \sigma \nabla u = f \text{ in } \Omega \end{cases}. \qquad \text{[P22]}$$

If v is a function in $L^2(\Omega)$, let $v_j = v_{|\Omega_j}$ be its restriction to Ω_j, $j = 1, 2$. Finally, denote $\Gamma_j = \partial\Omega_j \cap \partial\Omega$, $j = 1, 2$; assume that Γ_1, Γ_2 are both of non-zero measure.

a) Write down the variational formulation for the problem [P22]. In the following, denote by a the associated bilinear form:

$$a(u, v) = \int_\Omega \sigma \nabla u \cdot \nabla v \, d\Omega.$$

Is the form a coercive? Conclude that we cannot apply the Lax–Milgram theorem when studying the problem [P22].

In the following, we will use the Banach–Necas–Babuska theorem.

b) If v is a function in $L^2(\Omega)$ with the restrictions $v_j \in H^1(\Omega_j)$, $j = 1, 2$, under what condition do we have that $v \in H^1(\Omega)$? Deduce that the following are equivalent:

$$v \in H_0^1(\Omega) \iff v_j \in H_{0,\Gamma_j}^1(\Omega_j), \ j = 1, 2, \text{ and } \gamma_\Sigma v_1 = \gamma_\Sigma v_2$$

where γ_Σ is the trace on Σ and $H_{0,\Gamma_j}^1(\Omega_j) = \{w \in H^1(\Omega_j) \ : \ \gamma_{\Gamma_j} w = 0\}$, $j = 1, 2$.

c) Decompose $a(v, w)$ into the form $a(v, w) = a_{\Omega_1}(v_1, w_1) + a_{\Omega_2}(v_2, w_2)$.

d) For $j = 1, 2$, recall why the form a_{Ω_j} is coercive on $H_{0,\Gamma_j}^1(\Omega_j)$.

e) To establish the inf-sup condition [2.43], for any element v in $H_0^1(\Omega)$, we need to find a w of $H_0^1(\Omega)$ such that $a(v, w)$ "dominates" the product $\|v\|_{H^1(\Omega)} \|w\|_{H^1(\Omega)}$. Introduce an operator T_0, defined by:

$$T_0 v = \begin{cases} v_1 & \text{in } \Omega_1 \\ -v_2 & \text{in } \Omega_2. \end{cases}$$

By choosing $w = T_0 v$, can we obtain the inf-sup condition for the form a by using this operator?

From now on, we consider a *simplified* version of the problem. Assume that Ω_1 and Ω_2 are two cubes, and that the interface Σ between Ω_1 and Ω_2 forms a face, included in the plane given by the equation $\{x = 0\}$ (with $\Omega_1 \subset \{x < 0\}$ and $\Omega_2 \subset \{x > 0\}$).

f) In the following, denote $V_j = H^1_{0,\Gamma_j}(\Omega_j)$, $j = 1, 2$. Let $v_1 \in V_1$: introduce $(R_1 v_1)(x,y,z) = v_1(-x,y,z)$, for every $(x,y,z) \in \Omega_2$. Check that $\gamma_\Sigma(R_1 v_1) = \gamma_\Sigma v_1$ and deduce that R_1 is an operator on V_1 with values in V_2. In particular, what does the norm $|||R_1|||$ equal to?

g) Introduce:

$$Tv = \begin{cases} v_1 & \text{on } \Omega_1 \\ -v_2 + 2R_1 v_1 & \text{on } \Omega_2. \end{cases}$$

Show that T is an injective and continuous linear mapping from $H^1_0(\Omega)$ into itself. Using T, under what conditions on σ_1 and σ_2 can we establish the inf-sup condition for the form a, and therefore apply the Banach–Necas–Babuska corollary to the problem [P22]? It is possible to use Young's inequality:

$$\pm 2ab \geq -\eta^2 a^2 - \frac{1}{\eta^2} b^2 \quad \forall a, b, \forall \eta \neq 0$$

which also applies to inner products.

h) Given $v_2 \in V_2$, introduce $R_2 v_2(x,y,z) = v_2(-x,y,z)$, for every $(x,y,z) \in \Omega_1$, then construct a second bijective continuous linear mapping T from $H^1_0(\Omega)$ into itself, after modifying T_0 this time using R_2. To which new condition on σ_1 and σ_2 can we apply the Banach–Necas–Babuska corollary?

Exercise 23. Linear elasticity

The physics of linear elasticity is realized in $\mathbb{R}^3$, so let us consider $d = 3$ so that $\Omega \subset \mathbb{R}^3$[5]. Consider the following problem:

$$\begin{cases} \text{Find } \mathbf{u} \text{ such that} \\ \displaystyle\sum_{j=1}^{3} \frac{\partial}{\partial x_j} \sigma_{ij}(\mathbf{u}) = f_i \text{ in } \Omega & 1 \leq i \leq 3 \\ u_i = 0 & \text{on } \partial\Omega. \end{cases} \qquad \text{[P23]}$$

In the above, $\mathbf{f} \in \left(L^2(\Omega)\right)^3$, and $(\sigma_{ij})_{1 \leq i,j \leq 3}$ is the stress tensor, given by:

$$\sigma_{ij}(\mathbf{u}) = \lambda \operatorname{div} \mathbf{u}\, \delta_{ij} + 2\mu\, \varepsilon_{ij}(\mathbf{u}),\ \lambda, \mu \text{ the Lamé coefficients, with:}$$

$$\varepsilon_{ij}(\mathbf{u}) = \frac{1}{2}\left(\frac{\partial u_i}{\partial x_j} + \frac{\partial u_j}{\partial x_i}\right) \text{ the strain tensor.}$$

5 Mathematically, we can of course solve the same problem in $\Omega \subset \mathbb{R}^2$ using the tools given below. See section 5.3.3 for an illustration.

a) Considering $\mathbf{V} = \left(H_0^1(\Omega)\right)^3$, state the variational formulation [F23] associated with [P23] in $\mathbf{V}$ using the following bilinear form:

$$b(\mathbf{v}, \mathbf{w}) = \int_\Omega \Big(\sum_{i,j=1}^{3} \varepsilon_{ij}(\mathbf{v})\varepsilon_{ij}(\mathbf{w}) \Big) \, d\Omega.$$

Show that $\mathbf{u}$ is a solution to [P23] if, and only if, it is a solution to [F23].

b) *Korn's inequality*: For any element $\mathbf{v}$ in $\left(\mathcal{D}(\Omega)\right)^3$, check that the following equality holds:

$$\int_\Omega \frac{\partial v_i}{\partial x_j}\frac{\partial v_j}{\partial x_i} \, d\Omega = \int_\Omega \frac{\partial v_i}{\partial x_i}\frac{\partial v_j}{\partial x_j} \, d\Omega.$$

Does every function in $\mathbf{V}$ satisfy this identity?

From this, deduce Korn's inequality:

$$b(\mathbf{v}, \mathbf{v}) \geq \frac{1}{2}\sum_{i=1}^{3} \int_\Omega |\nabla v_i|^2 \, d\Omega \quad \forall \mathbf{v} \in \mathbf{V}.$$

c) When $\lambda \geq 0$ and $\mu > 0$, establish the existence, uniqueness and continuous dependence of the solution to the variational formulation [F23].

d) If $\mathbf{u}$ is a solution of [F23], show that $\mathbf{u}$ minimizes a functional defined on $\mathbf{V}$. What can be said about the converse?

Exercise 24. Solution to the plate equation

Consider the problem regarding the bending of a clamped plate. The displacement in the direction transverse to the plate, denoted by u, satisfies the following equations within the middle section Ω of the plate, which can be identified with a bounded open subset of $\mathbb{R}^2$:

$$\begin{cases} \text{Find } u \in H^2(\Omega) \text{such that} \\ \Delta^2 u = f \in L^2(\Omega) \text{ in } \Omega \\ u = 0 \text{ and } \dfrac{\partial u}{\partial n} = 0 \text{ on } \partial\Omega. \end{cases} \qquad \text{[P24]}$$

a) Recall the definitions of the Sobolev space $H^2(\Omega)$ and of the $H^2(\Omega)$ norm.

b) Introduce the subspace $H_0^2(\Omega) = \overline{\mathcal{D}(\Omega)}^{H^2(\Omega)}$ of $H^2(\Omega)$. Assume that:

$$H_0^2(\Omega) = \left\{ v \in H^2(\Omega) \text{ such that } v_{|\partial\Omega} = 0 \text{ and } \frac{\partial v}{\partial n}_{|\partial\Omega} = 0 \right\}.$$

Recall why $H_0^2(\Omega)$ is a Hilbert space when endowed with the $H^2(\Omega)$ norm.

c) Show that the solution u to [P24] satisfies a variational formulation in $H_0^2(\Omega)$ of the form:

$$\left| \begin{array}{l} \text{Find } u \in H_0^2(\Omega) \text{ such that} \\ a(u, v) = \ell(v), \quad \forall v \in H_0^2(\Omega). \end{array} \right. \qquad \text{[F24]}$$

d) Show that $a(\cdot, \cdot)$ is a symmetric continuous bilinear form on $H_0^2(\Omega)$, and that $\ell(\cdot)$ is a continuous linear form on $H_0^2(\Omega)$.

e) Let $v \in H_0^2(\Omega)$, and show that $\partial_1 v \in H_0^1(\Omega)$ and $\partial_2 v \in H_0^1(\Omega)$. For $i = 1, 2$, deduce that:

$$\left\| \frac{\partial v}{\partial x_i} \right\|_{L^2(\Omega)} \leq C_1 \, |v|_{2,\Omega} \, , \quad \forall v \in H_0^2(\Omega)$$

where $|\cdot|_{2,\Omega}$ is the semi-norm on $H^2(\Omega)$ (see [1.26]), and finally that:

$$\|v\|_{H^2(\Omega)} \leq C_2 \, |v|_{2,\Omega} \, , \quad \forall v \in H_0^2(\Omega),$$

where the constants C_1, C_2 are independent of v.

f) Show that:

$$a(v, v) = |v|_{2,\Omega}^2 \, , \quad \forall v \in H_0^2(\Omega)$$

which concludes our study of [F24].

3

Introduction to the Finite Element Method

There are mainly two techniques that are used to discretize elliptic problems: the finite difference method based upon the strong formulation of the problem (and therefore on classical solutions), and the finite element method based upon the variational formulation (and therefore on weak solutions).

As the fundamental principles that preside over finite difference discretizations of elliptic problems are similar to those encountered in the discretization of hyperbolic equations (Després 2010), we will only develop the theory of finite element approximation here.

The finite element method consists of replacing the solution and test-function spaces (which are infinite-dimensional) for the variational formulation, with those of solution and test-function approximation spaces, which are *finite-dimensional*. These approximation spaces are typically included within the initial space. We then will talk about either **internal**, **conformal** or **Galerkin**'s approximations so that, under certain conditions, the approximation space converges to the original space (in a sense that remains to be clarified).

The first section is generally dedicated to Galerkin's approximation. Then we emphasize the principle of finite element design.

3.1. Galerkin's approximation

Recall that the elliptic problems that we studied in the previous chapters (mainly Dirichlet, Neumann and Fourier) fit into the following abstract Hilbert space framework:

$$\left| \begin{array}{l} \text{Find } u \in V \text{ such that} \\ a(u,v) = \ell(v) \quad \forall v \in V \end{array} \right. \qquad [3.1]$$

where:

– V is a Hilbert space endowed with the norm $\|\cdot\|_V$;

– $a(\cdot,\cdot)$ is a continuous bilinear form on V, that is, that there exists a constant M such that:

$$|a(v,w)| \leq M\|v\|_V\|w\|_V \quad \forall(v,w) \in V \qquad [3.2]$$

– $\ell(\cdot)$ is a continuous form on V.

The objective of the following sections is to study how we may approximate problem [3.1]. In the last section, we will also consider cases involving different solution and test-function spaces.

Consider the *vector subspace* V_h of V, whose *dimension* $N = N(h)$ *is finite*, where h is a positive parameter that we will tend toward 0, and also that $\lim_{h\to 0} N(h) = \infty$ (it is implicitly assumed here and in the following that V is infinite-dimensional). Since $N(h) < \infty$, V_h is a *closed* vector subspace of V and therefore a Hilbert space, endowed with the norm $\|\cdot\|_V$. Denote by $(V_h)_h$ the family of subspaces of V, when h is allowed to vary. For a given h, V_h represents a finite-dimensional approximation for the space V.

We introduce the following approximate variational problem:

$$\left| \begin{array}{l} \text{Find } u_h \in V_h \text{ such that} \\ a(u_h,v_h) = \ell(v_h) \quad \forall v_h \in V_h. \end{array} \right. \qquad [3.3]$$

When $V_h \subset V$, then the approximation is called an **internal** or a **Galerkin** approximation. Then, we assume this for the rest of this chapter.

First, we will investigate the existence and uniqueness of the approximate solution u_h, depending on whether the abstract problem [3.1] can be solved or not using the Lax–Milgram theorem (along with a coercive form, a), or more generally using the Banach–Necas–Babuska theorem (with stability and solvability conditions on the form, a). Then we will study the behavior of the difference $u - u_h$.

In the following, we denote by $(\cdot|\cdot)$ the inner product on $\mathbb{R}^N$[1]. First, let us mention an immediate reinterpretation of the considered Problem [3.3].

1 From now and onwards, $\overrightarrow{x}$ will represent a vector in $\mathbb{R}^N$.

LEMMA 3.1.– *Let* $(\varphi_1, \cdots, \varphi_N)$ *be a basis for the vector space* V_h*. Then the approximate problem [3.3] is* equivalent *to the linear system based on* $\mathbb{R}^N$*:*

$$\textit{Find } \overrightarrow{\lambda} \in \mathbb{R}^N \textit{ such that } \quad \mathbb{A}\overrightarrow{\lambda} = \overrightarrow{L} \qquad [3.4]$$

where:

$$\mathbb{A}_{ij} = a(\varphi_j, \varphi_i), \; (\overrightarrow{\lambda})_j = \lambda_j, \; \textit{and} \; (\overrightarrow{L})_i = \ell(\varphi_i) \;\; 1 \leq i, j \leq N$$

and where the solutions are related to one another by:

$$u_h = \sum_{j=1}^{N} \lambda_j \varphi_j. \qquad [3.5]$$

Proof. We will first express u_h in terms of the basis $(\varphi_j)_j$ as in [3.5]. By choosing $v_h = \varphi_i$ for $i = 1, N$ in the variational formulation [3.3], we get the following equations:

$$\sum_{j=1}^{N} \lambda_j a(\varphi_j, \varphi_i) = \ell(\varphi_i) \quad \forall i = 1, N$$

and which make up the linear system [3.4] when written line by line. The converse is immediate by the *linearity* of the form a. ■

As [3.4] shows, an approximate problem is equivalent to solving a linear system, and it should be clear that it is this form that we will tackle numerically.

3.1.1. *Case of a coercive form*

Here, we assume that there exists a constant $\alpha > 0$ (see remark 2.11) such that:

$$a(v, v) \geq \alpha \|v\|_V^2 \quad \forall v \in V. \qquad [3.6]$$

By the Lax–Milgram theorem (theorem 2.4) and by assuming [3.2] and [3.6], the abstract problem [3.1] then has a unique solution $u \in V$.

3.1.1.1. *Existence and uniqueness of the approximate solution*

With the same assumptions, the approximate problem [3.3] also admits a unique solution $u_h \in V_h$ as the Lax–Milgram theorem can always be applied. Let us discuss another very instructive proof of this result, assuming the hypothesis [3.6].

PROPOSITION 3.1 (Existence and uniqueness of the approximate solution).– *With the hypothesis [3.6], the approximate variational problem [3.3] admits a unique solution* $u_h \in V_h$.

Proof. We will use an equivalent reformulation of lemma 3.1 and show that the matrix $\mathbb{A}$ is *invertible*. Indeed, let $\overrightarrow{\lambda}$ be such that $\mathbb{A}\overrightarrow{\lambda} = 0$. Then, after contracting $\mathbb{A}\overrightarrow{\lambda}$ with $\overrightarrow{\lambda}$ using the inner product and by using the bilinearity of the form a, we see that:

$$0 = (\mathbb{A}\overrightarrow{\lambda} \,|\, \overrightarrow{\lambda}) = \sum_{i=1}^{N} \Big(\sum_{j=1}^{N} \lambda_j a(\varphi_j, \varphi_i) \Big) \lambda_i = \sum_{i=1}^{N} \Big(a\Big(\sum_{j=1}^{N} \lambda_j \varphi_j, \varphi_i \Big) \Big) \lambda_i$$
$$= a\Big(\sum_{j=1}^{N} \lambda_j \varphi_j, \sum_{i=1}^{N} \lambda_i \varphi_i \Big).$$

Then by the *coercivity hypothesis* [3.6], which is valid since $\varphi_i \in V_h \subset V$ (internal approximation), we may deduce that:

$$\alpha \| \sum_{j=1}^{N} \lambda_j \varphi_j \|_V^2 \leq 0$$

that is:

$$\sum_{j=1}^{N} \lambda_j \varphi_j = 0$$

and, since $(\varphi_j)_j$ makes up a basis of V_h, we can conclude that $\lambda_j = 0$, $\forall j = 1, N$, and hence $\overrightarrow{\lambda} = 0$. ■

REMARK 3.1.– *By following exercise 16, we can see easily that the coercivity of the bilinear form* $a(\cdot, \cdot)$ *implies that the matrix* $\mathbb{A}$ *is* definite *(and therefore invertible): for all* $\overrightarrow{\lambda} \in \mathbb{R}^N \setminus \{0\}$, *we have that* $(\mathbb{A}\overrightarrow{\lambda} \,|\, \overrightarrow{\lambda}) \neq 0$. *Therefore* $\mathbb{A}\overrightarrow{\lambda} = 0$ *implies that* $\overrightarrow{\lambda} = 0$. *Lastly, since* $a(\cdot, \cdot)$ *is positive,* $\mathbb{A}$ *is* positive-definite.

REMARK 3.2.– *Note that the internal approximation assumption* $V_h \subset V$ *is essential. When* $V_h \not\subset V$, *the approximation is called either* ***external*** *or* ***non-conformal***. *This is a more complex case to study since the bilinear form* a *can no longer be defined on* V_h, *nor is it coercive on* V_h. *Thus, we must introduce an approximate bilinear form* a_h, *which complicates our analysis (see section 4.3).*

We now investigate what assumptions are necessary, in order for an internal approximation to converge.

3.1.1.2. *Convergence of Galerkin's approximation*

In order to obtain a convergence result with respect to how the approximate solution u_h (to the problem [3.3]) converges to the solution u (converges to the exact problem [3.1]), we will directly estimate its error:

$$e_h = \|u - u_h\|_V.$$

Note that this approach differs from the classical finite difference approach, for which error estimates are usually based upon the notions of stability and consistency (Després 2010).

Let us begin by proving Céa's fundamental lemma:

LEMMA 3.2 (Céa).– *If the bilinear form $a(\cdot,\cdot)$ is continuous and coercive, then:*

$$\|u - u_h\|_V \leq \frac{M}{\alpha} \inf_{v_h \in V_h} \|u - v_h\|_V \qquad [3.7]$$

where M is the modulus of continuity [3.2] and α is the coercivity constant [3.6].

Moreover, if $a(\cdot,\cdot)$ is symmetric, then:

$$\|u - u_h\|_V \leq \left(\frac{M}{\alpha}\right)^{1/2} \inf_{v_h \in V_h} \|u - v_h\|_V. \qquad [3.8]$$

Proof. From [3.6] (the coercivity of $a(\cdot,\cdot)$), we obtain the first bound:

$$\alpha\|u - u_h\|_V^2 \leq a\,(u - u_h, u - u_h)\,.$$

Now let $v_h \in V_h$ be arbitrary. By observing that $(v_h - u_h) \in V_h$ in taking the difference between [3.1] and [3.3], we obtain:

$$a\,(u - u_h, v_h - u_h) = 0.$$

It then implies:

$$a\,(u - u_h, u - u_h) = a\,(u - u_h, u - v_h) \quad \forall v_h \in V_h.$$

From [3.2] (the continuity of $a(\cdot,\cdot)$), we obtain the second bound:

$$a\,(u - u_h, u - v_h) \leq M\|u - u_h\|_V\|u - v_h\|_V.$$

It shows that:

$$\|u - u_h\|_V \leq \frac{M}{\alpha}\|u - v_h\|_V \qquad \forall v_h \in V_h$$

and thus [3.7] follows.

For the case in which the form $a(\cdot,\cdot)$ is symmetric, we can directly expand $a(u - v_h, u - v_h)$ by considering $a\,(u - u_h, u - u_h)$. To be precise:

$$\begin{aligned}
&a(u - v_h, u - v_h) = a(u - u_h + u_h - v_h, u - u_h + u_h - v_h)\\
&(a \text{ is symmetric}) = a(u - u_h, u - u_h) + 2a(u - u_h, u_h - v_h) + a(u_h - v_h,\\
&u_h - v_h)\\
&([3.1] \text{ and } [3.3]) = a(u - u_h, u - u_h) + a(u_h - v_h, u_h - v_h)\\
&(a \text{ is coercive}) \geq a(u - u_h, u - u_h).
\end{aligned}$$

Then, as a is both continuous and coercive, we may deduce that:

$$\alpha\,\|u - u_h\|_V^2 \leq M\,\|u - v_h\|_V^2 \qquad \forall v_h \in V_h$$

and so [3.8] follows. ■

REMARK 3.3.– *When the form a is symmetric, the calculation above shows that $a(u - v_h, u - v_h) > a(u - u_h, u - u_h)$ as soon as $v_h \neq u_h$.*

From the inequalities [3.7] and [3.8], we see that the error e_h is proportional to the approximation error $\inf_{v_h \in V_h} \|u - v_h\|_V$ between the two spaces V_h and V, with the latter error characterizing how "close" these spaces are to one another. In fact, we have:

$$\inf_{v_h \in V_h} \|u - v_h\|_V \leq \|u - u_h\|_V \leq C \inf_{v_h \in V_h} \|u - v_h\|_V \quad \forall h$$

where $C > 0$ is independent of both h and $\ell(\cdot)$.

Let us now provide a sufficient condition that guarantees that this error tends toward 0 as $h \to 0$ for the family $(V_h)_h$, and consequently the approximation converges.

DEFINITION 3.1.– *The family $(V_h)_h$ satisfies the* ***approximation property*** *for V if and only if:*

there exists a dense subspace W of V and, for each h,
there exists a mapping r_h from W into V_h, such that: [3.9]
$\lim_{h\to 0} \|v - r_h v\|_V = 0 \;\forall v \in W$.

PROPOSITION 3.2.– *Assume that the approximation property [3.9] holds. Then:*

$$\lim_{h\to 0}\left(\inf_{v_h\in V_h}\|v-v_h\|_V\right)=0\ \forall v\in V. \qquad [3.10]$$

Proof. Let $v \in V$ be given and consider $\eta > 0$. Since W is dense in V, there exists a $w \in W$ such that $\|v-w\|_V \leq \eta/2$. From [3.9], there exists a $h(\eta) > 0$ such that, for all $h \leq h(\eta)$, we have that $\|w - r_h w\|_V \leq \eta/2$. By using the triangle inequality, we see then that $\|v - r_h w\|_V \leq \eta$ for all $h \leq h(\eta)$. We can conclude that [3.10] must hold. ■

When the approximation property [3.9] is satisfied, then, by applying Céa's lemma, we can deduce that the error $\|u-u_h\|_V$ converges to 0 as h tends toward 0.

THEOREM 3.1 (Convergence of the internal approximation).– *Assume that the approximation property [3.9] holds. Then with the hypotheses [3.2] and [3.6], we have that:*

$$\lim_{h\to 0}\|u-u_h\|_V=0. \qquad [3.11]$$

REMARK 3.4.– *The space V is often equal to a Sobolev space of type $H^k(\Omega)$ ($k \geq 1$), in which a space of class $\mathcal{C}^m(\overline{\Omega})$ (for an appropriate $m \geq 0$) is dense (see proposition 1.10). Since spaces of classically differentiable functions $\mathcal{C}^m(\overline{\Omega})$ are easier to manipulate in the fundamental estimation $\|w - r_h w\|_V$, they usually play the role of W in [3.9]. See remark 4.5 for an example of obtaining [3.9] in the case where $V = H^1(\Omega)$ and $W = \mathcal{C}^2(\overline{\Omega})$.*

Equation [3.11] shows that the internal approximation converges with respect to the $\|\cdot\|_V$ norm. To specify the rate of convergence, we introduce the following definition; examples of using it are provided in Chapter 4.

DEFINITION 3.2 (Rate of convergence).– *We say that the internal approximation converges with order $k > 0$ if there exists a constant $C > 0$ that is independent of h, such that:*

$$\|u-u_h\|_V \leq C\,h^k.$$

3.1.2. *Case of a form satisfying the stability and solvability conditions*

In this case, it is the Banach–Necas–Babuska theorem that allows us to show that the abstract problem [3.1] admits a unique solution $u \in V$. The main difference between this case and the one with a coercive form is that the approximate variational problem [3.3] does not inherit any stability conditions, when set in V_h. Indeed, condition [2.43] only implies that $\alpha' > 0$ is such that, for all h and for all $v_h \in V_h$:

$$\sup_{v' \in V \setminus \{0\}} \frac{|a(v_h, v')|}{\|v'\|_V} \geq \alpha' \|v_h\|_V.$$

However, this obviously does not guarantee that:

$$\sup_{v'_h \in V_h \setminus \{0\}} \frac{|a(v_h, v'_h)|}{\|v'_h\|_V} \geq \alpha' \|v_h\|_V$$

since V_h is a *proper* subspace of V. This is why we impose a discrete uniform stability, or a discrete uniform inf-sup, condition. Compared to [2.43], this is an *additional* condition.

DEFINITION 3.3 (Discrete uniform inf-sup condition).– *We say that a bilinear form* $a(\cdot,\cdot)$ *on* $V \times V$ *satisfies a* discrete uniform stability condition, *or equivalently a* discrete uniform inf-sup condition, *on the subspaces* $(V_h)_h$, *if and only if there exists a constant* $\alpha'' > 0$, *such that:*

$$\sup_{v'_h \in V_h \setminus \{0\}} \frac{|a(v_h, v'_h)|}{\|v'_h\|_V} \geq \alpha'' \|v_h\|_V \qquad \forall h \qquad \forall v_h \in V_h. \qquad [3.12]$$

We will see (remark 3.5) why the *uniform* condition is preferred, that is, with a constant $\alpha'' > 0$ that is *independent* of h. In this case, we have the following result:

PROPOSITION 3.3 (Existence and uniqueness of the approximate solution).– *Under the hypothesis [3.12], the approximate variational problem [3.3] admits a unique solution* $u_h \in V_h$.

Proof. We will rely heavily on lemma 3.1 again to show that the matrix $\mathbb{A}$ is *invertible*. Let $\overrightarrow{\lambda}$ be such that $\mathbb{A}\overrightarrow{\lambda} = 0$, so that for every $\overrightarrow{\mu} \in \mathbb{R}^N$, we have that $(\mathbb{A}\overrightarrow{\lambda} | \overrightarrow{\mu}) = 0$. In setting $v_h = \sum_i \mu_i \varphi_i$, we see that:

$$0 = (\mathbb{A}\overrightarrow{\lambda} | \overrightarrow{\mu}) = \sum_{i=1}^{N} \Big(\sum_{j=1}^{N} \lambda_j a(\varphi_j, \varphi_i) \Big) \mu_i = a\Big(\sum_{j=1}^{N} \lambda_j \varphi_j, \sum_{i=1}^{N} \mu_i \varphi_i \Big) = a(u_h, v_h).$$

Then, according to the *uniform discrete stability condition* [3.12], we deduce that $u_h = 0$ and therefore conclude that $\lambda_j = 0$, $\forall j = 1, N$, since $(\varphi_j)_j$ is a basis of V_h, and thus $\overrightarrow{\lambda} = 0$. ■

Now we will concern ourselves with the convergence of Galerkin's approximation when applied to a form satisfying the stability and solvability conditions. Here, the same results from before apply: Céa's lemma regarding the proportionality between the error $e_h = \|u - u_h\|_V$ and the approximation error $\inf_{v_h \in V_h} \|u - v_h\|_V$ on V_h and V, independently of h, and then following that up with a convergence theorem applied to the internal approximation. The proof of the first result is slightly different from that of Céa's lemma, whereas the second is obtained in exactly the same way as in section 3.1.1.

LEMMA 3.3 (Céa's lemma).– *Assuming the hypotheses [3.2] and [3.12], there exists a constant $C > 0$, independent of $(V_h)_h$ and of $\ell(\cdot)$, such that:*

$$\|u - u_h\|_V \leq C \inf_{v_h \in V_h} \|u - v_h\|_V. \qquad [3.13]$$

Proof. Let $v_h \in V_h$. Then because of [3.12] (the discrete uniform stability condition for $a(\cdot, \cdot)$), we get the first bound:

$$\alpha'' \|u_h - v_h\|_V \|v_h'\|_V \leq |a(u_h - v_h, v_h')|$$

for an appropriate choice of $v_h' \in V_h \setminus \{0\}$. However, taking the difference between [3.1] and [3.3] with respect to the test-function v_h', we know that $a(u - u_h, v_h') = 0$, and so:

$$\alpha'' \|u_h - v_h\|_V \|v_h'\|_V \leq |a(u_h - v_h, v_h') + a(u - u_h, v_h')| = |a(u - v_h, v_h')|.$$

From [3.2] (the continuity of $a(\cdot, \cdot)$), we obtain the second bound:

$$|a(u - u_h, v_h')| \leq M \|u - u_h\|_V \|v_h'\|_V$$

which shows that:

$$\|u_h - v_h\|_V \leq \frac{M}{\alpha''} \|u - v_h\|_V.$$

The constant M/α'' does not depend on v_h, so the result is true for any element v_h in V_h. Using the triangle inequality, we conclude that:

$$\|u - u_h\|_V \leq \|u - v_h\|_V + \|u_h - v_h\|_V \leq \left(1 + \frac{M}{\alpha''}\right) \|u - v_h\|_V \quad \forall v_h \in V_h. \qquad \blacksquare$$

Finally, we have:

THEOREM 3.2 (Convergence of the internal approximation).– *Assume that the approximation property [3.9] holds. Then, assuming the hypotheses [3.2] and [3.12], we have that:*

$$\lim_{h \to 0} \|u - u_h\|_V = 0. \qquad [3.14]$$

A practical application of this is demonstrated in exercise 34.

REMARK 3.5.– *What would happen if the constant* $\alpha'' > 0$ *in [3.12] depends on* h *and is such that* $\lim_{h\to 0} \alpha''(h) = 0$*, that is, if the stability condition* is not uniform in *h? First of all, the result from proposition 3.3 still remains valid: the existence and uniqueness of* u_h *remains true for every* h*. On the other hand, we lose the (uniform) proportionality between the error* e_h *and the approximation error on* V_h *and* V*. Lastly, since the constant* C *equals* $(1 + M/\alpha''(h))$ *in [3.13], we cannot guarantee that* $\lim_{h\to 0} \|u - u_h\|_V = 0$*. In some cases, this result may not hold.*

3.1.3. *Approximation example: Hilbert bases*

A particular case of when an internal approximation can be used involves the use of a Hilbert basis, whose definition we recall below. Assume that V is separable.

Recall that $(w_j)_{j\geq 1}$ is a Hilbert basis for V if:

– $\forall i, j, \ (w_i, w_j)_V = \delta_{ij}$;

– the vector space generated by the finite linear combinations of the $(w_j)_{j\geq 1}$ is dense in V.

Set $V_h = Vect\,(w_1, \cdots, w_N)$ with $h = 1/N$, and let u_h be the unique solution to the approximation problem [3.3]. This approximation is called the Ritz–Galerkin approximation.

By applying the convergence theorems 3.1 and 3.2, we can prove that:

THEOREM 3.3 (Hilbert approximation).– *For all* $N \geq 1$*, the approximation problem [3.3] admits a unique solution* u_h *that converges to the solution* u *of the continuous problem [3.1].*

Proof. The existence and uniqueness of u_h follows from either proposition 3.1 or 3.3. To show the convergence, apply either theorem 3.1 or 3.2 to $W = V$ and with $r_h = P_N$ the orthogonal projection from V onto V_h. Indeed, we have that:

$$\lim_{h\to 0} \|v - P_N v\|_V = 0 \quad \forall v \in V. \qquad \blacksquare$$

For example, on the set $\Omega = (0,1) \times (0,1)$, the eigenfunctions $(w_{m,p})_{m,p\geq 1}$ of the homogeneous Dirichlet problem:

$$w_{m,p}(x, y) = a_{mn} \sin(m\pi x) \sin(p\pi y)$$

with a_{mn} such that $\|w_{mn}\|_{H_0^1(\Omega)} = 1$

make up a Hilbert basis of $H_0^1(\Omega)$. Taking $V_h = Vect\,(w_{m,p})_{m,p=1,n}$ with $h = 1/n^2$, we obtain a finite-dimensional approximation space $H_0^1(\Omega)$, which solves the approximate Dirichlet problem:

$$\left| \begin{array}{l} \text{Find } u_h \in V_h \text{ such that} \\ \displaystyle\int_\Omega k(x,y)\nabla u_h \cdot \nabla v_h \, d\Omega = \int_\Omega f v_h \, d\Omega \quad \forall v_h \in V_h. \end{array} \right.$$

In the above, $k \in L^\infty(\Omega)$ is an example such that $k(x,y) \geq k_{min} > 0$ for almost all $(x,y) \in \Omega$, ensuring the coercivity of the bilinear form.

Clearly, we could have used alternative Hilbert bases. In particular, let us point out that the bases formed up of orthogonal polynomials (e.g. Legendre polynomials) lead to the so-called spectral methods. These methods are very precise, but may give rise to the solution of a "full" linear system, a disadvantage of which the finite element method manages to avoid. For an introduction to spectral methods, see Shen et al. (2011).

3.1.4. *Different solution and test-function spaces*

When we are within the framework of the Banach–Necas–Babuska theorem, we have seen a priori that we can solve variational problems that are defined on two different Hilbert spaces V_1 and V_2:

$$\left| \begin{array}{l} \text{Find } u_1 \in V_1 \text{ such that} \\ a(u_1, v_2) = \ell_2(v_2) \quad \forall v_2 \in V_2. \end{array} \right. \qquad [3.15]$$

Here, consider two *vector subspaces* $V_{1,h}$ and $V_{2,h}$ of V_1 and V_2, respectively, whose *dimensions* $N_1 = N_1(h)$ and $N_2 = N_2(h)$ *are finite*, and where h is a positive parameter which we will eventually send to 0, with $\lim_{h\to 0} N_1(h) = \lim_{h\to 0} N_2(h) = \infty$. For a given h, consider the following approximate variational problem:

$$\left| \begin{array}{l} \text{Find } u_{1,h} \in V_{1,h} \text{ such that} \\ a(u_{1,h}, v_{2,h}) = \ell_2(v_{2,h}) \quad \forall v_{2,h} \in V_{2,h}. \end{array} \right. \qquad [3.16]$$

The approximate problem [3.16] can be reformulated as follows.

LEMMA 3.4.– *Let* $(\varphi_{1,1}, \cdots, \varphi_{1,N_1})$ *be a basis of the vector space* $V_{1,h}$, *and, respectively, let* $(\varphi_{2,1}, \cdots, \varphi_{2,N_2})$ *be a basis of the vector space* $V_{2,h}$. *Then the approximate problem [3.16] is* equivalent *to the linear system based in* $\mathbb{R}^{N_1}$*:*

$$\textit{Find } \overrightarrow{\lambda}_1 \in \mathbb{R}^{N_1} \textit{ such that } \mathbb{A}\overrightarrow{\lambda}_1 = \overrightarrow{L}_2 \qquad [3.17]$$

where $\mathbb{A}_{ij} = a(\varphi_{1,j}, \varphi_{2,i})$, $(\overrightarrow{\lambda}_1)_j = \lambda_{1,j}$, $(\overrightarrow{L}_2)_i = \ell_2(\varphi_{2,i})$ $1 \leq j \leq N_1$, $1 \leq i \leq N_2$. *Furthermore, we have the following relationship between both of the solutions:*

$$u_{1,h} = \sum_{j=1}^{N_1} \lambda_{1,j}\varphi_{1,j}. \qquad [3.18]$$

Proof. Write $u_{1,h}$ as in [3.18] and, for $i = 1, N_2$, set $v_{2,h} = \varphi_{2,i}$ in the variational formulation [3.3]. It leads us to the equations:

$$\sum_{j=1}^{N_1} \lambda_{1,j} a(\varphi_{1,j}, \varphi_{2,i}) = \ell_2(\varphi_{2,i}) \quad \forall i = 1, N_2$$

which makes up the linear system when written row by row. The converse is immediate by the linearity of the form a. ■

The approximate problem is equivalent to finding a solution of a linear system. Note that $\mathbb{A} \in \mathbb{R}^{N_2 \times N_1}$ by construction. Thus, a *necessary* condition to ensure the existence and uniqueness of $\overrightarrow{\lambda}_1$ – and hence the same properties for $u_{1,h}$ – is that $N_1 = N_2$. Moreover, one must impose a discrete uniform condition or the discrete uniform inf-sup condition, just like before.

DEFINITION 3.4 (Discrete uniform inf-sup condition).– *We say that a bilinear form* $a(\cdot,\cdot)$ *on* $V_1 \times V_2$ *satisfies a* discrete uniform condition, *or a* discrete uniform inf-sup condition, *on the subspaces* $(V_{1,h} \times V_{2,h})_h$, *if and only if there exists a constant* $\alpha'' > 0$ *such that:*

$$\sup_{v_{2,h} \in V_{2,h}\setminus\{0\}} \frac{|a(v_{1,h}, v_{2,h})|}{\|v_{2,h}\|_{V_2}} \geq \alpha'' \|v_{1,h}\|_{V_1} \qquad \forall h \qquad \forall v_{1,h} \in V_{1,h}. \qquad [3.19]$$

The results below can be proven in a similar fashion as before.

PROPOSITION 3.4 (Existence and uniqueness of the approximate solution).– *Under the hypothesis [3.19] and if* $N_1 = N_2$, *then the approximate variational problem [3.16] admits a unique solution* $u_{1,h} \in V_{1,h}$.

LEMMA 3.5 (Céa's lemma).– *Under the hypotheses [3.2] for* $V_1 \times V_2$ *and [3.19], and if* $N_1(h) = N_2(h)$ *for every* h, *then there exists a constant* $C > 0$, *independent of* $(V_{1,h} \times V_{2,h})_h$ *and of* $\ell_2(\cdot)$, *such that:*

$$\|u_1 - u_{1,h}\|_{V_1} \leq C \inf_{v_{1,h} \in V_{1,h}} \|u_1 - v_{1,h}\|_{V_1}. \qquad [3.20]$$

THEOREM 3.4 (Convergence of the internal approximation).– *Assume that the approximation property [3.9] holds for* V_1. *Then, under the hypotheses [3.2] in* $V_1 \times V_2$ *and [3.19], and if* $N_1(h) = N_2(h)$ *for all* h, *we have that:*

$$\lim_{h\to 0} \|u_1 - u_{1,h}\|_{V_1} = 0. \qquad [3.21]$$

REMARK 3.6.– *Let us mention finally that one could consider the case in which the two spaces* V_1 *and* V_2 *are identical, yet the approximation subspaces* $V_{1,h}$ *and* $V_{2,h}$ *are different, under the provision that* $N_1(h) = N_2(h)$ *for every* h.

3.2. Principles of the affine finite element method in two dimensions

First, let us use a simple example to identify the main principles of the finite element method. Recall that the Neumann problem can be expressed as the following variational formulation in $H^1(\Omega)$:

$$\left| \begin{array}{l} \text{Find } u \in H^1(\Omega) \text{such that} \\ \displaystyle\int_\Omega (\nabla u \cdot \nabla v + uv)\, d\Omega = \int_\Omega f v\, d\Omega + \int_{\partial\Omega} g v\, d\Gamma \quad \forall v \in H^1(\Omega). \end{array} \right. \qquad [3.22]$$

We are going to consider the case where Ω is a non-empty connected and bounded open subset of $\mathbb{R}^2$ with a polygonal boundary[2].

To use Galerkin's approximation with this problem, we need to construct a finite-dimensional approximation space $V_h \subset H^1(\Omega)$. Note that the principles are the same for $\Omega \subset \mathbb{R}^\mathrm{d}$, $\mathrm{d} = 1, 2, 3$. An analysis into the main differences with the finite difference method is discussed in exercises 25 and 26.

3.2.1. *Mesh*

The first natural idea consists of constructing an approximation space of *piecewise affine* functions, that is, functions whose gradients are *piecewise constant*. In order for these functions to belong to $H^1(\Omega)$, it is sufficient that they are *continuous* on $\overline{\Omega}$ (see property 1.1).

In order to build such a space, let us construct a **mesh** on the open subset Ω, for example, a partition consisting of L triangles (in two dimensions) (Figure 3.1).

2 This is a simplification that will have repercussions on the error analysis level (see section 4.3), as well as when implementing it.

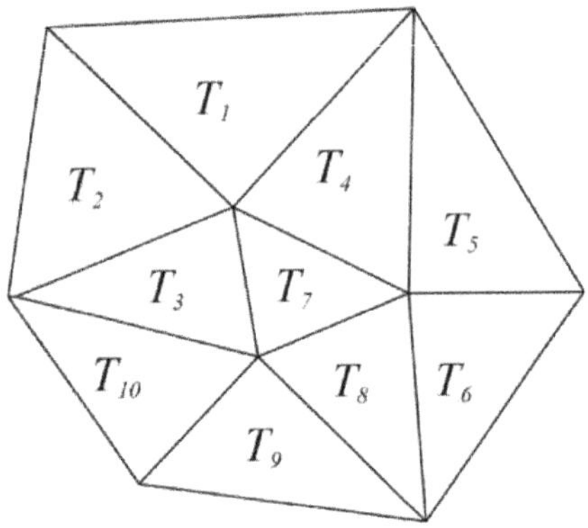

Figure 3.1. *A mesh*

We are going to assume that this partition satisfies the following natural properties:

$$\left|\begin{array}{l} -\text{ each triangle } T_\ell \text{ has a non-empty interior (i.e., } \overset{\circ}{T}_\ell \neq \emptyset) \\ -\ \overset{\circ}{T}_\ell \cap \overset{\circ}{T}_{\ell'} = \emptyset \text{ if } \ell \neq \ell' \\ -\ \bigcup_\ell T_\ell = \overline{\Omega} \\ -\text{ each edge of a triangle is either an edge of another triangle or} \\ \quad \text{an edge which forms part of the boundary } \partial\Omega. \end{array}\right. \qquad [3.23]$$

In particular, the last property excludes the case where a vertex M of the mesh is not a vertex of a triangle (Figure 3.2).

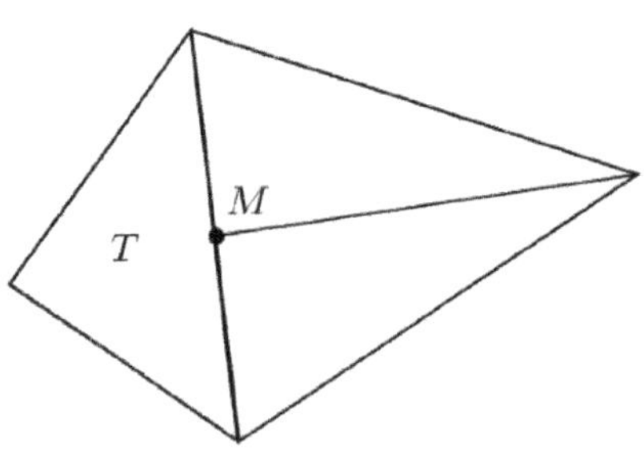

Figure 3.2. *An excluded configuration*

Denote by $(M_j)_{j=1,N}$ the set of vertices of the mesh. These vertices M_j are called the **nodes** of the mesh.

Write $h = \max_\ell h_\ell$, where h_ℓ is the radius of the circle circumscribing the triangle T_ℓ (Figure 3.3).

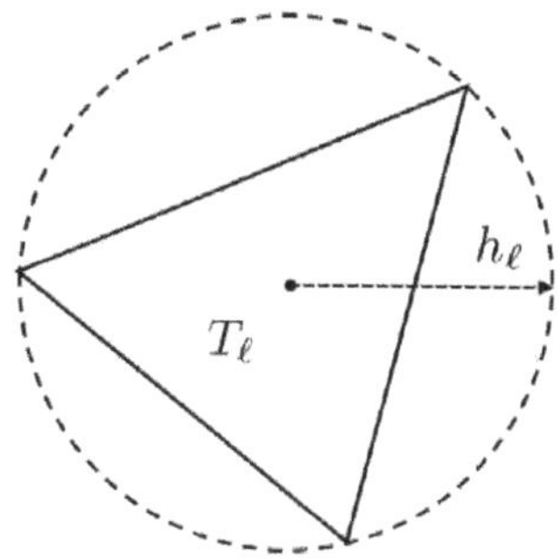

Figure 3.3. *The characteristic dimension of a triangle*

The length h is called the **mesh size**.

Now let us introduce the following space, indexed by the mesh size:

$$V_h = \left\{ v_h \in C^0(\overline{\Omega}) \text{ such that } v_{h|T_\ell} \text{ is affine}, \forall \ell = 1, L \right\}. \qquad [3.24]$$

Introduce the piecewise affine functions $(w_j)_{j=1,N}$ that satisfy the following relations:

$$w_j(M_i) = \delta_{ij} \quad \forall i, j = 1, N. \qquad [3.25]$$

We call these the **global basis functions**. Observe that these functions are *continuous on* $\overline{\Omega}$ provided that the partition satisfies [3.23]. Indeed, these functions are continuous on each triangle of the partition and, along any common edge to a pair of triangles, they are affine and completely determined by their values at the two extremities. Continuity when crossing each edge, and therefore on $\overline{\Omega}$, follows from this condition. Finally, they have a remarkable property, namely that:

$$\text{support } w_j = \bigcup_{\ell \text{ s.t. } M_j \in T_\ell} T_\ell. \qquad [3.26]$$

Therefore, the support of the w_j is bounded (see Figure 3.4).

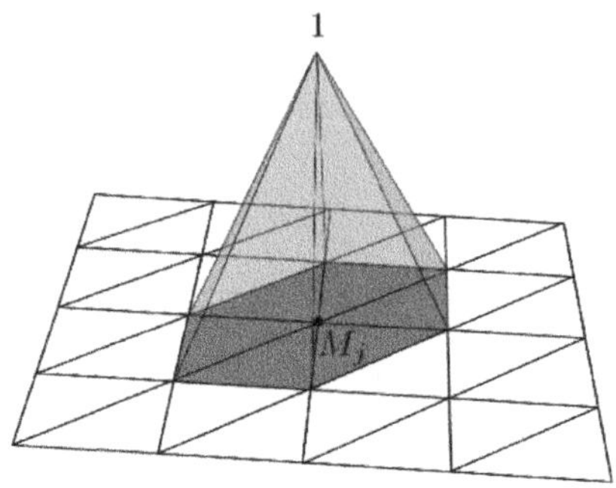

Figure 3.4. *Basis function* w_j

3.2.2. *Global basis functions properties*

We start with the result:

PROPOSITION 3.5.– *If a partition satisfies [3.23], then the space* V_h *is a vector space of dimension* N *with* $(w_j)_{j=1,N}$ *as a basis.*

Proof. By construction, the collection of functions $(w_j)_{j=1,N}$ satisfies:

– w_j is continuous on $\overline{\Omega}$, so $w_j \in V_h$;

– $(w_j)_{j=1,N}$ is a linearly independent subset of V_h, since:

$$\sum_j \alpha_j w_j = 0 \text{ on } \overline{\Omega} \Rightarrow \sum_j \alpha_j w_j (M_i) = 0,\ \forall i = 1, N \overset{[3.25]}{\Rightarrow} \alpha_i = 0,\ \forall i = 1, N.$$

– $(w_j)_{j=1,N}$ is a spanning set of V_h, since:

$$\text{if } v_h \in V_h \text{ then } v_h(M) = \sum_j v_h(M_j) w_j(M),\ \forall M \in \overline{\Omega}. \qquad [3.27]$$

Why is this so? For a given ℓ, on the one hand, consider $\left(\sum_j v_h(M_j) w_j\right)_{|T_\ell}$, and $v_{h|T_\ell}$, on the other hand. These two functions are affine and, if the three vertices of T_ℓ are $\{M_{\ell_1}, M_{\ell_2}, M_{\ell_3}\}$, then for $i = 1, 2, 3$, we have:

$$\left(\sum_j v_h(M_j) w_j\right)_{|T_\ell} (M_{\ell_i}) = \sum_j v_h(M_j) w_j(M_{\ell_i}) = v_h(M_{\ell_i}).$$

The values of the two affine functions coincide at the three vertices: therefore they are equal. Since this property is valid for any triangle T_ℓ, we have [3.27], given the fact that $\overline{\Omega} = \cup_\ell T_\ell$. ■

REMARK 3.7.– *Let us come back to the constraints that we established for a partition (see [3.23]). If a triangle T is such that one of its edges is included in the union of two (or more) edges from other triangles, then the trace – which is affine with respect to the curvilinear abscissa – from $v_{h|T}$ onto this edge is defined by three (or more) values: we run into problems when determining how this trace should be defined (see theorem 3.5 infra for a more in depth version of this reasoning).*

If T_ℓ and T_m are two different triangles such that $\overset{\circ}{T}_\ell \cap \overset{\circ}{T}_m \neq \emptyset$, then notice that the data of $v_{h|\overset{\circ}{T}_\ell \cap \overset{\circ}{T}_m}$ determines v_h completely on $T_\ell \cup T_m$ (by extension). Therefore, in this case, we will have six values to determine the affine function in two dimensions, which again poses a problem.

Finally, if a triangle T is such that $\overset{\circ}{T} = \emptyset$, then T is a line segment. In this case, since we know that $\overline{\Omega} = \cup_\ell T_\ell$, there exists an ℓ such that $T \subset T_\ell$ (recall that the intersection $T \cap T_\ell$ is either empty or equals an entire edge), and so we can simply delete T completely from the partition.

Equality [3.27] shows that the components of a function v_h, with respect to the $(w_j)_{j=1,N}$, have to coincide with its *values at the nodes*.

Now consider the approximate variational problem:

$$\int_\Omega (\nabla u_h \cdot \nabla v_h + u_h v_h)\, d\Omega = \int_\Omega f v_h\, d\Omega + \int_{\partial\Omega} g v_h\, d\Gamma \quad \forall v_h \in V_h. \qquad [3.28]$$

What are the interesting properties that follow from the choice of approximation space V_h defined by [3.24]?

Note that to begin with that, from lemma 3.1, the discretized problem [3.28] is equivalent to the following linear system based on $\mathbb{R}^N$:

$$\mathbb{A}\vec{U} = \vec{B}. \qquad [3.29]$$

$$\text{with} \quad \left| \begin{array}{l} \mathbb{A}_{ij} = \displaystyle\int_\Omega \nabla w_j \cdot \nabla w_i\, d\Omega + \int_\Omega w_j w_i\, d\Omega \\ B_i = \displaystyle\int_\Omega f w_i\, d\Omega + \int_{\partial\Omega} g w_i\, d\Omega \\ U_i = i^{th} \text{ component of the solution } u_h \text{ with} \\ \qquad\qquad \text{respect to the basis } (w_j)_j. \end{array} \right. \qquad [3.30]$$

By construction, $\mathbb{A}$ is also symmetric.

3.2.3. *Norm of the elements of the approximation space*

Let $\overrightarrow{V}$ be the vector in $\mathbb{R}^N$ whose components are equal to those of v_h relative to the basis $(w_j)_j$. Note that:

$$\begin{aligned}(u_h, v_h)_{H^1(\Omega)} &= \Big(\sum_i U_i w_i, \sum_j V_j w_j\Big)_{H^1(\Omega)} = \sum_{i,j} (w_i, w_j)_{H^1(\Omega)} U_i V_j \\ &= \sum_{i,j} \mathbb{A}_{ji} U_i V_j = \sum_j \Big(\sum_i \mathbb{A}_{ji} U_i\Big) V_j \\ &= \sum_j \left(\mathbb{A}\overrightarrow{U}\right)_j V_j = (\mathbb{A}\overrightarrow{U} \,|\, \overrightarrow{V}).\end{aligned}$$

Equally, we have that:

$$\|v_h\|^2_{H^1(\Omega)} = (\mathbb{A}\overrightarrow{V} \,|\, \overrightarrow{V}).$$

We see that the matrix $\mathbb{A}$ of $\mathbb{R}^{N\times N}$ is (symmetric) and definite-positive (see remark 3.1).

If we introduce the following matrices $\mathbb{K}$ and $\mathbb{M}$ in $\mathbb{R}^{N\times N}$, defined by:

$$\mathbb{K}_{ij} = \int_\Omega \nabla w_j \cdot \nabla w_i \, d\Omega, \qquad \mathbb{M}_{ij} = \int_\Omega w_j w_i \, d\Omega$$

then now we see that:

$$\|v_h\|^2_{L^2(\Omega)} = (\mathbb{M}\overrightarrow{V} \,|\, \overrightarrow{V}) \text{ and } |v_h|^2_{1,\Omega} = (\mathbb{K}\overrightarrow{V} \,|\, \overrightarrow{V}).$$

The matrix $\mathbb{K}$ is called the **stiffness matrix**: by construction, it is positive (symmetric), and satisfies the property:

$$(\mathbb{K}\overrightarrow{V} \,|\, \overrightarrow{V}) = 0 \iff v_h = cst \iff V_j = cst, \; j = 1, N.$$

The matrix $\mathbb{M}$ is called the **mass matrix**: by construction, it is (symmetric) positive-definite and clearly we have the equality $\mathbb{A} = \mathbb{K} + \mathbb{M}$. The terminology comes from mechanics, or more precisely from the linear elasticity equations of a body, where the stiffness refers to the partial derivative operator for the elasticity and, clearly, the mass refers to the mass of the body. Note in passing that, if $\overrightarrow{1}$ represents the vector whose components are all 1, we have the following relations which are easy to establish:

$$\mathbb{K}\overrightarrow{1} = \overrightarrow{0} \text{ and } (\mathbb{M}\overrightarrow{1} \,|\, \overrightarrow{1}) = \operatorname{vol}\Omega$$

and which are easy-to-implement validation tests.

3.2.4. *Sparse matrix property*

The property regarding the support [3.26] is essential. Indeed, according to this property, one can easily check that:

$$\mathbb{A}_{ij} = 0 \quad \text{if } M_i \text{ and } M_j \text{ do not belong to the same triangle.} \qquad [3.31]$$

This shows that the matrix $\mathbb{A}$ has a very large number of null components (zeros): we call such a matrix a **sparse matrix** (see Chapter 5, section 5.2). Consequently, on the one hand, storing the linear system occupies only a small amount of memory, and furthermore, on the other hand, one can apply specific solving methods, making it possible to drastically reduce the computation time (see Appendix 1; Meurant 1999).

3.2.5. *Lagrange interpolation*

The unknowns in the linear system are called the **degrees of freedom** for the approximation. Note from [3.27] that $U_i = u_h(M_i)$, for $i = 1, N$. Solving the linear system directly provides the approximation of the solution u for the nodes M_i of the mesh. In this case, such an approximation is said to be a **Lagrangian finite element** approximation, with each degree of freedom for the system supported on a node of the mesh.

3.2.6. *Local nature of the finite element method*

Consider a triangle T_ℓ with vertices $(M_{\ell_1}, M_{\ell_2}, M_{\ell_3})$. The only global basis functions $(w_j)_{j=1,N}$ that when restricted to T_ℓ are non-zero are w_{ℓ_1}, w_{ℓ_2} and w_{ℓ_3}.

Define the **local basis functions** on the triangle T_ℓ by:

$$\tau_{\ell_i} = w_{\ell_i|T_\ell} \qquad 1 \leq i \leq 3. \qquad [3.32]$$

Here, τ_{ℓ_i} is a polynomial of degree 1 in the (x, y)-coordinates since $w_{\ell_i|T_\ell}$ is affine by definition; thus, τ_{ℓ_i} belongs to $P^1(T_\ell)$, which is the space of degree 1 polynomials defined on T_ℓ with respect to the (x, y)-coordinates. It is easy to check that:

$$\tau_{\ell_i}(M) = \lambda_{\ell_i}(M) \qquad 1 \leq i \leq 3 \qquad [3.33]$$

where $(\lambda_{\ell_i})_{1\leq i\leq 3}$ denotes the barycentric coordinates[3] of M with respect to the vertices $(M_{\ell_i})_{1\leq i\leq 3}$. Moreover, $(\tau_{\ell_i})_{1\leq i\leq 3}$ forms a basis of $P^1(T_\ell)$.

3 Recall that the barycentric coordinates $(\lambda_i)_{i=1,2,3}$ associated to three non-colinear points $(M_i)_{i=1,2,3}$ (with the respective coordinates (x_i, y_i)) allow us to characterize any point M in $\mathbb{R}^2$. If we denote by (x, y) the coordinates of M, then $\lambda_1(x, y)$, $\lambda_2(x, y)$, $\lambda_3(x, y)$ are the three affine functions uniquely defined by the three linear equations: $\sum_{i=1,3} \lambda_i(x, y)x_i = x$, $\sum_{i=1,3} \lambda_i(x, y)y_i = y$, $\sum_{i=1,3} \lambda_i(x, y) = 1$.

Indeed, $M \mapsto \lambda_{\ell_i}(M)$ belongs to $P^1(T_\ell)$, and by definition $\lambda_{\ell_i}\left(M_{\ell_j}\right) = \delta_{ij} \quad 1 \leq i, j \leq 3$.

As τ_{ℓ_i} also satisfies (from [3.25] and [3.32]): $\tau_{\ell_i}\left(M_{\ell_j}\right) = \delta_{ij} \quad 1 \leq i, j \leq 3$, we obtain the relation [3.33] as $P^1(T_\ell)$ is three dimensional.

Reasoning locally, we can focus our attention to a reference configuration with respect to a fixed triangle. Indeed, the local basis functions (τ_{ℓ_i}) are derived from the basis functions $(\hat{\tau}_i)$, which are called reference functions, which are defined on the unit triangle $\hat{T}$ whose vertices are $\hat{M}_1(0,0)$, $\hat{M}_2(1,0)$ and $\hat{M}_3(0,1)$.

Let us introduce the **affine transformation** F_ℓ:

$$\begin{pmatrix} \hat{x} \\ \hat{y} \end{pmatrix} \mapsto \begin{pmatrix} x \\ y \end{pmatrix} = \begin{pmatrix} a_1\hat{x} + b_1\hat{y} + c_1 \\ a_2\hat{x} + b_2\hat{y} + c_2 \end{pmatrix}$$

which transforms[4] the triangle $\hat{T}$ into the triangle T_ℓ, by setting:

$$F_\ell\left(\hat{M}_i\right) = M_{\ell_i} \quad 1 \leq i \leq 3. \qquad [3.34]$$

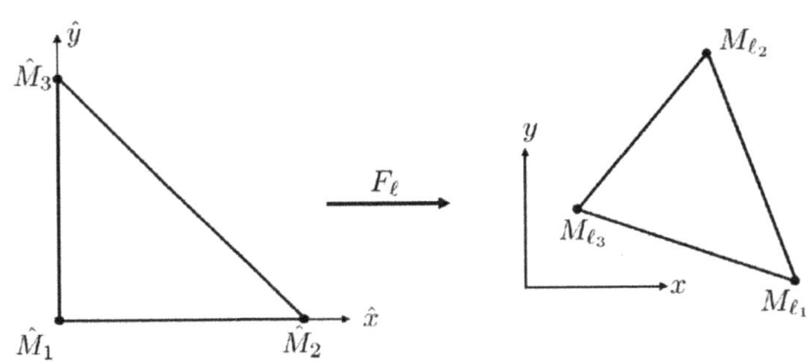

Figure 3.5. *A geometric transformation*

Since F_ℓ is affine, we have that:

$$M = F_\ell\left(\hat{M}\right) = \sum_{i=1}^{3} \hat{\lambda}_i\left(\hat{M}\right) M_{\ell_i} \qquad [3.35]$$

4 By an abuse of notation, if we mix up the point $\hat{M}$ with the vector $\overrightarrow{O\hat{M}}$, then the transformation F_ℓ can still be written as $F_\ell(\hat{M}) = \mathbb{B}_\ell \hat{M} + \vec{C}_\ell$ with $\mathbb{B}_\ell \in \mathbb{R}^{2\times 2}$ and $\vec{C}_\ell \in \mathbb{R}^2$.

where $\hat{\lambda}_i\left(\hat{M}\right)$ are the barycentric coordinates of $\hat{M}$ with respect to $\left(\hat{M}_i\right)_{1\leq i\leq 3}$:

$$\hat{\lambda}_1(\hat{x},\hat{y}) = 1 - \hat{x} - \hat{y}, \quad \hat{\lambda}_2(\hat{x},\hat{y}) = \hat{x}, \quad \hat{\lambda}_3(\hat{x},\hat{y}) = \hat{y}. \tag{3.36}$$

Then we set $\hat{\tau}_i = \hat{\lambda}_i$, $1 \leq i \leq 3$, because the local basis functions on the reference triangle are just the barycentric coordinates within this triangle.

It is easy to see that, since the points $(M_{\ell_i})_{1\leq i\leq 3}$ are not collinear – that is, T_{ℓ_i} is non-degenerate – that F_ℓ is invertible (with an affine inverse), and from [3.35]–[3.36] we have:

$$\left| \begin{array}{l} \tau_{\ell_i} = \hat{\tau}_i \circ F_\ell^{-1} \\ \hat{\tau}_i = \tau_{\ell_i} \circ F_\ell. \end{array} \right. \tag{3.37}$$

It follows then that one can construct the local basis functions τ_{ℓ_i}, and then subsequently the global basis functions w_j, using only the reference basis functions given by [3.36]. This provides a general method of constructing a basis for the approximation space, which is otherwise straightforward to implement both algorithmically and computationally.

This construction process also allows us to quickly and simply evaluate the linear system [3.29]. Indeed, introducing the term in this matrix:

$$\mathbb{K}_{ij} = \int_\Omega \nabla w_i \cdot \nabla w_j \, d\Omega$$

from the support property of $(w_j)_j$ ([3.26] and [3.32]), we have:

$$\mathbb{K}_{ij} = \sum_{T_\ell \text{ s.t. } M_i, M_j \in T_\ell} \int_{T_\ell} \nabla \tau_{\ell_i} \cdot \nabla \tau_{\ell_j} \, d\Omega. \tag{3.38}$$

REMARK 3.8.– *Obviously, when $i \neq j$, there exist at most two triangles with M_i and M_j as vertices (the triangles containing the edge $[M_i, M_j]$), while if $i = j$, then there may exist several triangles (all triangles whose vertex is M_i).*

Therefore we just need to calculate the so-called **elementary matrices**:

$$\mathbb{K}^\ell_{ij} = \int_{T_\ell} \nabla \tau_{\ell_i} \cdot \nabla \tau_{\ell_j} \, d\Omega. \tag{3.39}$$

Let $dF_\ell(\hat{M})$ be the differential of F_ℓ on $\hat{M}$, and let $[dF_\ell(\hat{M})]$ and $J_{F_\ell}(\hat{M})$ represent the Jacobian matrix of F_ℓ on $\hat{M}$ (i.e. the matrix representative of $dF_\ell(\hat{M})$), and its Jacobian on $\hat{M}$, equal to $det([dF_\ell(\hat{M})])$, respectively. Then we have the following formula (Ciarlet 2002) for the changes-of-variable $M = F_\ell(\hat{M})$:

$$\int_{T_\ell} v(M)\, d\Omega = \int_{\hat{T}} v \circ F_\ell(\hat{M})\, |J_{F_\ell}(\hat{M})|\, d\hat{\Omega}.$$

In the case that we are interested in, recall that we can derive the following relations $d\hat{\tau}_i(\hat{M}) = d\tau_{\ell_i}(M) \circ dF_\ell(\hat{M})$ and $\hat{\nabla}\hat{\tau}_i(\hat{M}) = [dF_\ell(\hat{M})]^t\, \nabla\tau_{\ell_i}\,(M)$ from the identity $\hat{\tau}_i(\hat{M}) = \tau_{\ell_i} \circ F_\ell(\hat{M})$. With this in mind, then:

$$\mathbb{K}^\ell_{ij} = \int_{\hat{T}} (\mathbb{C}^\ell(\hat{M})\nabla\hat{\tau}_i(\hat{M})) \cdot \hat{\nabla}\hat{\tau}_j(\hat{M}) |J_{F_\ell}(\hat{M})| d\hat{\Omega} \qquad [3.40]$$

where the (symmetric) matrix $\mathbb{C}^\ell(\hat{M})$ is given by:

$$\mathbb{C}^\ell(\hat{M}) = [dF_\ell(\hat{M})]^{-1}([dF_\ell(\hat{M})]^{-1})^t$$

Here, F_ℓ is affine[5], and therefore we have:

$$|J_{F_\ell}(\hat{M})| = \left| \overrightarrow{M_{\ell_1}M_{\ell_3}} \times \overrightarrow{M_{\ell_1}M_{\ell_2}} \right| = 2 \text{ mes } (T_\ell)$$

whereas, on the other hand, $\mathbb{C}^\ell(\hat{M})$ is a matrix independent of $\hat{M}$, and which allows us conclude that:

$$\mathbb{K}^\ell_{ij} = 2 \text{ vol } (T_\ell)\Big(\mathbb{C}^\ell_{11} \int_{\hat{T}} \partial_{\hat{x}}\hat{\tau}_i \partial_{\hat{x}}\hat{\tau}_j d\hat{x}d\hat{y} + \mathbb{C}^\ell_{22} \int_{\hat{T}} \partial_{\hat{y}}\hat{\tau}_i \partial_{\hat{y}}\hat{\tau}_j d\hat{x}d\hat{y} \\ + \mathbb{C}^\ell_{12}\Big(\int_{\hat{T}} \partial_{\hat{x}}\hat{\tau}_i \partial_{\hat{y}}\hat{\tau}_j d\hat{x}d\hat{y} + \int_{\hat{T}} \partial_{\hat{y}}\hat{\tau}_i \partial_{\hat{x}}\hat{\tau}_j d\hat{x}d\hat{y}\Big)\Big)$$

thereby showing us that the computation of $\mathbb{K}^\ell_{ij}$, and consequently that of $\mathbb{K}_{ij}$, requires us to only know the reference basis functions.

We can treat the term:

$$\mathbb{M}_{ij} = \int_\Omega w_i w_j \, d\Omega$$

in the same way, from which we obtain:

$$\mathbb{M}_{ij} = \sum_{T_\ell \text{ s.t. } M_i, M_j \in T_\ell} \mathbb{M}^\ell_{ij}$$

5 Note that $[dF_\ell(\hat{M})] = \mathbb{B}_\ell$, $J_{F_\ell}(\hat{M}) = det(\mathbb{B}_\ell)$ and $\hat{\nabla}\hat{\tau}_i(\hat{M}) = \mathbb{B}^t_\ell\, \nabla\tau_{\ell_i}(M)$.

this time with:

$$\mathbb{M}^{\ell}_{ij} = \int_{T_\ell} \tau_{\ell_i} \tau_{\ell_j} \, d\Omega = \int_{\hat{T}} \hat{\tau}_i(\hat{M}) \hat{\tau}_j(\hat{M}) |J_{F_\ell}(\hat{M})| \, d\hat{\Omega} \qquad [3.41]$$

or, in the affine case:

$$\mathbb{M}^{\ell}_{ij} = 2 \operatorname{vol}(T_\ell) \int_{\hat{T}} \hat{\tau}_i \hat{\tau}_j \, d\hat{x} \, d\hat{y}.$$

REMARK 3.9.– *In practice, one does not use the summation formula [3.38] but rather a different algorithm. Indeed, it is expensive (from a computational point of view) to detect triangles, which have both M_i and M_j as vertices. One proceeds by assembling elementary matrices. We will come back to this in detail when implementing the finite element method in Chapter 5 (section 5.1).*

The approximation space that we have just constructed and whose main properties we have just described make up the **P^1-Lagrangian finite element** approximation. This example highlights the general process used, when constructing a finite element method:

– the definition of the local basis functions for the reference triangle;

– the construction of global basis functions (by transforming and assembling local basis functions), which generate an internal approximation space.

3.3. General procedure of constructing finite elements

First, let us begin by generalizing the piecewise P^1-approximation. From now and onwards, we will assume that Ω is a connected and bounded non-empty open subset of $\mathbb{R}^{\mathrm{d}}$ ($\mathrm{d} = 1, 2, 3$), whose boundary is "sufficiently regular", that is, *polyhedral* if $\mathrm{d} = 3$ and *polygonal* if $\mathrm{d} = 2$.

3.3.1. *Lagrangian finite elements of order k*

DEFINITION 3.5.– *Let $k \in \mathbb{N}$. We say that a Lagrangian finite element of order k is a triple (K, Σ, P), where K is a bounded non-empty closed subset of $\mathbb{R}^{\mathrm{d}}$, Σ is a finite set of n_K points $\left(M_i^K\right)_{i=1,n_K}$ that belong to K, and P is a vector space of polynomials containing $P^k(K)$ (space of polynomials of degree at most k on K), such that:*

$$\forall (\alpha_1, \alpha_2, \cdots, \alpha_{n_K}) \in \mathbb{R}^{n_K}, \ \exists! p \in P \text{ such that } p(M_i^K) = \alpha_i, \ \forall i = 1, n_K. \qquad [3.42]$$

A point of Σ is called a **Lagrangian degree of freedom**, and when property [3.42] is satisfied we say that P is **Σ-unisolvent**, which expresses the fact that the mapping:

$$S : \left| \begin{array}{lcl} P & \longrightarrow & \mathbb{R}^{n_K} \\ p & \mapsto & (p(M_1^K), \cdots, p(M_{n_K}^K)) \end{array} \right.$$

is bijective, or, in other words, that any function in the approximation space can be uniquely determined by the values that it takes at each degree of freedom. Therefore, a simple way to prove the unisolvency property is to check that $\dim P = n_K$, and to show that the map S is either injective or surjective.

REMARK 3.10.– *In definition 3.5, K can take on any shape: triangle, quadrangle, etc., if $\mathtt{d} = 2$; tetrahedron, pentahedron, hexahedron, cube, etc., if $\mathtt{d} = 3$.*

Moreover, the points $(M_i^K)_i$ do not necessarily have to be vertices.

Finally, the order of the finite element appearing in the definition is connected to the accuracy of the approximation, a notion that we will discuss later (see section 4.1).

By the means of the unisolvency property [3.42], there exist n_K **local basis functions** such that:

$$\tau_i^K(M_j^K) = \delta_{ij} \quad \forall i, j = 1, n_K. \qquad [3.43]$$

Thus, $\left(\tau_i^K\right)_{i=1,n_K}$ forms a linearly independent set that also spans the vector space P.

3.3.2. *Examples of finite elements*

By the way of examples, let us highlight how finite elements can be used within some classical examples.

3.3.2.1. *In one dimension*

Let us denote the closed unit interval by $\hat{S} = [0, 1]$.

3.3.2.1.1. Zeroth-order Lagrangian finite element

$$\left| \begin{array}{l} \hat{K} = \hat{S} \\ \hat{\Sigma} = \{\frac{1}{2}\} \\ P^0 = \{p(\hat{x}) = a, \; a \in \mathbb{R}\} \end{array} \right.$$

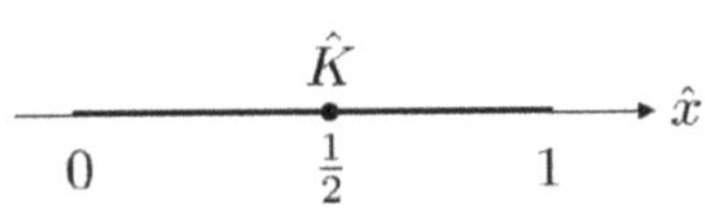

The unisolvency property [3.42] is satisfied trivially here, and for the basis function we have: $\hat{\tau}_1(\hat{x}) = 1$.

Note that we can choose any point of the segment $\hat{S}$ to be the degree of freedom. In this case, the approximation spaces V_h will contain the piecewise constant functions, which are therefore not continuous globally (apart from the everywhere constant function).

From the contrapositive of property 1.2, we see that $V_h \not\subset H^1(\Omega)$: V_h is not an internal approximation of $H^1(\Omega)$ here. On the other hand, $V_h \subset L^2(\Omega)$ and V_h does provide an internal approximation of $L^2(\Omega)$.

3.3.2.1.2. First-order Lagrangian finite element

$$\left|\begin{array}{l}\hat{K} = \hat{S} \\ \hat{\Sigma} = \{0, 1\} \\ P^1 = \left\{p(\hat{x}) = a\hat{x} + b,\ (a, b) \in \mathbb{R}^2\right\}\end{array}\right.$$

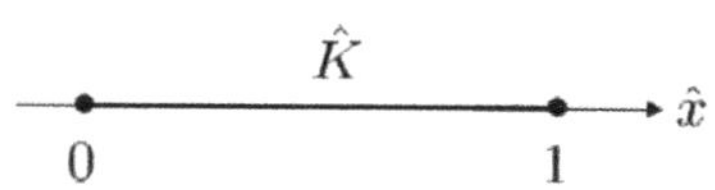

Property [3.42] is still satisfied, and we see that:

$$\hat{\tau}_1(\hat{x}) = 1 - \hat{x}$$
$$\hat{\tau}_2(\hat{x}) = \hat{x}$$

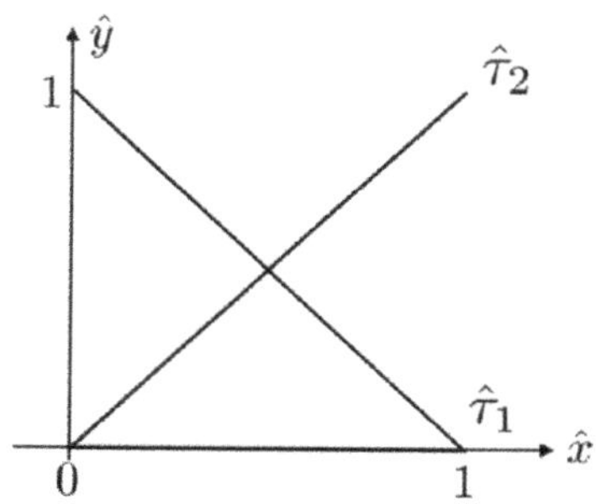

3.3.2.1.3. Second-order Lagrangian finite element

$$\left|\begin{array}{l}\hat{K} = \hat{S} \\ \hat{\Sigma} = \left\{0, 1, \frac{1}{2}\right\} \\ P^2 = \left\{p(\hat{x}) = a\hat{x}^2 + b\hat{x} + c,\ (a, b, c) \in \mathbb{R}^3\right\}\end{array}\right.$$

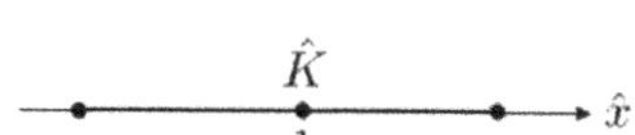

Property [3.42] is satisfied and this time the basis functions are given by:

$$\hat{\tau}_1(\hat{x}) = (1 - \hat{x})(1 - 2\hat{x})$$
$$\hat{\tau}_2(\hat{x}) = \hat{x}(2\hat{x} - 1)$$
$$\hat{\tau}_3(\hat{x}) = 4\hat{x}(1 - \hat{x})$$

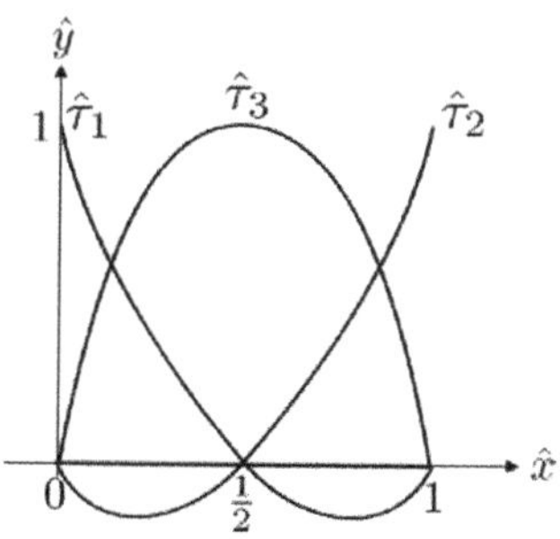

3.3.2.2. *In two dimensions*

With respect to the reference triangle $\hat{T}$, we may classically construct approximations P^0, P^1 and P^2. Denote the vertices of $\hat{T}$ by $\hat{M}_1(0,0)$, $\hat{M}_2(1,0)$, $\hat{M}_3(0,1)$.

3.3.2.2.1. P^0-finite element

$$\left|\begin{array}{l}\hat{K}=\hat{T}\\ \hat{\Sigma}=\left\{\hat{M}_{123}=\left(\frac{1}{3},\frac{1}{3}\right)\right\}\\ \hat{P}=P^0(\hat{T})=\left\{p\left(\hat{x},\hat{y}\right)=a\right\}\end{array}\right.$$

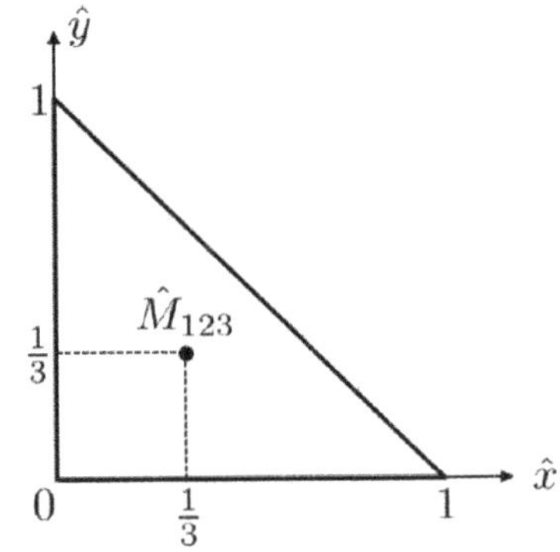

It is clear that $\hat{P}$ is $\hat{\Sigma}$-unisolvent and that $\hat{\tau}_{123}\left(\hat{M}\right)=1$. Again, this choice leads to the approximation spaces V_h containing functions which, although piecewise regular, are not continuous globally. Thus, according to the contrapositive of property 1.2, we see that $V_h \not\subset H^1(\Omega)$, and therefore V_h does not let us construct an internal approximation of $H^1(\Omega)$. On the other hand, these elements are suitable for an approximation of $L^2(\Omega)$.

3.3.2.2.2. P^1-finite element

$$\left|\begin{array}{l}\hat{K}=\hat{T}\\ \hat{\Sigma}=\left\{\hat{M}_1,\hat{M}_2,\hat{M}_3\right\}\\ \hat{P}=P^1(\hat{T})=\left\{p\left(\hat{x},\hat{y}\right)=a\hat{x}+b\hat{y}+c\right\}\end{array}\right.$$

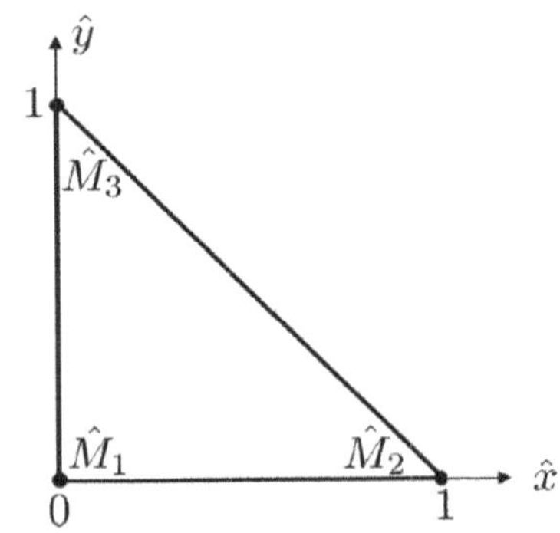

Since the three vertices of the triangle are not collinear, the unisolvency property [3.42] holds, and we have:

$$\hat{\tau}_1\left(\hat{x},\hat{y}\right)=1-\hat{x}-\hat{y},\qquad \hat{\tau}_2\left(\hat{x},\hat{y}\right)=\hat{x},\qquad \hat{\tau}_3\left(\hat{x},\hat{y}\right)=\hat{y}.$$

Note that we could very well have chosen the points located at the midpoint of each edge for $\hat{\Sigma}$, namely:

$$\hat{\Sigma} = \left\{ \hat{M}_{12}\left(\frac{1}{2}, 0\right), \hat{M}_{23}\left(\frac{1}{2}, \frac{1}{2}\right), \hat{M}_{31}\left(0, \frac{1}{2}\right) \right\}$$

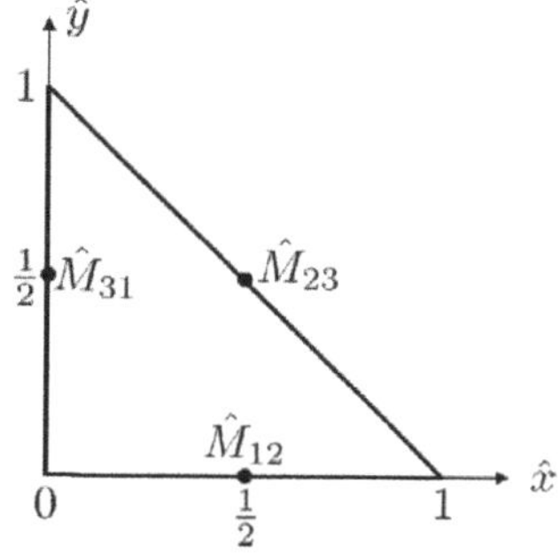

which also leads to a P^1-type finite element, with:

$$\hat{\tau}_{12}(\hat{x}, \hat{y}) = 1 - 2\hat{y}, \qquad \hat{\tau}_{23}(\hat{x}, \hat{y}) = 2(\hat{x} + \hat{y}) - 1, \qquad \hat{\tau}_{31}(\hat{x}, \hat{y}) = 1 - 2\hat{x}.$$

Within the internal approximation framework, this element is not relevant, since after assembling the local basis functions, it does not allow us to construct the functions in $H^1(\Omega)$ (continuity at each interface is not guaranteed).

3.3.2.2.3. P^2-finite element

$$\left| \begin{array}{l} \hat{K} = \hat{T} \\ \hat{\Sigma} = \left\{ \hat{M}_1, \hat{M}_2, \hat{M}_3, \hat{M}_{12}, \hat{M}_{23}, \hat{M}_{31} \right\} \\ \hat{P} = P^2(\hat{T}) \end{array} \right.$$

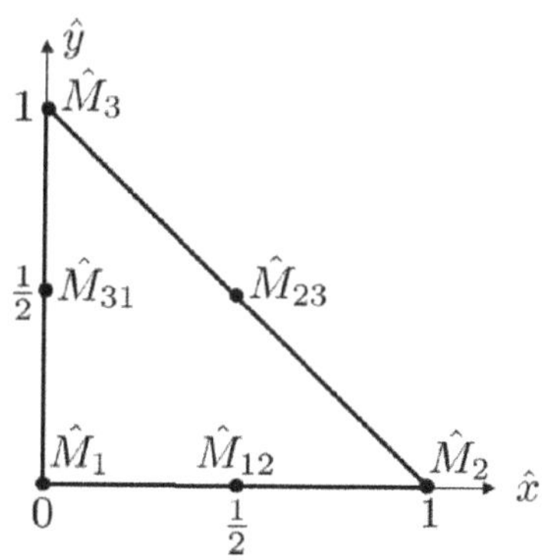

The polynomial space is $P^2\left(\hat{T}\right) = \{p(\hat{x}, \hat{y}) = a\hat{x}^2 + b\hat{y}^2 + c\hat{x}\hat{y} + d\hat{x} + e\hat{y} + f\}$. When $\dim \hat{P} = \operatorname{card} \hat{\Sigma}$, to show that $\hat{P}$ is $\hat{\Sigma}$-unisolvent it is sufficient to find the local basis functions $\hat{\tau}_i$. A quick technique for calculating the basis functions $\hat{\tau}_i$ consists of writing down the equations for the straight lines, along which the basis functions which we are after should cancel.

For example, to calculate $\hat{\tau}_1$, we have the points $\hat{M}_3$, $\hat{M}_2$ and $\hat{M}_{23}$, which lie along the line with the equation: $1 - \hat{x} - \hat{y} = 0$, whereas $\hat{M}_{31}$ and $\hat{M}_{12}$ lie along the line with the equation: $2\hat{x} + 2\hat{y} - 1 = 0$, from which we find:

$$\hat{\tau}_1(\hat{x}, \hat{y}) = \alpha (1 - \hat{x} - \hat{y})(2\hat{x} + 2\hat{y} - 1).$$

In setting $\hat{\tau}_1(0,0) = 1$, we get the constant $\alpha = -1$.

Proceeding similarly for the other basis functions, we subsequently obtain:

$$\begin{array}{ll} \hat{\tau}_1(\hat{x}, \hat{y}) = (1 - \hat{x} - \hat{y})(1 - 2\hat{x} - 2\hat{y}) & \hat{\tau}_{12}(\hat{x}, \hat{y}) = 4\hat{x}(1 - \hat{x} - \hat{y}) \\ \hat{\tau}_2(\hat{x}, \hat{y}) = \hat{x}(2\hat{x} - 1) & \hat{\tau}_{23}(\hat{x}, \hat{y}) = 4\hat{x}\hat{y} \\ \hat{\tau}_3(\hat{x}, \hat{y}) = \hat{y}(2\hat{y} - 1) & \hat{\tau}_{31}(\hat{x}, \hat{y}) = 4\hat{y}(1 - \hat{x} - \hat{y}). \end{array}$$

Care should be taken when using this technique, since it is essential that the constructed functions indeed do belong to $\hat{P}$, which will be the case provided that we use the minimum number of lines (here 2).

For geometric reasons, it is sometimes more appropriate to form a mesh for a computational domain using quadrangles instead. Naturally, this will introduce quadrangular finite elements. Let us introduce them for the unit square $\hat{C}$ whose vertices are $\hat{M}_1(0,0)$, $\hat{M}_2(1,0)$, $\hat{M}_3(1,1)$ and $\hat{M}_4(0.1)$. The associated polynomial space $\hat{P}$ no longer coincides with $P^k(\hat{C})$, but rather with a space consisting of polynomials whose degrees are at most k in each variable, which we denote by $Q^k(\hat{C})$:

$$Q^k\left(\hat{C}\right) = \left\{p(\hat{x}, \hat{y}) = \sum_{0 \le i,j \le k} \alpha_{ij}\hat{x}^i\hat{y}^j,\ \alpha_{ij} \in \mathbb{R}\right\}.$$

3.3.2.2.4. Q^1-finite element

$$\left|\begin{array}{l} \hat{K} = \hat{C} \\ \hat{\Sigma} = \left\{\hat{M}_1, \hat{M}_2, \hat{M}_3, \hat{M}_4\right\} \\ \hat{P} = Q^1(\hat{C}) = \{p(\hat{x}, \hat{y}) = a\hat{x}\hat{y} + b\hat{x} + c\hat{y} + d\} \end{array}\right.$$

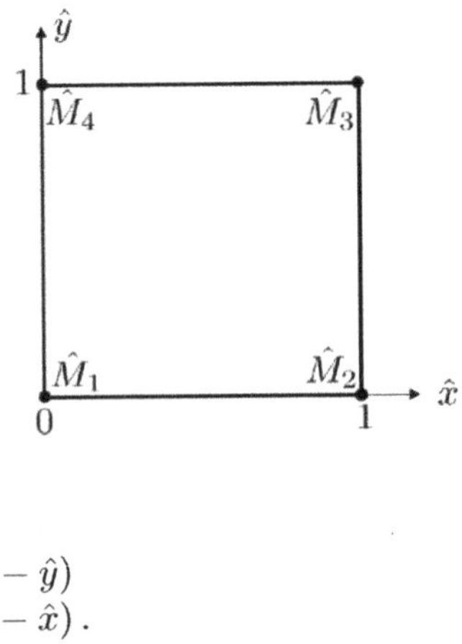

The basis functions are given by:

$$\begin{array}{ll} \hat{\tau}_1(\hat{x}, \hat{y}) = (1 - \hat{x})(1 - \hat{y}) & \hat{\tau}_2(\hat{x}, \hat{y}) = \hat{x}(1 - \hat{y}) \\ \hat{\tau}_3(\hat{x}, \hat{y}) = \hat{x}\hat{y} & \hat{\tau}_4(\hat{x}, \hat{y}) = \hat{y}(1 - \hat{x}). \end{array}$$

3.3.2.2.5. Q^2-finite element

$$\left|\begin{array}{l}\hat{K}=\hat{C}\\ \hat{\Sigma}=\left\{\hat{M}_1,\hat{M}_2,\hat{M}_3,\hat{M}_4,\hat{M}_{12},\right.\\ \qquad\left.\hat{M}_{23},\hat{M}_{34},\hat{M}_{41},\hat{M}_{1234}\right\}\\ \hat{P}=Q^2\left(\hat{C}\right)=\left\{p(\hat{x},\hat{y})=\sum\limits_{0\le i,j\le 2}\alpha_{ij}\hat{x}^i\hat{y}^j\right\}.\end{array}\right.$$

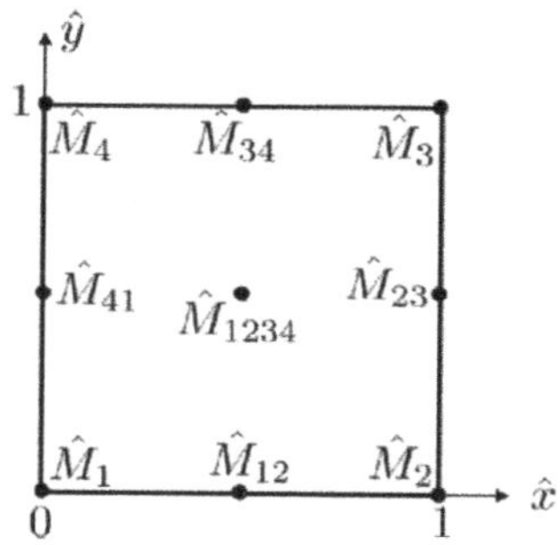

The information for the first basis function is:

$$\hat{\tau}_1(\hat{x},\hat{y})=(1-\hat{x})(1-2\hat{x})(1-\hat{y})(1-2\hat{y}).$$

REMARK 3.11.– *For practical reasons (to reduce the size of the matrices), one may opt to "delete" the central node* $\hat{M}_{1234}$, *by determining a hyperplane in* Q^2 *containing* P^2, *along which the unisolvency property is satisfied for the set of nodes* $\left\{\hat{M}_1,\hat{M}_2,\hat{M}_3,\hat{M}_4,\hat{M}_{12},\hat{M}_{23},\hat{M}_{34},\hat{M}_{41}\right\}$. *This technique gives rise to a new type of finite element, called* Q^2*-Serendipity, and is also a second-order Lagrangian finite element (see exercise 27).*

To pass from the unit square $\hat{C}$ to any *convex* quadrilateral Q_ℓ (by passing from the reference plane to the "physical" plane), a transformation F_ℓ is used just as was done for triangles. This time, however, F_ℓ is not necessarily affine[6], though it will be of the form $F_\ell \in Q^1(\hat{C}) \times Q^1(\hat{C})$, namely:

$$F_\ell\begin{pmatrix}\hat{x}\\ \hat{y}\end{pmatrix}=\begin{pmatrix}a_1^\ell\hat{x}\hat{y}+b_1^\ell\hat{x}+c_1^\ell\hat{y}+d_1^\ell\\ a_2^\ell\hat{x}\hat{y}+b_2^\ell\hat{x}+c_2^\ell\hat{y}+d_2^\ell\end{pmatrix}.$$

Indeed, a priori we require 2×4 parameters to fully determine the transformation F_ℓ, since the quadrilateral Q_ℓ is defined by the coordinates of its four vertices. We can check (Girault and Raviart 1986) that there exists a unique and invertible $F_\ell \in Q^1(\hat{C}) \times Q^1(\hat{C})$, such that:

$$F_\ell\left(\hat{M}_i\right)=M_{\ell_i}\quad 1\le i\le 4. \qquad [3.44]$$

(to be compared with [3.34] for triangles).

6 We can check using a calculation that F_ℓ is affine if, and only if, Q_ℓ is a parallelogram.

Moreover, each side of $\hat{C}$, $[\hat{M}_i, \hat{M}_{i+1}]$ for $1 \leq i \leq 4$ (with the index convention $\hat{M}_5 = \hat{M}_1$) will be carried onto the corresponding side of Q_ℓ, $[M_{\ell_i}, M_{\ell_{i+1}}]$ for $1 \leq i \leq 4$, and finally the *restriction* of F_ℓ to any one of these sides is *affine*.

The basis functions on Q_ℓ are then defined by "pulling back" polynomials from $Q^k(\hat{C})$ via F_ℓ. More precisely, they are defined by:

$$Q^k(Q_\ell) = \left\{p_\ell = \hat{p} \circ F_\ell^{-1},\ \hat{p} \in Q^k(\hat{C})\right\}.$$

In general[7], $Q^k(Q_\ell)$ *is not a polynomial space*, although we do have the inclusion $Q^k(Q_\ell) \subset \mathcal{C}^\infty(Q_\ell)$. It is easier therefore to represent its elements by pulling them back via F_ℓ (and F_ℓ^{-1}), and referring to the square $\hat{C}$.

Using functions which are locally $(P^k)_{k\geq 1}$ or $(Q^k)_{k\geq 1}$, and are *globally continuous*, we may construct an internal approximation of $H^1(\Omega)$ (see section 3.3.3).

3.3.2.3. *In three dimensions*

Let us emphasize that in three dimensions, the P^1- and Q^1-finite elements are constructed essentially in an identical way to those in two dimensions.

3.3.2.3.1. P^1-finite element

Denote by $\hat{T}'$ the reference tetrahedron with the vertices $\hat{M}_1(0,0,0)$, $\hat{M}_2(1,0,0)$, $\hat{M}_3(0,1,0)$ and $\hat{M}_4(0,0,1)$.

$$\left|\begin{array}{l} \hat{K} = \hat{T}' \\ \hat{\Sigma} = \left\{\hat{M}_1, \hat{M}_2, \hat{M}_3, \hat{M}_4\right\} \\ \hat{P} = P^1(\hat{T}') = \left\{p(\hat{x},\hat{y},\hat{z}) = a\hat{x} + b\hat{y} + c\hat{z} + d\right\} \end{array}\right.$$

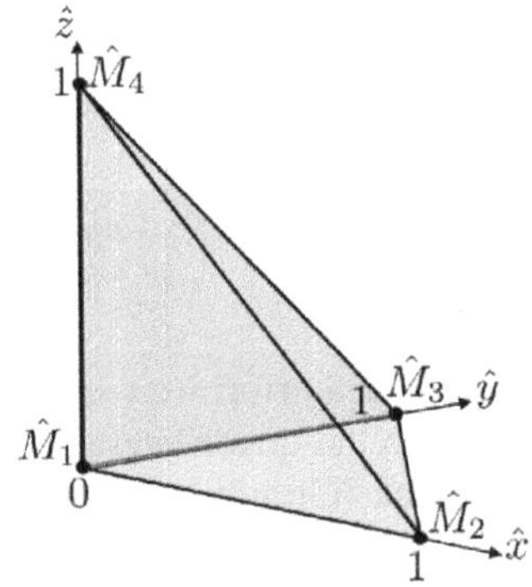

The basis functions associated with this element are:

$$\hat{\tau}_1(\hat{x},\hat{y},\hat{z}) = 1 - \hat{x} - \hat{y} - \hat{z},\ \hat{\tau}_2(\hat{x},\hat{y},\hat{z}) = \hat{x},\ \hat{\tau}_3(\hat{x},\hat{y},\hat{z}) = \hat{y},\ \hat{\tau}_4(\hat{x},\hat{y},\hat{z}) = \hat{z}.$$

7 The space $Q^k(Q_\ell)$ is a polynomial space with respect to (x, y), only in the particular case when Q_ℓ is a parallelogram. Indeed, in general, F_ℓ^{-1} will be a *rational fraction*.

3.3.2.3.2. Q^1-finite element

Denote by $\hat{C}$ the unit cube with the vertices $\hat{M}_1, \hat{M}_2, \hat{M}_3, \hat{M}_4$, and $\hat{M}_5(1,1,0)$, $\hat{M}_6(1,0,1)$, $\hat{M}_7(1,1,1)$, $\hat{M}_8(0,1,1)$.

$$\left| \begin{array}{l} \hat{K} = \hat{C} \\ \hat{\Sigma} = \left\{ \hat{M}_i, i = 1,8 \right\} \\ \hat{Q}^1(\hat{C}) = \left\{ p(\hat{x},\hat{y},\hat{z}) = \sum_{0 \le i,j,l \le 1} \alpha_{ijk} \hat{x}^i \hat{y}^j \hat{z}^l \right\} \end{array} \right.$$

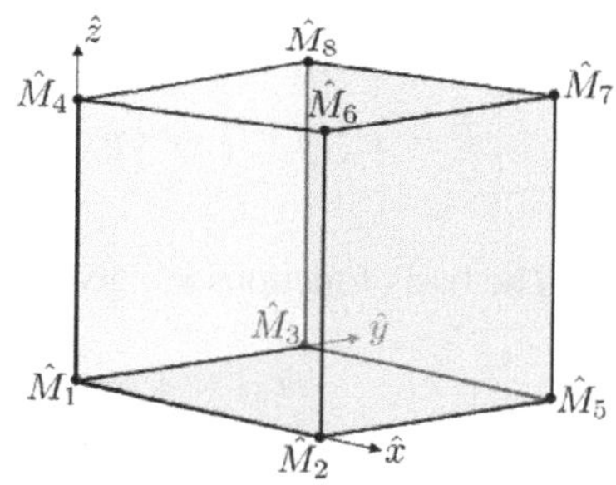

For example, we have that:

$$\hat{\tau}_1(\hat{x},\hat{y},\hat{z}) = (1-\hat{x})(1-\hat{y})(1-\hat{z}) \text{ and } \hat{\tau}_2(\hat{x},\hat{y},\hat{z}) = \hat{x}(1-\hat{y})(1-\hat{z}).$$

We leave it to the reader to determine the other basis functions.

It is not difficult to construct the P^2-finite elements on $\hat{T}'$ and the Q^2-finite elements on $\hat{C}$, which have 10 and 27 degrees of freedom, respectively. Let us point out that frequently one encounters first- and second-order finite elements on the prism (see exercise 29):

$$\left| \begin{array}{l} \hat{K} = \hat{T} \otimes \hat{S} \\ \hat{\Sigma} = \left\{ \hat{M}_i, i = 1,6 \right\} \\ \hat{P} = \left\{ p(\hat{x},\hat{y},\hat{z}) = a\hat{x}\hat{z} + b\hat{y}\hat{z} + c\hat{x} + d\hat{y} + e\hat{z} + f \right\} \end{array} \right.$$

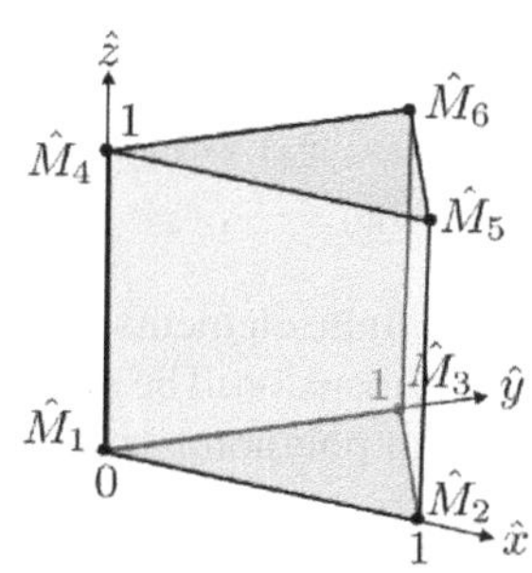

Again, using functions that are locally defined by "carrying" them from the reference finite element and are *globally continuous*, one can construct internal approximation spaces for $H^1(\Omega)$ (see section 3.3.3).

3.3.2.4. *In* d *dimensions*

Sometimes, we resort to solving problems in $\mathrm{d} \geq 4$ dimensions numerically. The principle of constructing the P^1- and Q^1-finite elements remains the same as before.

3.3.2.4.1. P^1-finite element

Let $\hat{T}^{(\mathrm{d})}$ be the reference simplex with the vertices $\hat{M}_1(0,0,0,\cdots,0)$, $\hat{M}_2(1,0,0,\cdots,0)$, $\hat{M}_3(0,1,0,\cdots,0)$, $\cdots$, $\hat{M}_{\mathrm{d}+1}(0,0,0,\cdots,1)$. Introduce:

$$\left|\begin{array}{l}\hat{K}=\hat{T}^{(\mathrm{d})}\\ \hat{\Sigma}=\left\{\hat{M}_i,\ i=1,\mathrm{d}+1\right\}\\ \hat{P}=P^1(\hat{T}^{(\mathrm{d})})=\left\{p\left(\hat{x}_1,\cdots,\hat{x}_{\mathrm{d}}\right)=\displaystyle\sum_{k=1,\mathrm{d}}a_k\hat{x}_k+a_{\mathrm{d}+1}\right\}.\end{array}\right.$$

The basis functions are given by:

$$\hat{\tau}_1(\hat{x}_1,\cdots,\hat{x}_{\mathrm{d}})=1-\sum_{k=1,\mathrm{d}}\hat{x}_k,\quad \hat{\tau}_i(\hat{x}_1,\cdots,\hat{x}_{\mathrm{d}})=\hat{x}_{i-1},\ 2\leq i\leq \mathrm{d}+1.$$

3.3.2.4.2. Q^1-finite element

Let $\hat{C}^{(\mathrm{d})}$ be the unit hypercube with the vertices $(\hat{M}_i)_{1\leq i\leq 2^{\mathrm{d}}}$. The coordinates of $\hat{M}_i$ correspond to the value of $i-1$ when expressed in base 2. For example: $\hat{M}_1(0,\cdots,0)$, $\hat{M}_2(0,\cdots,0,1)$, $\hat{M}_3(0,\cdots,0,1,0)$, $\ldots$, $\hat{M}_{2^{\mathrm{d}}}(1,\cdots,1)$.

Introduce:

$$\left|\begin{array}{l}\hat{K}=\hat{C}^{(\mathrm{d})}\\ \hat{\Sigma}=\left\{\hat{M}_i,\ i=1,2^{\mathrm{d}}\right\}\\ \hat{P}=Q^1(\hat{C}^{(\mathrm{d})})=\left\{p\left(\hat{x}_1,\cdots,\hat{x}_{\mathrm{d}}\right)=\displaystyle\sum_{0\leq i_1,i_2,\cdots,i_{\mathrm{d}}\leq 1}a_{i_1,i_2,\cdots,i_{\mathrm{d}}}\hat{x}_1^{i_1}\hat{x}_2^{i_2}\cdots\hat{x}_{\mathrm{d}}^{i_{\mathrm{d}}}\right\}.\end{array}\right.$$

. Then, the basis functions will be:

$\hat{\tau}_1(\hat{x}_1,\cdots,\hat{x}_{\mathrm{d}})=\prod_{k=1,\mathrm{d}}(1-\hat{x}_k)$, $\hat{\tau}_2(\hat{x}_1,\cdots,\hat{x}_{\mathrm{d}})=\hat{x}_{\mathrm{d}}\prod_{k=1,\mathrm{d}-1}(1-\hat{x}_k)$, $\cdots$, $\hat{\tau}_{2^{\mathrm{d}}}(\hat{x}_1,\cdots,\hat{x}_{\mathrm{d}})=\prod_{k=1,\mathrm{d}}\hat{x}_k$.

The finite elements that we have just presented generalize to any order k, regardless of the dimension. The spaces P just constructed equal either P^k, Q^k or "intermediate" spaces of polynomials.

One may "transport" finite elements onto any domain, using bijections from $\mathbb{R}^{\mathrm{d}}$ onto $\mathbb{R}^{\mathrm{d}}$ to help us. Let us introduce the following definition:

DEFINITION 3.6.– *The Lagrangian finite elements* $\left(\hat{K},\hat{\Sigma},\hat{P}\right)$ *and* (K,Σ,P) *are said to be equivalent if there exists a bijection* F *from* $\hat{K}$ *onto* K*, such that:*

$$\Sigma=F\left(\hat{\Sigma}\right) \quad [3.45]$$

$$P=\left\{p:K\to\mathbb{R}\ \text{such that}\ p\circ F\in\hat{P}\right\}. \quad [3.46]$$

When F is an affine mapping, the finite elements $\left(\hat{K}, \hat{\Sigma}, \hat{P}\right)$ and (K, Σ, P) are said to be affinely equivalent.

REMARK 3.12.– *In practice, the geometric transformation F is used to reduce the setting to the reference finite element, whose basis functions are intrinsic, that is, defined in a unique and natural way (see the preceding examples).*

3.3.3. *Constructing the approximation space*

We will now describe the process of constructing the internal approximation spaces V_h of $H^1(\Omega)$ using Lagrangian finite elements.

3.3.3.1. *Mesh*

Consider a mesh $\mathcal{T}_h$ of $\overline{\Omega}$, that is, a finite partition of the computational domain into polyhedra, if $\mathrm{d} = 3$ (line segments, if $\mathrm{d} = 1$; polygons, if $\mathrm{d} = 2$):

$$\overline{\Omega} = \bigcup_{K_\ell \in \mathcal{T}_h} K_\ell. \tag{3.47}$$

$$\text{such that} \quad \left| \begin{array}{l} \text{– each polyhedron } K_\ell \text{ has a non-empty interior } (\overset{\circ}{K_\ell} \neq \emptyset) \\ \text{– } \overset{\circ}{K_\ell} \cap \overset{\circ}{K_m} = \emptyset \quad \forall \ell \neq m \\ \text{– each face of a polyhedron } K_\ell \in \mathcal{T}_h \text{ is :} \\ \quad \text{a face of another polyhedron } K_m \in \mathcal{T}_h \\ \quad \text{or a subset of the boundary } \partial\Omega. \end{array} \right. \tag{3.48}$$

The parameter h denotes the characteristic size of the mesh:

$$h = \max_{K_\ell \in \mathcal{T}_h} h_{K_\ell} \tag{3.49}$$

where h_{K_ℓ} is the radius of the smallest sphere containing the polyhedron K_ℓ.

REMARK 3.13.– *It is not necessary for every polyhedron K_ℓ to be of the same type. In $\mathbb{R}^2$, we can therefore obtain a mixture of triangles and quadrangles, provided that properties [3.47] and [3.48] are satisfied.*

Assume that each polyhedron $K_\ell \in \mathcal{T}_h$ is associated with a Lagrangian finite element of order k: $E_\ell = (K_\ell, \Sigma_\ell, P_\ell)$. For the sake of simplicity, assume that k is the same for each polyhedron. Denote by n_ℓ the number of degrees of freedom that the finite element E_ℓ ($n_\ell = \mathrm{card}\Sigma_\ell$) has, and denote by $\left(\tau_i^\ell\right)_{i=1,n_\ell}$ its corresponding local basis functions. We will now see how we can construct the approximation space:

$$V_h = \left\{ v_h \in \mathcal{C}^0\left(\overline{\Omega}\right) \text{ such that } v_{h|K_\ell} \in P_\ell,\ \forall K_\ell \in \mathcal{T}_h \right\}. \tag{3.50}$$

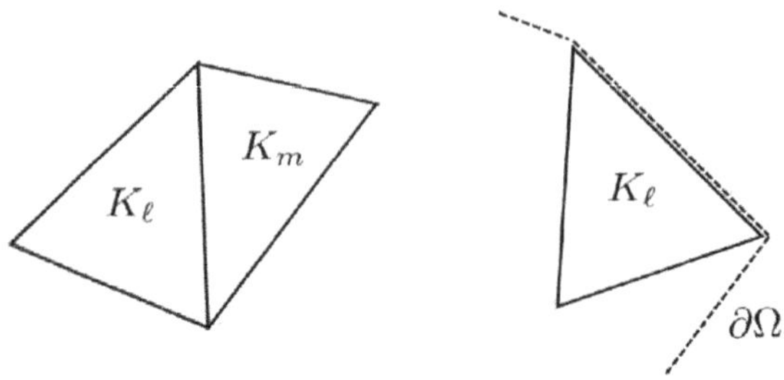

Figure 3.6. *Two adjacent "polyhedra" and one boundary "polyhedron"*

3.3.3.2. *Construction of the global basis functions*

Let us denote by L the number of finite elements E_ℓ associated with the mesh $\mathcal{T}_h$, and define the set of all the nodes of the mesh to be:

$$\Sigma = \bigcup_{\ell=1,L} \Sigma_\ell$$

which is equipped with the global index: $I = 1, N$ with $N = \text{card}\Sigma$. Starting from *each* finite element E_ℓ, define the map $\ell_g(\ell, \cdot)$

$$\left| \begin{array}{ll} \{1, 2, \cdots, n_\ell\} & \to \{1, 2, \cdots, N\} \\ i & \mapsto I = \ell_g(\ell, i) \end{array} \right. \qquad [3.51]$$

where $\ell_g(\ell, i)$ is the global number of the ith local degree of freedom of the ℓth finite element.

$$\ell_g(\ell,1) = 10$$
$$\ell_g(\ell,2) = 15$$
$$\ell_g(\ell,3) = 30$$
$$\ell_g(\ell,4) = 2$$

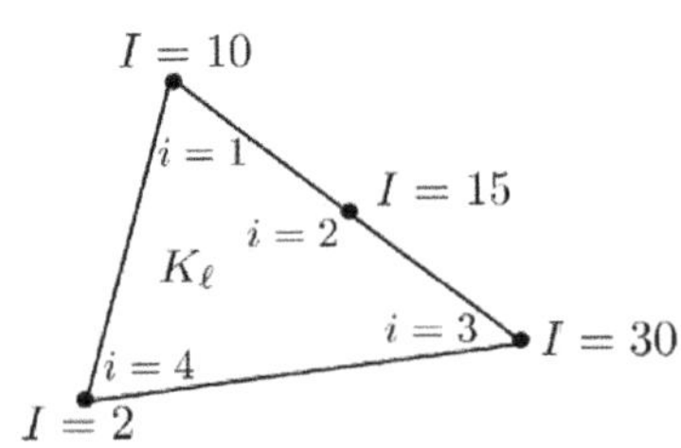

Consider the function w_I associated with the node M_I in the mesh, that is constructed by combining each local basis function τ_i^ℓ – defined on K_ℓ – that is associated with the node M_I, that is, those satisfying $\tau_i^\ell(M_I) = 1$. The function ℓ_g then lets us express the global basis function w_I on K_ℓ as:

$$w_I = \begin{cases} \tau_i^\ell & \text{if } \exists i \text{ such that } \ell_g(\ell, i) = I \\ 0 & \text{otherwise.} \end{cases} \qquad [3.52]$$

By construction, we immediately have the following properties:

PROPOSITION 3.6.– *The function w_I is a piecewise polynomial satisfying:*

$$w_I(M_J) = \delta_{IJ} \qquad \forall I, J = 1, N \tag{3.53}$$

and the support of w_I equals the union of all the polyhedra K_ℓ having the node M_I as a degree of freedom, in other words:

$$\text{support}(w_I) = \bigcup_{\ell \text{ s.t. } \exists i \text{ s.t. } \ell_g(\ell,i)=I} K_\ell. \tag{3.54}$$

The collection of functions $(w_I)_{I=1,N}$ just constructed forms a linearly independent set according to property [3.53], and therefore it constitutes a basis for the space $V_h = Vect(w_I)_{I=1,N}$. This is why we introduce the definition:

DEFINITION 3.7 (Global basis functions).– *We say that a global basis function associated with the node M_I is the function w_I defined by [3.52].*

REMARK 3.14.– *Recall that the support property is fundamental as it leads to sparse matrices.*

This construction process is called the **assembly principle**. We are going to illustrate it more concretely with two examples, and also highlight the continuous nature (or lack thereof) of the global basis functions.

3.3.4. *Two-dimensional examples of assembly*

– Let (K_1, Σ_1, P_1) and (K_2, Σ_2, P_2) be two adjacent finite elements with:

$$P_1 = P^1(K_1), \ P_2 = Q^1(K_2), \ \Sigma_1 = \{M_4, M_5, M_3\}, \\ \text{and } \Sigma_2 = \{M_2, M_1, M_4, M_3\}.$$

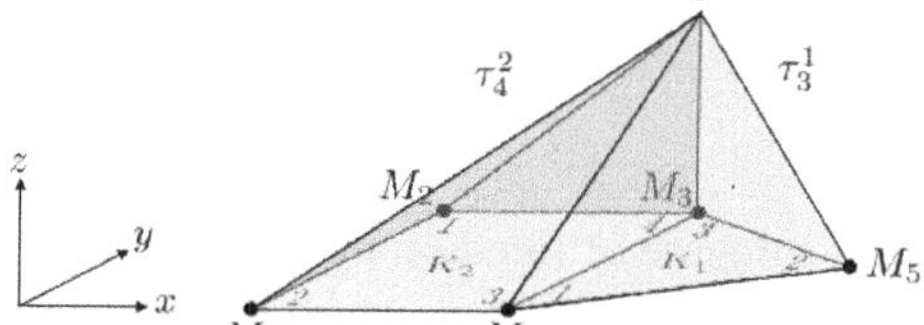

Figure 3.7. $P^1 - Q^1$ *assembly*

We have that $\tau_3^1 \in P^1(K_1)$ and $\tau_4^2 \in Q^1(K_2)$ (but they do not belong to $P^1(K_2)$). By the virtue of [3.52], we have:

$$w_3 = \begin{cases} \tau_3^1 & \text{on } K_1 \\ \tau_4^2 & \text{on } K_2. \end{cases}$$

However, the restrictions of τ_4^2 and τ_3^1 to the line segment $[M_3, M_4]$ are degree 1 polynomials in the variable s (the parameter along $[M_3, M_4]$) taking the values of 1 and 0 at the endpoints M_3 and M_4, respectively. Indeed, for $\tau^1_{3\,|[M_3,M_4]}$, this is immediate. Since τ_3^1 belongs to $P^1(K_1)$: $\tau_3^1(x,y) = ax + by + c$, then $\tau^1_{3\,|[M_3,M_4]}(s) = ax(s) + by(s) + c$. Moreover, $x(s)$ and $y(s)$ are both affine with respect to s: $x(s) = \alpha s + \beta$, $y(s) = \gamma s + \delta$. Then $\tau^1_{3\,|[M_3,M_4]}(s) = (a\alpha + b\gamma)s + (a\beta + b\delta + c)$. For $\tau^2_{4\,|[M_3,M_4]}$, we reason using the reference square $\hat{C}$ (remember: $K_2 = F_2(\hat{C})$). By definition, we have $\tau_4^2 = \hat{\tau}_4 \circ F_2^{-1}$, and we use the property that F_2^{-1} is affine from $[M_3, M_4]$ into $[\hat{M}_3, \hat{M}_4]$, knowing that $\hat{\tau}_4$ is also affine on $[\hat{M}_3, \hat{M}_4]$. Hence, the restrictions of τ_4^2 and τ_3^1 to the line segment $[M_3, M_4]$ are identical, proving that w_3 is continuous along the interface $\partial K_1 \cap \partial K_2$ between the two polygons, and therefore is continuous on $K_1 \cup K_2$.

– Now consider the case where $P_1 = P^2(K_1)$ and $P_2 = Q^1(K_2)$, with:

$$\Sigma_1 = \{M_3, M_4, M_5, M_6, M_7, M_8\} \text{ and } \Sigma_2 = \{M_2, M_1, M_4, M_3\}.$$

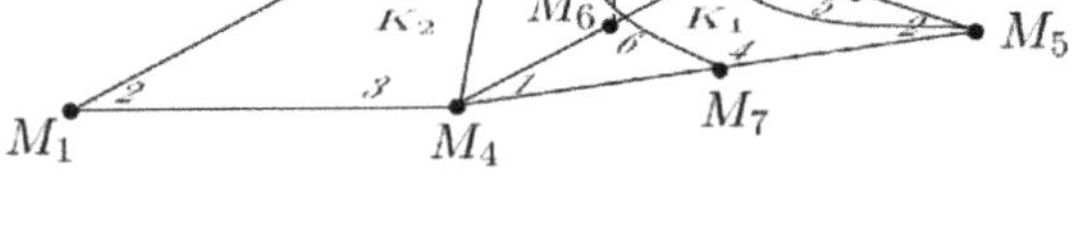

Figure 3.8. *$P^2 - Q^1$ assembly (non-conformal H^1)*

As the node M_6 is not a degree of freedom of K_2, we have:

$$w_6 = \begin{cases} 0 & \text{on } K_2 \\ \tau_6^1 & \text{on } K_1. \end{cases}$$

which shows that w_6 is not continuous on $K_1 \cup K_2$.

This last example shows that any degree of freedom located along a face F of a polygon K must *necessarily* also be a degree of freedom for any polygon adjacent to K along this face F, so that the global basis function associated with this degree of freedom is continuous on F (this is a *necessary* condition, but it is not a *sufficient* condition). See also exercise 28 for an example of assembling finite elements with various degrees and conforming in H^1.

Moreover, if a finite element (K, Σ, P) has a degree of freedom M_I not belonging to any face of K (in other words, a degree of freedom situated within the interior of K), then by construction we will have:

$$w_I = \begin{cases} \tau_i & \text{on } K \\ 0 & \text{otherwise.} \end{cases}$$

Therefore, for the function w_I to be continuous across each face of K, it is necessary and sufficient that:

$$w_{I|F} = 0 \quad \forall \text{ faces } F \text{ of } K. \qquad [3.55]$$

These observations lead us to the following continuity result, which ensures that the approximation space V_h, defined by [3.50], exists.

PROPOSITION 3.7.– *Let $\mathcal{T}_h$ be a mesh of Ω that satisfies [3.48]. Then every global basis function w_I defined by [3.52] is continuous on $\overline{\Omega}$, if:*

Any local degree of freedom that belongs to a face F of a polyhedron K is also a degree of freedom for each polyhedron adjacent to K via the face F. [3.56]

The restriction (F, Σ_F, P_F) of any finite element (K, Σ, P) to one of its faces F: $\Sigma_F = \Sigma \cap F$ and $P_F = \{p_{|F},\ p \in P\}$ is a Lagrangian finite element. [3.57]

If two elements $(K_\ell, \Sigma_\ell, P_\ell)$ and (K_m, Σ_m, P_m) are adjacent via the face F, then $P_{\ell|F} = P_{m|F}$. [3.58]

Condition [3.57] implies [3.55]. Indeed, let M_I be a degree of freedom that does not belong to any face of K, and let τ_I be its corresponding basis function. By definition, τ_I satisfies:

$$\tau_I(M) = 0 \quad \forall M \in \Sigma_F.$$

and since $\tau_{I|F} \in P_F$ from [3.57], we have: $\tau_{I|F} \equiv 0$ on F as (F, Σ_F, P_F) is a finite element that satisfies [3.42] by definition.

We are now able to state the following construction theorem:

THEOREM 3.5.– *Let $\mathcal{T}_h$ be a mesh of $\overline{\Omega}$ satisfying [3.47] and [3.48], on which the finite elements $(K_\ell, \Sigma_\ell, P_\ell)_{\ell=1,L}$ that satisfy the hypotheses [3.56], [3.57] and [3.58] are defined.*

Then the space $V_h = \left\{v_h \in \mathcal{C}^0(\overline{\Omega}) \text{ s.t. } v_{h|K_\ell} \in P_\ell,\ \forall K_\ell \in \mathcal{T}_h\right\}$ has $(w_I)_{I=1,N}$ as a basis, where the w_I are defined by [3.52], with $N = \text{card}\Sigma$, and we have that:

$$v_h(M) = \sum_{I=1,N} v_h(M_I) w_I(M) \quad \forall v_h \in V_h. \tag{3.59}$$

Additionally, the following inclusions are satisfied:

$$V_h \subset H^1(\Omega) \text{ and } V_h \subset \mathcal{C}^0(\overline{\Omega}).$$

Proof. According to proposition 3.7, the global basis functions w_I are continuous. By construction, see [3.52], they therefore belong to V_h. Moreover, according to [3.53], they form a linearly independent subset within V_h, and clearly they generate V_h. The expression [3.59] then follows from [3.53], because:

$$v_h(M_I) = \sum_{J=1,N} \alpha_J w_J(M_I) = \sum_{J=1,N} \alpha_J \delta_{IJ} = \alpha_I.$$

Finally, the inclusion $V_h \subset \mathcal{C}^0(\overline{\Omega})$ follows from the definition of V_h, and the inclusion $V_h \subset H^1(\Omega)$ follows from property 1.1. ■

Lagrangian finite elements also allow us to construct internal approximations of $H_0^1(\Omega)$.

It is a question of taking the essential condition (vanishing trace on $\partial\Omega$) into account when constructing the approximation space. See exercise 30 for an example of another essential condition (the average value).

Denote by $\mathcal{I}$ an indexing set of $\{1, 2, \cdots, N\}$ such that $M_I \notin \partial\Omega$ and set $q = \text{card}\,\mathcal{I}$. Then we have the following corollary:

COROLLARY 3.1.– *Under the same assumptions as those in theorem 3.5, the space:*

$$V_h^0 = \left\{v_h = \sum_{I \in \mathcal{I}} \alpha_I w_I,\ (\alpha_I) \in \mathbb{R}^q\right\}$$

is an internal approximation space of $H_0^1(\Omega)$.

Proof. V_h^0 is a subspace of V_h, and from theorem 3.5, we have that $V_h^0 \subset H^1(\Omega)$. It remains to check that $w_{I|\partial\Omega} = 0, \forall I \in \mathcal{I}$.

Let $I \in \mathcal{I}$, and let M be a point on the boundary $\partial\Omega$, which we assume to be located along the face F of a polyhedron K. By the definition of $\mathcal{I}$, M_I does not belong to $\partial\Omega$. If $M_I \notin K$, then according to the support property [3.54], $w_I = 0$ on F. If $M_I \in K$, we have that $w_I(M_s) = 0$ for any degree of freedom M_s belonging to F. But according to [3.57], (F, Σ_F, P_F) defines a finite element, and hence $w_I(M) = 0$. ■

REMARK 3.15.– *The basis functions w_I are not differentiable on $\overline{\Omega}$ in general, and therefore the Lagrangian finite elements do not let us construct an internal approximation space for $H^2(\Omega)$.*

Indeed, the functions in $H^2(\Omega)$, which are piecewise regular, must necessarily be regular globally in $C^1(\overline{\Omega})$ (the proof is analogous to that of property 1.2: all one has to do is reason for the gradient components that belong to $H^1(\Omega)$) (see section 3.4.3).

3.4. Extension of finite elements

The Lagrangian finite elements do not let us treat every situation. In particular, they do not lead to an internal approximation for $H^2(\Omega)$, which is useful when solving fourth-order elliptic problems (see exercise 24). This is why we generalize the notion of a degree of freedom, and of a finite element, in the following way.

Let K be a closed subset of $\mathbb{R}^{\mathrm{d}}$, $\mathrm{d} = 1, 2, 3$ with a non-empty interior (and K is not necessarily a polyhedron).

– We say that the local **degree of freedom** on K is any *distribution*[8] on K.

– Then, we say that a **finite element** on K is any triple (K, Σ, P) with $\Sigma = \{\varphi_i, i = 1, n_K,\ \varphi_i$ local degree of freedom on $K\}$, and where P is a vector space of functions on K, such that:

$$\forall(\alpha_1, \alpha_2, \cdots, \alpha_{n_K}) \in \mathbb{R}^{n_K},\ \exists! p \in P \text{ such that } \varphi_i(p) = \alpha_i,\ \forall i = 1, n_K. \quad [3.60]$$

This is indeed a generalization of a Lagrangian finite elements. Indeed, a Lagrangian degree of freedom M_i can be represented equivalently by the distribution:

$$\varphi_i = \delta_{M_i} \qquad \text{(Dirac mass at the point } M_i).$$

8 Warning: the domain K is closed. In this case, a distribution acts on the elements in $C^\infty(K)$, which in particular may be non-zero on the boundary ∂K.

Since the equality $\varphi_i(p) = \alpha_i$ then becomes $p(M_i) = \alpha_i$, property [3.60] therefore coincides with property [3.42] that defines a Lagrangian finite element. Moreover, property [3.60] ensures both the existence and uniqueness of local basis functions $(\tau_i)_{i=1,n_K}$ satisfying:

$$\varphi_j(\tau_i) = \delta_{ij} \quad \forall i, j = 1, n_K. \tag{3.61}$$

Two interesting examples that can be tackled now that this definition has been generalized are the Hermitian finite elements, as well as those of "moment" type.

3.4.1. *Hermitian finite elements*

These finite elements involve the values of the derivative of a function at just a few points in a domain K. In other words, we consider:

$$\Sigma \subset \left\{\delta'_{M_i}, \delta_{M_i},\ i = 1, n_K\right\}.$$

As a general rule, the Lagrangian degrees of freedom are kept fixed at the same points from before.

3.4.1.1. *One dimension*

Along the line interval $\hat{S} = [0,1]$, choose:

$$\left|\begin{array}{l} \hat{K} = \hat{S} \\ \hat{\Sigma} = \{\delta_0, \delta_1, \delta'_0, \delta'_1\} \\ \hat{P} = P^3([0,1]) \end{array}\right.$$

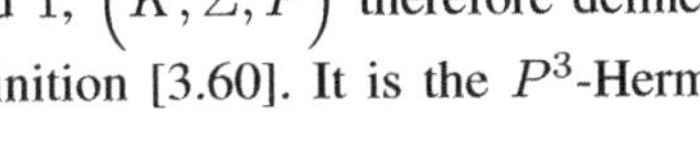

Since a third-order polynomial is determined completely by its values and derivatives that it takes at the points 0 and 1, $\left(\hat{K}, \hat{\Sigma}, \hat{P}\right)$ therefore defines a finite element that satisfies the statement of definition [3.60]. It is the P^3-Hermite finite element.

For its basis functions, it has:

$$\begin{array}{l} \hat{\tau}_1(\hat{x}) = (\hat{x} - 1)^2 (2\hat{x} + 1) \\ \hat{\tau}_2(\hat{x}) = \hat{x}^2 (3 - 2\hat{x}) \\ \tilde{\tau}_1(\hat{x}) = \hat{x}(\hat{x} - 1)^2 \\ \tilde{\tau}_2(\hat{x}) = \hat{x}^2 (\hat{x} - 1) \end{array}$$

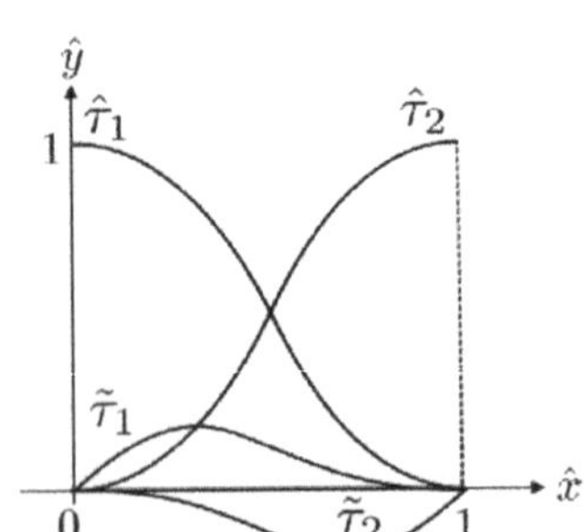

3.4.1.2. *Two dimensions*

We can also define the P^3-Hermite finite element when d $= 2$ for the triangle $\hat{T}$:

$$\left| \begin{array}{l} \hat{K} = \hat{T} \\ \hat{\Sigma} = \{\delta_{M_1}, \delta_{M_2}, \delta_{M_3}, \delta_{M_{123}}, \varphi_{12}, \varphi_{21}, \varphi_{23}, \varphi_{32}, \varphi_{13}, \varphi_{31}\}, \\ \text{where } \varphi_{ij} \text{ denotes the distribution } \varphi_{ij} : f \mapsto \nabla f(M_i) \cdot \overrightarrow{M_i M_j} \\ \hat{P} = P^3\left(\hat{K}\right) \end{array} \right.$$

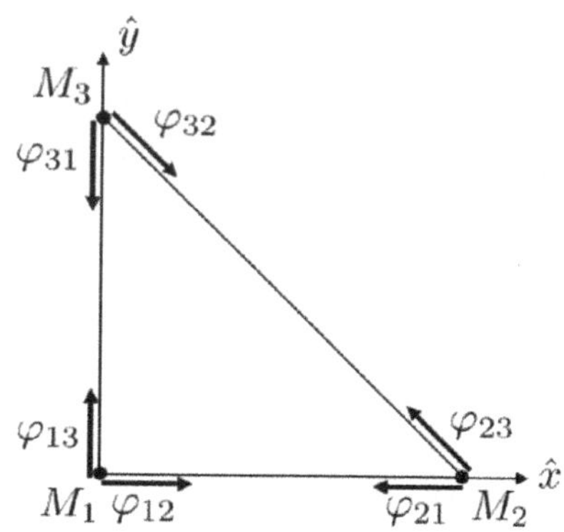

Property [3.60] is easy to verify, and it shows that $\left(\hat{K}, \hat{\Sigma}, \hat{P}\right)$ does indeed define a finite element.

3.4.2. *"Moment"-type finite elements*

Moment-type degrees of freedom are represented by the distribution:

$$\varphi = (-1)^{|\alpha|}\partial_\alpha 1_K \qquad (1_K \text{ indicator function of } K)$$

where α is a multi-index, which is possibly zero. For any function $v \in C^\infty(K)$, we have:

$$\langle \varphi, v \rangle = \langle 1_K, \partial_\alpha v(x) \rangle = \int_K \partial_\alpha v(x)\, d\Omega.$$

In d dimensions, one can thus consider zeroth-order finite elements that are associated with a zeroth-order moment. Let K be a closed domain with a non-empty interior. Then for an example, we may choose:

$$\Sigma = \{1_K\} \quad \text{and} \quad P = P^0(K)$$

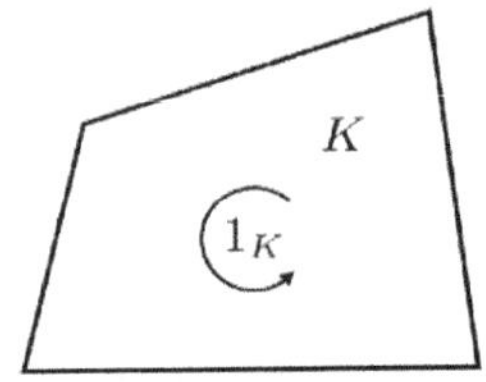

Such a degree of freedom is not associated to any node of K, since it represents the mean-value over K of the solution. See section 3.4.4 for a moment-type degree of freedom, which is defined over a face or along an edge.

3.4.3. *Constructing the approximation space*

The principle of construction is similar to that used for Lagrangian finite elements. Take a mesh $\mathcal{T}_h$ satisfying the hypotheses [3.47] and [3.48]. On each domain K_ℓ, consider a finite element $(K_\ell, \Sigma_\ell, P_\ell)$ with the basis functions $\left(\tau_i^\ell\right)_{i=1,n_\ell}$ (here, n_ℓ will always mean cardΣ_ℓ).

We also impose the following conditions for $m \geq 0$.

– The local basis functions τ_i^ℓ are of class $\mathcal{C}^{m+1}$ on K_ℓ.

– As soon as the support of a degree of freedom φ_i^ℓ is not included within a face F (which is a subset of the boundary of K_ℓ), then:

$$\partial_\alpha \tau^\ell_{i|F} = 0 \quad \forall |\alpha| \leq m.$$

– Let F be a face common to two domains K_ℓ and $K_{\ell'}$, then any degree of freedom of K_ℓ, whose support is included in F, is also a degree of freedom of $K_{\ell'}$. In denoting $D^\alpha P_\ell$ to be the set of derivatives of functions of order α on P_ℓ, we have:

$$D^\alpha P_{\ell|F} = D^\alpha P_{\ell'|F}, \quad \forall |\alpha| \leq m.$$

Under these assumptions, we can construct the global basis functions w_I by deciding whether or not to consider every single local degree of freedom that is shared between several finite elements, as a (single) global degree of freedom, which then, using the mapping ℓ_g defined by [3.51], leads to the definition:

$$w_I = \begin{cases} \tau_i^\ell & \text{on } K_\ell \text{ if } \exists i \text{ such that } \ell_g(\ell, i) = I \\ 0 & \text{otherwise.} \end{cases} \qquad [3.62]$$

We then introduce the following approximation space (with $\Sigma = \cup_{\ell=1,L}\Sigma_\ell$ and $N = \text{card }\Sigma$):

$$V_h = \text{vect}\left(w_I\right)_{I=1,N}.$$

With the previous assumptions, we can show that the global basis functions w_I belong to $\mathcal{C}^m(\overline{\Omega})$ and, since they are piecewise regular, further deduce that V_h is an internal approximation space for $H^{m+1}(\Omega)$. By reasoning by induction on m: typically, for $m = 1$, we noted that:

$$H^2(\Omega) = \{v \in H^1(\Omega) \text{ such that } \frac{\partial v}{\partial x_i} \in H^1(\Omega),\ 1 \leq i \leq \mathrm{d}\}.$$

Furthermore, when $v \in \mathcal{C}^1(\overline{\Omega})$ is piecewise regular, we have that $\partial_i v \in \mathcal{C}^0(\overline{\Omega})$ and that $\partial_i v$ is piecewise regular for $1 \leq i \leq \mathrm{d}$. According to property 1.1, we can conclude that $\partial_i v \in H^1(\Omega)$ for $1 \leq i \leq \mathrm{d}$, etc.

For example, it is easy to show that a P^3*-Hermite* finite element in *one dimension* leads to an internal approximation space for $H^2(\Omega)$ (see exercise 33). On the other hand, in *two dimensions*, we can check that the P^3*-Hermite* finite element is neither globally in $\mathcal{C}^1(\overline{\Omega})$ nor in $H^2(\Omega)$, because of the central degree of freedom, $\delta_{M_{123}}$.

3.4.4. *Vectorial finite elements*

Finally, if we consider problems with vector-valued indeterminates, then we can adapt the techniques presented above (there are two applications of this in exercises 36 and 37). Thus, if we consider $\mathbf{u}$ with values in $\mathbb{R}^\mathrm{d}$ for $\mathrm{d} \geq 2$, then we can choose Lagrangian finite elements for each component of $\mathbf{u}$. Or, one can choose moment-type finite elements for one or more components of $\mathbf{u}$. For example, if K is a polyhedron in $\mathbb{R}^3$, and F is one of the faces of ∂K or e is one of its edges, then we may define:

$$\langle \varphi_F, \mathbf{u} \rangle = \int_F \mathbf{u} \cdot \mathbf{n}\, dS, \text{ where } \langle \varphi_e, \mathbf{u} \rangle = \int_e \mathbf{u} \cdot \mathbf{t}\, dl.$$

where $\mathbf{n}$ is a unit normal to F and $\mathbf{t}$ is a unit tangent vector to e.

3.5. Exercises – statements

Exercise 25. Discretization in one dimension

Consider the boundary value problem posed on the open interval $(0, 1)$, with $f \in L^2(0, 1)$:

$$\begin{cases} \text{Find } u \in H^1(0,1) \text{ such that} \\ -u'' = f \text{ in } (0,1) \\ u(0) = u(1) = 0. \end{cases} \qquad \text{[P25]}$$

a) State the associated variational formulation [F25] of [P25].

b) Discretize [F25] using P^1-finite elements: label the approximate solution by u_h.

c) For a uniform mesh, compare the linear system $\mathbb{A}_1 \vec{U} = \vec{L}$, which is equivalent to the approximate variational formulation, with the system that comes from its discretization via the finite differences of [P25], which we write as $\mathbb{B}_1 \vec{U}' = \vec{L}'$.

d) What are the fundamental properties possessed by the two matrices: $\mathbb{A}_1$ and $\mathbb{B}_1$?

Exercise 26. P^1-finite elements in two dimensions on a structured mesh

Consider the square $\Omega = (0,1)\times(0,1)$, along with the grid points $(M_{i_1 i_2})$, $i_1, i_2 = 0, n$ whose coordinates are $(i_1 h, i_2 h)$ where $h = 1/n$. A mesh $\mathcal{T}_h$ can be constructed from each elementary square of side length h, by cutting it into two triangles after bisecting it along the first bisector. To solve the Dirichlet problem:

$$\left| \begin{array}{l} \text{Find } u \in H_0^1(\Omega) \text{ such that} \\ \displaystyle\int_\Omega \nabla u \cdot \nabla v \, d\Omega = \int_\Omega f v d\Omega \ \forall v \in H_0^1(\Omega). \end{array} \right. \quad \text{[F26]}$$

we can use P^1-finite elements on the triangulation $\mathcal{T}_h$.

a) What is the size of the linear system that we wish to solve?

b) Calculate the elementary stiffness matrix $(\mathbb{K}^\ell_{ij})_{1\leq i,j\leq 3}$ for the triangle T_ℓ, and from it deduce the total stiffness matrix $\mathbb{K}$.

c) Calculate the elementary mass matrix $(\mathbb{M}^\ell_{ij})_{1\leq i,j\leq 3}$ for the triangle T_ℓ, and from it deduce the total mass matrix $\mathbb{M}$.

d) If we approximate the data f using the decomposition:

$$f_h = \sum_{I=1}^{N} f(M_I) w_I$$

state the general equations of the linear system, from which we may obtain the discretization via P^1-finite elements. Compare it with that obtained by employing the finite difference method.

Exercise 27. Q^2-Serendipity finite elements

On the unit square $\hat{C}$, consider the set $\hat{\Sigma}$ of the degrees of freedom coming from the points: $\hat{M}_1(0,0)$, $\hat{M}_2(1,0)$, $\hat{M}_3(1,1)$, $\hat{M}_4(0,1)$, $\hat{M}_5(\frac{1}{2},0)$, $\hat{M}_6(1,\frac{1}{2})$, $\hat{M}_7(\frac{1}{2},1)$ and $\hat{M}_8(0,\frac{1}{2})$. Let $\hat{M}_9$ denote the coordinates $(\frac{1}{2},\frac{1}{2})$.

a) Prove that any polynomial p, whose degree is at most two in $(\hat{x},\hat{y})$, satisfies the relation:

$$4p(\hat{M}_9) + \sum_{i=1}^{4} p(\hat{M}_i) - 2\sum_{i=5}^{8} p(\hat{M}_i) = 0. \qquad \text{[E27]}.$$

(Use a second-order Taylor expansion.)

b) Denoting by $(\hat{\tau}_i)_{i=1,9}$ the usual basis functions of the element $Q^2(\hat{C})$, introduce the functions:

$$\tilde{\tau}_i = \hat{\tau}_i - \frac{1}{4}\hat{\tau}_9 \quad 1 \leq i \leq 4$$
$$\tilde{\tau}_i = \hat{\tau}_i + \frac{1}{2}\hat{\tau}_9 \quad 5 \leq i \leq 8.$$

Check that the functions $(\tilde{\tau}_i)_{i=1,8}$ form a linearly independent set, and that they also satisfy relation [E27].

c) Set $\hat{P} = \{p \in Q^2(\hat{C})$ such that p satisfies [E27]$\}$. Deduce from the above that the Q^2-Serendipity finite element $(\hat{C},\hat{\Sigma},\hat{P})$ is a second-order Lagrangian finite element $(\mathbb{P}^2(\hat{C}) \subset \hat{P})$.

d) In the case of a square mesh made up of $n \times n$ quadrangular elements, estimate the number of non-zero terms in the stiffness matrix when $Q^2(\hat{C})$-finite elements are used, and also when Q^2-Serendipity finite elements are used. Estimate the increase in the asymptotic (i.e., for large n) space (i.e. memory) complexity that occurs when Q^2-Serendipity finite elements are used.

Exercise 28. Finite element for the $P^1 - P^2$ coupling

On the triangle $T_1 = \hat{T}$ of $\mathbb{R}^2$, consider the following four Lagrangian degrees of freedom: $M_1(0,0)$, $M_2(1,0)$, $M_3(0,1)$ and $M_4(\frac{1}{2},\frac{1}{2})$, that form the set Σ, and also consider the set of polynomials $P = Q^1(T_1)$.

a) Show that (T_1,Σ,P) is a first-order Lagrangian finite element, whose local basis functions will be given.

b) Consider the open set Ω that arises from taking the union of three triangles T_1, T_2, T_3, on which we define the following finite elements:

$$(T_1, \Sigma_1, P_1) : \begin{cases} T_1 \text{ with vertices } M_1, M_2, M_3 \\ \Sigma_1 = \{M_1, M_2, M_3, M_4\} \\ P_1 = Q^1(T_1) \end{cases}$$

$$(T_2, \Sigma_2, P_2) : \begin{cases} T_2 \text{ with vertices } M_2, M_5, M_3 \\ \Sigma_2 = \{M_i, i = 2, 7\}\,; \\ \qquad M_6, M_7 \text{ the midpoints of } [M_3, M_5] \text{ and } [M_2, M_5] \\ P_2 = P^2(T_2) \end{cases}$$

$$(T_3, \Sigma_3, P_3) : \begin{cases} T_3 \text{ with vertices } M_1, M_3, M_8 \\ \Sigma_3 = \{M_1, M_3, M_8\} \\ P_3 = P^1(T_3) \end{cases}$$

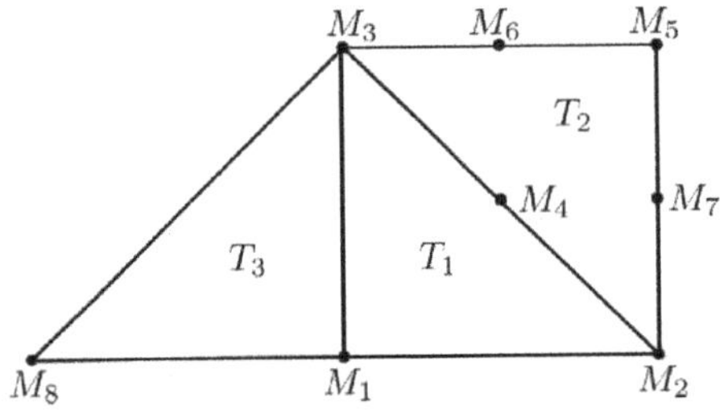

Construct the approximation space V_h generated by the three finite elements.

c) Do we have that $V_h \subset C^0(\overline{\Omega})$? Or $V_h \subset H^1(\Omega)$?

Exercise 29. Three-dimensional prismatic finite element

Consider the unit prism $\hat{K} = \hat{T} \otimes \hat{S}$ with the vertices:

$$\begin{array}{lll} \hat{M}_1 = (0,0,0), & \hat{M}_2 = (1,0,0), & \hat{M}_3 = (0,1,0) \\ \hat{M}_4 = (0,0,1), & \hat{M}_5 = (1,0,1), & \hat{M}_6 = (0,1,1). \end{array}$$

Denote $\hat{\Sigma} = \{\hat{M}_i,\ i = 1, 6\}$, and denote by $\hat{P}$ the space generated by polynomials of the form:

$$p(\hat{x}_1, \hat{x}_2, \hat{x}_3) = a_0 + a_1\hat{x}_1 + a_2\hat{x}_2 + (b_0 + b_1\hat{x}_1 + b_2\hat{x}_2)\hat{x}_3.$$

a) Show that $(\hat{K}, \hat{\Sigma}, \hat{P})$ is a Lagrangian finite element and compute the local basis functions $\{\hat{\tau}_i,\ i = 1, 6\}$ for it. These should be expressed using the local basis functions for the affine P^1-finite elements, which are defined either on the unit triangle $\hat{T}$ in the plane $O\hat{x}_1\hat{x}_2$, or on the unit line segment $\hat{S} = [0, 1]$ along the line $O\hat{x}_3$, respectively.

b) Show that the elementary mass and stiffness matrices can be expressed using the elementary matrices for the two affine elements mentioned in the previous question.

Exercise 30. Mean-type essential condition[9]

Let α and β be two real numbers, and suppose that $f \in L^2(0,1)$. We wish to solve the following problem:

$$\begin{cases} \text{Find } u \in H^1(0,1) \text{ such that} \\ -u'' = f \text{ in } (0,1) \\ u'(0) = \alpha, \quad u'(1) = \beta \\ \displaystyle\int_0^1 u(x)dx = 0. \end{cases} \qquad \text{[P30]}$$

a) How regular is the solution to problem [P30]? Show that it satisfies:

$$\int_0^1 u'v'dx = \int_0^1 f\,v\,dx - \alpha v(0) + \beta v(1), \quad \forall v \in H^1(0,1). \qquad \text{[E30]}$$

b) For problem [P30] to admit a solution, deduce that the data f, α and β must satisfy the following compatibility condition:

$$\int_0^1 f\,dx - \alpha + \beta = 0. \qquad \text{[C30]}$$

c) Show that $V = \left\{v \in H^1((0,1)) \text{ such that } \displaystyle\int_0^1 v\,dx = 0\right\}$ is a closed subspace of $H^1(0,1)$.

d) Let u be a solution of [P30]. Deduce from the above that u is the solution for the following variational problem:

$$\left| \begin{array}{l} \text{Find } u \in V \text{ such that} \\ \displaystyle\int_0^1 u'v'dx = \int_0^1 f\,v\,dx - \alpha v(0) + \beta v(1), \qquad \forall v \in V. \end{array} \right. \qquad \text{[F30]}$$

and show that the problem is well-posed (we will use the Poincaré–Wirtinger inequality [2.23] to prove that the bilinear form is coercive).

e) Let u be the solution of [F30]. If condition [C30] is satisfied, show that u satisfies [E30]. Thus deduce that u is a solution of [P30].

---

9 The abstract general case was dealt with in exercise 19 (Neumann condition).

f) Let N be a strictly positive integer, and set:

$$h = \frac{1}{N} \text{ and } x_i = (i-1)h, \quad i = 1, N+1.$$

For $i = 1, N+1$, denote by w_i the unique continuous function on $[0, 1]$ such that:

$$\begin{cases} w_i\left(x_j\right) = \delta_{ij}, \quad j = 1, N+1 \\ \text{the restriction of } w_i \text{ to each interval } [x_j, x_{j+1}],\ j = 1, N, \text{ is affine.} \end{cases}$$

Finally, denote by X_h the vector space that is generated by the functions $(w_i)_{i=1,N+1}$, and by V_h by the subspace of X_h which is defined by:

$$V_h = \left\{ v_h \in X_h \text{ such that } \int_0^1 v_h dx = 0 \right\}.$$

Then consider the following discrete variational formulation:

$$\left| \begin{array}{l} \text{Find } u_h \in V_h \text{ such that} \\ \displaystyle\int_0^1 u_h' v_h' dx = \int_0^1 f v_h dx - \alpha v_h(0) + \beta v_h(1), \qquad \forall v_h \in V_h. \end{array} \right. \qquad [\text{F30}_h]$$

Show that this discrete variational formulation is well posed.

g) Set:

$$\hat{w}_1 = 2w_1 - w_2,\ \hat{w}_2 = w_2 - w_3,\ \cdots,\ \hat{w}_{N-1} = w_{N-1} - w_N,$$
$$\hat{w}_N = w_N - 2w_{N+1}.$$

- Show that the functions $(\hat{w}_i)_{i=1,N}$ form a basis of V_h.

- Show that solving the discrete problem [F30$_h$] is equivalent to solving a linear system of the form

$$\mathbb{A}\overrightarrow{U} = \overrightarrow{L}$$

where $\mathbb{A}$ denotes a square matrix of order N, defined as follows:

$$\mathbb{A}_{ij} = \int_0^1 \hat{w}_j' \hat{w}_i' dx.$$

- What is the form of the matrix $\mathbb{A}$ (we will explicitly calculate a central line of the matrix)? Comment on it, and then calculate the matrix $\mathbb{A}$.

- Calculate the term $\overrightarrow{L}$ on the right-hand side, given the data $f = constant$.

4

Numerical Analysis of the Finite Element Method

In the previous chapter, we introduced a method for approximating a variational formulation, based on the definition of finite-dimensional vector subspaces of solution and test-function spaces. When these spaces are equal to $H^1(\Omega)$, where $\Omega \subset \mathbb{R}^{\mathrm{d}}$ ($\mathrm{d} = 1, 2, 3$) is an open and bounded polygon ($\mathrm{d} = 2$) or polyhedron ($\mathrm{d} = 3$), we used the finite element method. This method allows us to build a family of subspaces from the meshes of $\overline{\Omega}$, whose mesh size tends toward 0. We are now going to study how a sequence of approximate solutions converges toward the exact solution more precisely, when the mesh size tends toward 0, and in particular the speed of this convergence. For this, we will need to use an **interpolation operator** and the local interpolation errors. Afterward, we will introduce the mesh refinement technique that improves the quality of the calculated approximate solution. Finally, we will discuss approximation methods involving approximate calculations of the data and/or of the (bi)linear forms – for example, via numerical quadrature formulas – as well as external, or non-conforming, methods, for which the approximation spaces are not included within the solution and/or test-function spaces.

4.1. Convergence of finite element methods

In this section, we will examine some aspects of finite element convergence analysis. We will only be interested in the case when the finite elements are Lagrangian. On the one hand, the convergence proofs are based upon Céa's lemma (see lemma 3.2), when the finite elements are a particular case of Galerkin's approximation and, on the other hand, upon very technical estimates for the interpolation error, of which we present just the broad outlines.

We place ourselves in the case of an elliptic problem on $H^1(\Omega)$, which is of the form:

$$\left| \begin{array}{l} \text{Find } u \in H^1(\Omega) \text{ such that} \\ a(u,v) = \ell(v) \quad \forall v \in H^1(\Omega). \end{array} \right. \qquad [4.1]$$

In the following, we suppose for the sake of example that $a(\cdot,\cdot)$ and $\ell(\cdot)$ satisfy the hypotheses for the Lax–Milgram theorem (see theorem 2.4). The Neumann problem (stated in section 2.3.1) falls within this framework.

On a connected and bounded non-empty open subset Ω with a "sufficiently regular" boundary, which is always assumed to be either polygonal (in $\mathbb{R}^2$) or polyhedral (in $\mathbb{R}^3$), we are given the mesh $\mathcal{T}_h : (K_\ell)_{\ell=1,L}$ satisfying the hypotheses [3.47] and [3.48]. To each polyhedron K_ℓ, one can associate to it the Lagrangian finite element of order k, that is given by: $(K_\ell, \Sigma_\ell, P_\ell)$, with local basis functions $\left(\tau_i^\ell\right)_{i=1,n_\ell}$.

We assume that the finite elements satisfy the compatibility conditions [3.56], [3.57] and [3.58], so that the global basis functions $(w_I)_{I=1,N}$ defined in [3.52] generate an N-dimensional vector space V_h included in both $\mathcal{C}^0\left(\overline{\Omega}\right)$ and $H^1(\Omega)$.

Introduce the discrete problem:

$$\left| \begin{array}{l} \text{Find } u_h \in V_h \text{ such that} \\ a\left(u_h, v_h\right) = \ell\left(v_h\right) \quad \forall v_h \in V_h \end{array} \right. \qquad [4.2]$$

which constitutes Galerkin's approximation for the problem [4.1]. We can repeat the process for a set of meshes whose mesh size h becomes smaller and smaller. We denote by $(V_h)_h$ the family of subspaces of $H^1(\Omega)$ generated by this.

Recall that from Céa's lemma (lemma 3.2), we have:

$$\|u - u_h\|_{H^1(\Omega)} \leq C \inf_{v_h \in V_h} \|u - v_h\|_{H^1(\Omega)} \qquad [4.3]$$

where C is independent of $(V_h)_h$ and $\ell(\cdot)$. For a given h, it is necessary therefore to estimate:

$$\inf_{v_h \in V_h} \|u - v_h\|_{H^1(\Omega)}.$$

To do so, introduce the **interpolation operator**:

$$\pi_h : \left| \begin{array}{lll} \mathcal{C}^0(\overline{\Omega}) & \longrightarrow & V_h \\ v & \mapsto & \pi_h v = \sum_{I=1,N} v(M_I) w_I \end{array} \right. \qquad [4.4]$$

where M_I represents the node associated with the global basis function w_I. In other words, $\pi_h v$ is the unique function on V_h that takes the same values as v at the nodes $(M_I)_{I=1,N}$.

From [4.3], it is clear that for every $w \in \mathcal{C}^0(\overline{\Omega})$:

$$\|u - u_h\|_{H^1(\Omega)} \leq C\|u - \pi_h w\|_{H^1(\Omega)}. \qquad [4.5]$$

Which function should we choose for w? The choice $w = u$ seems natural.

Given the fact that the interpolation operator π_h is defined for continuous functions on $\overline{\Omega}$, in the calculations we will assume that $u \in \mathcal{C}^0(\overline{\Omega})$. Typically, if u belongs to $\Psi_D(\Omega)$, $\Psi_N(\Omega)$ (see [1.43]), or $H^2(\Omega)$, then according to propositions 1.11 or 1.16 (d $\leq$ 3), we can indeed choose $w = u$ in [4.5].

REMARK 4.1.– *In the cases that interest us, it is fundamental to note that it is not necessary to estimate the quantity* $\inf_{v_h \in V_h} \|u - v_h\|_{H^1(\Omega)}$ *for* any $u \in H^1(\Omega)$, *but for u belonging to a subspace of* $H^1(\Omega)$, *and even only* $H^1(\Omega) \cap \mathcal{C}^0(\overline{\Omega})$ *if we use the interpolation operator [4.4].*

Thus, we are going to estimate $\|u - \pi_h u\|_{H^1(\Omega)}$, which we call the **interpolation error**.

REMARK 4.2.– *Indeed, there is no reason for us to have that* $u_h = \pi_h u$.

This error decomposes according to each polyhedron in $\mathcal{T}_h$:

$$\|u - \pi_h u\|_{H^1(\Omega)} = \Big(\sum_{\ell=1,L} \|u - \pi_h u\|^2_{H^1(K_\ell)}\Big)^{1/2}.$$

Therefore, we must estimate a local interpolation error for each polyhedron K_ℓ.

4.1.1. *Local interpolation error*

Let $1 \leq \ell \leq L$ be given. By the virtue of definition [3.52] of the global basis functions (w_I), we have:

$$(\pi_h u)_{|K_\ell} = \sum_{i=1,n_\ell} u\left(M_i^\ell\right) \tau_i^\ell = \pi_{K_\ell}(u_{|K_\ell})$$

where M_i^ℓ represents the local degree of freedom associated with each local basis function τ_i^ℓ.

Suppose that the finite element $(K_\ell, \Sigma_\ell, P_\ell)$ is affinely equivalent to the reference finite element $\left(\hat{K}, \hat{\Sigma}, \hat{P}\right)$ (see definition 3.6), whose size is independent of h, and let h_{K_ℓ} be the radius of the smallest ball circumscribing K_ℓ, and let ρ_{K_ℓ} be the radius of the largest ball inscribed within K_ℓ (for quadrilaterals, ρ_{K_ℓ} is the minimum between the four radii of the largest balls inscribing each of the four triangles, which are formed from choosing three out of the four vertices of the quadrilateral). By definition, we have the relation $h_{K_\ell} \geq \rho_{K_\ell}$.

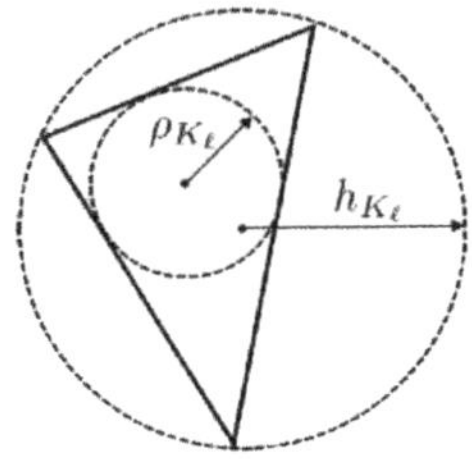

Figure 4.1. *Characteristic dimensions of the triangle*

For $m \geq 0$, let us introduce the **semi-norm** $|v|_{m,K_\ell}$ of $H^m(K_\ell)$, defined by:

$$|v|_{m,K_\ell} = \Big(\sum_{\alpha \in \mathbb{N}^d, |\alpha| = m} \int_{K_\ell} (\partial_\alpha v(x))^2 \, d\Omega \Big)^{1/2}.$$

We have the following fundamental result:

PROPOSITION 4.1 (Local interpolation error).– *Let $(K_\ell, \Sigma_\ell, P_\ell)$ be a Lagrangian finite element of order k, that is to say:*

$$P^k(K_\ell) \subset P_\ell \subset H^{k+1}(K_\ell)$$

and which is affinely equivalent to the finite element $\left(\hat{K}, \hat{\Sigma}, \hat{P}\right)$. Then there exists a constant C, depending only on $(\hat{K}, \hat{\Sigma}, \hat{P})$, such that for each integer $0 \leq m \leq k+1$, we have the following estimate:

$$|v - \pi_{K_\ell} v|_{m,K_\ell} \leq C \frac{\left(h_{K_\ell}\right)^{k+1}}{\left(\rho_{K_\ell}\right)^m} |v|_{k+1,K_\ell} \qquad \forall v \in H^{k+1}(K_\ell). \qquad [4.6]$$

To establish this result, we will proceed in several steps. Of course, as we can simplify the problem by using a reference domain, the first step is to make a change-of-variables to go from K_ℓ to $\hat{K}$. Then, we can make a direct estimate for the reference domain $\hat{K}$, which by definition will be independent of K_ℓ, and then we can conclude by returning to K_ℓ.

First, we establish estimates between the semi-norms evaluated on K_ℓ and on $\hat{K}$, respectively. As stated, we assume that $(K_\ell, \Sigma_\ell, P_\ell)$ is affinely equivalent to $\left(\hat{K}, \hat{\Sigma}, \hat{P}\right)$. In other words, the transformation $F_\ell : \hat{K} \to K_\ell$ is affine, and we recall that we can write:

$$M = F_\ell(\hat{M}) = \mathbb{B}_\ell \hat{M} + \vec{C}_\ell, \quad \mathbb{B}_\ell \in \mathbb{R}^{\mathrm{d}\times\mathrm{d}} \text{ is invertible, } \vec{C}_\ell \in \mathbb{R}^{\mathrm{d}}.$$

In this case, we check that $[dF_\ell(\hat{M})] = \mathbb{B}_\ell$, $J_{F_\ell}(\hat{M}) = det(\mathbb{B}_\ell)$, and finally $\hat{M} = \mathbb{B}_\ell^{-1} M - \mathbb{B}_\ell^{-1} \vec{C}_\ell$. To start with, note that we can easily bound the norm of $\mathbb{B}_\ell$ and that of its inverse.

LEMMA 4.1.– *We have the following bounds:*

$$||\mathbb{B}_\ell|| \leq \frac{h_{K_\ell}}{\hat{\rho}}, \; ||\mathbb{B}_\ell^{-1}|| \leq \frac{\hat{h}}{\rho_{K_\ell}}. \tag{4.7}$$

Proof. As usual, for an arbitrary $R > 0$ we have:

$$||\mathbb{B}_\ell|| = \frac{1}{R} \sup_{\hat{y},\, |\hat{y}|=R} |\mathbb{B}_\ell \hat{y}|.$$

But as $\hat{\rho}$ is equal to the radius of the largest ball inscribed within $\hat{K}$ by definition, for any vector $\hat{y}$ with $2\hat{\rho}$ for its norm, there exist two points $\hat{M}_1$ and $\hat{M}_2$ of $\hat{K}$ such that $\hat{y} = \overrightarrow{\hat{M}_1 \hat{M}_2}$. The points $M_1 = F_\ell(\hat{M}_1)$ and $M_2 = F_\ell(\hat{M}_2)$ belong to K_ℓ of course, and, by the definition of h_{K_ℓ}, we have that $|\overrightarrow{M_1 M_2}| \leq 2h_{K_\ell}$. If we apply the previous formula to calculate $||\mathbb{B}_\ell||$, we obtain the left result of [4.7]. For the result on the right-hand side, we proceed in a similar fashion though this time passing from K_ℓ to $\hat{K}$. ■

To be complete, note that $|det(\mathbb{B}_\ell)| = \text{vol}(K_\ell)/\text{vol}(\hat{K})$.

Now, to a function $\hat{w}$ that is defined on $\hat{K}$, we associate the function w defined on K by $\hat{w}(\hat{M}) = w(M)$ (and conversely). We can write $w = \hat{w} \circ F_\ell^{-1}$ or $\hat{w} = w \circ F_\ell$. As the transformations F_ℓ and F_ℓ^{-1} are, respectively, of the class C^∞ on $\overline{\hat{K}}$ and $\overline{K}_\ell$, then in particular it follows that $w \in H^p(K_\ell) \Longleftrightarrow \hat{w} \in H^p(\hat{K})$, for all $p \geq 0$.

LEMMA 4.2.– *For each $q \geq 0$, we have the following estimates:*

$$|v|_{q,K_\ell} \leq C_q \left(\rho_{K_\ell}\right)^{-q} |det(\mathbb{B}_\ell)|^{1/2} \left|v \circ F_\ell\right|_{q,\hat{K}}, \quad \forall v \in H^q(K_\ell) \tag{4.8}$$

$$\left|\hat{v}\right|_{q,\hat{K}} \leq C_q' \left(h_{K_\ell}\right)^q |det(\mathbb{B}_\ell)|^{-1/2} \left|\hat{v} \circ F_\ell^{-1}\right|_{q,K_\ell} \quad \forall \hat{v} \in H^q(\hat{K}). \tag{4.9}$$

In the above, the constants C_q and C'_q depend only on q (and on the dimension d*).*

Proof. We begin by establishing the bound [4.8] in the $q = 0$ and $q = 1$ cases. Recall the change-of-variables formula:

$$\int_{K_\ell} w \, d\Omega = |det(\mathbb{B}_\ell)| \int_{\hat{K}} \hat{w} \, d\hat{\Omega}.$$

For $q = 0$, it suffices to choose $w = v^2$ (and $C_0 = 1$). For $q = 1$, we replace w with $|\nabla v|^2$. Recall that $\nabla v(M) = (\mathbb{B}_\ell^{-1})^t \, \hat{\nabla} \hat{v}(\hat{M})$. Then in this case, the change-of-variables formula gives:

$$\begin{aligned} |v|_{1,K_\ell}^2 = \int_{K_\ell} |\nabla v|^2 \, d\Omega &= |det(\mathbb{B}_\ell)| \int_{\hat{K}} |(\mathbb{B}_\ell^{-1})^t \hat{\nabla} \hat{v}|^2 \, d\hat{\Omega} \\ &\leq |det(\mathbb{B}_\ell)| \, \|\mathbb{B}_\ell^{-1}\|^2 \int_{\hat{K}} |\hat{\nabla} \hat{v}|^2 \, d\hat{\Omega} \\ &\leq \hat{h}^2 \, |det(\mathbb{B}_\ell)| \, (\rho_{K_\ell})^{-2} \, |\hat{v}|_{1,\hat{K}}^2. \end{aligned}$$

We obtain [4.8] for $q = 1$, with $C_1 = \hat{h}$. For $q \geq 2$, we reason partial derivative by partial derivative. Let us detail the method for $q = 2$, which then easily generalizes to the case $q > 2$.

For $1 \leq i, j \leq$ d, note that:

$$\begin{aligned} \left| \frac{\partial^2 v}{\partial x_i \partial x_j}(M) \right| &= |d^2 v(M).(\mathbf{e}_i, \mathbf{e}_j)| \leq \sup_{\mathbf{x}_1, \mathbf{x}_2, \, |\mathbf{x}_1| = |\mathbf{x}_2| = 1} |d^2 v(M).(\mathbf{x}_1, \mathbf{x}_2)| \\ &= \|d^2 v(M)\|. \end{aligned}$$

Otherwise, for every pair of vectors $(\mathbf{x}_1, \mathbf{x}_2)$, we have the following relation when differentiating a composition of functions:

$$d^2 v(M).(\mathbf{x}_1, \mathbf{x}_2) = d^2 \hat{v}(\hat{M}).(\mathbb{B}_\ell^{-1} \mathbf{x}_1, \mathbb{B}_\ell^{-1} \mathbf{x}_2)$$

so that:

$$\|d^2 v(M)\| \leq \|d^2 \hat{v}(\hat{M})\| \, \|\mathbb{B}_\ell^{-1}\|^2.$$

For each partial derivative, we have thus obtained an increase by the same amount. However, the number of second-order partial derivatives is finite: let D_2 be their number (which depends on the dimension d).

Then we find:

$$
\begin{aligned}
|v|^2_{2,K_\ell} &= \sum_{\alpha\in\mathbb{N}^d, |\alpha|=2} \int_{K_\ell} |\partial_\alpha v(M)|^2 \, d\Omega \leq D_2 \, \|\mathbb{B}_\ell^{-1}\|^4 \int_{K_\ell} \|d^2\hat{v}(F_\ell^{-1}(M))\|^2 \, d\Omega \\
&\leq D_2 \, \hat{h}^4 \, |det(\mathbb{B}_\ell)| \, (\rho_{K_\ell})^{-4} \int_{\hat{K}} \|d^2\hat{v}(\hat{M})\|^2 \, d\hat{\Omega}.
\end{aligned}
$$

To finish the proof, it suffices to note that we have an inequality of the form:

$$
\|d^2\hat{v}(\hat{M})\| \leq D_2' \max_{\alpha\in\mathbb{N}^d, |\alpha|=2} |\partial_\alpha \hat{v}(\hat{M})|
$$

for a constant $D_2' > 0$ (which depends on d), and which leads us to the bound:

$$
|v|^2_{2,K_\ell} \leq D_2 (D_2')^2 \, \hat{h}^4 \, |det(\mathbb{B}_\ell)| \, (\rho_{K_\ell})^{-4} \, |\hat{v}|^2_{2,\hat{K}}
$$

that is to say [4.8] for $q = 2$, with $C_2 = D_2^{1/2} D_2' \, \hat{h}^2$. We proceed in the same way for $q > 2$.

To establish the bounds [4.9], it suffices to reason by using the change-of-variables formula for the inverse. ■

The last result, established in exercise 9 and recalled below, allows us to compare the semi-norm of a function in $H^{r+1}(\hat{K})$, with the total norm of the same function, with a correction polynomial term of degree at most r.

LEMMA 4.3 (Deny-Lions).– *For $r \in \mathbb{N}$, there exists a $C_r > 0$ which only depends on $\hat{K}$ and on r, such that:*

$$
\inf_{\hat{p}\in P^r(\hat{K})} \|\hat{w} + \hat{p}\|_{H^{r+1}(\hat{K})} \leq C_r |\hat{w}|_{r+1,\hat{K}}, \; \forall \hat{w} \in H^{r+1}(\hat{K}). \qquad [4.10]
$$

We can now move on to:

Proof of Proposition 4.1. As the transformation F_ℓ is affine, we have that $\hat{\pi}\hat{v}(\hat{M}) = \pi_{K_\ell} v(M)$, where $\hat{\pi}$ is the interpolation operator on the reference domain $\hat{K}$. In applying [4.8] to $q = m$, we see immediately that:

$$
|v - \pi_{K_\ell} v|_{m,K_\ell} \leq C \, (\rho_{K_\ell})^{-m} \, |det(\mathbb{B}_\ell)|^{1/2} \, |\hat{v} - \hat{\pi}\hat{v}|_{m,\hat{K}}.
$$

Since $m \leq k+1$, we get that $|\hat{v} - \hat{\pi}\hat{v}|_{m,\hat{K}} \leq \|\hat{v} - \hat{\pi}\hat{v}\|_{H^{k+1}(\hat{K})}$. Then, as we are dealing with a Lagrangian finite element of order k, we have $\hat{\pi}\hat{p} = \hat{p}$ for every $\hat{p} \in P^k(\hat{K})$:

$$
\|\hat{v} - \hat{\pi}\hat{v}\|_{H^{k+1}(\hat{K})} = \inf_{\hat{p}\in P^k(\hat{K})} \|(\hat{v} - \hat{p}) - \hat{\pi}(\hat{v} - \hat{p})\|_{H^{k+1}(\hat{K})}.
$$

And furthermore:

$$\begin{aligned}&\|(\hat{v}-\hat{p})-\hat{\pi}(\hat{v}-\hat{p})\|_{H^{k+1}(\hat{K})}=\|(Id-\hat{\pi})(\hat{v}+\hat{p})\|_{H^{k+1}(\hat{K})}\\&\leq C_{k+1}\,\|\hat{v}+\hat{p}\|_{H^{k+1}(\hat{K})}\end{aligned}$$

with C_{k+1} the continuity modulus of the operator $(Id-\hat{\pi})$ from $H^{k1}(\hat{K})$ into itself. We can therefore use lemma 4.3 with $r=k$, to get (we have merged all of the previous constants into C):

$$|v-\pi_{K_\ell}v|_{m,K_\ell}\leq C\,(\rho_{K_\ell})^{-m}\,|det(\mathbb{B}_\ell)|^{1/2}|\hat{v}|_{k+1,\hat{K}}.$$

The proof concludes by applying the bound [4.9] for $q=k+1$:

$$|v-\pi_{K_\ell}v|_{m,K_\ell}\leq C\,(h_{K_\ell})^{k+1}(\rho_{K_\ell})^{-m}\,|v|_{k+1,K_\ell}.$$ ■

The minimal case for which the proposition applies corresponds to the situation involving function $v\in H^2(K_\ell)$ and which uses first-order finite elements (P^1 or Q^1-Lagrangian), which comes up frequently. In which case, the following estimates ($m=0$ or 1 and $k=1$ in [4.6]) apply:

$$\begin{aligned}&\|v-\pi_{K_\ell}v\|_{L^2(K_\ell)}\leq Ch^2_{K_\ell}|v|_{2,K_\ell}\\&|v-\pi_{K_\ell}v|_{1,K_\ell}\leq C\frac{h^2_{K_\ell}}{\rho_{K_\ell}}|v|_{2,K_\ell}.\end{aligned}\qquad[4.11]$$

In the second estimate of [4.11], note the presence of the coefficient ρ_{K_ℓ} in the denominator, which is meant to tend toward zero. To obtain an estimate to use in practice, we place ourselves in the following situation, where:

$$\exists\sigma>0,\ \forall h,\ \forall K_\ell\in\mathcal{T}_h,\ \frac{h_{K_\ell}}{\rho_{K_\ell}}\leq\sigma.\qquad[4.12]$$

In such a situation, we say that the family of meshes $(\mathcal{T}_h)_h$ is **regular**.

Then second bound then becomes:

$$|v-\pi_{K_\ell}v|_{1,K_\ell}\leq C\,\sigma\,h_{K_\ell}|v|_{2,K_\ell}.\qquad[4.13]$$

REMARK 4.3.– *The $L^2(\Omega)$ norm estimate is much better than that of the $H^1(\Omega)$ semi-norm. This comes from the fact that the derivatives of a function are poorly controlled by their values at the nodes. Furthermore, the estimate [4.6] shows that the order of the error is limited by the regularity of the function v, on the one hand, and by the order of finite elements, on the other hand. Thus, it is useless to use finite elements of an order greater than one when the sought-after solution is only H^2-regular in each polyhedron (see section 4.1.3 for more comments).*

At the end of Chapter 2, we observed that a solution u of the Laplace problem may not belong to $H^2(\Omega)$, in particular when the open subset Ω is a non-convex polyhedron or polygon. Typically, u belongs to $H^{1+s}(\Omega)$, for $s \in (1/2, 1)$. If we use definition [1.28] for the semi-norm of $H^{1+s}(\Omega)$, which involves double integrals, then by performing a change-of-variables in each one of them, we obtain the more precise result below.

PROPOSITION 4.2 (Local interpolation error).– *Let $(K_\ell, \Sigma_\ell, P_\ell)$ be a Lagrangian finite element of order k defined on a sequence of meshes satisfying [4.12], and affinely equivalent to the finite element $\left(\hat{K}, \hat{\Sigma}, \hat{P}\right)$. Let $s \in (1/2, 1)$. Then there is a constant C, only depending on $\left(\hat{K}, \hat{\Sigma}, \hat{P}\right)$ and on s, such that we have the bounds:*

$$\begin{aligned} \|v - \pi_{K_\ell} v\|_{L^2(K_\ell)} &\leq C h_{K_\ell}^{1+s} |v|_{s+1,K_\ell} \quad \forall v \in H^{s+1}(K_\ell) \\ |v - \pi_{K_\ell} v|_{1,K_\ell} &\leq C\,\sigma\, h_{K_\ell}^{s} |v|_{s+1,K_\ell} \quad \forall v \in H^{s+1}(K_\ell). \end{aligned} \qquad [4.14]$$

REMARK 4.4.– *If we go back to definition [1.28], we can easily check that the general sub-additivity property is obtained:*

$$\sum_{\ell=1,L} |u|^2_{s+1,K_\ell} \leq |u|^2_{s+1,\Omega}.$$

If $\Omega \subset \mathbb{R}^2$ and considering triangles, condition [4.12] is equivalent to the following angular condition:

$$\exists \theta_0 > 0,\ \forall h,\ \forall K_\ell \in \mathcal{T}_h,\ \theta_{K_\ell} \geq \theta_0 \quad (\theta_{K_\ell} \text{ smallest angle of } K_\ell).$$

It reflects that the triangles cannot be completely collapsed.

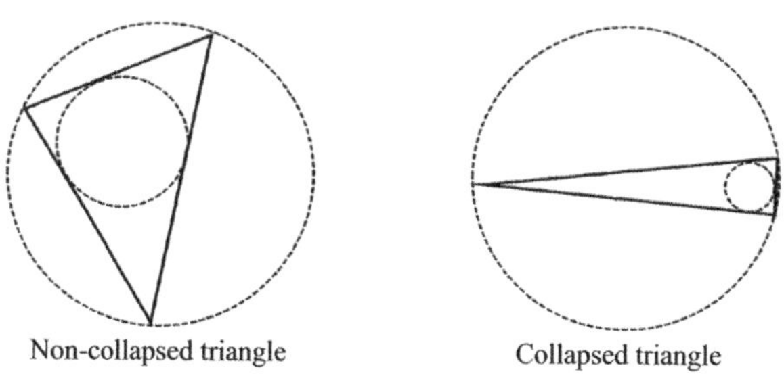

Figure 4.2. *Geometric condition for the triangle case*

When $\Omega \subset \mathbb{R}^3$, condition [4.12] is equivalent to the following condition on the solid angles at the vertex as well as a condition on the dihedral angles between two faces[1]:

$$\left| \begin{array}{l} \exists \omega_0 > 0,\ \forall h,\ \forall K_\ell \in \mathcal{T}_h,\ \omega_{K_\ell} \geq \omega_0 \quad (\omega_{K_\ell} \text{ smallest solid angle in } K_\ell) \\ \exists \iota_0 > 0,\ \forall h,\ \forall K_\ell \in \mathcal{T}_h,\ \iota_{K_\ell} \geq \iota_0 \quad (\iota_{K_\ell} \text{ smallest dihedral angle in } K_\ell). \end{array} \right.$$

4.1.2. Estimating errors in the finite element method

From the estimate of the local interpolation error [4.6], we can immediately deduce the convergence theorem for approximating Lagrangian finite elements:

THEOREM 4.1 (Convergence of finite element methods – the regular case).– *For every h, suppose that the mesh $\mathcal{T}_h$, that is made up of Lagrangian finite elements of order $k \geq 1$, satisfies the hypotheses [3.47], [3.48], [3.56], [3.57] and [3.58], and that the family of meshes $(\mathcal{T}_h)_h$ is regular (see [4.12]). If the solution u to the problem [4.1] belongs to $H^{k+1}(\Omega)$, then there exists a $C > 0$ such that, for every h:*

$$\|u - u_h\|_{H^1(\Omega)} \leq C h^k |u|_{k+1,\Omega}. \tag{4.15}$$

Proof. From [4.6] and hypothesis [4.12], we have:

$$\|u - \pi_{K_\ell} u\|_{L^2(K_\ell)} \leq C_1 h_{K_\ell}^{k+1} |u|_{k+1,K_\ell}$$

$$|u - \pi_{K_\ell} u|_{1,K_\ell} \leq C_2 \sigma h_{K_\ell}^k |u|_{k+1,K_\ell}$$

and which leads to the bound:

$$\begin{aligned} \|u - \pi_{K_\ell} u\|_{H^1(K_\ell)} &\leq \left(C_1^2 h_{K_\ell}^2 + C_2^2 \sigma^2\right)^{1/2} h_{K_\ell}^k |u|_{k+1,K_\ell} \\ &\leq C(\sigma) h^k |u|_{k+1,K_\ell} \end{aligned}$$

since, by definition, $h = \max_{K_\ell \in \mathcal{T}_h} h_{K_\ell}$ (see [3.49]). After summing over each of the polyhedra, we obtain:

$$\|u - \pi_h u\|_{H^1(\Omega)} \leq C(\sigma) h^k \Big(\sum_{\ell=1,L} |u|_{k+1,K_\ell}^2 \Big)^{1/2} = C(\sigma) h^k |u|_{k+1,\Omega}$$

which yields [4.15], by the virtue of [4.5] with $w = u$. ■

---

1 A solid angle at a vertex ranges from 0 to 4π; a dihedral angle between two faces ranges from 0 to 2π.

REMARK 4.5.– *From the proof above, we see in particular that:*

$$\lim_{h\to 0} \|w - \pi_h w\|_{H^1(\Omega)} = 0 \qquad \forall w \in \mathcal{C}^2(\overline{\Omega}).$$

From [3.9], this is an approximation property with $V = H^1(\Omega)$ *and* $W = \mathcal{C}^2(\overline{\Omega})$, *since* $\mathcal{C}^2(\overline{\Omega})$ *is dense in* $H^1(\Omega)$.

Even if $u \notin H^2(\Omega)$, one can establish whether the finite element approximation converges or not within the space $H^1(\Omega)$ by using proposition 4.2 (see also section 4.1.3). Typically, for the homogeneous Neumann problem with the data $f \in L^2(\Omega)$ and $g = 0$ (see section 2.3.1), we know that the sought-after solution will belong to $\Psi_N(\Omega)$ (see [1.43]). We now arrive at the following *minimal* result that involves the parameter σ_N, defined in section 2.5.2.

THEOREM 4.2 (Convergence of the finite element method).– *Let* $(\mathcal{T}_h)_h$ *be a family of meshes made up of first-order Lagrangian finite elements satisfying the hypotheses of theorem 4.1. Let* u *be the solution to the homogeneous Neumann problem, with the right-hand side* f *in* $L^2(\Omega)$ *and with* u_h *the solution to the discrete problem. Then, for every* $s < \sigma_N$, *or* $s = 1$ *if* $\sigma_N = 1$, *there exists a* $C_s > 0$ *such that, for all* h:

$$\|u - u_h\|_{H^1(\Omega)} \le C_s h^s \|f\|_{L^2(\Omega)}. \qquad [4.16]$$

Proof. We will use the same method as for the proof of theorem 4.1, though this time using the estimate [4.14] for both the $L^2(K_\ell)$ norm and for the $H^1(K_\ell)$ semi-norm, and also using the sub-additivity property from remark 4.4 if $s \neq 1$. Therefore, first we have:

$$\|u - u_h\|_{H^1(\Omega)} \le C_s h^s |u|_{s+1,\Omega}.$$

Then the continuous inclusion from $\Psi_N(\Omega)$ into $H^{1+s}(\Omega)$ (see section 2.5.2) allows us to write:

$$|u|_{s+1,\Omega} \le C_1 \|u\|_{H^1(\Delta;\Omega)}$$

where C_1 is a constant independent of u. Finally by definition, observe that:

$$\|u\|^2_{H^1(\Delta;\Omega)} = \|u\|^2_{H^1(\Omega)} \|\Delta u\|^2_{L^2(\Omega)} \le C_2 \|f\|^2_{L^2(\Omega)}$$

where the constant C_2 is independent of u. Indeed, on the one hand, according to the Lax–Milgram theorem, $\|u\|_{H^1(\Omega)}$ depends continuously on $\|f\|_{L^2(\Omega)}$ and thus allows us to bound the first term, whereas, on the other hand, the relation $-\Delta u = f - u$ allows us to bound the second term. ■

REMARK 4.6.– *From section 2.5.2, the parameter σ_N lies within $(\frac{1}{2}, 1]$, depending on whether the open subset Ω is convex or not. By the previous proposition, using first-order Lagrangian finite elements is sufficient to obtain the minimal order of convergence.*

Using Céa's lemma only leads to an estimate with respect to the $H^1(\Omega)$ norm, and does not allow us to get a better estimate by using the $L^2(\Omega)$ norm, as was suggested by the estimate regarding the interpolation error. However, with an additional assumption, one can estimate the error with respect to the $L^2(\Omega)$ norm for greater orders by using a duality type argument.

Let us introduce the following problem, which is called the **adjoint** problem:

$$\left| \begin{array}{l} \text{Find } \varphi \in H^1(\Omega) \text{ such that} \\ a(v, \varphi) = \displaystyle\int_\Omega g v \, d\Omega \quad \forall v \in H^1(\Omega) \end{array} \right. \qquad [4.17]$$

where $g \in L^2(\Omega)$. From the Lax–Milgram theorem, the problem [4.17] admits a unique solution.

DEFINITION 4.1.– *The problem [4.17] is said to be* ***regular*** *if the mapping $g \mapsto \varphi$ is continuous and linear from $L^2(\Omega)$ into $H^2(\Omega)$, which is to say (C is independent of g):*

$$g \in L^2(\Omega) \implies \varphi \in H^2(\Omega) \text{ and } \|\varphi\|_{H^2(\Omega)} \leq C\|g\|_{L^2(\Omega)}.$$

When the problem [4.17] is regular, we then obtain the following convergence result regarding the $L^2(\Omega)$ norm:

LEMMA 4.4 (Aubin-Nitsche).– *Under the assumptions of theorem 4.1, if the adjoint problem [4.17] is regular, then there exists a constant C, independent of h, such that:*

$$\|u - u_h\|_{L^2(\Omega)} \leq C h^{k+1} |u|_{k+1,\Omega}. \qquad [4.18]$$

Proof. Let $g \in L^2(\Omega)$, then we have:

$$\begin{aligned} \left| \int_\Omega g(u - u_h) \, d\Omega \right| &= |a(u - u_h, \varphi)| && \text{from [4.17]} \\ &= |a(u - u_h, \varphi - \psi_h) + a(u - u_h, \psi_h)| && \forall \psi_h \in V_h \\ &= |a(u - u_h, \varphi - \psi_h)| && \text{as } a(u, \psi_h) = \ell(\psi_h) = a(u_h, \psi_h) \\ &\leq M \, \|u - u_h\|_{H^1(\Omega)} \, \|\varphi - \psi_h\|_{H^1(\Omega)} && \text{by the continuity of } a(\cdot, \cdot). \end{aligned}$$

From which:

$$\left| \int_\Omega g(u - u_h)\, d\Omega \right| \leq M \, \|u - u_h\|_{H^1(\Omega)} \inf_{\psi_h \in V_h} \|\varphi - \psi_h\|_{H^1(\Omega)} .$$

Since the problem [4.17] is regular, we get that $\varphi \in H^2(\Omega)$ and $\|\varphi\|_{H^2(\Omega)} \leq C \, \|g\|_{L^2(\Omega)}$. By applying theorem 4.1, we get the bound:

$$\inf_{\psi_h \in V_h} \|\varphi - \psi_h\|_{H^1(\Omega)} \leq \|\varphi - \varphi_h\|_{H^1(\Omega)} \leq Ch\|\varphi\|_{H^2(\Omega)} \leq Ch\|g\|_{L^2(\Omega)}$$

where φ_h is the solution to the discrete problem associated with the adjoint problem [4.17]. On the other hand, if $u \in H^{k+1}(\Omega)$, then from the error estimate [4.15]:

$$\|u - u_h\|_{H^1(\Omega)} \leq Ch^k |u|_{k+1,\Omega}$$

which, after taking into account the above, yields to the bounds:

$$\left| \int_\Omega g(u - u_h)\, d\Omega \right| \leq Ch^{k+1} \|g\|_{L^2(\Omega)} |u|_{k+1,\Omega} \quad \forall g \in L^2(\Omega).$$

To conclude the proof, it suffices to pick $g = u - u_h$:

$$\begin{aligned} &\|u - u_h\|^2_{L^2(\Omega)} \leq Ch^{k+1} \|u - u_h\|_{L^2(\Omega)} |u|_{k+1,\Omega} \\ &\Longrightarrow \|u - u_h\|_{L^2(\Omega)} \leq Ch^{k+1} |u|_{k+1,\Omega}. \end{aligned}$$

■

When the solution belongs to $H^2(\Omega)$, lemma 4.4 shows that the first-order finite elements (i.e. P^1 or Q^1) converge to the first-order in the $H^1(\Omega)$ norm, and to second-order in the $L^2(\Omega)$ norm, once the adjoint problem becomes regular.

REMARK 4.7.– *For the homogeneous Neumann problem with data belonging to $L^2(\Omega)$, the solution belongs to $\Psi_N(\Omega)$. If $\Psi_N(\Omega) \subset H^2(\Omega)$, we gain a Ch factor between the estimate in the $H^1(\Omega)$ norm and that in the $L^2(\Omega)$ norm, since the solution to the adjoint problem belongs to $\Psi_N(\Omega)$. In the case where $\Psi_N(\Omega) \not\subset H^2(\Omega)$, the previous proof allows for a factor of $C_s h^s$ to be obtained, where $s < \sigma_N$.*

The analysis just carried out in the case that involves the space $V = H^1(\Omega)$ can be adapted rather effortlessly to the case that involves the space $V = H^1_0(\Omega)$ for the homogeneous Dirichlet problem.

THEOREM 4.3 (Convergence of the finite element method).– *Let $(\mathcal{T}_h)_h$ be a family of meshes composed of first-order Lagrangian finite elements satisfying the hypotheses of theorem 4.1. Let u be the solution to the homogeneous Dirichlet problem with* *right-hand side f in $L^2(\Omega)$, and let u_h be the solution to the discrete problem. Then, for all $s < \sigma_D$ and $s = 1$, if $\sigma_D = 1$, then there exists a $C_s > 0$ such that, for all h:*

$$\|u - u_h\|_{H^1(\Omega)} \leq C_s h^s \|f\|_{L^2(\Omega)}. \quad [4.19]$$

We also have an improved convergence result in the $L^2(\Omega)$ norm of the Aubin–Nitsche lemma with a $C_s h^s$ factor gain, if the adjoint problem (with homogeneous Dirichlet boundary condition) is regular. Finally, the analysis can be adapted to cases involving Dirichlet problems with non-homogeneous boundary conditions (exercise 32) as well as mixed boundary condition problem cases.

In more general cases, for example those with less regular data, or for mixed boundary condition problems, it is no longer guaranteed that the solution belongs to a type $H^{1+s}(\Omega)$ space with $s > \frac{1}{2}$. The solution u is therefore not systematically continuous on $\overline{\Omega}$, which is the condition for obtaining local error estimates using $\pi_h u$, and then concluding using theorem 4.1. Let us look at two specific cases. First of all, when we have a second term $f \in H^{-1}(\Omega)$ for the homogeneous Dirichlet problem, we only know that the solution belongs to $H^1(\Omega)$. Then, with regard to mixed Dirichlet–Neumann or mixed Dirichlet–Fourier problems, these will not necessarily be regular (Grisvard 1985, 1992) in the sense that, even when posed on a convex open subset Ω, the solution may not automatically belong to $H^2(\Omega)$; this is due to the change in the boundary conditions. The solution will however belong to a type $H^{1+s}(\Omega)$ space for $s < \sigma_M$, with $\sigma_M > 0$ (which should be compared with the results of section 2.5.2, where we always have $\sigma_D, \sigma_N > \frac{1}{2}$). In these more general cases, it will nevertheless be possible to establish the convergence of the finite element approximation, by relying on new interpolation operators that make it possible to approximate functions which are not globally continuous. Let us quote the interpolation operators of Clément or Scott-Zhang (Ciarlet 2002; Ern and Guermond 2004; Brenner and Scott 2008).

4.1.3. *"Piecewise regular" solutions*

Let us return to the problem [2.10] with a varying coefficient k, solution u, and with the data $f \in L^2(\Omega)$. To be precise, suppose that k is a constant piecewise function[2] on Ω. There is a partition $\mathcal{P} = \{\Omega_j\}_{j=1,J}$ of Ω into polyhedral subdomains:

– the Ω_j, $j = 1, J$, are "sufficiently regular" open subsets of the boundary;

– $\Omega_j \cap \Omega_{j'} = \emptyset$ if $j \neq j'$, $\overline{\Omega} = \cup_{j=1,J}\overline{\Omega_j}$;

– the restrictions of k to each subdomain Ω_j are constant and pairwise distinct.

2 The same could be done by assuming that the function is piecewise continuous on Ω.

Denote $k_j = k_{|\Omega_j}$ and $u^{(j)} = u_{|\Omega_j}$, for $j = 1, J$. Define Σ to be the interface between the subdomains, namely $\Sigma = \cup_{j \neq j'}(\partial\Omega_j \cap \partial\Omega_{j'})$ and we say that two subdomains Ω_j and $\Omega_{j'}$ are adjacent if $\partial\Omega_j \cap \partial\Omega_{j'} \neq \emptyset$ is a $\mathrm{d}-1$ dimensional manifold. We say that the $(\Sigma \cap \partial\Omega_j)_{j=1,J}$ are the components of the interface.

Since the coefficient k is discontinuous, we have that $u \notin H^2(\Omega)$. Let us explain this negative result with a simple example, that is, by considering the case when the partition $\mathcal{P}$ consists solely of two subdomains ($J = 2$), with a plane interface that is perpendicular to the axis (Ox_1). By construction, $k\nabla u \in H(\mathrm{div};\Omega)$ and, by adapting the proof of exercise 15, we find:

$$k_1\nabla u^{(1)} \cdot \mathbf{e}_1 - k_2\nabla u^{(2)} \cdot \mathbf{e}_1 = 0 \text{ on } \Sigma, \quad \text{where} \quad k_1\frac{\partial u^{(1)}}{\partial x_1} = k_2\frac{\partial u^{(2)}}{\partial x_1} \text{ on } \Sigma.$$

Moreover, if we had $u \in H^2(\Omega)$, then in particular we would have $\partial_1 u \in H^1(\Omega)$. From exercise 11 on the equality of traces, this would imply that:

$$\frac{\partial u^{(1)}}{\partial x_1} = \frac{\partial u^{(2)}}{\partial x_1} \text{ on } \Sigma.$$

However $k_1 \neq k_2$, since k is discontinuous when crossing the interface, and consequently the two equalities on Σ are *incompatible* in general, as soon as $\partial_1 u^{(1)} \neq 0$ or $\partial_1 u^{(2)} \neq 0$. Then indeed, we have that $u \notin H^2(\Omega)$.

Note that, to obtain the convergence results for the finite element method (see theorems 4.1, 4.2 or 4.3), we deduce a result *global* – on all of Ω – from the *local* results that are obtained on each polyhedron of the mesh. Thus, one could apply the same type of proof in the case of a piecewise regular solution u to the problem [2.10]. Let us specify: we associate to the partition $\mathcal{P}$ the Sobolev space, for $m \geq 2$:

$$\mathcal{P}H^m(\Omega) = \left\{v \in H^1(\Omega) \text{ such that } v^{(j)} \in H^m(\Omega_j),\ j = 1, J\right\}.$$

It is a Hilbert space, endowed with the inner product:

$$(u, v)_{\mathcal{P}H^m(\Omega)} = (u, v)_{H^1(\Omega)} + \sum_{j=1,J} (u^{(j)}, v^{(j)})_{H^m(\Omega_j)}.$$

We also define the semi-norm:

$$|v|_{m,\mathcal{P}} = \Big(\sum_{j=1,J}\ \sum_{\alpha\in\mathbb{N}^d,|\alpha|=m} \int_{\Omega_j} \left(\partial_\alpha v^{(j)}(x)\right)^2 d\Omega\Big)^{1/2} = \Big(\sum_{j=1,J} |v^{(j)}|^2_{m,\Omega_j}\Big)^{1/2}.$$

Then, to be able to apply the typical polyhedron-by-polyhedron reasoning, we choose families of meshes that *conform* to the partition $\mathcal{P}$, that is to say that they are made up of polyhedra which are never cut into two by the interface Σ. Mathematically, we write:

$$\forall h, \quad \forall K \in \mathcal{T}_h, \quad \exists ! j \quad \text{such that} \quad \overset{\circ}{K} \cap \Omega_j \neq \emptyset.$$

In this case, we have the following result.

THEOREM 4.4 (Convergence of the finite element method – "piecewise regular" solution).– *Suppose that, for all h, the mesh $\mathcal{T}_h$ conforms to the partition $\mathcal{P}$ and is made up of Lagrangian finite elements of order $k \geq 1$, and that satisfies the hypotheses of [3.47], [3.48], [3.56], [3.57] and [3.58], and that the mesh family $(\mathcal{T}_h)_h$ is regular (see [4.12]). Then, if the solution u to the problem [2.10] belongs to $\mathcal{P}H^{k+1}(\Omega)$, we have that there exists a $C > 0$ such that, for every h:*

$$\|u - u_h\|_{H^1(\Omega)} \leq C h^k |u|_{k+1,\mathcal{P}}. \quad [4.20]$$

Proof. Identical to that of theorem 4.1, except for the last bound that becomes:

$$\begin{aligned} \|u - \pi_h u\|_{H^1(\Omega)} &\leq C(\sigma) h^k \Big(\sum_{j=1,J} \sum_{K_\ell \text{ st } K_\ell \subset \overline{\Omega_j}} |u|^2_{k+1,K_\ell} \Big)^{1/2} \\ &\leq C(\sigma) h^k \Big(\sum_{j=1,J} |u^{(j)}|^2_{k+1,\Omega_j} \Big)^{1/2} \end{aligned}$$

which gives [4.20]. ■

In the *minimal* "piecewise regular" case, we have the following results. For $s \in (1, 2)$, we define:

$$\mathcal{P}H^s(\Omega) = \Big\{ v \in H^1(\Omega) \text{ such that } v^{(j)} \in H^s(\Omega_j),\ j = 1, J \Big\}.$$

For the homogeneous Neumann problem [2.10] for a given $f \in L^2(\Omega)$, by construction we have that $u \in \Psi_N(\mathcal{P})$, where:

$$\Psi_N(\mathcal{P}) = \{ v \in H^1(\Omega) \text{ such that } k\nabla u \in H(\text{div}; \Omega),\ \gamma_n(k\nabla u) = 0 \}.$$

We can prove the properties generalizing those found in section 2.5.2, namely the *continuous* injections:

– $\Psi_N(\mathcal{P}) \subset \mathcal{P}H^2(\Omega)$ in the *regular* case described below ($\sigma'_N = 1$);

– $\Psi_N(\mathcal{P}) \subset \bigcap_{s<\sigma'_N} \mathcal{P}H^{1+s}(\Omega)$ *if* $\sigma'_N \in (0, 1)$.

The classification above, as well as the value of σ'_N, depend on Ω, the partition $\mathcal{P}$, and the values $(k_j)_{j=1,J}$. In dimension one, we are always in the regular case (see remark 2.18).

As the subdomains $(\Omega_j)_{j=1,J}$ are polyhedral, each of the J components of Σ is therefore a surface equal to the union of one or more plane faces in three dimensions, a curve equal to the union of one or more segments in two dimensions. In dimensions two or three, we are in the *regular* case if and only if:

– the open subset Ω is convex;

– the interface Σ and the boundary $\partial\Omega$ are perpendicular at each of their intersection points;

– at every point of Σ, the number of adjacent subdomains equals two;

– the components of Σ are all reduced to a sole segment or face.

This result is valid regardless of the values $(k_j)_{j=1,J}$, so it depends only on the geometric configuration, denoted (G_0). If we are not in the regular case, the classification is more complex than in section 2.5.2. Let us mention three frequent cases that come up in practice:

– (G_1) the open subset Ω is convex and at every point of Σ, the number of adjacent subdomains equals two;

– (G_2) there exists a j such that $\partial\Omega \subset \partial\Omega_j$ and, at every point of Σ, the number of adjacent subdomains equals two;

– (G_3) there exists a point of Σ for which the number of adjacent subdomains is greater than or equal to three.

In the (G_1) and (G_2) cases, we always have that $\sigma'_N \in (1/2, 1)$ regardless of the values $(k_j)_{j=1,J}$. On the other hand, in the (G_3) case, σ'_N can be arbitrarily close to 0, depending on the values $(k_j)_{j=1,J}$.

To conclude, we have the following result, which applies to the homogeneous Neumann problem with varying coefficients [2.10], and which is proved in the same way as theorem 4.2.

THEOREM 4.5 (Convergence of finite element methods).– *Let $(\mathcal{T}_h)_h$ be a family of meshes made up of first-order Lagrangian finite elements that satisfy the hypotheses of theorem 4.4. Let u be the solution to the problem [2.10] with the second term f in $L^2(\Omega)$ and u_h the solution to the discrete problem. Then, for every $s < \sigma'_N$ and $s = 1$, if $\sigma'_N = 1$, there exists a $C_s > 0$ such that, for every h:*

$$\|u - u_h\|_{H^1(\Omega)} \leq C_s h^s \|f\|_{L^2(\Omega)}. \qquad [4.21]$$

The same type of reasoning can be adapted to the case of a homogeneous Dirichlet problem with a varying coefficient.

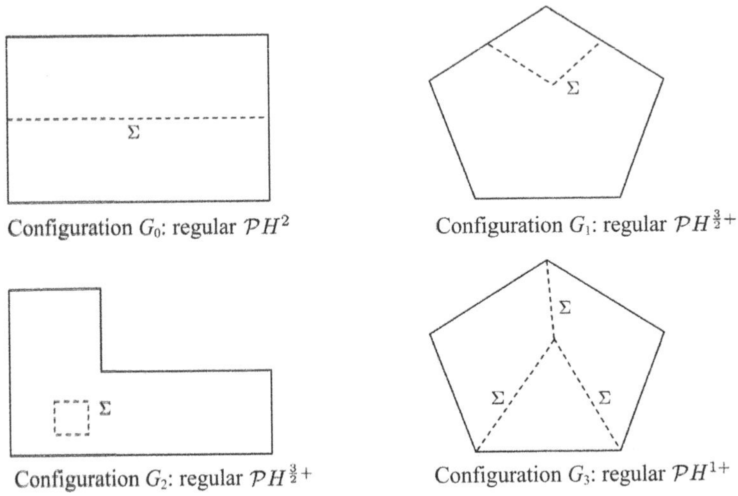

Figure 4.3. *Geometric configurations and regularity*

4.2. Error estimators and mesh refinement

Theorems 4.1–4.5 of convergence provide global error estimates. In reality, as the interpolation error estimates [4.14] suggest, the error for the solution is not uniform over the computational domain. It depends on the local mesh size, the local regularity, or the local variation of the solution. It seems clear that in an area where the solution varies rapidly, it is worthwhile to use a finer mesh than in an area where it varies just a little. Therefore, we can consider a **local refinement** of the mesh. One can also choose to use an approximation locally of a higher order k, where the solution presents large variations. We speak then of refining the order of approximation. In the following, we focus on local mesh refinement. One can implement mesh refinement techniques from a priori information, knowledge of the singularity of the solution, for example, from a posteriori information from the calculated solution, obtained by **error estimators**.

For example, if we recall section 2.5.2, we see that the solution to the Neumann or Dirichlet problems are H^2 regular, except in a neighborhood ω of the reentrant corners (and edges) of $\partial\Omega$. If we want to improve the rate of convergence, then we can make the following remark: for each triangle (or tetrahedron) K_ℓ such that $K_\ell \cap \omega = \emptyset$, we can use the standard estimate [4.13] and, for every other K_ℓ, we may use the refined estimate [4.14]. In the case of the uniform mesh shown in Figure 4.4 (i.e. with each triangle of a similar size between one another), this second estimate (in $h^s_{K_\ell}$) is a priori worse than the first one (respectively, in h_{K_ℓ}).

For the second estimate, the idea here is to construct smaller and smaller triangles or tetrahedra in the vicinity of the reentrant corners (or edges), so that h_{K_ℓ} decreases when we approach them. By decreasing the radii h_{K_ℓ} quickly enough (Apel 1999), we arrive at the result:

$$\sum_{\ell \text{ s.t. } K_\ell \cap \omega \neq \emptyset} h_{K_\ell}^{2s} |u|_{s,K_\ell}^2 \leq Ch^2 \|f\|_{L^2(\omega)}^2.$$

This makes it possible to finally recover a global rate of convergence in $O(h)$. In Figure 4.4, we give an example of a mesh of the same open subset, albeit refined in the vicinity of the reentrant corner.

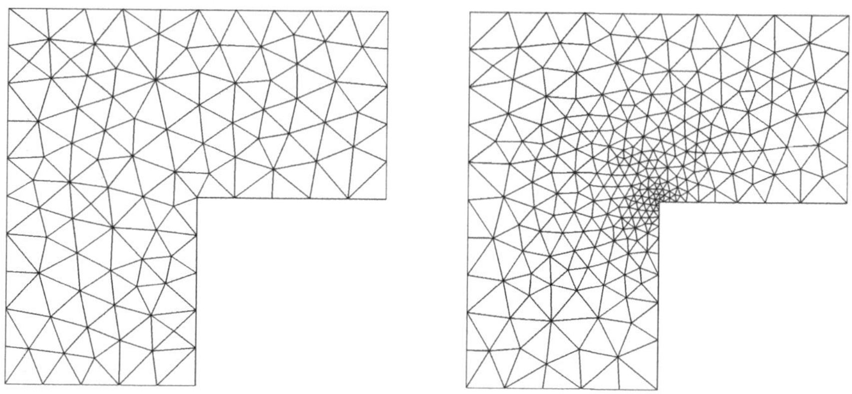

Figure 4.4. *Uniform and locally refined meshes*

In some cases when it is difficult to obtain a priori information regarding the solution, we can use error estimators that allow us to locally estimate the quality of the discrete solution obtained. If this quality is considered to be insufficient with regard to the estimator, one locally refines the mesh further and recalculates a new approximate solution on this new mesh. We iterate this process until we find an approximate solution satisfying a quality criterion that is sufficient in the sense of the estimator. A simple local refinement process consists of subdividing a triangle (respectively, a tetrahedron) into four triangles (respectively, eight tetrahedra).

There are several error estimators. The simplest and most commonly used one is that based upon the volume residual. We are going to describe it in the case of the Dirichlet problem posed on an open and bounded polygonal or polyhedral subset Ω of $\mathbb{R}^d$ (d $= 2, 3$):

$$\begin{cases} \text{Find } u \in H^1(\Omega) & \text{such that} \\ -\Delta u = f & \text{on } \Omega \\ u = 0 & \text{on } \partial\Omega. \end{cases}$$

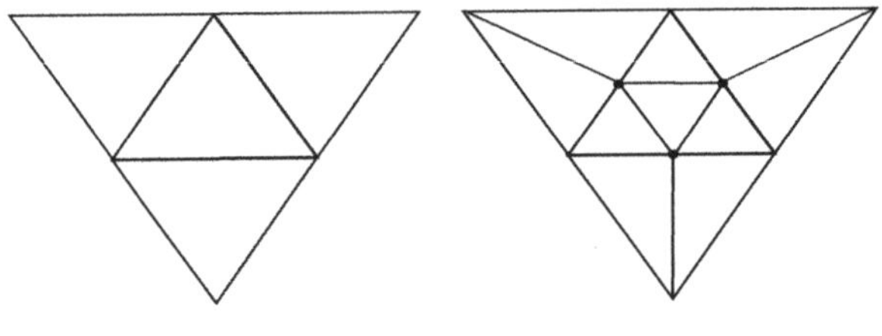

Figure 4.5. *Local refinement of a triangle*

Let u_h be the approximate solution obtained by an approximation using P^k-Lagrangian finite elements on a mesh $\mathcal{T}_h$, composed of polyhedra $(K_\ell)_{\ell=1,L}$. Denote by $\mathcal{E}_h$ the set of all inner faces of the mesh. For any inner face $e = K_\ell \cap K_m$, $\ell \neq m$, we denote by $\mathbf{n}_e$ the unit normal vector to e, oriented from K_ℓ to K_m, and by $[\cdot]_e$ the jump in value when crossing e, compatible with the orientation of $\mathbf{n}_e$ (i.e. $[v]_e = v_{|K_m} - v_{|K_\ell}$). The local residuals are defined below:

$$\left|\begin{array}{ll} R_\ell(u_h) = f + \Delta u_h \text{ on } \overset{\circ}{K_\ell} & \ell = 1, L \\ R_e(u_h) = \left[\dfrac{\partial u_h}{\partial n_e}\right]_e = [\nabla u_h \cdot \mathbf{n}_e]_e & e \in \mathcal{E}_h. \end{array}\right.$$

Note that, in the framework of the Lagrangian approximation, the basis functions are only H^1 regular, and not H^2 regular, and Δu_h is only defined (in the classical sense) on the interior of each triangle or tetrahedron. Note that, in the case of the P^1 approximation, $R_\ell(u_h) = f$ because $u_{h|K_\ell}$ is an affine function. On the other hand, the residual $R_e(u_h)$ indeed measures an approximation defect since the exact solution u satisfies $\left[\frac{\partial u}{\partial n_e}\right]_e = 0$ when the data f belongs to $L^2(\Omega)$. In practice, these residues are relatively easy to calculate. From these residuals, we can construct the local estimator:

$$E_\ell(u_h) = \left(h_{K_\ell}^{\mathrm{d}} \|R_\ell(u_h)\|^2_{L^2(K_\ell)} + \sum_{e \in \partial K_\ell \cap \mathcal{E}_h} h_e^{\mathrm{d}-1} \|R_e(u_h)\|^2_{L^2(e)}\right)^{\frac{1}{2}}$$

where h_e represents the radius of the face e.

We have the following estimates (see, for example, Verfürth (2013) for the proof), which guarantee that these estimators *asymptotically* characterize the approximation error. There exists a constant c, depending uniquely on Ω and the shape factor σ of the mesh (see [4.12]), such that:

$$\|u - u_h\|_{H^1(\Omega)} \leq c\Big(\sum_{\ell=1,L} E_\ell^2(u_h)\Big)^{\frac{1}{2}}$$

$$E_\ell(u_h) \leq c \left\| u - u_h \right\|_{H^1(V_\ell)}, \quad \ell = 1, L$$

where V_ℓ is the geometric domain made up from the union of K_ℓ and its neighbors K_m that share a face with K_ℓ.

On the one hand, these estimates show that if the local residuals tend toward 0, then the global error for the solution tends toward 0 also and, on the other hand, that the local residuals indeed tend toward 0 when the local approximation error tends toward 0. They do not prove that the local approximation error is proportional to the local estimator E_ℓ, and therefore do not guarantee that a local refinement step will improve the local precision of the solution. On the other hand, it is true asymptotically, that is, if one refines the mesh more and more. Let us point out that, for second-order elliptic problems that are more general than the Laplacian, for example those with varying coefficients, the constant c which plays a role in these estimates is related to the continuity and coercivity constants for the problem. In particular, in the case of a weakly coercive problem (where the constant α is small in [2.29]), this constant is large and the estimator quality may be poor. Nevertheless, in practice, this simple estimator can be put to work and provides satisfactory results.

There are other estimators, with some of them leading to estimates where the constants are independent of the coefficients (see Verfürth (2013) for more details). Let us cite three of them as examples based on:

– the solution to local problems and using a finite element approximation of an order greater than that used to solve the initial problem;

– the calculation of an approximate solution on a finer mesh;

– estimators that use conformal finite elements in $H(\mathrm{div};\Omega)$ on a dual mesh.

4.3. Non-polyhedral domains and approximations of the data

In the error analysis that we have just presented, situations that are more complicated but which are often encountered in practice were excluded, in order to simplify things: the *non-polyhedral open subset* (i.e. curvilinear polyhedra) case and the *data approximation* case. We will discuss them here.

4.3.1. *Non-polyhedral open subsets*

So far, we have assumed that the open subset Ω is polyhedral and that $\overline{\Omega} = \bigcup_{\ell=1,L} K_\ell$. However, in practice, meshed objects are rarely polyhedral. When meshing such objects using flat-faced finite elements leads to an open subset Ω_h that is different from Ω:

$$\overline{\Omega_h} = \bigcup_{\ell=1,L} K_\ell \neq \overline{\Omega}$$

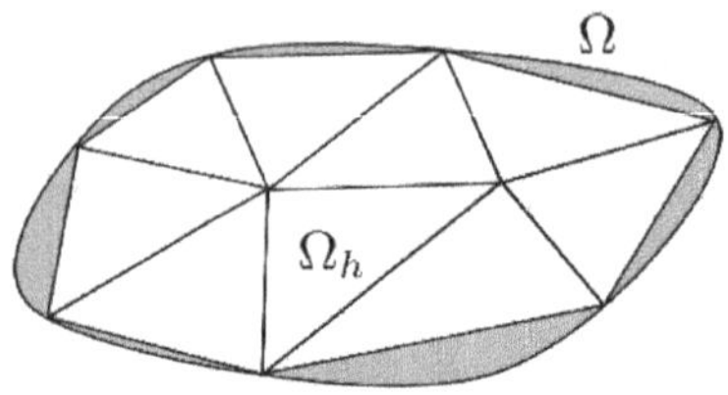

This gap[3] between the open subsets Ω and Ω_h gives rise to approximations of the bilinear form $a(\cdot,\cdot)$ and the linear form $\ell(\cdot)$. Moreover, if we call $\widetilde{V_h}$ the finite-dimensional space based on the mesh, we will generally have $\widetilde{V_h} \not\subset V$. We speak of an **external**, or **nonconforming**, **approximation** in this case (see remark 3.2). The discretized problem then takes the following form:

$$\left|\begin{array}{l}\text{Find } \widetilde{u_h} \in \widetilde{V_h} \text{ such that} \\ \widetilde{a_h}\left(\widetilde{u_h}, v_h\right) = \widetilde{\ell_h}\left(v_h\right) \quad \forall v_h \in \widetilde{V_h}.\end{array}\right. \qquad [4.22]$$

The question is whether the error estimate [4.15] remains true with $\widetilde{u_h}$ replacing u_h. In order to measure both the elements of V and those of $\widetilde{V_h}$, we must start by introducing a norm on $V + \widetilde{V_h}$, denoted $\|\cdot\|_h$, chosen such that $\|v\|_h = \|v\|_V$ for all $v \in V$. Similarly, the (bi)linear forms a and ℓ must be extended to $V + \widetilde{V_h}$: we therefore consider $\widetilde{a_h}$ and $\widetilde{\ell_h}$ defined on $V + \widetilde{V_h}$ and such that:

$$\widetilde{a_h}(v,w) = a(v,w) \text{ and } \widetilde{\ell_h}(v) = \ell(v) \qquad \forall v, w \in V.$$

DEFINITION 4.2.– *The bilinear forms $(\widetilde{a_h})_h$ are* uniformly $(V + \widetilde{V_h})$-continuous *if there exists a constant $\widetilde{M}$, independent from $(\widetilde{V_h})_h$ and from $(\widetilde{a_h})_h$, such that:*

$$|\widetilde{a_h}(v,w)| \leq \widetilde{M}\, \|v\|_h \, \|w\|_h \qquad \forall h, \quad \forall v, w \in (V + \widetilde{V_h}).$$

The bilinear forms $(\widetilde{a_h})_h$ are uniformly $\widetilde{V_h}$-coercive *if there exists a constant $\widetilde{\alpha} > 0$, independent of $(\widetilde{V_h})_h$ and $(\widetilde{a_h})_h$, such that:*

$$|\widetilde{a_h}(v_h, v_h)| \geq \widetilde{\alpha}\, \|v_h\|_h^2 \qquad \forall h, \quad \forall v_h \in \widetilde{V_h}.$$

In the non-conforming case, the result is similar to that of proposition 3.1.

PROPOSITION 4.3.– *If the forms $(\widetilde{a_h})_h$ are uniformly $\widetilde{V_h}$-coercive, then the approximate variational problem [4.22] admits a unique solution $\widetilde{u_h} \in \widetilde{V_h}$.*

3 If the open subset Ω is convex, then $\Omega_h \subset \Omega$. If, on the contrary, Ω is not convex, then $\Omega_h \not\subset \Omega$.

For every h, we introduce a **consistency error**, which lets us measure the differences between the extended forms:

$$\widetilde{Cons_h}(\ell) = \sup_{w_h \in \widetilde{V_h} \setminus \{0\},\, \|w_h\|_h = 1} |\widetilde{a_h}(u, w_h) - \widetilde{\ell}_h(w_h)|.$$

As u is the solution of the exact problem, recall that we may write it in the form $u = A^{-1}L$, where L is the unique element of V such that $(L, v)_V = \ell(v)$ for all $v \in V$ (see section 2.4). Therefore, the formula defining $\widetilde{Cons_h}(\ell)$ depends only on $\ell(\cdot)$ (and the approximation technique).

LEMMA 4.5 (Strang).– *If the bilinear forms $(\widetilde{a_h})_h$ are uniformly $(V + \widetilde{V_h})$-continuous and uniformly $\widetilde{V_h}$-coercive, then there exists a constant $C > 0$, independent of h and $\ell(\cdot)$, such that:*

$$\|u - \widetilde{u_h}\|_h \leq C \inf_{v_h \in \widetilde{V_h}} \left(\|u - v_h\|_h + \widetilde{Cons_h}(\ell) \right). \qquad [4.23]$$

Proof. Let $v_h \in \widetilde{V_h}$. From the uniform coercivity, there exists a $\widetilde{\alpha} > 0$ independent of h and v_h such that:

$$\begin{aligned} \widetilde{\alpha}\|\widetilde{u_h} - v_h\|_h^2 &\leq |\widetilde{a_h}(\widetilde{u_h} - v_h, \widetilde{u_h} - v_h)|, \quad \text{with} \\ \widetilde{a_h}(\widetilde{u_h} - v_h, \widetilde{u_h} - v_h) &= \widetilde{a_h}(\widetilde{u_h} - u, \widetilde{u_h} - v_h) + \widetilde{a_h}(u - v_h, \widetilde{u_h} - v_h) \\ &= \widetilde{\ell}_h(\widetilde{u_h} - v_h) - \widetilde{a_h}(u, \widetilde{u_h} - v_h) + \widetilde{a_h}(u - v_h, \widetilde{u_h} - v_h). \end{aligned}$$

From this, we deduce that:

$$\|\widetilde{u_h} - v_h\|_h \leq \frac{1}{\widetilde{\alpha}}(\widetilde{Cons_h}(\ell) + \widetilde{M}\|u - v_h\|_h)$$

with $\widetilde{M}$ independent of h by uniform continuity. With the help of the triangle inequality $\|u - \widetilde{u_h}\|_h \leq \|u - v_h\|_h + \|\widetilde{u_h} - v_h\|_h$, finally we obtain [4.23]. ■

REMARK 4.8.– *For [4.23] to be satisfied, it is sufficient for the uniform continuity property to be true on $(V + \widetilde{V_h}) \times \widetilde{V_h}$.*

Thus, in the case of an external, or non-conforming, approximation, the error $\|u - \widetilde{u_h}\|_h$ tends to 0 if, on the one hand, the approximability property [3.9] with respect to the norm $\|\cdot\|_h$ is satisfied and if, on the other hand, the consistency error tends toward 0:

$$\lim_{h \to 0} \left(\sup_{\ell \in V',\, \|\ell\|_{V'} = 1} \widetilde{Cons_h}(\ell) \right) = 0.$$

For example, if we solve the Dirichlet problem, then $V = H_0^1(\Omega)$ and we have the inclusion $\widetilde{V_h} \subset H_0^1(\Omega_h)$. To define the norm $\|\cdot\|_h$ and the extended forms $\widetilde{a_h}$ and $\widetilde{\ell_h}$, one can choose to extend $v \in V$ defined on Ω by 0 to $\Omega \cup \Omega_h$; the same can be done for $v_h \in \widetilde{V_h}$ defined on Ω_h, which is extended by 0 to $\Omega \cup \Omega_h$. If we label $\widetilde{v}$ and $\widetilde{v_h}$ to be the extensions, then we have $\widetilde{v}, \widetilde{v_h} \in H_0^1(\Omega \cup \Omega_h)$ (see exercise 11). Thus, $V + \widetilde{V_h} \subset H_0^1(\Omega \cup \Omega_h)$. By the previous formalism, we introduce the norm $\|\cdot\|_h = \|\cdot\|_{H^1(\Omega\cup\Omega_h)}$, as well as the bilinear form:

$$\widetilde{a_h}(u, v) = \int_{\Omega\cup\Omega_h} \nabla\widetilde{u} \cdot \nabla\widetilde{v}\, d\Omega.$$

We then do the same to extend the linear form ℓ.

As an application of the theory stated above, one will find in Lenoir (1987) the proof that the error coming from the approximation by plane faces is of the order $\frac{3}{2}$ in the approximation of the homogeneous Dirichlet problem case. This indicates that using[4] Lagrangian finite elements of orders greater than or equal to 2 is not optimal, since any better rate of convergence than $\frac{3}{2}$ is "lost".

In order to "recover" the expected orders for finite elements of order k, one must use **curved finite elements**, combined with a technique called **isoparametrization**.

As one can imagine, a curved finite element is an element whose faces are not planar, and clearly can approximate a non-polyhedral open subset better than an element with planar faces. A standard technique used to construct curved finite elements is to use a transform from a reference element, whose boundary is composed of planar faces, using a map F_ℓ which is no longer affine. Let $\left(\hat{K}, \hat{\Sigma}, \hat{P}\right)$ be a reference finite element of order k, of basis functions $(\hat{\tau}_i)_{i=1,n_k}$, and denote $(K_\ell, \Sigma_\ell, P_\ell)$ its image by the transformation F_ℓ.

Among all the possible transformations, those constructed from polynomials are used: the order of the transformation is then equal to the degree of the chosen polynomial. When the geometric transformation F_ℓ is:

– of the same order as the reference finite element, $(K_\ell, \Sigma_\ell, P_\ell)$ is called **isoparametric**;

– of a lower order, $(K_\ell, \Sigma_\ell, P_\ell)$ is called **subparametric**;

– of a greater order, $(K_\ell, \Sigma_\ell, P_\ell)$ is called **superparametric**.

This last case does not present any interest (except within the framework of discretizing finite elements of integral equations).

4 For example, from theorem 4.1, the choice of P^2-Lagrangian finite element allows for an order of 2 when $u \in H^3(\Omega)$.

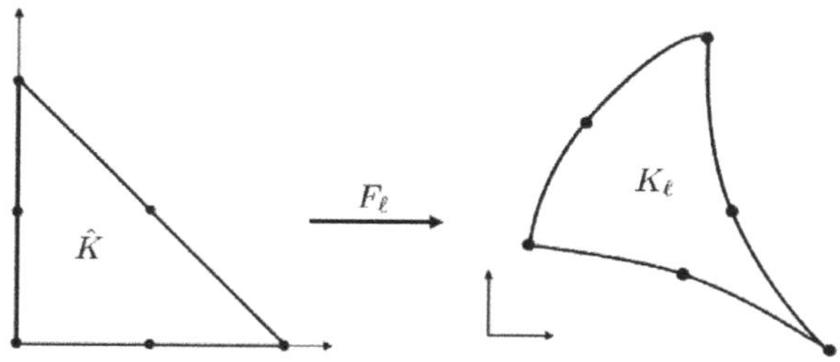

Figure 4.6. *Geometric transformation*

4.3.2. *Isoparametric finite elements*

We present an example here, the isoparimetric P^2-Lagrangian finite element in d = 2. We denote $(\hat{M}_i)_{i=1,3}$ the vertices of the reference triangle $\hat{K}$, and $(\hat{M}_{ij})_{i,j=1,3,\ i\neq j}$ the midpoints of $[\hat{M}_i, \hat{M}_j]$. To return to mono-index notation, one sets, for example $\hat{M}_4 = \hat{M}_{12}$, $\hat{M}_5 = \hat{M}_{23}$ and $\hat{M}_6 = \hat{M}_{31}$, as well as $\hat{\Sigma} = \{\hat{M}_1, \cdots, \hat{M}_6\}$. In the same fashion, $(M_i^\ell)_{i=1,3}$ represents the vertices of K_ℓ, and $(M_{ij}^\ell)_{i,j=1,3,\ i\neq j}$ are the "midpoints" of $[M_i^\ell, M_j^\ell]$.

Thus, we aim to construct an invertible mapping $F_\ell : \mathbb{R}^2 \to \mathbb{R}^2$, of degree 2, associating each vertex of $\hat{K}$ to a vertex of K_ℓ, and to each midpoint of the edge $\hat{M}_{ij}$ the "midpoint" of the corresponding M_{ij}^ℓ:

$$F_\ell(\hat{M}_i) = M_i^\ell,\ i = 1,3, \quad F_\ell(\hat{M}_{ij}) = M_{ij}^\ell,\ i,j = 1,3,\ i \neq j.$$

Therefore, we set $\Sigma_\ell = F_\ell(\hat{\Sigma})$ and $K_\ell = F_\ell(\hat{K})$. Resume using the mono-index notation. As the family $(\hat{\tau}_i)_{i=1,6}$ generates $\hat{P} = P^2(\hat{K})$, we have the relation:

$$F_\ell(\hat{x}, \hat{y}) = \sum_{i=1}^{6} \hat{\tau}_i(\hat{x}, \hat{y})\, M_i^\ell, \qquad \forall (\hat{x}, \hat{y}) \in \hat{K}$$

which lets us construct the finite element $(K_\ell, \Sigma_\ell, P_\ell)$ from $(\hat{K}, \hat{\Sigma}, \hat{P})$, as was done for the Q^k-finite elements (see section 3.3.2).

The advantage here lies in the fact that a parabola, which can be described by a transformation of degree 2, fits a curved boundary better than a straight line. More generally, the use of **isoparametric finite elements of order** k makes it possible to approximate the open subset Ω by an open subset Ω_h up to order k. Using these isoparametric finite elements of order k provides a finite element approximation, which *also converges to order* k (Lenoir 1987).

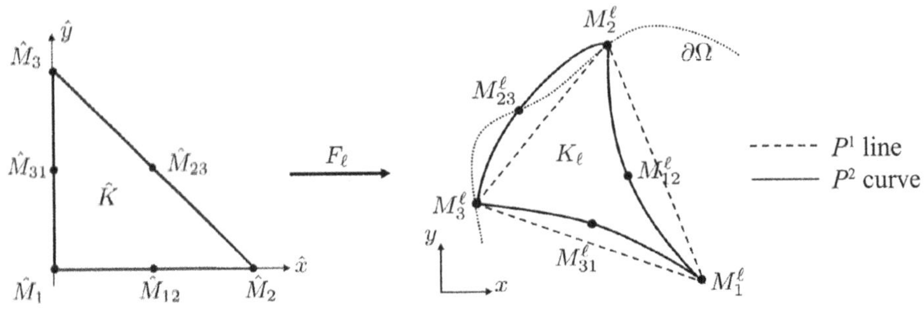

Figure 4.7. *Isoparametric P^2-transformation*

4.3.3. *Approximation of the data*

When solving a linear system, two situations can occur when one must calculate the integrals used to construct the matrix and the right-hand side. Either we know how to calculate these integrals exactly or if this is not the case then we must use approximate integral formulas.

If we can express the quantity we wish to integrate as a product of barycentric coordinates and their integer powers, then we can use a general formula (see exercise 31). We place ourselves in a d-simplex S_d in $\mathbb{R}^\mathrm{d}$, that is to say, in the convex hull of $\mathrm{d}+1$ points not belonging to the same hyperplane: in practice, it is a line interval for $\mathrm{d} = 1$, a triangle for $\mathrm{d} = 2$, and a tetrahedron for $\mathrm{d} = 3$. Let $(\lambda_m)_{1\leq m\leq \mathrm{d}+1}$ be the barycentric coordinates associated with these $\mathrm{d}+1$ points. Then, for each positive or zero integers $k_1, \cdots, k_{\mathrm{d}+1}$, we have the formula:

$$\int_{S_\mathrm{d}} \left(\prod_{m=1}^{m=\mathrm{d}+1} \lambda_m^{k_m} \right) dx_1 \cdots dx_\mathrm{d} = \frac{\mathrm{d}! \prod_{m=1}^{m=\mathrm{d}+1} k_m!}{(\mathrm{d} + \sum_{m=1}^{m=\mathrm{d}+1} k_m)!} \operatorname{vol}(S_\mathrm{d}). \qquad [4.24]$$

For example, if one uses P^1-Lagrangian finite elements for an equation with constant coefficients of the form $-u + \Delta u = f$, the elementary stiffness and mass matrices can then be written in this form (see [3.39] and [3.41]).

In general, we do not know how to calculate the right-hand side $\ell(v_h)$ exactly, which is involved in the discrete problem [4.2]. For example, since

$$\ell(v_h) = \int_\Omega f\, v_h\, d\Omega \quad \text{(with } f \text{ being given)}$$

one has to calculate the elementary integrals on each K_ℓ:

$$I_i^\ell(f) = \int_{K_\ell} f \tau_i^\ell\, d\Omega.$$

When the function f is sufficiently regular, for example if it is piecewise continuous[5], a classical technique consists of approximating these integrals with the help of quadrature formulas, say with nq_ℓ points:

$$I_i^\ell(f) \simeq \sum_{q=1}^{nq_\ell} \omega_q^\ell f\left(x_q^\ell\right) \tau_i^\ell\left(x_q^\ell\right)$$

where $\left(x_q^\ell\right)_{q=1,nq_\ell}$ are the quadrature points in K_ℓ and $\left(\omega_q^\ell\right)_{q=1,nq_\ell}$ are the positive weights associated with each quadrature point.

There are many types of quadratures (Gauss-Lobatto, etc.) (see, for example, Solin et al. 2004; Cohen and Perret 2017), which provide approximations for the integrals:

$$\int_{K_\ell} g\, d\Omega.$$

These quadrature methods differ in that they integrate exactly the polynomials of a given degree k_{quad}, which characterizes their accuracy.

In two dimensions, we can use, for example, the seven-point quadrature formula (of Gauss–Lobatto) on the reference triangle $\hat{T}$, which integrates degree five polynomials exactly (Table 4.1) with the coordinates $s_0 = \frac{1}{3}$, $s_1 = \frac{6-\sqrt{15}}{21}$, $s_2 = \frac{6+\sqrt{15}}{21}$, $s_3 = \frac{9+2\sqrt{15}}{21}$, $s_4 = \frac{9-2\sqrt{15}}{21}$, and the weights $\eta_0 = \frac{9}{80}$, $\eta_1 = \frac{155-\sqrt{15}}{2400}$, $\eta_2 = \frac{155+\sqrt{15}}{2400}$.

$\hat{\boldsymbol{x}}_q$	(s_0, s_0)	(s_1, s_1)	(s_1, s_3)	(s_3, s_1)	(s_2, s_2)	(s_2, s_4)	(s_4, s_2)
$\hat{\omega}_q$	η_0	η_1	η_1	η_1	η_2	η_2	η_2

Table 4.1. *Quadrature of seven points in two dimensions: points and weights*

In one dimension (calculations of boundary terms in dimension two), we will use, for example, the three-point quadrature formula (of Gauss–Legendre) on the line interval $[-1, 1]$, which integrates degree three polynomials exactly (Table 4.2).

The same question is asked when one wants to evaluate:

$$a(u_h, v_h) = \int_\Omega k(x) \nabla u_h \cdot \nabla v_h \, d\Omega$$

5 Beware of any singular or hyperoscillating functions, for which special numerical quadrature techniques should be used.

when the coefficient $k(x)$ varies and is piecewise continuous. In this case, quadrature formulas will also be needed.

$\hat{\boldsymbol{x}}_q$	$-\sqrt{3/5}$	0	$\sqrt{3/5}$
$\hat{\boldsymbol{\omega}}_q$	$5/9$	$8/9$	$5/9$

Table 4.2. *Quadrature of three points in one dimension: points and weights*

Using such methods leads to discretized problems of the form:

$$\left|\begin{array}{l}\text{Find } u_h' \in V_h \text{ such that}\\ a_h(u_h', v_h) = \ell_h(v_h) \quad \forall v_h \in V_h.\end{array}\right. \qquad [4.25]$$

This time, we remain within the internal approximation framework, but an additional error is introduced for the approximate evaluation of the forms. First, we adapt the notion of uniform coercivity.

DEFINITION 4.3.– *The bilinear forms $(a_h)_h$ are* uniformly V_h-coercive *if there is a constant $\alpha^\star > 0$, independent of $(V_h)_h$ and of $(a_h)_h$, such as:*

$$|a_h(v_h, v_h)| \geq \alpha^\star \, \|v_h\|_V^2 \qquad \forall h, \quad \forall v_h \in V_h.$$

PROPOSITION 4.4.– *If the forms $(a_h)_h$ are uniformly V_h-coercive, then the approximate variational problem [4.25] admits a unique solution $u_h' \in V_h$.*

For all h and $v_h \in V_h$, we introduce two **consistency errors** this time, allowing us to measure the differences between exact and approximate forms:

$$Cons_h(\ell) = \sup_{w_h \in V_h, \ \|w_h\|_V = 1} |\ell(w_h) - \ell_h(w_h)| \qquad [4.26]$$

$$Cons_{a,h}(v_h) = \sup_{w_h \in V_h, \ \|w_h\|_V = 1} |a(v_h, w_h) - a_h(v_h, w_h)|. \qquad [4.27]$$

LEMMA 4.6 (Strang).– *If the bilinear forms $(a_h)_h$ are uniformly coercive, then there exists a constant $C > 0$, independent of h and $\ell(\cdot)$, such that:*

$$\|u - u_h'\|_V \leq C \inf_{v_h \in V_h} \left(\|u - v_h\|_V + Cons_h(\ell) + Cons_{a,h}(v_h)\right). \qquad [4.28]$$

Proof. Let $v_h \in V_h$, $v_h \neq u'_h$; we set $w_h = u'_h - v_h$. By uniform coercivity, there exists an $\alpha^\star > 0$, independent of h and v_h, such that:

$$\begin{aligned} \alpha^\star \|u'_h - v_h\|_V \, \|w_h\|_V &\leq |a_h(u'_h - v_h, w_h)|, \text{ with} \\ a_h(u'_h - v_h, w_h) &= \ell_h(w_h) - a_h(v_h, w_h) \\ &= (\ell_h - \ell)(w_h) + \ell(w_h) - a_h(v_h, w_h) \\ &= (\ell_h - \ell)(w_h) + a(u, w_h) - a_h(v_h, w_h) \\ &= (\ell_h - \ell)(w_h) + a(u - v_h, w_h) + (a - a_h)(v_h, w_h). \end{aligned}$$

Therefore, we have:

$$\|u'_h - v_h\|_V \leq \frac{1}{\alpha^\star}(Cons_h(\ell) + M \, \|u - v_h\|_V + Cons_{a,h}(v_h))$$

with M greater than or equal to the continuity modulus of a. From this, [4.28] follows because of the triangle inequality $\|u - u'_h\|_V \leq \|u - v_h\|_V + \|u'_h - v_h\|_V$. ■

In the case of an internal approximation representing approximate forms, the error $\|u - u'_h\|_V$ tends to 0 if, on the one hand, the approximation property [3.9] is satisfied, and if, on the other, the consistency errors tend to 0:

$$\lim_{h \to 0} \sup_{\ell \in V', \ \|\ell\|_{V'} = 1} Cons_h(\ell) = 0 \text{ and } \lim_{h \to 0} \sup_{v_h \in V_h, \ \|v_h\|_V = 1} Cons_{a,h}(v_h) = 0.$$

In Raviart and Thomas (1983), one will find a proof of the following result:

PROPOSITION 4.5.– *Suppose that the quadrature formula used integrates the degree* k_{quad} *polynomials exactly, where* $k_{quad} \geq 2k - 2$*, and where* k *is the order of the finite elements that are associated with the mesh* $\mathcal{T}_h$ *for the open subset* Ω*. Then, if* $u \in C^{k+2}\left(\overline{\Omega}\right)$*:*

$$\|u - u'_h\|_{H^1(\Omega)} \leq C h^k \|u\|_{C^{k+2}\left(\overline{\Omega}\right)}.$$

More general results can be found in Lenoir (1987).

Lastly, one should not overestimate the precision of integration because then the calculation becomes unnecessarily expensive. When one uses first-order finite elements, then one observes that it is enough to use a quadrature formula that integrates the constants exactly (that is to say, $k_{quad} = 0$).

4.4. Exercises – statements

Exercise 31. Integrating barycentric coordinates

We want to prove the two-dimensional integration formula for barycentric coordinates [4.24]. For each triangle T_ℓ, and in denoting $\lambda_{\ell_1}, \lambda_{\ell_2}, \lambda_{\ell_3}$, the three barycentric coordinates associated with the three vertices, the formula can be written as:

$$\int_{T_\ell} \lambda_{\ell_1}^{k_1} \lambda_{\ell_2}^{k_2} \lambda_{\ell_3}^{k_3}\, dx_1 dx_2 = \frac{2\, k_1!\, k_2!\, k_3!}{(2 + k_1 + k_2 + k_3)!} \,\text{Area}(T_\ell) \;\; \forall (k_1, k_2, k_3) \in \mathbb{N}^3. \qquad \text{[E31]}$$

a) Pass to the reference triangle $\hat{T}$ by changing the variables.

b) Establish formula [E31] by successively integrating by parts, and by noting that:

$$\hat{T} = \{(\hat{x}_1, \hat{x}_2) \text{ such that } \hat{x}_1 \in [0, 1],\ \hat{x}_2 \in [0, 1 - \hat{x}_1]\}.$$

c) How can formula [E31] be generalized to dimensions $\mathrm{d} \geq 3$?

Exercise 32. Electrostatics: potentials and capacitance

Let Ω be a connected and bounded open subset of $\mathbb{R}^3$, with a "sufficiently regular" boundary, occupied by a dielectric material with a constant permittivity ε_0, whose boundary $\partial\Omega$ can be split into $K + 1$ disjoint connected components: $\partial\Omega = \cup_{k=0,\cdots,K} \Gamma_k$. Consider the boundary value problem:

$$\begin{cases} \text{Find } u \in H^1(\Omega) & \text{such that} \\ -\Delta u = f & \text{in } \Omega \\ u = V_k & \text{on } \Gamma_k,\ 0 \leq k \leq K. \end{cases} \qquad \text{[P32]}$$

The given data consists of $f = \rho/\varepsilon_0 \in L^2(\Omega)$ where ρ is the charge density, in addition to the potentials $(V_k)_{0 \leq k \leq K} \in \{0\} \times \mathbb{R}^K$.

a) Explain why we can decompose u into $u = u_0 + u_R$, with $u_0 \in H_0^1(\Omega)$ and u_R depending only on $(V_k)_{1 \leq k \leq K}$. State the variational formulation [F32] associated to [P32]. Recall why [F32] is well-posed and why [F32] is equivalent to [P32].

b) *Conductors in electrostatic equilibrium.* Now consider the case when there is an absence of any charge, that is, $f = 0$.

- Consider the space:

$$W = \left\{ u \text{ is a solution to [P32] when } f = 0 \text{and } (V_k)_{1 \leq k \leq K} \in \mathbb{R}^K \right\}.$$

What is the dimension of the vector space W? Find a basis for it, which will be denoted by $(\chi_k)_{1 \leq k \leq K}$ in the following questions.

- We say that a capacitance matrix is the matrix determined by the elements:

$$\mathbb{C}_{kl} = \int_\Omega \nabla\chi_l \cdot \nabla\chi_k \, d\Omega \quad 1 \le k, l \le K.$$

What are the properties of the matrix $\mathbb{C}$?

- Let u be a solution to [P32]. Express:

$$Q_k = \int_{\Gamma_k} \frac{\partial u}{\partial n} \, d\Gamma, \; 0 \le k \le K$$

as a function of the $(V_k)_{1 \le k \le K}$ and $\mathbb{C}$.

c) *Discretization*. Suppose that the boundary $\partial\Omega$ is polyhedral.

- Discretize the equivalent variational formulation [F32] using P^1 finite elements: label u_{0h} the approximate solution. Write the equivalent linear system in the form of $\mathbb{A}\vec{U} = \vec{L} + \vec{G}$.

- Estimate the error $\|u_0 - u_{0h}\|_{H^1(\Omega)}$.

- Denote by $(\chi_{k,h})_{1 \le k \le K}$ the approximations to the functions $(\chi_k)_{1 \le k \le K}$, and denote by $\mathbb{C}_h$ the approximate capacitance matrix. Estimate the error of the capacitance matrix, namely $\|\mathbb{C} - \mathbb{C}_h\|$.

Exercise 33. Bending of a beam

On the open interval (a, b) $(a < b)$, consider the following variational problem:

$$\left| \begin{array}{l} \text{Find } u \in H_0^2(a,b) \text{ such that} \\ \displaystyle\int_a^b u''v'' \, dx = \int_a^b fv \, dx \qquad \forall v \in H_0^2(]a,b[) \end{array} \right. \qquad \text{[P33]}$$

with the data $f \in L^2(]a, b[)$.

a) Why can one not choose Lagrangian finite elements to approximate the problem [P33]?

b) One may propose another type of finite element: Hermitian elements defined in section 3.4. Denote $\hat{K} = [0,1]$, $\hat{\Sigma} = \{\delta_0, \delta_1, \delta_0', \delta_1'\}$ and $\hat{P} = P^3([0,1])$. Show that, for every $(\alpha_0, \alpha_1, \widetilde{\alpha}_0, \widetilde{\alpha}_1)$, there exists a unique polynomial $p \in \hat{P}$, such that:

$$\delta_0(p) = \alpha_0, \; \delta_1(p) = \alpha_1, \; \delta_0'(p) = \widetilde{\alpha}_0, \; \delta_1'(p) = \widetilde{\alpha}_1.$$

Determine the local basis functions for $(\hat{K}, \hat{\Sigma}, \hat{P})$, that is to say, $\tau_1, \tau_2, \widetilde{\tau}_1, \widetilde{\tau}_2$ such that:

$$\begin{aligned}
\delta_0(\tau_1) &= 1, \ \delta_1(\tau_1) = \delta'_0(\tau_1) = \delta'_1(\tau_1) = 0 \\
\delta_1(\tau_2) &= 1, \ \delta_0(\tau_2) = \delta'_0(\tau_2) = \delta'_1(\tau_2) = 0 \\
\delta'_0(\widetilde{\tau}_1) &= 1, \ \delta_0(\widetilde{\tau}_1) = \delta_1(\widetilde{\tau}_1) = \delta'_1(\widetilde{\tau}_1) = 0 \\
\delta'_1(\widetilde{\tau}_2) &= 1, \ \delta_0(\widetilde{\tau}_2) = \delta_1(\widetilde{\tau}_2) = \delta'_0(\widetilde{\tau}_2) = 0.
\end{aligned}$$

c) Decompose the line segment $[a, b]$ into L intervals $K_\ell = [x_\ell, x_{\ell+1}]$, $\ell = 0, L-1$, on which the family of Hermitian finite elements $(K_\ell, \Sigma_\ell, P_\ell)$ is based. Denote by $(\tau_1^\ell, \tau_2^\ell, \widetilde{\tau}_1^\ell, \widetilde{\tau}_2^\ell)$ the local basis functions on the element $(K_\ell, \Sigma_\ell, P_\ell)$ and define the global basis functions $(w_i, \widetilde{w}_i)_{i=1,L-1}$ by the following formulae:

$$w_i = \begin{cases} \tau_2^{i-1} & \text{on } K_{i-1} \\ \tau_1^i & \text{on } K_i \\ 0 & \text{otherwise} \end{cases} \quad \text{and } \widetilde{w}_i = \begin{cases} \widetilde{\tau}_2^{i-1} & \text{on } K_{i-1} \\ \widetilde{\tau}_1^i & \text{on } K_i \\ 0 & \text{otherwise.} \end{cases}$$

Introduce the approximation space:

$$V_h = \text{vect}(w_i, \widetilde{w}_i)_{i=1,L-1}.$$

Show that the family $(w_i, \widetilde{w}_i, \ i = 1, L-1)$ is linearly independent, then show that $V_h \subset H_0^2(a, b)$.

d) Interpret the coefficients in the decomposition of an element $v_h \in V_h$.

e) Write the discretization for the problem [P33] in the space V_h and describe the equivalent linear system to the approximate variational problem.

f) What is the matrix structure for this linear system?

Exercise 34. Helmholtz problem

Consider the Helmholtz equation ($\omega > 0$) with a homogeneous Dirichlet boundary condition posed in a bounded open subset Ω of $\mathbb{R}^\mathrm{d}$ with a Lipschitz boundary:

$$\begin{cases} \text{Find } u \in H_0^1(\Omega) \text{ such that} \\ -\Delta u - \omega^2 u = f \quad \text{in } \Omega \end{cases} \qquad \text{[P34]}$$

with the data $f \in L^2(\Omega)$. In what follows, set $V = H_0^1(\Omega)$ endowed with the $\|v\|_V = |v|_{1,\Omega}$ norm, and $(\lambda_j, e_j)_{j \geq 1}$ the pairs of eigenvalues and eigenfunctions of the Laplacian with the Dirichlet condition. We assume that $\|e_j\|_{L^2(\Omega)} = 1$. These pairs are arranged in the order of increasing eigenvalue.

a) Recall the variation formulation [F34] associated with the Helmholtz problem. Denote the bilinear form by a. Give an expression for $a(v, w)$ by decomposing $v, w \in V$ with respect to the basis formed from the eigenvectors.

Assume that ω^2 is not equal to any eigenvalue, or more precisely that there exists an index J_0 such that $\lambda_{J_0} < \omega^2 < \lambda_{J_0+1}$.

b) Using the orthogonal projection P from V onto $\text{vect}(e_j)_{j=1,J_0}$, show that $a(v, v)$ can be written as the sum of a positive term and a negative term.

Since P is a projection, it satisfies $P^2 = P$ and $\|Pv\|_V \leq \|v\|_V, \forall v \in V$.

c) Using P, suggest a linear map $T \in \mathcal{L}(V)$ such that:

$$a(v, Tv) > 0 \quad \forall v \in V \setminus \{0\}.$$

d) Show that T is an isometry and that there exists $\alpha > 0$ such that:

$$a(v, Tv) \geq \alpha \|v\|_V^2 \quad \forall v \in V.$$

Establish the well-posed character of the problem [P34] using the Banach–Nečas–Babuška corollary.

Assume below that Ω is polygonal (d = 2) or polyhedral (d = 3), and that we have a regular family of meshes for Ω. Discretize the variational formulation [F34] using P^1-Lagrangian finite elements, and denote by $(V_h)_h$ the vector subspaces of $H_0^1(\Omega)$ thus constructed.

e) Construct the corresponding discrete variational formulation (denote the approximate solution by u_h).

f) To obtain a uniform discrete inf-sup condition, we define a linear map $T_h \in \mathcal{L}(V_h)$ for all h, by transposing the previous approach to the discrete case.

- Define P_h that approximates P.
- Show that the P_h thus constructed converges to P, that is:

$$\lim_{h \to 0} \left(\sup_{v_h \in V_h \setminus \{0\}} \frac{\|(P - P_h)v_h\|_V}{\|v_h\|_V} \right) = 0.$$

- How should T_h be defined?
- For $v_h \in V_h$, calculate $(T - T_h)v_h$.
- Conclude.

g) If u is piecewise regular, for example if $u \in \mathcal{P}H^2(\Omega)$, what is the error estimate obtained from $\|u - u_h\|_{H^1(\Omega)}$?

Exercise 35. Regularization of the Dirichlet condition

Let Ω be a *convex* and bounded open subset of $\mathbb{R}^{\mathrm{d}}$, $\mathrm{d} = 2, 3$, with boundary $\partial\Omega$. Given two functions $f \in L^2(\Omega)$ and $g \in H^2(\Omega)$, consider the following problem:

$$\begin{cases} \text{Find } u \in H^1(\Omega) & \text{such that} \\ -\Delta u + u = f & \text{in } \Omega \\ u = g_{|\partial\Omega} & \text{on } \partial\Omega. \end{cases} \qquad \text{[P35]}$$

According to the regularity results from section 2.5.2, the solution u to [P35] belongs to $H^2(\Omega)$. Therefore, the function $p = \dfrac{\partial u}{\partial n}_{|\partial\Omega}$ belongs to $L^2(\partial\Omega)$.

a) Let $\varepsilon > 0$ be a "small" parameter. Consider a "penalized" version of [P35], stated in variational form:

$$\left| \begin{array}{l} \text{Find } u_\varepsilon \in H^1(\Omega) \text{ such that} \\ \displaystyle\int_\Omega (\nabla u_\varepsilon \cdot \nabla v + u_\varepsilon v)\, d\Omega + \frac{1}{\varepsilon}\int_{\partial\Omega} u_\varepsilon v d\Gamma \\ \displaystyle\qquad = \int_\Omega f v d\Omega + \frac{1}{\varepsilon}\int_{\partial\Omega} g v d\Gamma \quad \forall v \in H^1(\Omega). \end{array} \right. \qquad [\mathrm{F35}_\varepsilon]$$

Show that the variational formulation [F35$_\varepsilon$] is indeed well posed and interpret it in terms of partial differential equations.

b) Show that for any test function $v \in H^1(\Omega)$, we have that:

$$\int_\Omega (\nabla u \cdot \nabla v + uv) d\Omega + \frac{1}{\varepsilon}\int_{\partial\Omega} uv d\Gamma = \int_\Omega f v d\Omega + \int_{\partial\Omega} p v d\Gamma + \frac{1}{\varepsilon}\int_{\partial\Omega} g v d\Gamma.$$

c) Set $v_\varepsilon = u - u_\varepsilon \in H^1(\Omega)$. Show that:

$$\int_\Omega \left(|\nabla v_\varepsilon|^2 + v_\varepsilon^2\right) d\Omega + \frac{1}{\varepsilon}\int_{\partial\Omega} v_\varepsilon^2 d\Gamma = \int_{\partial\Omega} p v_\varepsilon d\Gamma$$

and deduce the following two inequalities:

$$||v_\varepsilon||^2_{H^1(\Omega)} \le ||p||_{L^2(\partial\Omega)}\, ||v_\varepsilon||_{L^2(\partial\Omega)} \text{ and } ||v_\varepsilon||_{L^2(\partial\Omega)} \le \varepsilon ||p||_{L^2(\partial\Omega)}.$$

d) Deduce from this the estimate:

$$||u - u_\varepsilon||_{H^1(\Omega)} \le \sqrt{\varepsilon}||p||_{L^2(\partial\Omega)}. \qquad \text{[I35]}$$

Comment on this result.

Now we propose to calculate an approximation u_ε^h for u_ε using finite elements. Because of the estimation [I35], u_ε^h also makes up an approximation of u, if ε is chosen to be small enough. Now we assume that Ω is polygonal (d $=2$) or polyhedral (d $=3$), and that we have a regular family of meshes for Ω. One discretizes the variational formulation [F35$_\varepsilon$] using P^1-Lagrangian finite elements. Denote by $(V_h)_h$ the vector subspaces of $H^1(\Omega)$ hence generated (h is the mesh size).

e) Write out the discrete problem associated with the variational formulation [F35$_\varepsilon$] and show that it is well posed.

f) For a given f and g, assume that $(|u_\varepsilon|_{2,\Omega})_\varepsilon$ is *uniformly* bounded in ε. Show that there exists a constant C, depending only on f and g, such that:

$$\|u - u_\varepsilon^h\|_{H^1(\Omega)} \leq C\left(\frac{h}{\sqrt{\varepsilon}} + \sqrt{\varepsilon}\right). \qquad [\text{I35}_b]$$

Using the estimate (I35$_b$), how should the penalization parameter ε be chosen?

g) Write down the discretized problem in matrix form. How does solving this problem differ from solving the problem [P35] numerically?

Exercise 36. Stokes' problem

Let Ω be a connected bounded open subset of $\mathbb{R}^2$ with a Lipschitz boundary. To describe the stationary motion of a incompressible viscous fluid, we introduce Stokes' problem:

$$\begin{cases} \text{Find } \mathbf{u} \text{ and } p & \text{such that} \\ -\nu\Delta\mathbf{u} + \nabla p = \mathbf{f} & \text{in } \Omega \\ \operatorname{div}\mathbf{u} = 0 & \text{in } \Omega \\ \mathbf{u} = 0 & \text{on } \partial\Omega. \end{cases} \qquad [\text{P36}]$$

Here, $\nu > 0$ is the viscosity coefficient, and:

– the vector $\mathbf{u} = \begin{pmatrix} u_1 \\ u_2 \end{pmatrix}$ denotes the fluid velocity;

– the scalar p denotes the fluid pressure;

– the data $\mathbf{f} = \begin{pmatrix} f_1 \\ f_2 \end{pmatrix} \in (L^2(\Omega))^2$ is the applied force density.

We are looking for $\mathbf{u}$ in $(H^1(\Omega))^2$ and p in $L^2(\Omega)$. To take into account the essential boundary condition, introduce the space $(H_0^1(\Omega))^2$ equipped with the norm:

$$|\cdot|_1 : \mathbf{v} = \begin{pmatrix} v_1 \\ v_2 \end{pmatrix} \mapsto |\mathbf{v}|_1 = \left(|v_1|_{1,\Omega}^2 + |v_2|_{1,\Omega}^2\right)^{1/2}.$$

From Poincaré's inequality, $\left((H_0^1(\Omega))^2, |\cdot|_1\right)$ is a Hilbert space.

a) Show that the variational formulation associated with [P36] is:

$$\left| \begin{array}{l} \text{Find } (\mathbf{u}, p) \in (H_0^1(\Omega))^2 \times L^2(\Omega) \text{ such that:} \\ \nu \displaystyle\sum_{i=1,2} \int_\Omega \nabla u_i \cdot \nabla v_i \, d\Omega - \int_\Omega p \operatorname{div} \mathbf{v} \, d\Omega = \int_\Omega \mathbf{f} \cdot \mathbf{v} \, d\Omega \\ \displaystyle\int_\Omega q \operatorname{div} \mathbf{u} \, d\Omega = 0 \qquad \forall (\mathbf{v}, q) \in (H_0^1(\Omega))^2 \times L^2(\Omega). \end{array} \right. \quad \text{[F36]}$$

For this, we can use the approach developed in section 2.2.4.

b) Note that p is known up to a constant. To remove the indeterminacy, we want to find a p whose mean value is zero, that is to say:

$$p \in L^2_{zmv}(\Omega) = \left\{ q \in L^2(\Omega) \text{ such that } \int_\Omega q \, d\Omega = 0 \right\}$$

. Show that we can just consider a test function $q \in L^2_{zmv}(\Omega)$ in [F36].

In what follows, we assume that the more "general" variational formulation:

$$\left| \begin{array}{l} \text{Find } (\mathbf{u}^+, p^+) \in (H_0^1(\Omega))^2 \times L^2_{zmv}(\Omega) \text{ such that} \\ \nu \displaystyle\sum_{i=1,2} \int_\Omega \nabla u_i^+ \cdot \nabla v_i \, d\Omega - \int_\Omega p^+ \operatorname{div} \mathbf{v} \, d\Omega \\ \qquad = \left\langle \mathbf{f}^+, \mathbf{v} \right\rangle_{((H_0^1(\Omega))')^2, (H_0^1(\Omega))^2} \quad \forall \mathbf{v} \in (H_0^1(\Omega))^2 \\ \displaystyle\int_\Omega q \operatorname{div} \mathbf{u}^+ \, d\Omega = \int_\Omega q g^+ \, d\Omega \quad \forall q \in L^2_{zmv}(\Omega) \end{array} \right. \quad \text{[F36}^+\text{]}$$

is *well posed*. For every $(\mathbf{f}^+, g^+) \in ((H_0^1(\Omega))')^2 \times L^2_{zmv}(\Omega)$, there exists a unique solution $(\mathbf{u}^+, p^+)$ to (F36$^+$), and additionally:

$$|\mathbf{u}^+|_1 + \|p^+\|_{L^2(\Omega)} \leq C^+ \left(\|\mathbf{f}^+\|_{((H_0^1(\Omega))')^2} + \|g^+\|_{L^2(\Omega)} \right)$$

with $C^+ > 0$ independent of $\mathbf{f}^+$ and g^+.

c) Show that the space:

$$\mathbf{V} = \left\{ \mathbf{v} = \begin{pmatrix} v_1 \\ v_2 \end{pmatrix} \in (H_0^1(\Omega))^2 \text{ such that } \operatorname{div} \mathbf{v} = 0 \right\}$$

is a Hilbert space with the norm $|\cdot|_1$.

d) Deduce from the above that if $(\mathbf{u}, p)$ satisfies equation [P36], then $\mathbf{u}$ is a solution to:

$$\left| \begin{array}{l} \text{Find } \mathbf{u} \in \mathbf{V} \text{ such that} \\ \nu \sum_{i=1,2} \int_\Omega \nabla u_i \cdot \nabla v_i \, d\Omega = \int_\Omega \mathbf{f} \cdot \mathbf{v} d\Omega \quad \forall \mathbf{v} \in \mathbf{V}. \end{array} \right. \qquad [\text{F36}_a]$$

e) Prove that the problem [F36$_a$] admits a unique solution that depends continuously on the data $\mathbf{f}$.

Assume below that Ω is a polygon.

f) Explain why we cannot use a P^1-Lagrangian finite element approximation for the problem [F36$_a$].

g) Let $\varepsilon > 0$. We introduce the following "penalized" problem:

$$\left| \begin{array}{l} \text{Find } \mathbf{u}^\varepsilon \in (H_0^1(\Omega))^2 \text{ such that} \\ \nu \sum_{i=1,2} \int_\Omega \nabla u_i^\varepsilon \cdot \nabla v_i \, d\Omega + \dfrac{1}{\varepsilon} \int_\Omega \text{div}\, \mathbf{u}^\varepsilon \, \text{div}\, \mathbf{v} \, d\Omega \\ \qquad = \int_\Omega \mathbf{f} \cdot \mathbf{v} d\Omega \quad \forall \mathbf{v} \in (H_0^1(\Omega))^2. \end{array} \right. \qquad [\text{F36}_\varepsilon]$$

Show that the problem [F36$_\varepsilon$] admits a unique solution.

h) Show that $\|\text{div}\, \mathbf{u}^\varepsilon\|_{L^2(\Omega)} \leq C\sqrt{\varepsilon}\, \|\mathbf{f}\|_{(L^2(\Omega))^2}$, with $C > 0$ a constant independent of $\mathbf{f}$ and ε.

i) By using the well-posed character of [F36$^+$], show that:

$$|\mathbf{u} - \mathbf{u}^\varepsilon|_1 \leq C'\sqrt{\varepsilon}\, \|\mathbf{f}\|_{(L^2(\Omega))^2}$$

with $C' > 0$ independent of both $\mathbf{f}$ and ε.

Let $\mathcal{T}_h$ be a mesh for the open subset Ω, consisting of the triangles $(T_\ell)_{\ell=1,L}$ with vertices $(M_I)_{I=1,N}$. Assume that the vertices not belonging to the boundary are the vertices $(M_I)_{I=1,M}$. We use P^1-Lagrangian finite elements and we denote by w_I the global basis function attached to the vertex M_I. Set $V_0^h = \text{vector}\,(w_I)_{I=1,M}$, and recall that $V_0^h \subset H_0^1(\Omega)$. We are looking for a **vectorial** approximation for $\mathbf{u}^\varepsilon$ in the form:

$$\mathbf{u}_h = \begin{pmatrix} \sum_{I=1,M} u_1^I w_I \\ \sum_{I=1,M} u_2^I w_I \end{pmatrix}, \text{ represented by } \vec{U} = \begin{pmatrix} \vec{U}_1 \\ \vec{U}_2 \end{pmatrix} \in \mathbb{R}^{2M}$$

with $\vec{U}_1$ (respectively, $\vec{U}_2$) the vector in $\mathbb{R}^M$ with the components $\left(u_1^I\right)_{I=1,M}$ (respectively, $\left(u_2^I\right)_{I=1,M}$).

j) Construct the associated discrete variational formulation.

k) Given $\mathbf{f}$, suppose that $\mathbf{u}^\varepsilon \in (H^2(\Omega))^2$ for every $\varepsilon > 0$, and that $(|u_1^\varepsilon|_{2,\Omega})_\varepsilon$ and $(|u_2^\varepsilon|_{2,\Omega})_\varepsilon$ are *uniformly* bounded in ε. Show that there exists a constant C, depending only on $\mathbf{f}$, such that:

$$|\mathbf{u} - \mathbf{u}_h|_1 \leq C\left(\frac{h}{\sqrt{\varepsilon}} + \sqrt{\varepsilon}\right) \qquad [I36]$$

where h is the mesh size.

l) Show that the discretization of the problem [F36$_\varepsilon$] in $\left(V_0^h\right)^2$ is equivalent to a matrix system of order $2M$ that is of the form:

$$\begin{bmatrix} \mathbb{A}_{11} & \mathbb{A}_{12} \\ \mathbb{A}_{21} & \mathbb{A}_{22} \end{bmatrix} \begin{pmatrix} \vec{U}_1 \\ \vec{U}_2 \end{pmatrix} = \begin{pmatrix} \vec{B}_1 \\ \vec{B}_2 \end{pmatrix}$$

where one should describe the matrices $\mathbb{A}_{11}$, $\mathbb{A}_{21}$, $\mathbb{A}_{21}$, $\mathbb{A}_{22}$ of order M and the vectors $\vec{B}_1$, $\vec{B}_2$ in $\mathbb{R}^M$.

m) Denote $p_h = \operatorname{div} \mathbf{u}_h$. For $\ell = 1, L$, introduce the functions $\mathcal{X}_\ell$ defined by:

$$\mathcal{X}_\ell = \begin{cases} 1 & \text{on } T_\ell \\ 0 & \text{otherwise.} \end{cases}$$

Show that $p_h = \sum_{\ell=1,L} p_\ell \mathcal{X}_\ell$, where we will express p_ℓ as a function in the components of the vectors $\vec{U}_1$ and $\vec{U}_2$.

Exercise 37. Solving the Maxwell equations

We place ourselves on the open square $\Omega = (0,1) \times (0,1)$. We want to solve the following problem:

$$\begin{cases} \text{Find } \mathbf{E} & \text{such that} \\ \operatorname{curl} \mathbf{E} = f & \text{in } \Omega \\ \operatorname{div} \mathbf{E} = 0 & \text{in } \Omega \\ \mathbf{E} \cdot \mathbf{t} = 0 & \text{on } \partial\Omega \end{cases} \qquad [P37]$$

where the datum f is an element of $L^2(\Omega)$ whose mean value is zero, and $\mathbf{t}$ denotes the unit tangent vector to $\partial\Omega$ such that $(\mathbf{t}, \mathbf{n})$ is positively oriented: $\mathbf{t} = \begin{pmatrix} n_2 \\ -n_1 \end{pmatrix}$.

$\mathbf{E} = \begin{pmatrix} E_1 \\ E_2 \end{pmatrix}$ is the electric field to be determined. Recall the following notations, valid for all $\mathbf{q} \in (\mathcal{D}(\Omega))^2$ and all $q \in \mathcal{D}(\Omega)$:

$$\operatorname{curl} \mathbf{q} = \frac{\partial q_2}{\partial x_1} - \frac{\partial q_1}{\partial x_2}, \quad \mathbf{curl}\, q = \begin{pmatrix} \dfrac{\partial q}{\partial x_2} \\ -\dfrac{\partial q}{\partial x_1} \end{pmatrix}$$

as well as the elementary formulae:

$$\Delta q = \operatorname{div}(\nabla q) = -\operatorname{curl}(\mathbf{curl}\, q)\,;\ \operatorname{div}(\mathbf{curl}\, q) = 0;\ \operatorname{curl}(\nabla q) = 0.$$

Subsequently, we will look for $\mathbf{E}$ in the space:

$$\mathbf{V} = \left\{ \mathbf{v} = \begin{pmatrix} v_1 \\ v_2 \end{pmatrix} \in (H^1(\Omega))^2 \text{ such that } \mathbf{v} \cdot \mathbf{t}_{\,|\partial\Omega} = 0 \right\}$$

endowed with the norm: $\|\mathbf{v}\|_{(H^1(\Omega))^2} = \left(\|v_1\|^2_{H^1(\Omega)} + \|v_2\|^2_{H^1\Omega)} \right)^{1/2}$.

a) Explain the boundary condition for each side of $\partial\Omega$.

b) Show that $\mathbf{E}$ satisfies:

$$\left| \begin{array}{l} \text{Find } \mathbf{E} \in \mathbf{V} \text{ such that} \\ \displaystyle\int_\Omega \operatorname{curl} \mathbf{E} \operatorname{curl} \mathbf{v}\, d\Omega + \int_\Omega \operatorname{div} \mathbf{E} \operatorname{div} \mathbf{v}\, d\Omega = \int_\Omega f \operatorname{curl} \mathbf{v}\, d\Omega \quad \forall\, \mathbf{v} \in \mathbf{V}. \end{array} \right. \quad \text{[F37]}$$

c) Prove that if $\mathbf{E}$ satisfies [F37], then $\operatorname{div} \mathbf{E} = 0$.

d) Indicate the principle of a method to prove that $\operatorname{curl} \mathbf{E} = f$.

We will now show that the variational formulation [F37] is well posed. For this, introduce the following notation:

$$(\mathbf{u}, \mathbf{v})_{\mathbf{V}} = \sum_{i,j=1}^{2} \int_\Omega \partial_i u_j \partial_i v_j\, d\Omega.$$

e) Let $\mathbf{u}, \mathbf{v} \in \mathbf{V} \cap (H^2(\Omega))^2$. Using the integration-by-parts formula:

$$\int_\Omega \operatorname{curl} \mathbf{u} \operatorname{curl} \mathbf{v}\, d\Omega = \int_\Omega \left(\left(\partial_1 \partial_2 u_2 - \partial_2^2 u_1 \right) v_1 + \left(\partial_1 \partial_2 u_1 - \partial_1^2 u_2 \right) v_2 \right) d\Omega$$

show that:

$$\int_\Omega \text{curl}\, \mathbf{u}\, \text{curl}\mathbf{v}\, d\Omega + \int_\Omega \text{div}\, \mathbf{u}\, \text{div}\mathbf{v}\, d\Omega$$

$$= - \int_\Omega ((\Delta u_1)\, v_1 + (\Delta u_2)\, v_2)\, d\Omega + \int_{\partial\Omega} \text{div}\, \mathbf{u}(\mathbf{v} \cdot \mathbf{n}) d\Gamma$$

$$= (\mathbf{u}, \mathbf{v})_{\mathbf{V}} + \int_{\partial\Omega} \text{div}\, \mathbf{u}(\mathbf{v} \cdot \mathbf{n}) d\Gamma - \int_{\partial\Omega} \Big(\sum_{i=1}^{2} \frac{\partial u_i}{\partial n}\, v_i\Big) d\Gamma.$$

By explaining the boundary terms, show that:

$$\int_C \text{div}\, \mathbf{u}(\mathbf{v} \cdot \mathbf{n}) d\Gamma - \int_C \Big(\sum_{i=1}^{2} \frac{\partial u_i}{\partial n}\, v_i\Big) d\Gamma = 0, \text{ on either side } C \text{ of } \partial\Omega$$

and conclude that:

$$\int_\Omega (\text{curl}\, \mathbf{u}\, \text{curl}\mathbf{v} + \text{div}\, \mathbf{u}\, \text{div}\mathbf{v})\, d\Omega = (\mathbf{u}, \mathbf{v})_{\mathbf{V}}\,.$$

In the sequel, assume that the result holds for all $\mathbf{u}, \mathbf{v} \in \mathbf{V}$.

f) We also assume the following generalization to Poincaré's inequality, called *Weber's inequality*: there exists a $C_W > 0$, depending only on Ω, such that:

$$\|\mathbf{v}\|^2_{(L^2(\Omega))^2} \le C_W\, (\mathbf{v}, \mathbf{v})_{\mathbf{V}} \quad \forall \mathbf{v} \in \mathbf{V}.$$

Prove that the variational formulation [F37] admits a unique solution.

We can now consider a P^1-finite element approximation. Choose a triangulation $(T_\ell)_{\ell=1,L}$ of the square Ω with vertices $(M_i)_{i=1,N}$ along with their attached basis functions $(w_i)_{i=1,N}$. Introduce the **vectorial** basis functions: for each i in $\{1, ..., N\}$, let:

$$\mathbf{w}_1^i = \begin{pmatrix} w_i \\ 0 \end{pmatrix} \quad \text{and} \quad \mathbf{w}_2^i = \begin{pmatrix} 0 \\ w_i \end{pmatrix}.$$

g) Determine the largest internal approximation space for $\mathbf{V}$ of the form:

$$\mathbf{V}_h = \text{vect}(\mathbf{w}_1^i)_{i\in\mathcal{I}_1} \oplus \text{vect}(\mathbf{w}_2^j)_{j\in\mathcal{I}_2}.$$

h) Write the approximate variational formulation.

i) Show that it can be written equivalently in the form:

$$\begin{bmatrix} \mathbb{A}_{11} & \mathbb{A}_{12} \\ \mathbb{A}_{21} & \mathbb{A}_{22} \end{bmatrix} \begin{pmatrix} \vec{E}_1 \\ \vec{E}_2 \end{pmatrix} = \begin{pmatrix} \vec{F}_1 \\ \vec{F}_2 \end{pmatrix}. \quad [S37]$$

j) What is $\left(\mathbf{w}_1^k, \mathbf{w}_2^l\right)_{\mathbf{V}}$ for $k \in \mathcal{I}_1$ and $l \in \mathcal{I}_2$? Deduce a method for solving the matrix system [S37].

Hint: In Question (c), one can use the gradient of $q \in H_0^1(\Omega)$ that satisfies $\Delta q = \operatorname{div} \mathbf{E}$ in Ω (q is a solution to the Dirichlet problem). On the one hand, since $q_{|\partial\Omega} = 0$, we will assume that we automatically have $\nabla q \cdot \mathbf{t}_{|\partial\Omega} = 0$. On the other hand, since Ω is convex, note that $q \in H^2(\Omega)$ (see section 2.5.2). In Question (d), we can use a similar technique, though this time by solving a Neumann problem.

5

Concrete Aspects of the Finite Element Method

This chapter is dedicated to the practical aspects used in the finite element method. We start by asking questions about the computer implementation of the method before continuing onto the algorithmic considerations, and then concluding with some numerical illustrations obtained via Matlab. The Matlab scripts presented here are easily translatable into other scripting languages (Python, Julia, etc.).

5.1. Implementation

The realization of finite element code presents some peculiarities that are linked, on the one hand, to the properties of the basis functions (their local support and definition) and, on the other hand, to the variational formulations – in particular when treating essential conditions.

Solving an elliptic problem numerically using a finite element method proceeds according to following seven key steps:

- the generation of a mesh for the computational domain $\overline{\Omega}$;
- the creation of the elementary matrices;
- the assembly of the global matrices;
- the construction of the right-hand side;
- the elimination of the essential condition(s) (if any);
- the solution to the linear system;
- any numerical and/or graphical post-processing.

Note that the automatic generation of the mesh is usually performed by specialized software, so we will not discuss the principles of this here and instead only indicate what information is required for the mesh to implement a finite element method. Toward the end of this chapter, snippets of Matlab code are presented, allowing one to realize the meshes of some elementary 2D domains, that is, when $\mathrm{d} = 2$.

Throughout this section, we will describe the different steps, with more or less detail, in the case of the following mixed Dirichlet–Neumann problem that is posed on a bounded and open subset Ω of $\mathbb{R}^{\mathrm{d}}$, along with a "sufficiently regular" boundary:

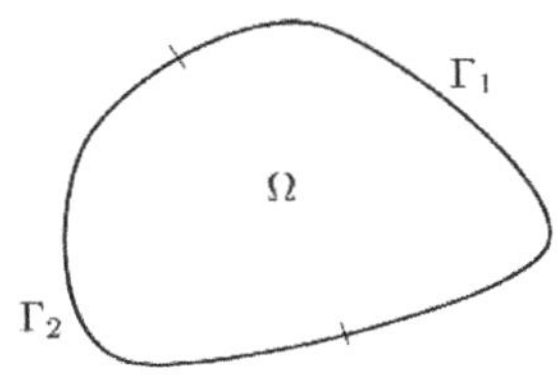

$$\left\{\begin{array}{lll}\text{Find } u \in H^1(\Omega) \text{ such that} & & \\ -\Delta u = f & \text{in } \Omega & \\ \dfrac{\partial u}{\partial n} = g & \text{on } \Gamma_1 & \\ u = \alpha & \text{on } \Gamma_2 & \end{array}\right. \qquad [5.1]$$

with: $\Gamma_1 \cap \Gamma_2 = \emptyset$ (Γ_1 and Γ_2 are open) and $\overline{\Gamma}_1 \cup \overline{\Gamma}_2 = \partial\Omega$, $f \in L^2(\Omega)$, $g \in L^2(\Gamma_1)$ and $\alpha \in L^2(\Gamma_2)$ such that there exists a lifting $\tilde{\alpha} \in H^1(\Omega)$, that is, $\tilde{\alpha} = \alpha$ over Γ_2, that is to say, $\alpha \in H^{1/2}(\Gamma_2)$, as in Chapter 1.

The variational formulation of problem [5.1] in the space:

$$H^1_{0,\Gamma_2}(\Omega) = \left\{v \in H^1(\Omega) \text{ such that } v_{|\Gamma_2} = 0\right\}$$

is then:

$$\left|\begin{array}{l}\text{Find } u = \tilde{u} + \tilde{\alpha} \text{ with } \tilde{u} \in H^1_{0,\Gamma_2}(\Omega) \text{ such that} \\ \displaystyle\int_\Omega \nabla\tilde{u} \cdot \nabla v \, d\Omega = \int_\Omega f v \, d\Omega \\ \displaystyle\qquad\qquad + \int_{\Gamma_1} g v \, d\Gamma - \int_\Omega \nabla\tilde{\alpha} \cdot \nabla v \, d\Omega \quad \forall v \in H^1_{0,\Gamma_2}(\Omega).\end{array}\right. \qquad [5.2]$$

From theorem 1.7, $H^1_{0,\Gamma_2}(\Omega)$ is a closed subset of $H^1(\Omega)$. In the following, we denote by Σ the interface between Γ_1 and Γ_2: $\Sigma = \partial\Gamma_1 \cap \partial\Gamma_2$.

Furthermore, we will restrict our focus to cases involving Lagrangian finite elements.

5.1.1. *Finite element meshes*

A mesh is described by:

– the coordinates of the Lagrangian degrees of freedom $(M_I)_{I=1,N}$;

– for each finite element $E_\ell = (K_\ell, \Sigma_\ell, P_\ell)$, $\ell = 1, L$:

a) its type (triangle, quadrangle, cube, prism, etc.);

b) its order (0, 1, 2, etc.);

c) the list of global degrees of freedom associated to this element, in a pre-established order (e.g., given a P^2 triangle in 2D: the three vertices and then the three midpoints of each edge);

d) the so-called reference numbers: for the domains K_ℓ, faces, edges and vertices.

The information in (a) and (b) tells us the number of the local degrees of freedom n_ℓ, as well as the local basis functions $\left(\tau_i^\ell\right)$, via the geometric transformation F_ℓ constructed from the coordinates of each degree of freedom. The information in (c) corresponds to the data regarding the indexing function ℓ_g defined in [3.51]. The information in (d) is optional, but nonetheless allows us to quickly identify specific objects: faces, edges and nodes located along a boundary (processing of the boundary conditions, for example), or even the areas that form part of the computational domain (processing of any discontinuous coefficients, for example). In particular, explicitly knowing each node along the boundary that belongs to the interface Σ is desirable.

The data coming from the information in (a), (b) and (c) form the mesh $\mathcal{T}_h$: $(K_\ell, \Sigma_\ell, P_\ell)_{\ell=1,L}$, which must clearly satisfy each assumption of [3.48]. Therefore, we will denote: $\overline{\Omega}_h = \cup_{\ell=1,L} K_\ell$ (here, we consider the case where Ω is an open subset that is not necessarily polyhedral).

To divide up the boundary $\partial\Omega_h$, we let Γ_1^h (respectively, Γ_2^h) equal the set of boundary faces[1] that connects the nodes along $\overline{\Gamma}_1$ (respectively, $\overline{\Gamma}_2$). With this definition, Γ_1^h and Γ_2^h are two closed sets and their intersection equals $\Sigma = \Gamma_1^h \cap \Gamma_2^h$, whereas their union equals the boundary $\partial\Omega_h = \Gamma_1^h \cup \Gamma_2^h$.

Furthermore, let us introduce the following sets of indices:

$$
\begin{aligned}
\mathcal{N} &= \left\{I \text{ such that } M_I \in \overset{\circ}{\Gamma_1^h}\right\} && (N_1 = \operatorname{card} \mathcal{N}) \\
\mathcal{D} &= \left\{I \text{ such that } M_I \in \Gamma_2^h\right\} && (N_2 = \operatorname{card} \mathcal{D}) \\
\mathcal{I} &= \left\{1, 2, ..., N\right\} \setminus \mathcal{D} && (M = \operatorname{card} \mathcal{I})
\end{aligned}
$$

which in practice are constructed using the information coming from (d). The indexing set $\mathcal{N}$ identifies the nodes belonging to the boundary Γ_1^h and which possess

1 To quickly identify the boundaries Γ_1^h and Γ_2^h (different from Γ_1 and Γ_2, if Ω is not polyhedral), one starts with the following observation. Each domain K_ℓ has a boundary that is made up from a set of faces. Among these faces, those forming part of the boundary $\partial\Omega_h$ are exactly those which belong only to K_ℓ: they are called boundary faces. Conversely, any internal face belongs to exactly two domains, K_ℓ and $K_{\ell'}$, with $\ell \neq \ell'$.

the Neumann condition. In reality, as we will have to evaluate an integral along this boundary, just the location of these faces in the boundary suffices. Obviously, the knowledge of these nodes located along the boundary Γ_1^h tells us where to find the faces located along Γ_1^h. The indexing set $\mathcal{D}$ identifies the nodes along the boundary Γ_2^h that possess the Dirichlet condition, and we introduce the indexing set $\mathcal{I}$ to represent the nodes that do not possess the Dirichlet condition, since they correspond to the set of global basis functions that generate the approximation space for which the Dirichlet condition is taken into account.

REMARK 5.1.– *A priori, any node along the interface Σ that forms part of the boundary with a Dirichlet-type boundary condition is included.*

Example: A mesh consisting of two P^1-Lagrangian finite elements.

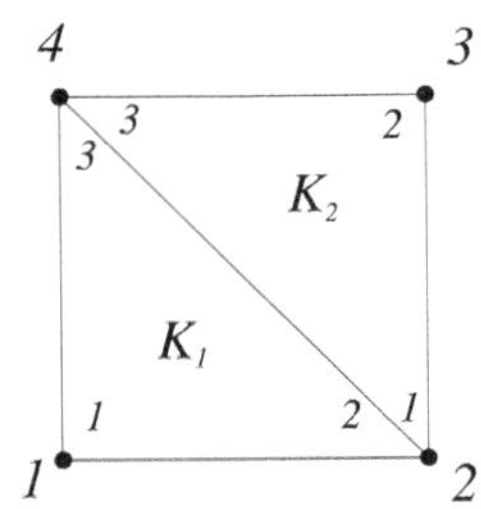

Figure 5.1. *A mesh of two triangles*

In this example, the vertex coordinate arrays and local $\rightarrow$ global numbering are coded as:

coor(1,1)=0 coor(1,2)=0	lg(1,1)=1 lg(2,1)=2
coor(2,1)=1 coor(2,2)=0	lg(1,2)=2 lg(2,2)=3
coor(3,1)=1 coor(3,2)=1	lg(1,3)=4 lg(2,3)=4.
coor(4,1)=0 coor(4,2)=1	

It is easy to create meshing programs for simple objects such as rectangles, circles, cubes, spheres, or even geometric transformations of such objects. To form meshes from more complex objects, such as those often encountered in industry, general meshing software is used; from just a few points on the surface of the object[2], it produces a mesh using sophisticated algorithms whose descriptions are beyond the scope of this book. Unfortunately, only a few of these meshing software are able to automatically generate *isoparametric meshes of order* k ($k \geq 2$).

2 In our "abstract" framework, these will be points located along the boundary $\partial\Omega$.

We are now able to discretize the variational formulation [5.2]. Let us introduce the spaces:

$$V_h = Vect\,(w_I)_{I=1,N} \text{ and } V_h^0 = Vect\,(w_I)_{I\in\mathcal{I}}$$

along with the discrete variational formulation in V_h^0:

$$\left|\begin{array}{l}\text{Find } u_h = \tilde{u}_h + \tilde{\alpha} \text{ with } \tilde{u}_h \in V_h^0 \text{ such that} \\ \displaystyle\int_{\Omega_h} \nabla\tilde{u}_h \cdot \nabla\tilde{v}_h d\Omega = \int_{\Omega_h} f\tilde{v}_h d\Omega \\ \displaystyle\qquad\qquad + \int_{\Gamma_1^h} g\tilde{v}_h d\Gamma - \int_{\Omega_h} \nabla\tilde{\alpha}\cdot\nabla\tilde{v}_h\, d\Omega \quad \forall\tilde{v}_h \in V_h^0.\end{array}\right. \tag{5.3}$$

By lemma 3.1, this variational formulation is equivalent to the linear system of order M:

$$\mathbb{K}_{\mathcal{I}\mathcal{I}}\vec{U}_{\mathcal{I}} = \vec{S}_{\mathcal{I}} \tag{5.4}$$

where:

$$\begin{aligned}(\mathbb{K}_{\mathcal{I}\mathcal{I}})_{IJ} &= \int_{\Omega_h} \nabla w_I \cdot \nabla w_J d\Omega \qquad I, J \in \mathcal{I} \\ (S_{\mathcal{I}})_I &= \int_{\Omega_h} f\, w_I d\Omega + \int_{\Gamma_1^h} g\, w_I d\Gamma - \int_{\Omega_h} \nabla\tilde{\alpha}\cdot\nabla w_I d\Omega \qquad I \in \mathcal{I}\end{aligned}$$

and also:

$$u_h = \sum_{I\in\mathcal{I}} (U_{\mathcal{I}})_I w_I + \tilde{\alpha}. \tag{5.5}$$

5.1.2. *Basic calculations*

As previously seen, it is preferable to calculate the elementary matrices:

$$\mathbb{K}_{ij}^\ell = \int_{K_\ell} \nabla\tau_i^\ell \cdot \nabla\tau_j^\ell d\Omega \quad \text{and} \quad \mathbb{M}_{ij}^\ell = \int_{K_\ell} \tau_i^\ell \tau_j^\ell d\Omega \qquad \forall i, j = 1, n_\ell$$

on each domain K_ℓ (see [3.38] and [3.39]). The elementary matrices $\mathbb{K}^\ell$ are used to construct the (global) matrix $\mathbb{K}_{\mathcal{I}\mathcal{I}}$, and the elementary matrices $\mathbb{M}^\ell$ are used to construct the right-hand side (more precisely, when constructing an approximation to the right-hand side in a particular way).

Recall that the finite element $(K_\ell, \Sigma_\ell, P_\ell)$ is linked to the reference finite element $(\hat{K}, \hat{\Sigma}, \hat{P})$ by the geometric transformation F_ℓ, that sends $\hat{K}$ into K_ℓ. In performing the change in variable $M = F_\ell(\hat{M})$, the previous integrals become integrals over $\hat{K}$. We find (see [3.40] and [3.41]):

$$\mathbb{K}^\ell_{ij} = \int_{\hat{K}} (\left\{[dF_\ell(\hat{M})]^{-1}([dF_\ell(\hat{M})]^{-1})^t\right\} \nabla\hat{\tau}_i(\hat{M})) \cdot \hat{\nabla}\hat{\tau}_j(\hat{M})\, |J_{F_\ell}(\hat{M})|\, d\hat{\Omega}$$

$$\mathbb{M}^\ell_{ij} = \int_{\hat{K}} \hat{\tau}_i(\hat{M})\hat{\tau}_j(\hat{M})\, |J_{F_\ell}(\hat{M})|\, d\hat{\Omega}$$

which uses only the knowledge of the basis functions on the reference finite element. These integrals are:

– calculated both analytically and exactly (see [4.24]) in simple cases (finite elements with low degrees);

– either evaluated (exact evaluation of finite element matrices) or estimated (calculating integrals of finite elements with varying coefficients, or the right-hand side) via numerical quadrature formulas; let us recall that, when calculating the right-hand side, it is enough to integrate the polynomials of degree $2k - 2$ exactly, when the finite element has degree k itself (see proposition 4.5).

Note that the quadrature points $((\hat{x}^\ell_q)_{q=1,nq_\ell})$ and their weights $((\omega^\ell_q)_{q=1,nq_\ell})$ are defined once, and for all, relative to the reference domain $\hat{K}$.

For each domain K_ℓ, when using a quadrature formula we execute the following algorithm 5.1.

$\mathbb{K}^\ell = 0$	
for $q = 1, nq_\ell$	*loop over quadrature points*
$\quad DF = [dF_\ell(\hat{x}^\ell_q)]$	*calculate Jacobian matrix at* $\hat{x}^\ell_q$
$\quad J = \|\det DF\|$	*calculate Jacobian at* $\hat{x}^\ell_q$
$\quad \mathrm{C} = J\,(DF^{-1}\,(DF^{-1})^t)$	
$\quad$ **for** $i, j = 1, n_\ell$	*loop over dof's*
$\qquad \mathbb{K}^\ell_{ij} = \mathbb{K}^\ell_{ij} + \omega^\ell_q(\mathrm{C}\hat{\nabla}\hat{\tau}_i(\hat{x}^\ell_q)) \cdot \hat{\nabla}\hat{\tau}_j(\hat{x}^\ell_q)$	*"calculate" integral*

Algorithm 5.1. *Algorithm that calculates the elementary matrices*

REMARK 5.2.– *When calculating the elementary matrices, note that a choice of isoparametric finite element results in the explicit appearance of the geometric transformation* F_ℓ.

5.1.3. *Assembly of the global matrices and the right-hand side*

To construct the linear system [5.4], it is necessary to construct the (global) matrix $\mathbb{K}_{\mathcal{I}\mathcal{I}}$ from the elementary matrices $\mathbb{K}^{\ell}$, on the one hand, and the right-hand vector term $\overrightarrow{S}_{\mathcal{I}}$, on the other hand.

5.1.3.1. *Assembling the matrices*

For reasons which will appear later (when taking Dirichlet boundary conditions into account), it is preferable and more straightforward to initially assemble the matrix $\mathbb{K}$ of order N by taking *every* degree of freedom of the mesh into account, including those located along Γ_2^h (including the interface Σ). Interestingly, $\mathbb{K}$ "contains" $\mathbb{K}_{\mathcal{I}\mathcal{I}}$. In the second step, we will "extract" the matrix $\mathbb{K}_{\mathcal{I}\mathcal{I}}$.

Recall that, by the definition of the global basis functions (w_I), we have:

$$w_I = \tau_i^{\ell} \quad \text{on } K_{\ell} \text{ if } \exists i \text{ such that } \ell_g(\ell, i) = I$$

where the index i, if it exists, is unique. Introduce the set:

$$\mathcal{C}_{IJ} = \{\ell \text{ such that there exists } i_{\ell}, j_{\ell} \text{ such that } \ell_g(\ell, i_{\ell}) = I \text{ and } \ell_g(\ell, j_{\ell}) = J\}$$

which represents the collection of mesh elements *simultaneously* possessing the nodes M_I and M_J as their degrees of freedom. We have then, with i_{ℓ}, j_{ℓ} the indices appearing in the definition of $\mathcal{C}_{IJ}$, that:

$$\mathbb{K}_{IJ} = \int_{\Omega_h} \nabla w_I \cdot \nabla w_J d\Omega = \sum_{\ell \in \mathcal{C}_{IJ}} \int_{K_{\ell}} \nabla \tau_{i_{\ell}}^{\ell} \cdot \nabla \tau_{j_{\ell}}^{\ell} d\Omega = \sum_{\ell \in \mathcal{C}_{IJ}} \mathbb{K}_{i_{\ell} j_{\ell}}^{\ell}$$

$$\mathbb{M}_{IJ} = \int_{\Omega_h} w_I w_J d\Omega = \sum_{\ell \in \mathcal{C}_{IJ}} \int_{K_{\ell}} \tau_{i_{\ell}}^{\ell} \tau_{j_{\ell}}^{\ell} d\Omega = \sum_{\ell \in \mathcal{C}_{IJ}} \mathbb{M}_{i_{\ell} j_{\ell}}^{\ell}.$$

The sets $(\mathcal{C}_{IJ})_{I,J=1,N}$ are never constructed explicitly because that would require us to "inverse" the mapping ℓ_g with respect to its second argument i, which is computationally very expensive. This is why we proceed in the opposite direction. Indeed, one can easily verify that each elementary term $\mathbb{K}_{ij}^{\ell}$ occurs *once and only once*, when constructing the global matrix $\mathbb{K}$ (and likewise when constructing $\mathbb{M}$) since, given (i, j), in a domain K_{ℓ} we know the position of the elementary term $\mathbb{K}_{ij}^{\ell}$ explicitly within the global matrix: $(I, J) = (\ell_g(\ell, i), \ell_g(\ell, j))$.

After initializing the matrix $\mathbb{K}$ to 0, we continue with the process for $\ell = 1, L$:

$$\mathbb{K}_{\ell_g(\ell,i),\ell_g(\ell,j)} = \mathbb{K}_{\ell_g(\ell,i),\ell_g(\ell,j)} + \mathbb{K}_{ij}^{\ell} \quad i, j = 1, n_{\ell}$$

which constitutes the **assembly** phase of the global matrix $\mathbb{K}$.

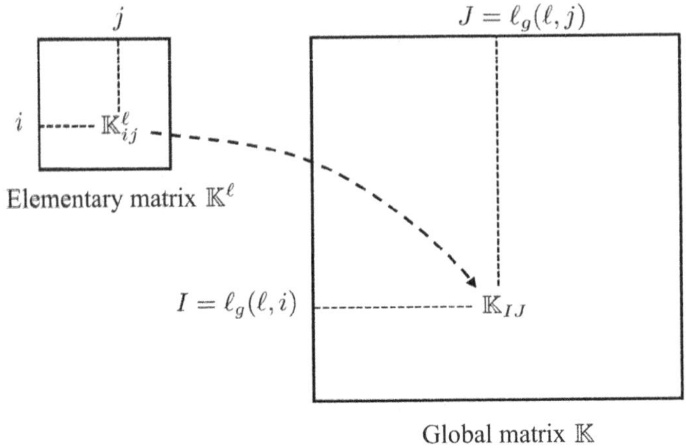

Figure 5.2. *Principle of assembly*

In practice, one computes the elementary matrices and also assembles the global matrix simultaneously, leading to assembly algorithm 5.2.

$\mathbb{K} = 0, \mathbb{M} = 0$	*initialization*
for $\ell = 1, L$	*loop over the domains*
calculate the elementary matrices $\mathbb{K}^\ell$, $\mathbb{M}^\ell$	
for $i, j = 1, n_\ell$	*loop over local dof's*
$I = \ell_g(\ell, i),\ J = \ell_g(\ell, j)$	*global numbers*
$\mathbb{K}_{IJ} = \mathbb{K}_{IJ} + \mathbb{K}^\ell_{ij}$	*assembling matrices*
$\mathbb{M}_{IJ} = \mathbb{M}_{IJ} + \mathbb{M}^\ell_{ij}$	

Algorithm 5.2. *The assembly algorithm*

5.1.3.2. *Constructing the right-hand side*

The right-hand side in the variational formulation [5.3] is made up of the following three terms:

– the volume term: $\int_{\Omega_h} f w_I d\Omega$;

– the surface term: $\int_{\Gamma_1^h} g w_I d\Gamma$, for $I \in \mathcal{N}$, from the Neumann condition;

– the volume term: $\int_\Omega \nabla\tilde{\alpha} \cdot \nabla w_I d\Omega$, from the Dirichlet condition.

There are essentially two methods when it comes to calculating these terms.

5.1.3.2.1. The interpolation method

If the functions f, g, and $\tilde{\alpha}$ belong to $\mathcal{C}^0(\overline{\Omega})$, then we may approximate them using their respective interpolates (see [4.4]) $\pi_h f$, $\pi_h g$ and $\pi_h \tilde{\alpha}$, namely:

$$\pi_h f = \sum_{J=1,N} f(M_J)\, w_J, \quad \pi_h g = \sum_{J\in\mathcal{N}} g(M_J)\, w_J, \quad \pi_h \tilde{\alpha} = \sum_{J=1,N} \tilde{\alpha}(M_J)\, w_J.$$

Then the volume integrals are approached in the following way, for $I = 1, N$:

$$\int_{\Omega_h} f\, w_I d\Omega \simeq \int_{\Omega_h} \pi_h f\, w_I d\Omega = \sum_{J=1,N} f(M_J) \int_{\Omega_h} w_I w_J d\Omega$$

$$\int_{\Omega_h} \nabla\tilde{\alpha}\cdot\nabla w_I d\Omega \simeq \int_{\Omega_h} \nabla(\pi_h\tilde{\alpha})\cdot\nabla w_I d\Omega = \sum_{J=1,N} \tilde{\alpha}(M_J) \int_{\Omega_h} \nabla w_I\cdot\nabla w_J d\Omega$$

and then the surface integrals:

$$\int_{\Gamma_1^h} g\, w_I d\Gamma \simeq \int_{\Gamma_1^h} \pi_h g\, w_I d\Gamma = \sum_{J\in\mathcal{N}} g(M_J) \int_{\Gamma_1^h} w_I w_J d\Gamma \qquad \forall I \in \mathcal{N}.$$

These approximations lead us to the following expression for the right-hand side appearing in [5.4]:

$$\overrightarrow{S} = \mathbb{M}\overrightarrow{F} + P_{\mathcal{N}}\left(\mathbb{M}_{\mathcal{N}}\overrightarrow{G}\right) - \mathbb{K}\overrightarrow{\alpha} \qquad [5.6]$$

where:

– $\overrightarrow{F}$ is the vector in $\mathbb{R}^N$ with components $(f(M_I))_I$;

– $\overrightarrow{G}$ is the vector in $\mathbb{R}^{N_1}$ with components $(g(M_I))_{I\in\mathcal{N}}$;

– $\overrightarrow{\alpha}$ is the vector in $\mathbb{R}^N$ with components $(\tilde{\alpha}(M_I))_I$;

– $\mathbb{M}_{\mathcal{N}}$ is the square matrix of rank N_1, called the boundary mass matrix, and is defined by:

$$(\mathbb{M}_{\mathcal{N}})_{IJ} = \int_{\Gamma_1^h} w_I w_J d\Gamma \qquad I, J \in \mathcal{N}.$$

– $P_{\mathcal{N}}$ is the canonical extension operator from $\mathbb{R}^{N_1}$ into $\mathbb{R}^N$.

The boundary mass matrix is constructed via the aforementioned assembly process, taking care to keep just the degrees of freedom located along the boundary $\overset{\circ}{\Gamma_1^h}$. For this technique, we must extract the actual right-hand vector $\overrightarrow{S}_{\mathcal{I}}$ from the vector $\overrightarrow{S}$, by eliminating the terms that correspond to nodes located along Γ_2^h (again, including those along the interface Σ).

5.1.3.2.2. The quadrature method

The other technique consists of forming the right-hand side locally, with the help of some quadrature formulas. Indeed, we have:

$$\int_{\Omega_h} f\, w_I d\Omega = \sum_{\ell \in C_{II}} \int_{K_\ell} f\, \tau^{\ell}_{i_\ell} d\Omega.$$

We arrive back at the reference finite element via the formula:

$$F_i^\ell = \int_{K_\ell} f\, \tau_i^\ell d\Omega = \int_{\hat{K}} f\left(F_\ell\left(\hat{x}\right)\right) \hat{\tau}_i \left|J_{F_\ell}\left(\hat{x}\right)\right| d\hat{\Omega}$$

and we calculate this integral using the quadrature points $(\hat{x}_q^\ell)_{q=1,nq_\ell}$ with the weights $(\hat{\omega}_q^\ell)_{q=1,nq_\ell}$. By performing the following assembly operation, and after initialization at 0:

$$(S_f)_{\ell_g(\ell,i)} = (S_f)_{\ell_g(\ell,i)} + F_i^\ell \quad \ell = 1, L,\ i = 1, n_\ell$$

thereby constructing the contribution from the volume $\overrightarrow{S}_f$ on the right-hand side. In a similar fashion, we may construct the other contributions $\overrightarrow{S}_g$ and $\overrightarrow{S}_\alpha$.

As indicated during our error analysis (see proposition 4.5), the quadrature formulas that we use have to integrate the polynomials of degree $k_{quad} \geq 2k - 2$ exactly.

REMARK 5.3.– *When using the quadrature formulae (instead of using interpolation), the components of the right-hand side are calculated* directly, *without needing the mass matrix that appears in [5.6] when using the interpolation method.*

REMARK 5.4.– *With the quadrature method, one may choose either to calculate $\overrightarrow{S}_{\mathcal{I}}$ directly, or instead calculate $\overrightarrow{S}$ prior to extracting $\overrightarrow{S}_{\mathcal{I}}$. The second approach does not introduce any additional manipulations, since one has to extract $\mathbb{K}_{\mathcal{II}}$ from $\mathbb{K}$ regardless.*

These two calculation techniques coexist. The first requires more memory space as soon as the additional mass matrix is not required elsewhere. On the other hand, the second requires more calculations. Lastly, when the data f or g possess singularities (discontinuities or unboundedness), the second method using adapted quadrature formulae is preferred, due to the nature of the singularities.

REMARK 5.5.– *If we use Lobatto's quadrature formula to approximate the mass matrix $\mathbb{M}$:*

$$\int_{K_\ell} v d\Omega \simeq \frac{vol(K_\ell)}{\mathrm{d}+1} \sum_{i=1,\mathrm{d}+1} v(\bar{x}_i^\ell) \qquad (v \textit{ is a regular function on } K_\ell)$$

where $\left(\bar{x}_i^\ell\right)_{i=1,\mathrm{d}+1}$ *are the vertices of the* d*-simplex, then we obtain the approximated matrix* $\widetilde{\mathbb{M}}$*, which is diagonal.*

Furthermore, one can also construct another approximation $\hat{\mathbb{M}}$ to the mass matrix $\mathbb{M}$, by using a technique called "mass lumping":

$$\hat{\mathbb{M}}_{II} = \sum_{J=1}^{N} \mathbb{M}_{IJ} \text{ and } \hat{\mathbb{M}}_{IJ} = 0 \text{ if } I \neq J$$

and which also leads to a diagonal matrix.

These two techniques allow a substantial saving of time and space complexity. Furthermore, it can be shown that they do not deteriorate the error estimates of first-order Lagrangian finite elements (Cohen and Perret 2017).

5.1.4. *Eliminating the essential boundary conditions*

Recall that the essential boundary conditions are those that explicitly intervene within the space that the variational formulation is defined in. In the case of problem [5.1], we have to deal with the Dirichlet condition:

$$u = \alpha \text{ on } \Gamma_2.$$

We refer to [5.3] for the discrete variational formulation. As one can see, without risking efficiency when assembling the matrices and the right-hand side, it is not possible to eliminate the Dirichlet degrees of freedom at this stage. In order to form the linear system [5.4], we must extract the matrix $\mathbb{K}_{\mathcal{I}\mathcal{I}}$ from the matrix $\mathbb{K}$, and the right-hand side $S_{\mathcal{I}}$ from the vector $\overrightarrow{S}$ once their assembly has been completed.

5.1.4.1. *Actual elimination techniques*

This operation consists of eliminating the *rows and columns* that are indexed by $I, J \in \mathcal{D}$ in the matrix $\mathbb{K}$, as well as the *rows* that are indexed by $I \in \mathcal{D}$ in the vector $\overrightarrow{S}$. In practice, one creates a new way to classify the degrees of freedom not belonging to Γ_2^h (reminder: $\Sigma \subset \Gamma_2^h$). A pointer $\mathcal{P}_{\mathcal{I}}$: $I' \mapsto I$ is used that associates to the number I' (between 1 and M = card $\mathcal{I}$) the previous I which lies between 1 and N. Then we have:

$$\begin{aligned} (\mathbb{K}_{\mathcal{I}\mathcal{I}})_{I'J'} &= \mathbb{K}_{\mathcal{P}_{\mathcal{I}}(I')\mathcal{P}_{\mathcal{I}}(J')} \qquad I', J' = 1, M \\ (S_{\mathcal{I}})_{I'} &= S_{\mathcal{P}_{\mathcal{I}}(I')} \qquad I' = 1, M. \end{aligned}$$

This technique has some drawbacks since the finite element matrices have very particular storage structures, given the fact that they are sparse. It turns out to be both complicated and expensive to carry out the row and column elimination when finite element matrices are involved. This is why we prefer to use pseudo-elimination techniques, which will be presented next.

5.1.4.2. *Pseudo-elimination techniques*

Here, the case that we shall consider is when the right-hand side has been obtained via interpolation, so that its expression is given by [5.6].

Partition the matrix $\mathbb{K}$ and the right-hand side $\vec{S}$ into blocks, first using $\mathcal{I}$ and then using $\mathcal{D}$ (recall that $N = M + N_2$ and card $\mathcal{D} = N_2$):

$$\mathbb{K} = \begin{bmatrix} \mathbb{K}_{\mathcal{I}\mathcal{I}} & \mathbb{K}_{\mathcal{I}\mathcal{D}} \\ \mathbb{K}_{\mathcal{D}\mathcal{I}} & \mathbb{K}_{\mathcal{D}\mathcal{D}} \end{bmatrix} \qquad \vec{S} = \begin{pmatrix} \vec{S}_{\mathcal{I}} \\ \vec{S}_{\mathcal{D}} \end{pmatrix}.$$

Then proceed in the same way for each $N \times N$ matrix and for each vector in $\mathbb{R}^N$. Set:

$$\vec{T} = \mathbb{M}\vec{F} + P_{\mathcal{N}}\left(\mathbb{M}_{\mathcal{N}}\vec{G}\right).$$

Then, by the virtue of [5.6], we have:

$$\vec{S} = \vec{T} - \mathbb{K}\vec{\alpha} = \begin{pmatrix} \vec{T}_{\mathcal{I}} \\ \vec{T}_{\mathcal{D}} \end{pmatrix} - \begin{pmatrix} \mathbb{K}_{\mathcal{I}\mathcal{I}}\vec{\alpha}_{\mathcal{I}} + \mathbb{K}_{\mathcal{I}\mathcal{D}}\vec{\alpha}_{\mathcal{D}} \\ \mathbb{K}_{\mathcal{D}\mathcal{I}}\vec{\alpha}_{\mathcal{I}} + \mathbb{K}_{\mathcal{D}\mathcal{D}}\vec{\alpha}_{\mathcal{D}} \end{pmatrix}$$

from which we may see that:

$$\vec{S}_{\mathcal{I}} = \vec{T}_{\mathcal{I}} - \mathbb{K}_{\mathcal{I}\mathcal{I}}\vec{\alpha}_{\mathcal{I}} - \mathbb{K}_{\mathcal{I}\mathcal{D}}\vec{\alpha}_{\mathcal{D}}.$$

Now introduce the diagonal matrix: $\mathbb{D} = \text{Diag}\left(\mathbb{K}_{\mathcal{D}\mathcal{D}}\right)$ of order N_2 which is invertible since the diagonal components of $\mathbb{K}$ are strictly positive. After taking into account our previous discussion, the vector:

$$\vec{U} = \begin{pmatrix} \vec{U}_{\mathcal{I}} \\ \vec{U}_{\mathcal{D}} \end{pmatrix} \quad \text{in } \mathbb{R}^N \text{where } \vec{U}_{\mathcal{I}} \text{ is the solution of } [5.4], \text{ and } \vec{U}_{\mathcal{D}} = \vec{\alpha}_{\mathcal{D}}$$

satisfies:

$$\begin{bmatrix} \mathbb{K}_{\mathcal{I}\mathcal{I}} & 0 \\ 0 & \mathbb{D} \end{bmatrix} \begin{pmatrix} \vec{U}_{\mathcal{I}} \\ \vec{U}_{\mathcal{D}} \end{pmatrix} = \begin{pmatrix} \vec{S}_{\mathcal{I}} \\ \mathbb{D}\vec{\alpha}_{\mathcal{D}}. \end{pmatrix} \qquad [5.7]$$

We choose to interpolate $\tilde{\alpha}$ on $\overline{\Gamma}_2$ in order to define $\vec{\alpha}_{\mathcal{D}}$. By the definition of $\mathcal{D}$ and of the Lagrangian finite element basis functions w_I, we have therefore:

$$(\alpha_{\mathcal{D}})_I = \alpha(M_I) \quad \forall I \in \mathcal{D}$$

which shows that the vector $\overrightarrow{U}_{\mathcal{D}}$ is the vector formed from the nodes coming from the Dirichlet data $\tilde{\alpha}$ along the boundary $\overline{\Gamma}_2$.

The pseudo-elimination technique therefore consists of substituting the linear system [5.4] of rank M with the linear system [5.7] of rank N. No renumbering is required in doing this, and hence neither the storage structure of the matrix $\mathbb{K}$ nor of the vector $\overrightarrow{S}$ are changed – only new assignments of certain components are undertaken.

Before describing this operation in full detail, let us introduce a simplification first.

5.1.4.2.1. Choice of lifting

We are going to choose the following lifting:

$$\alpha_h = \sum_{I \in \mathcal{D}} \alpha(M_I) w_I \tag{5.8}$$

which belongs to $H^1(\Omega_h)$, since $w_I \in H^1(\Omega_h)$, for every I, and satisfies:

$$\alpha_h(M_I) = \begin{cases} \alpha(M_I) & \text{if } I \in \mathcal{D} \\ 0 & \text{if } I \notin \mathcal{D} \text{ (that is to say, } I \in \mathcal{I}). \end{cases} \tag{5.9}$$

This lifting depends on the approximation, although this is of no consequence. Indeed, the variational formulation [5.2] is equivalent to the following one:

$$\left| \begin{array}{l} \text{Find } u \in H^1(\Omega) \text{ such that} \\ \displaystyle\int_\Omega \nabla u \cdot \nabla v d\Omega = \int_\Omega f v d\Omega + \int_{\Gamma_1} g v d\Gamma \quad \forall v \in H^1_{0,\Gamma_2}(\Omega) \\ u = \alpha \quad \text{on } \Gamma_2 \end{array} \right.$$

which discretizes into:

$$\left| \begin{array}{l} \text{Find } u_h \in V_h \text{ such that} \\ \displaystyle\int_{\Omega_h} \nabla u_h \cdot \nabla \tilde{v}_h d\Omega = \int_{\Omega_h} f \tilde{v}_h d\Omega + \int_{\Gamma_1^h} g \tilde{v}_h d\Gamma \quad \forall \tilde{v}_h \in V_h^0 \\ u_h(M_I) = \alpha(M_I) \quad \forall I \in \mathcal{D} \end{array} \right.$$

hence showing that the lifting can be put into the form [5.8].

From our choice [5.9], we immediately see that the superior block of $\overrightarrow{\alpha}$ satisfies $\overrightarrow{\alpha}_{\mathcal{I}} = 0$, and consequently:

$$\overrightarrow{S}_{\mathcal{I}} = \overrightarrow{T}_{\mathcal{I}} - \mathbb{K}_{\mathcal{I}\mathcal{D}} \overrightarrow{\alpha}_{\mathcal{D}} = \left(\overrightarrow{T} - \mathbb{K}\overrightarrow{\alpha}\right)_{\mathcal{I}} \quad \text{where } \overrightarrow{\alpha} = \begin{pmatrix} 0 \\ \overrightarrow{\alpha}_{\mathcal{D}} \end{pmatrix}.$$

Therefore, it is not necessary to partition the matrix $\mathbb{K}$ effectively in order to calculate the right-hand side.

From [5.5] and the Dirichlet data interpolation, the discrete solution u_h is given by:

$$u_h = \sum_{I \in \mathcal{I}} (U_{\mathcal{I}})_I w_I + \alpha_h = \sum_{I \in \mathcal{I}} (U_{\mathcal{I}})_I w_I + \sum_{I \in \mathcal{D}} \alpha(M_I) w_I \quad \text{(from [5.8])}$$

which shows that:

$$u_h = \sum_{I=1,N} U_I w_I \quad \text{with } U_I = \begin{cases} (U_{\mathcal{I}})_I & \text{if } I \in \mathcal{I} \\ \alpha(M_I) & \text{si } I \in \mathcal{D}. \end{cases}$$

Finally, the solution to the linear system [5.7] is indeed the vector $\overrightarrow{U}$, whose components are the coordinates of the discrete solution u_h, relative to the finite element basis.

5.1.4.2.2. Practical realization of the pseudo-elimination technique

In order to perform the pseudo-elimination technique, one can either construct the numbering pointer for the Dirichlet degrees of freedom directly:

$$\mathcal{P}_{\mathcal{D}} : \left| \begin{array}{ll} \{1, \cdots, N_2\} & \rightarrow \{1, \cdots, N\} \\ I_{\mathcal{D}} & \mapsto I \ : \ \text{number of the } I_{\mathcal{D}}^{th} \text{ Dirichlet degree of freedom} \end{array} \right.$$

or instead use a test that tells us if a degree of freedom is a Dirichlet one or not, for example, by referencing each degree of freedom with a specific number, should it be a Dirichlet one.

REMARK 5.6.– *If one were to "forget" to take the Dirichlet conditions into account, one would obtain a non-invertible linear system since the pure Neumann problem is ill-posed (see section 2.4.1), and also since kernel of the matrix* $\mathbb{K}$ *is the vector space generated by the vector* $\overrightarrow{1}$, *whose components are all* 1 *(constant functions on* V_h*). Note that the property* $\mathbb{K}\overrightarrow{1} = \overrightarrow{0}$ *can be exploited as a validation test when calculating the matrix* $\mathbb{K}$*. Similarly, calculating* $(\mathbb{M}\overrightarrow{1} \,|\, \overrightarrow{1})$ *yields us the area and/or the volume of the domain, since:*

$$vol(\Omega_h) = \int_{\Omega_h} d\Omega = \int_{\Omega_h} \sum_I w_I \sum_J w_J d\Omega = (\mathbb{M}\overrightarrow{1} \,|\, \overrightarrow{1}).$$

In cases when the Dirichlet pointer $\mathcal{P}_{\mathcal{D}}$ is used, a pseudo-elimination algorithm (Algorithm 5.3) is then realized.

$\overrightarrow{\alpha} = \begin{pmatrix} 0 \\ \overrightarrow{\alpha}_{\mathcal{D}} \end{pmatrix}$	*Dirichlet data*
$\overrightarrow{S} = \overrightarrow{T} - \mathbb{K}\overrightarrow{\alpha}$	*calculate right-hand side*
for $I_{\mathcal{D}} = 1, N_2$	*loop over Dirichlet nodes*
$I = \mathcal{P}_{\mathcal{D}}(I_{\mathcal{D}})$	
$\left(\overrightarrow{S}\right)_I = \mathbb{K}_{II}\alpha(M_I)$	
for $J = 1, N$	*loop over dof's*
if $J \neq I$ **then**	
$\mathbb{K}_{IJ} = 0$ and $\mathbb{K}_{JI} = 0$	*"elimination"*

Algorithm 5.3. *Pseudo-elimination algorithm with Dirichlet conditions*

The pseudo-elimination technique can be generalized to other types of essential conditions: transmission or periodic conditions, for example. But then it is more complex to implement as these types of conditions couple together several degrees of freedom.

There are other procedures for taking into account Dirichlet conditions. For example, in many computational codes, the technique of making the diagonal coefficients $\mathbb{K}_{II}$ "very large", for each index I corresponding to a Dirichlet node, is used (penalization). This technique has the advantage of being easy to implement and fast. Nevertheless, it has two flaws: first is that the concept of "very large" must be defined, and second is that the presence of very large coefficients deteriorates the conditioning of the matrix, which in some situations can be a nuisance. It is also possible to deal with Dirichlet conditions via a penalization-regularization type of approach (see exercise 35) that introduces the "approximate" boundary condition:

$$u_\varepsilon + \varepsilon \frac{\partial u_\varepsilon}{\partial n} = \alpha \text{ on } \Gamma_2.$$

We can show that when $\varepsilon \to 0$, the obtained solution u_ε converges to the initial value problem: $\|u - u_\varepsilon\|_{H^1(\Omega)} \leq C\sqrt{\varepsilon}$ with $C > 0$ depending solely on the data. This approach no longer requires any algebraic treatment of the matrices, but still presents defects of a similar nature to those of the previous approach: a choice of ε has to be made and the ill-conditioning of the linear system (again, see exercise 35).

5.2. Algorithmic considerations

In addition to the fact that finite element methods allow us to approach weak solutions with great generality, there is also the advantage that *sparse linear systems* arise when using them. Indeed, recall that $\mathbb{K}_{IJ} = 0$ as soon as M_I and M_J do not belong to the same domain, K_ℓ (see [3.31]).

5.2.1. *A particular case*

Let us consider the case of a regular mesh of the unit square composed of $N = n^2$ degrees of freedom. Each vertex belongs to (at most) four squares (with exactly four squares per interior vertex), and whose union includes (at most) nine vertices.

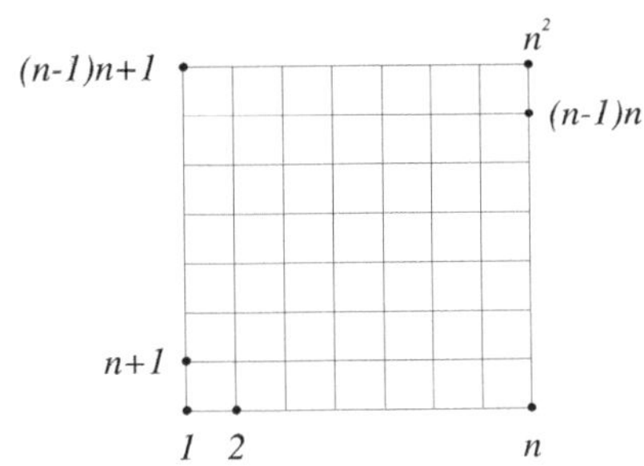

Figure 5.3. *Quadrangular unit square mesh*

We choose to use a Q^1 finite element approximation. From the mesh structure, we deduce that, for a given I, card $\{J \text{ s.t. } \mathbb{K}_{IJ} \neq 0\} \leq 9$. There are therefore less than $9N$ non-zero terms among the N^2 terms of the matrix, that is, it has a filling rate less than $9N^{-1}$. However, in the case we are considering, the mesh size is $h \sim n^{-1}$ and thus $N = n^2 \to \infty$ when $h \to 0$. The filling rate of the matrix therefore tends toward 0 when $h \to 0$: in this case, we have a **sparse matrix**. Also note that the data of the pointer ℓ_g, which allows us to pass from the local degrees of freedom numbering to their global numbering (see [3.51]), a priori tells us the matrix structure.

The main storage techniques and algorithms used to solve these linear systems are described in Appendix 1.

5.2.2. *The general 2D case*

In this section, we are going to show that matrices resulting from finite element discretization are *sparse* in a general way, by always referring to the idea that their filling rate tends toward 0 when the mesh size h tends toward 0. To this end, we are going to treat the case that uses a discretization generated by P^1-*Lagrangian finite elements*, for a problem defined on an *polygonal* open subset of $\mathbb{R}^2$ (i.e. a triangular mesh) in detail, before moving on to other cases set in $\mathbb{R}^2$ or $\mathbb{R}^3$. To fix ideas, suppose that we wish to solve a Neumann problem. Let $(\mathcal{T}_h)_h$ be a sequence of triangulations. After choosing h, the triangulation $\mathcal{T}_h$ consists of N vertices, A edges and L triangles.

By construction, the dimension of the discrete space V_h is equal to N.

Next, let us examine the number of non-zero components Z_1 there are in the stiffness matrix $\mathbb{K}$ (or in the mass matrix $\mathbb{M}$), which has size $N \times N$.

The **fill rate** τ_1 is then defined by:

$$\tau_1(N) = \frac{Z_1}{N^2}. \tag{5.10}$$

We have seen in section 3.2 that if M_I and M_J are two vertices that do not belong to the same triangle of $\mathcal{T}_h$, then $\mathbb{K}_{IJ} = 0$. Thus, for $\mathbb{K}_{IJ}$ to be non-zero, it is *necessary*[3] for:

$$M_I = M_J \quad \text{or} \quad [M_I, M_J] \text{ is a side of a triangle}$$

to hold.

The first case occurs exactly N times, whereas the second occurs exactly $2A$ times, with an edge being represented either by $[M_I, M_J]$, or by $[M_J, M_I]$. To conclude:

$$Z_1 \leq N + 2A. \tag{5.11}$$

To evaluate the filling rate $\tau_1(N)$, we will express A (as well as L) as a function of N. In the following, we will denote N_b (respectively, A_b) to be the number of vertices (respectively, number of edges) of $\partial\Omega$, and $N_i = N - N_b$ (respectively, $A_i = A - A_b$).

PROPOSITION 5.1.– *Let Ω be a polygon of $\mathbb{R}^2$ without any holes, and $\mathcal{T}_h$ be a triangulation of Ω. Then we have the relations:*

$$L = 2N - N_b - 2. \tag{5.12}$$

$$A = 3N - N_b - 3. \tag{5.13}$$

3 This condition *is not sufficient*. Indeed, on certain meshes (e.g. those composed of isosceles right-angled triangles), it can happen that, although $[M_I, M_J]$ equals one side of a triangle, yet $\nabla w_I(x) \perp \nabla w_J(x)$ for every $x \in \Omega$, and in this case $\mathbb{K}_{IJ} = 0$ (see exercise 26 for an example). Conversely, $\mathbb{K}_{II} > 0$ for each index in I.

Proof.

Step 1: Let us begin by *counting the angles.*

i) In each triangle, the sum of the internal angles equals π radians: we count $\pi \times L$ radians, summing over each internal angle of each triangle.

ii) Reasoning per vertex in the decomposition above:

- the sum of the angles about an *internal* vertex equals 2π radians;

- for the boundary vertices, we *do not know* the value of the angles about any *given* vertex. Conversely, we can recursively prove that the sum over the internal angles about any vertex of a polygon, with K sides or vertices, will equal $(K-2)\pi$.

To recap: $\pi L = 2\pi N_i + (N_b - 2)\pi$, which gives us $L = 2N_i + N_b - 2 = 2N - N_b - 2$.

Step 2: *Edge count.*

i) In each triangle, there are *three* edges: so $3 \times L$ edges traversing every triangle.

ii) Reasoning per edge in the count above:

- an edge *of the boundary* is counted exactly *once*, because it belongs to a single triangle;

- an *internal* edge is counted exactly *twice*, because it is shared between two triangles.

To recap: $3L = A_b + 2A_i$, or $3L = 2A - A_b$.

Note that $N_b = A_b$, since a polygon has the same number of vertices and edges. We then take the previous results, and we eliminate the number of triangles L to find that:

$$2A - A_b \overset{Step\ 2}{=} 3L \overset{Step\ 1}{=} 6N - 3N_b - 6$$

and finally that $A = 3N - N_b - 3$. ■

REMARK 5.7.– *The presence of holes only changes the angle counting step. It is easy to check that the formulas [5.12] and [5.13] only marginally differ, with the constant terms (respectively,* -2 *and* -3*) having to be modified to take into account the number of holes.*

COROLLARY 5.1.– *As h tends to 0, the dimension of V_h tends to infinity:*

$$\lim_{h \to 0} N = \infty. \qquad [5.14]$$

REMARK 5.8.– *This result is of course strongly desired in order to be able to apply the general framework used in the approximation methods (see section 3.1), but it first needs to be proven...*

Proof of corollary 5.1. From [3.49], we have that $\text{Area}(T_\ell) \leq \pi h^2$, for $\ell = 1, L$. Since $\text{Area}(\Omega) = \sum_\ell \text{Area}(T_\ell)$, we deduce from this that $\text{Area}(\Omega) \leq \pi h^2 L$ and consequently:

$$\lim_{h \to 0} L = \infty.$$

Then, from [5.12], $L < 2N$, the conclusion follows. ■

THEOREM 5.1.– *Let Ω be a polygon in $\mathbb{R}^2$. If a discretization is made up of P^1 finite elements, then the matrices $\mathbb{K}$ and $\mathbb{M}$ are sparse.*

Proof. From [5.11] and [5.13], we have that $N < Z_1 < 7N$. The filling rate is therefore bounded between:

$$\frac{1}{N} < \tau_1(N) < \frac{7}{N}.$$

However, according to corollary 5.1, N tends toward infinity when h tends to 0, and the conclusion follows. ■

Remarkably, this result generalizes to higher order Lagrangian finite elements, as well as to other types of finite elements. The fundamental results that allow us to reach this conclusion are based upon counting, and are the contents of proposition 5.1 and corollary 5.1. Let us take two examples that use P^2-Lagrangian finite elements and the P^3-Hermitian finite elements.

For *P^2-Lagrangian finite elements*, there are six degrees of freedom per triangle: one for each vertex and one for each edge midpoint. The total number of degrees of freedom (and the dimension of V_h) is therefore equal to $N_2 = N + A$, and $\mathbb{K}$ and $\mathbb{M}$ are $N_2 \times N_2$ matrices. For $\mathbb{K}_{IJ}$ to be non-zero, it is necessary that:

– if M_I and M_J are two vertices: $M_I = M_J$, or otherwise $[M_I, M_J]$ is a side of a triangle;

– if M_I and M_J are two edge midpoints: $M_I = M_J$, or otherwise M_I and M_J belong to the same triangle;

– if M_I and M_J are of different types: M_I and M_J belong to the same edge, otherwise they are in the same triangle.

By counting each set of configurations, we then find that the number of non-zero elements Z_2 is bounded above by:

$$Z_2 \leq (N + 2A) + (A + 6L) + (4A + 6L) = N + 7A + 12L.$$

From [5.12] and [5.13], $Z_2 < 46N$ and $N_2 > 3N$. Thus, the filling rate $\tau_2(N_2)$ for the P^2-Lagrangian finite element is less than:

$$\tau_2(N_2) < \frac{46N}{9N^2} = \frac{46}{9} \times \frac{1}{N}$$

and tends to 0 when h tends to 0.

Let us compare the filling rates, τ_1 and τ_2 (for a *same* array size). To do so, we make the "reasonable" assumption that $N_b = o(N)$ when $N \to \infty$, that is, the number of boundary vertices is negligible relative to the total number of vertices as $N \to \infty$. If the case described in footnote 3 (p. 203) does not occur, then we have the following equivalence results:

$$\tau_1(N) \sim \frac{7}{N} \quad \text{for the } P^1 \text{ finite element}$$

$$\tau_2(N_2) \sim \frac{46N}{N_2^2} \sim \frac{11.5}{N_2} \quad \text{for the } P^2 \text{ finite element.}$$

Thus, for a given dimension of the discretization space, the filling rate grows as the order increases (with the multiplicative factor transitioning from 7 to 11.5). This property holds in general.

For the P^3-*Hermitian finite element*, there are 10 degrees of freedom per triangle: one for each vertex, two for each edge and one for the middle. The total number of degrees of freedom is equal to $N_3 = N + 2A + L$ this time, and $\mathbb{K}$ and $\mathbb{M}$ are $N_3 \times N_3$ matrices. By counting the configurations for which $\mathbb{K}_{IJ}$ is non-zero, we find that the total number of non-zero elements Z_3 is bounded above by:

$$Z_3 \leq N + 10A + 31L.$$

From [5.12] and [5.13], we have $Z_3 < 93N$ and $N_3 > 6N$, and also the filling rate $\tau_3(N_3)$ for the P^3-Hermitian finite element is less than:

$$\tau_3(N_3) < \frac{31}{12} \times \frac{1}{N}.$$

Thus, τ_3 tends toward 0 as h tends toward 0.

In the general case, by listing the degrees of freedom per triangle, we obtain the count $N_{tot} = c_N + N + c_A + A + c_L + L$. The matrices $\mathbb{K}$ and $\mathbb{M}$ have size $N_{tot} \times N_{tot}$. After counting the configurations in which $\mathbb{K}_{IJ} \neq 0$, we deduce that there exist the positive constants, C_N, C_A and C_L, which are independent of the triangulation $\mathcal{T}_h$, such that:

$$Z_{tot} \leq C_N N + C_A A + C_L L.$$

From [5.12] and [5.13]:

$$Z_{tot} \leq (C_N + 3C_A + 2C_L)N \quad \text{and} \quad N_{tot} > (c_N + 2c_A + c_L)N$$

the filling rate is bounded above by:

$$\tau(N_{tot}) < \frac{(C_N + 3C_A + 2C_L)}{(c_N + 2c_A + c_L)^2} \times \frac{1}{N}$$

which tends toward 0 as h tends toward 0.

Of course, we could reason in the same way for quadrangular meshes, with the only differences coming from the quadrilaterals and edges numbers for this type of mesh (which would modify [5.12] and [5.13] accordingly). Hence, the general result below.

THEOREM 5.2 (Sparse matrices for 2D finite elements).– *Let Ω be a polygon in $\mathbb{R}^2$. Then the matrices $\mathbb{K}$ and $\mathbb{M}$, obtained using finite element discretization, are sparse.*

5.2.3. *The general 3D case*

One can reason along the same lines for a problem (e.g. a Neumann problem) defined on an open *polyhedral* subset of $\mathbb{R}^3$ (a tetrahedral mesh).

Let $(\mathcal{T}_h)_h$ be a sequence of triangulations of the polyhedron. Let the triangulation $\mathcal{T}_h$ (h fixed) contain N vertices, A edges, F faces and L tetrahedra.

To establish that the finite element matrices $\mathbb{K}$ and $\mathbb{M}$ are sparse, one would like to determine the number of edges, faces and tetrahedra, in terms of the number of vertices, to obtain the 3D versions of [5.12] and [5.13]. As we will see later, while one cannot ascertain exact formulas, frameworks can be created that nevertheless let us conclude the desired result. We denote the number of vertices (respectively, number of edges, faces) by N_b (respectively, by A_b, F_b) in $\partial\Omega$, and $N_i = N - N_b$ (respectively, $A_i = A - A_b$, $F_i = F - F_b$).

First, A_i, A_b, F_i and F_b can be determined as functions of N_i, N_b and L. For this, we will use the four relations below:

– boundary edge count: $3F_b = 2A_b$;

– face count: $4L = 2F_i + F_b$;

– Euler's formula for the boundary mesh: $N_b - A_b + F_b = 2$;

– Euler's formula for polyhedra: $N - A + F - L = 1$.

To fix ideas, we assume that the polyhedron is connected, simply connected and without holes. If this were not the case, then the values of each constant in the formulas must be modified. From above, we get the formulae:

$$A_b = 3N_b - 6, \quad F_b = 2N_b - 4, \quad A_i = N_i - N_b + L + 3, \quad F_i = 2L - N_b + 2.$$

Now it remains to eliminate L as a function of N_i and N_b. For this, a natural idea here is to count the *solid angles* at each vertex of the tetrahedra, which we denote by $(\omega^i_{K_\ell})_{i=1,4}$. For each tetrahedron, recall:

$$0 \leq \sum_{i=1,4} \omega^i_{K_\ell} \leq 2\pi. \tag{[5.15]}$$

Indeed, the sum of the solid angles of a tetrahedron is neither constant nor bounded from below by a strictly positive value.

The sum of the solid angles at an interior vertex of $\mathcal{T}_h$ is 4π. For a vertex of $\mathcal{T}_h$ located on the boundary, the sum is greater than $\omega_{\partial\Omega}$, which is the smallest solid angle value for any vertex of $\partial\Omega$. By summing the inequalities [5.15] over each tetrahedron, we arrive at the *lower bound*:

$$4N_i + \frac{\omega_{\partial\Omega}}{\pi} N_b \leq 2L.$$

Unfortunately, we do not know how to *bound L from above* as a function of N_i and N_b. To achieve this, an *additional assumption* on the triangulations must be made: assume that the family of triangulations $(\mathcal{T}_h)_h$ is *regular*. Indeed, returning to the discussion after the condition [4.12] was stated, then in this case we see that:

$$\exists \omega_0 > 0,\ \forall h,\ \forall K_\ell \in \mathcal{T}_h,\ \omega_{K_\ell} \geq \omega_0 \quad (\omega_{K_\ell} \text{ is the smallest solid angle of } K_\ell).$$

In this case, [5.15] becomes:

$$\forall h,\ \forall K_\ell \in \mathcal{T}_h,\ 4\omega_0 \leq \sum_{i=1,4} \omega^i_{K_\ell} \leq 2\pi. \tag{[5.16]}$$

By summing the inequalities [5.16] over each tetrahedron, we arrive at:

$$L \leq \frac{\pi}{\omega_0}(N_i + N_b)$$

providing us with the desired upper bound. As stated so far, we no longer know how to express the relations between the various quantities using equalities, but we still have uniform upper and lower bounds. Continuing from here, one may reason as in the 2D case to obtain the general result below.

THEOREM 5.3 (Sparse matrices for 3D finite elements).– *Let* Ω *be a polyhedron in* $\mathbb{R}^3$, *and let* $(\mathcal{T}_h)_h$ *be a family of regular triangulations of* Ω. *Then the matrices* $\mathbb{K}$ *and* $\mathbb{M}$, *obtained using finite element discretization, are sparse.*

In conclusion, for us to use this advantage (i.e. the a priori knowledge regarding sparse matrices that we need to manipulate in 2D and in 3D), it is imperative for us to use storage structures that allow to only store the non-zero terms, in addition to using algorithms that solve linear systems using only the non-zero terms, in order to reduce the calculation time: these aspects are presented and then treated in Appendix 1. Finally, for effective computer implementation, software and/or programming languages that can deal with sparse structures effectively, and can perform the required operations on them, are necessary.

5.3. Some numerical illustrations

The purpose of this section is to illustrate concepts that implement what we have previously discussed. We will present two examples successively, which reveal the relative simplicity when implementing P^1 and P^2 finite elements for two-dimensional problems within a Matlab environment. In the first, we consider the Laplace equation, showing that the errors obtained are completely in agreement with the theoretical estimates presented in Chapter 4, Theorem 4.1 and Lemma 4.4. Also, we will show how easy it is to generalize the calculation to include an inhomogeneous medium, in which may appear transmission conditions that the variational formulation takes into account naturally. We then tackle a problem of a *vectorial* nature, namely dealing with homogeneous linear isotropic elasticity. The techniques remain fundamentally the same, with the difficulty residing in the vectorial aspect of the problem, which requires a little more complicated management.

For each example, we will recall its theoretical background and discuss its implementation aspects, as well as the associated Matlab codes that can easily be translated into other programming languages (Scilab, Python, Julia, etc.).

The finite element method is based upon mesh data (triangulation for P^1 and P^2 finite elements). There is general open-access meshing software (e.g., *gmsh, grummp, netgen, mefisto*)[4], but here we have only used elementary meshing tools – more than enough for our purposes – that have been developed in Matlab. These tools essentially produce four tables for either P^1 or P^2 instances:

– S: $ns \times 2$ table containing the coordinates of the mesh nodes; ns total number of nodes;

4 See http//www.robertschneiders.de/meshgeneration/software.html for a very complete list of what meshing tools are available.

– T: $nt \times nd$ table containing the triangle labeling; nt number of triangles; nd number of nodes per triangle: $nd = 3$ in P^1 and $nd = 6$ in P^2 (for the second case, with the convention that the first three numbers correspond to the vertices of a triangle);

– BR: $nt \times 3$ table providing a reference number for each edge of the triangles (allows for the boundary conditions to be processed);

– RT: $nt \times 1$ table providing one reference number per triangle (allows for different media to be treated).

The conventions for the last two tables are that, on the one hand, any boundary edge is subject to a single boundary condition (Dirichlet, Neumann, etc.), and, on the other hand, each triangle contains a homogeneous material.

In the following, denote by $(M_I)_{I=1,ns}$ the nodes of the mesh, by $(T_\ell)_{\ell=1,nt}$ the triangles of the mesh and, for either P^1 or P^2, by w_I the global basis function attached to node M_I. We assume that the open subset Ω is polygonal, in which case $\overline{\Omega} = \bigcup_{\ell=1,nt} T_\ell$ (see [3.47]), and that the mesh is admissible (see [3.48]), so that the ns-dimensional vector space:

$$V_h = \text{vect}(w_I)_{I=1,ns}$$

is contained in $H^1(\Omega)$ (internal approximation). Finally, assume that the family of triangulations is regular (see [4.12]).

5.3.1. *The Laplace equation*

Let us consider the Dirichlet problem on a connected and bounded open subset Ω of $\mathbb{R}^2$ whose boundary is denoted by Γ, with the data $f \in L^2(\Omega)$ and $g \in H^{1/2}(\Gamma)$:

$$\begin{cases} \text{Find } u \in H^1(\Omega) & \text{such that} \\ -\Delta u = f & \text{in } \Omega \\ u = g & \text{on } \Gamma \end{cases}$$

whose variational formulation is:

$$\left| \begin{array}{l} \text{Find } u \in H^1(\Omega) \text{ such that} \\ u = g \text{ on } \Gamma \text{ and } \displaystyle\int_\Omega \nabla u \cdot \nabla v \, d\Omega = \int_\Omega f \, v \, d\Omega \quad \forall v \in H_0^1(\Omega). \end{array} \right.$$

Recall that this problem is well posed. Moreover, if we denote by $\mathcal{D}$ the indexing set of the nodes located along the boundary ($\mathcal{D} = \{I$ such that $M_I \in \Gamma\}$) and $\mathcal{I} = \{1, ..., ns\} \setminus \mathcal{D}$, an approximation in V_h of this formulation is:

$$\left| \begin{array}{l} \text{Find } u_h \in V_h \text{ such that} \\ u_h(M_I) = g(M_I)\ \forall I \in \mathcal{D} \text{ and} \\ \displaystyle\int_\Omega \nabla u_h \cdot \nabla v_h \, d\Omega = \int_\Omega f\, v_h \, d\Omega \qquad \forall v_h \in V_h^0 \end{array} \right.$$

where $V_h^0 = \text{vect}(w_I)_{I \in \mathcal{I}}$. We can reformulate the approximate problem into a matrix format, by denoting $u_h = \sum_{J=1,ns} U_J w_J$, $\mathbb{K}_{IJ} = \int_\Omega \nabla w_J \cdot \nabla w_I \, d\Omega$, $G_I = g(M_I)$ and $S_I = \int_\Omega f\, w_I \, d\Omega$ for $1 \le I, J \le ns$, and by partitioning the set of indices into $\mathcal{I}$ and $\mathcal{D}$. With the help of the pseudo-elimination technique (see [5.7] with $\mathbb{D} =$ Diag $(\mathbb{K}_{\mathcal{D}\mathcal{D}})$), we end up with:

$$\begin{bmatrix} \mathbb{K}_{\mathcal{I}\mathcal{I}} & 0 \\ 0 & \mathbb{D} \end{bmatrix} \begin{pmatrix} \vec{U}_\mathcal{I} \\ \vec{U}_\mathcal{D} \end{pmatrix} = \begin{pmatrix} \vec{S}_\mathcal{I} - \mathbb{K}_{\mathcal{I}\mathcal{D}} \vec{G}_\mathcal{D} \\ \mathbb{D}\vec{G}_\mathcal{D}. \end{pmatrix}$$

If the data f are regular ($f \in \mathcal{C}^0(\overline{\Omega})$), it is possible to approximate S_I by using the interpolate of f ($\pi_h f = \sum_{J=1,ns} f(M_J) w_J$):

$$S_I \simeq \int_\Omega \pi_h f\, w_I \, d\Omega = \left(\mathbb{M}\vec{F}\right)_I \quad \text{with } F_I = f(M_I) \text{ and } \mathbb{M}_{IJ} = \int_\Omega w_J\, w_I \, d\Omega.$$

In summary, we must:

– calculate the mass matrix $\mathbb{M}_{IJ} = \displaystyle\int_\Omega w_J\, w_I \, d\Omega$;

– calculate the stiffness matrix $\mathbb{K}_{IJ} = \displaystyle\int_\Omega \nabla w_J \cdot \nabla w_I \, d\Omega$;

– assemble the system $\begin{bmatrix} \mathbb{K}_{\mathcal{I}\mathcal{I}} & 0 \\ 0 & \mathbb{D} \end{bmatrix} \begin{pmatrix} \vec{U}_\mathcal{I} \\ \vec{U}_\mathcal{D} \end{pmatrix} = \begin{pmatrix} (\mathbb{M}\vec{F})_\mathcal{I} - \mathbb{K}_{\mathcal{I}\mathcal{D}} \vec{G}_\mathcal{D} \\ \mathbb{D}\vec{G}_\mathcal{D} \end{pmatrix}$.

In practice, we calculate the complete matrices $\mathbb{K}$ and $\mathbb{M}$ without distinguishing between the indices of $\mathcal{I}$ and $\mathcal{D}$, before applying the *pseudo-elimination* procedure which provides the previous system. As we have already seen, the *pseudo-elimination* technique does not modify the index order as the previous presentation might suggest.

The following Matlab script solves the Dirichlet problem on the square of side length 4, whose solution is $u(x, y) = \cos(xy)$:

```
%finite element calculation
[S,T,BR,RT]=rectangle_triangle([0 4 0 4],40,40,1);          %mesh
[K,M]=FE_calc_2D(S,T,RT);                                   %FE matrices
```

```
%test of U=cos(xy)
X=S(:,1);Y=S(:,2);
Uex=cos(X.*Y);                                    %exact sol.
B=M*(Uex.*(X.*X+Y.*Y));                           %right-hand side
%take Dirichlet cond. into account
Node_dir=node_bound(S,T,BR,[2 3]);                %Dirichlet nodes
G=Node_dir.*Uex;                                  %Dirichlet data
[Ke,Be]=cd_Dirichlet(K,B,Node_dir,G);             %pseudo-elimination
%solve system, calculate L2 and H1 errors
U=Ke\B;                                           %solve system
E=U-Uex;                                          %difference
el2=sqrt(E'*M*E);eh1=sqrt(E'*K*E);                %L2 and H1 error
%plot
figure; trisurf(T,S(:,1),S(:,2),U);shading interp;
```

It is interesting to examine the code for the function `FE_calc_2D.m` carefully, as it constitutes the core ideas behind of the finite element method. This function constructs the mass $\mathbb{M}$ and stiffness $\mathbb{K}$ matrices by forming a loop over each triangle in the mesh. On each triangle T_ℓ, the so-called elementary matrices are then calculated:

$$\mathbb{K}^\ell_{ij} = \int_{T_\ell} \nabla\tau^\ell_i \cdot \nabla\tau^\ell_j \, d\Omega \quad \text{and} \quad \mathbb{M}^\ell_{ij} = \int_{T_\ell} \tau^\ell_i \tau^\ell_j \, d\Omega \qquad \forall i,j = 1, nd.$$

In the code proposed, we have adopted a general point of view that consists of calculating the integrals via a change in variable, bringing the integral back to the unit triangle $\hat{T}$, and a numerical quadrature formula (see Chapter 4). The quadrature formula used involves seven points, and is exact for degree five polynomials. This point of view lets us treat cases involving equations with varying coefficients, such as $-\operatorname{div} k(x)\nabla u + q(x)u = f$.

Of course, if we are interested in constant coefficient problems that are approximated by P^1 finite elements, then it is more efficient to use the explicit formulae for the elementary matrices, that is, without going through $\hat{T}$.

```
function [K,M]=FE_calc_2D(S,T,RT,fK,fM)
if(nargin==3) fK=@un;fM=@un;end,                  %fK coeff. rel. to K
if(nargin==4) fM=@un;end,                         %fM coeff. rel. to M
%7 point quadrature formula
os=sqrt(15);s3=1./3.;
pp1=(6.-os)/21.;pp2=(6.+os)/21.;
pp3=(9.+2.*os)/21.;pp4=(9.-2.*os)/21}.;
quadT_pts=[s3 s3;pp1 pp1;pp1 pp3;pp3 pp1;...      %quadrature points
           pp2 pp2;pp2 pp4;pp4 pp2];
pp1=(155.-os)/2400.;pp2=(155.+os)/2400.;
quadT_wgts=[9./80.;pp1;pp1;pp1;pp2;pp2;pp2];      %quadrature weights
%initialization
nt=size(T,1);ns=size(S,1);q=size(T,2);            %various sizes
nbq=length(quadT_wgts);
K=sparse(ns,ns);M=sparse(ns,ns);                  %def. sparse matrices
%loop over each triangle
for t=1:nt,                                       %index of triangle t
    St=[S(T(t,:),:)];                             %vertices of triangle t
    S21=St(2,:)-St(1,:);S31=St(3,:)-St(1,:);
    delta=S21(1)*S31(2)-S21(2)*S31(1);
    Jfmt=[S31(2) -S21(2);-S31(1) S21(1)]/delta;   %inverse Jacobian
```

```
        Mt=zeros(q,q);Kt=zeros(q,q);                                  %elementary matrices
        for k=1:nbq,                                                  %loop over quadrature
            x=quadT_pts(k,1);y=quadT_pts(k,2);
            if(q==3) w=[1-x-y x y];gw=[-1 1 0;-1 0 1];                %P1 basis fct.
            else a=1-x-y;b=2*x-1;c=2*y-1;                             %P2 basis fct.
                 w=[a*(1-2*x-2*y),b*x*,c*y*,4*x*a,4*x*y,4*y*a];
                 gw=[4*(x+y)-3 4*x-1 0 4*(1-2*x-y) 4*y -4*y;
                     4*(x+y)-3 0 4*y-1 -4*x 4*x 4*(1-x-2*y)];
            end,
        P=St'*[1-x-y;x;y];pk=quadT_wgts(k)*abs(delta);                %pt. of physical plane
        Mt=Mt+fM(P(1),P(2),RT(t))*pk*w'*w;                            %elem. mass matrix
        jg=Jfmt*gw;
        Kt=Kt+fK(P(1),P(2),R(t))*pk*jg'*jg;                           %elem. stiff. matrix
    end,
    %assembly of M and K
    I=T(t,:);
    K(I,I)=K(I,I)+Kt;                                                 %assembly of K
    M(I,I)=M(I,I)+Mt;                                                 %assembly of M
    function [z]=un(x,y,r) z=1}.;                                     %function f=1
```

Taking the Dirichlet conditions into account constitutes the other important part of the finite element code. We must first determine the list of nodes that are subject to a Dirichlet condition:

```
function [Nodes]=bound_nodes(S,T,Boundary,ref_bound)
ns=size(S,1);                                                 %number of vertices
nt=size(T,1);                                                 %number of triangles
nd=size(T,2);                                                 %number of nodes per triangle
Nodes=zeros(ns,1);
for t=1:nt,
    for a=1:3;
        as=a+1;am=a+3;if(a==3)as=1;end,
        if(ismember(Bound(t,a),ref_bound))                    %vertex on a Dirichlet boundary
            Nodes(T(t,a))=1;
            Nodes(T(t,as))=1;
            if(nd==6) Nodes(T(t,am))=1;end,                   %for the P2
        end,
    end,
end
```

Then once we have this list of nodes, it is straightforward to generate the final linear system:

```
function [Ae,Be] = cd_Dirichlet(A,B,Node_dir,G)
ns=size(A,1);Ae=A;Be=B;                                       %initialization
if(nargin==3) G=zeros(ns,1);                                  %homog. cond.
else Be=B-A*(Node_dir.*G);end,                                %rhs correction
%pseudo-elimination
for I=1:ns,
    if(Node_dir(I)==1),
        Ae(I,:)=0;Ae(:,I)=0;
        Ae(I,I)=A(I,I);Be(I)=A(I,I)*G(I);                     %elimination
    end,
end
```

Figure 5.4 presents the result obtained by executing the previous Matlab script.

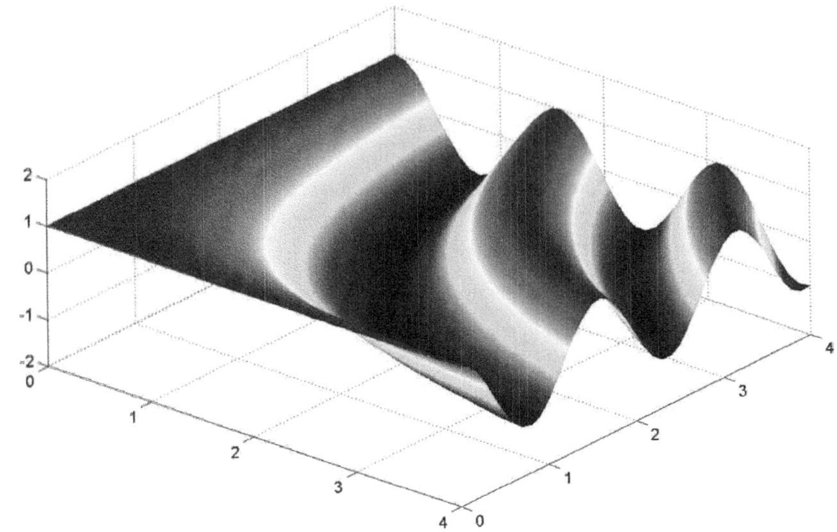

Figure 5.4. *Approximate solution for the Laplace equation using finite elements*

One can convince themselves of the quality that the finite element approximation possesses, by viewing the $L^2(\Omega)$ and the $H^1(\Omega)$ norms of the error, that is, by looking at the quantities:

$$\begin{aligned}\|u_h - u\|_{L^2(\Omega)} &= \Big(\int_\Omega |u_h - u|^2 d\Omega\Big)^{1/2} \\ &\simeq \Big(\int_\Omega |u_h - \pi_h u|^2 d\Omega\Big)^{1/2} = \big(\mathbb{M}(\vec{U} - \vec{U}_\pi)|(\vec{U} - \vec{U}_\pi)\big)^{1/2} \\ |u_h - u|_{1,\Omega} &= \Big(\int_\Omega |\nabla u_h - \nabla u|^2 d\Omega\Big)^{1/2} \\ &\simeq \Big(\int_\Omega |\nabla u_h - \nabla \pi_h u|^2 d\Omega\Big)^{1/2} = \big(\mathbb{K}(\vec{U} - \vec{U}_\pi)|(\vec{U} - \vec{U}_\pi)\big)^{1/2}\end{aligned}$$

where $\vec{U}_\pi$ represents the component vector of $\pi_h u$ relative to the finite element basis $(w_I)_{I=1,ns}$. Using the interpolated $\pi_h u$ for the solution u is only valid if the function u is regular enough (again, if $u \in \mathcal{C}^0(\overline{\Omega})$).

For the exact solution $u(x, y) = \cos(xy)$, the error curves are represented in Figure 5.5 as a function of how fine (which is characterized by h) the mesh is, represented logarithmically. The theoretically predicted asymptotic behaviors are obtained when $h \to 0$, namely that the standard error $H^1(\Omega)$ is of the same order as the finite element degree, and that the $L^2(\Omega)$ error is one order greater. The gradients of each dotted line equals 1, 2, or 3, respectively.

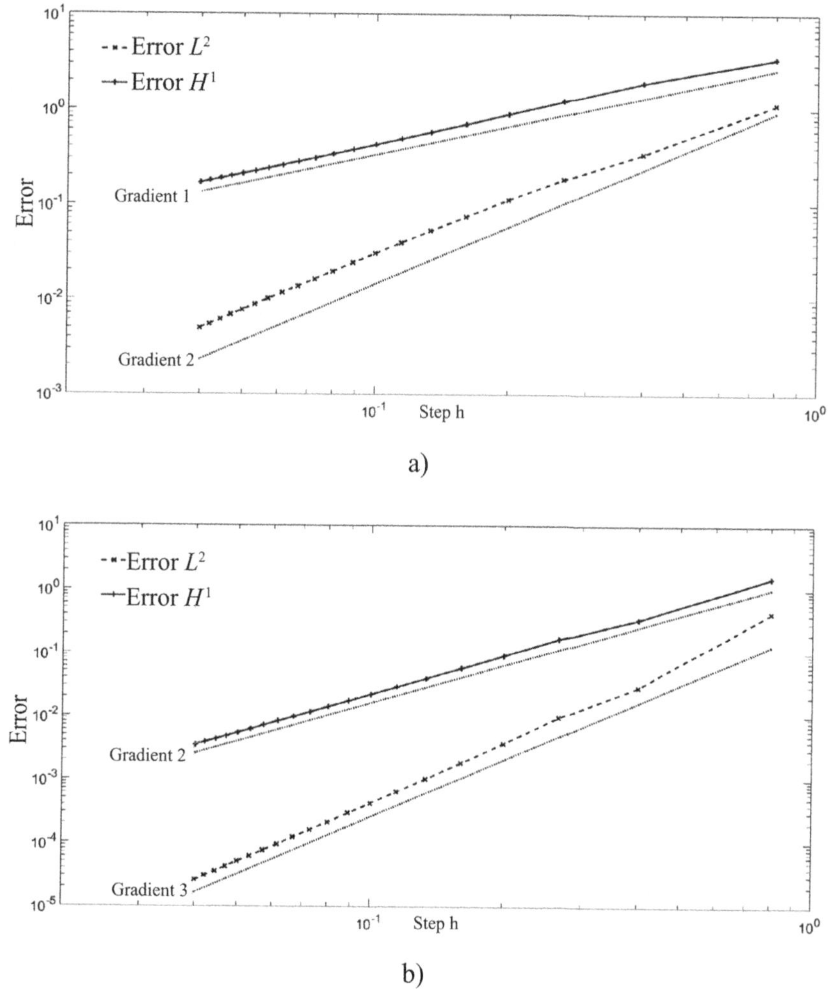

Figure 5.5. *Finite element errors with: (a) P^1 and (b) P^2*

5.3.2. *The diffusion equation*

The previous Matlab scripts allow us to process the diffusion equation (with varying coefficients). For example, the diffusion problem posed upon the open subset

Ω, with $\overline{\Omega} = \overline{\Omega}_1 \cup \overline{\Omega}_2$ and where Ω_1 and Ω_2 are two disjoint open subsets, with $k \in L^\infty(\Omega)$ and $f \in L^2(\Omega)$:

$$\begin{cases} \text{Find } u \in H^1(\Omega) & \text{such that} \\ -\operatorname{div}(k(x)\nabla u) = f & \text{on } \Omega_1 \cup \Omega_2 \\ u = 0 & \text{on } \Gamma_d \\ \dfrac{\partial u}{\partial n} = 0 & \text{on } \Gamma_n \\ \left[k \dfrac{\partial u}{\partial n}\right] = 0 & \text{on } \Sigma \end{cases} \qquad \text{where } k(x) = \begin{cases} k_1 \text{ on } \Omega_1 \\ k_2 \text{ on } \Omega_2 \end{cases}$$

whose variational formulation is:

$$\left| \begin{array}{l} \text{Find } u \in H^1_{0,\Gamma_d}(\Omega) \text{ such that} \\ \displaystyle\int_\Omega k \nabla u \cdot \nabla v \, d\Omega = \int_\Omega f\, v \, d\Omega \quad \forall v \in H^1_{0,\Gamma_d}(\Omega). \end{array} \right.$$

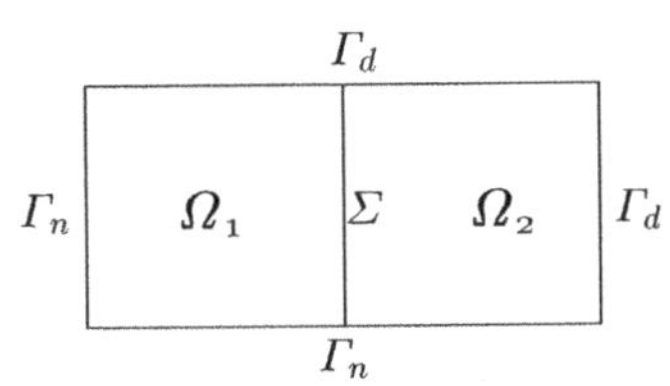

Figure 5.6. *The diffusion problem*

If there exists a $k_- > 0$ such that $k(x, y) \geq k_-$ for almost all $(x, y) \in \Omega$, then this variational problem admits a unique solution.

The following script calculates the approximate solution using P^2 finite elements with the data $k_1 = 10$, $k_2 = 1$, and also the data f where:

$$f(x, y) = \begin{cases} 1 \text{ if } \left(x - \frac{3}{4}\right)^2 + \left(y - \frac{1}{2}\right)^2 \leq \frac{1}{10} \\ 0 \text{ otherwise.} \end{cases}$$

```
%finite element calculation
  [S,T,BR,RT]=triangle_rectangle([-1 1 0 1],20,10);              %P1 mesh
  [S,T,BR,RT]=meshP2(S,T,BR,RT);[T,S]=renum(T,S);                %P2 mesh
  [K,M]=FE_calc_2D(S,T,RT,@k);                                   %FE matrices
%right-hand side
  X=S(:,1);Y=S(:,2);X1=X-0.75;Y1=Y-0.5;
  B=M*((X1.*X1+Y1.*Y1)<0.1);}                                    %right-hand side
```

```
%take Dirichlet cond. into account
Node_dir=node_bound(S,T,BR,[2 3]);                    %list Dirichlet dof's
[Ke,Be]=cd_Dirichlet(K,B,Node_dir);                   %pseudo-elimination
%plot
T1=isop2(T);
figure; trisurf(T,S(:,1),S(:,2),U);shading interp;
%k(x,y) function
function z= k(x,y,r)
z=1;if(x>0)z=10;end,
```

The solution obtained using this Matlab script is presented in Figure 5.7. If the solution is globally continuous, observe that it is not differentiable when passing via the interface $\Sigma : u \in \mathcal{C}^0(\overline{\Omega})$, $\nabla u \notin (\mathcal{C}^0(\overline{\Omega}))^2$.

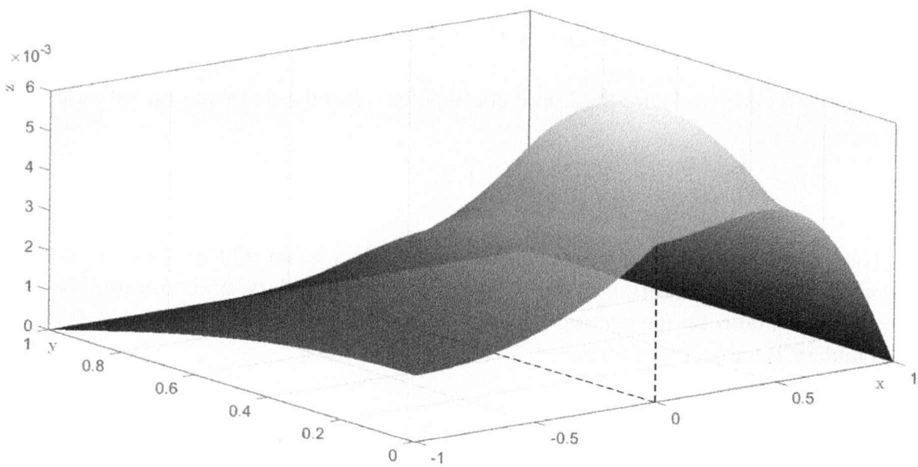

Figure 5.7. *Solution to a diffusion problem approximated using finite elements*

5.3.3. *Bidimensional elasticity*

In this example, we are interested in implementing the finite element method with regard to a vector problem, namely that of homogeneous linear isotropic elasticity in two dimensions. Recall that, for a solid Ω (a connected and bounded open subset of $\mathbb{R}^2$ with a "sufficiently regular" boundary) that is supposed to be homogeneously isotropic, this becomes a question of finding the displacement field $\mathbf{u} = \begin{pmatrix} u_1 \\ u_2 \end{pmatrix}$ that solves:

$$\begin{cases} \text{Find } \mathbf{u} \in (H^1(\Omega))^2 \quad \text{such that} \\ -\sum_{j=1,2} \frac{\partial}{\partial x_j} \sigma_{ij}(\mathbf{u}) = f_i \quad \text{in } \Omega \\ \sum_{j=1,2} \sigma_{ij}(\mathbf{u}) n_j = h_i \quad \text{on } \Gamma_c \qquad \text{for } i = 1,2 \qquad (\mathcal{E}) \\ \sum_{j=1,2} \sigma_{ij}(\mathbf{u}) n_j = 0 \quad \text{on } \Gamma_l \\ \mathbf{u} = 0 \quad \text{on } \Gamma_e \end{cases}$$

where the stiffness tensor is given by:

$$\sigma_{ij}(\mathbf{u}) = \lambda(\operatorname{div} \mathbf{u})\, \delta_{ij} + 2\mu\varepsilon_{ij}(\mathbf{u}) \quad i,j = 1,2$$

(where $\lambda > 0$ and $\mu > 0$ are the Lamé coefficients), and the deformation tensor by:

$$\varepsilon_{ij}(\mathbf{u}) = \frac{1}{2}(\frac{\partial u_i}{\partial x_j} + \frac{\partial u_j}{\partial x_i}) \quad i,j = 1,2.$$

The case considered here is that of a solid subject to an interior force $\mathbf{f}$, with a clamping boundary condition at the section Γ_e of its boundary, and additionally it is subject to a force $\mathbf{h}$ on the section Γ_c of its boundary, whilst the remaining part Γ_l of its boundary is subject to no force:

$$\mathbf{f} = \begin{pmatrix} f_1 \\ f_2 \end{pmatrix}, \ \mathbf{h} = \begin{pmatrix} h_1 \\ h_2 \end{pmatrix}.$$

The variational formulation of $(\mathcal{E})$ in the space:

$$\mathbf{V}_0 = \{\mathbf{v} \in (H^1(\Omega))^2 \text{ such that } \mathbf{v} = 0 \text{ on } \Gamma_e\}$$

is:

$$\left| \begin{array}{l} \text{Find } \mathbf{u} \in \mathbf{V}_0 \text{ such that} \\ a(\mathbf{u},\mathbf{v}) = \ell(\mathbf{v}) \quad \forall \mathbf{v} \in \mathbf{V}_0 \end{array} \right.$$

where we have set:

$$a(\mathbf{u},\mathbf{v}) = \lambda \int_\Omega \operatorname{div} \mathbf{u} \operatorname{div} \mathbf{v}\, d\Omega + 2\mu \sum_{i,j=1,2} \int_\Omega \varepsilon_{ij}(\mathbf{u})\, \varepsilon_{ij}(\mathbf{v})\, d\Omega$$
$$\ell(\mathbf{v}) = \int_{\Gamma_c} \mathbf{h} \cdot \mathbf{v}\, d\Gamma + \int_\Omega \mathbf{f} \cdot \mathbf{v}\, d\Omega.$$

One can show that this problem is well posed (see exercise 23, by reasoning in $\mathbb{R}^2$ here rather than in $\mathbb{R}^3$). Within the context of P^1- or P^2-Lagrangian finite elements, we denote the usual finite element basis by $(w_I)_{I=1,ns}$, and we introduce the approximation space (included within $(H^1(\Omega))^2$) of dimension $2ns$ by:

$$\mathbf{V}_h = \text{vect}(\mathbf{w}_1^I, \mathbf{w}_2^I)_{I=1,ns} \quad \text{where } \mathbf{w}_1^I = \begin{pmatrix} w_I \\ 0 \end{pmatrix} \text{ and } \mathbf{w}_2^I = \begin{pmatrix} 0 \\ w_I \end{pmatrix}, \qquad I = 1, ns$$

and additionally the space (where $\mathcal{D}$ represents the indexing set for the clamped nodes, $M_i \in \overline{\Gamma}_e$, and $\mathcal{I} = \{1, \cdots, ns\} \setminus \mathcal{D}$ represents the non-clamped nodes):

$$\mathbf{V}_h^0 = \text{vect}(\mathbf{w}_1^I, \mathbf{w}_2^I)_{I\in\mathcal{I}} \subset \mathbf{V}_0.$$

We then introduce the approximate variational problem:

$$\left|\begin{array}{l} \text{Find } \mathbf{u}_h \in \mathbf{V}_h^0 \text{ such that} \\ \lambda \displaystyle\int_\Omega \text{div}\, \mathbf{u}_h \,\text{div}\, \mathbf{v}_h \, d\Omega + 2\mu \sum_{i,j=1,2} \int_\Omega \varepsilon_{ij}(\mathbf{u}_h)\, \varepsilon_{ij}(\mathbf{v}_h)\, d\Omega = \\ \qquad\qquad \displaystyle\int_{\Gamma_c} \mathbf{h}\cdot \mathbf{v}_h \, d\Gamma + \int_\Omega \mathbf{f}\cdot \mathbf{v}_h \, d\Omega \qquad \forall \mathbf{v}_h \in \mathbf{V}_h^0. \end{array}\right.$$

After decomposing $\mathbf{u}_h$ relative to the basis $(\mathbf{w}_1^I, \mathbf{w}_2^I)_{I\in\mathcal{I}}$:

$$\mathbf{u}_h = \sum_{J\in\mathcal{I}} U_1^J \mathbf{w}_1^J + U_2^J \mathbf{w}_2^J$$

and after considering each equation obtained from $\mathbf{v}_h = \mathbf{w}_1^I$ and $\mathbf{v}_h = \mathbf{w}_2^I$ for each $I \in \mathcal{I}$, we obtain a linear system of dimension 2card$\mathcal{I}$, though now with the unknown $\vec{U} = \begin{pmatrix} \vec{U}_1 \\ \vec{U}_2 \end{pmatrix}$. The entries are now in terms of card$\mathcal{I} \times 1$ blocks for the vectors, and card$\mathcal{I}\times$card$\mathcal{I}$ blocks for the matrices:

$$\begin{bmatrix} \mathbb{A}_{11} & \mathbb{A}_{12} \\ \mathbb{A}_{21} & \mathbb{A}_{22} \end{bmatrix} \begin{pmatrix} \vec{U}_1 \\ \vec{U}_2 \end{pmatrix} = \begin{pmatrix} \vec{B}_1 \\ \vec{B}_2 \end{pmatrix}, \text{ that is } \mathbb{A}\vec{U} = \vec{B} \qquad [5.17]$$

with $k, l = 1, 2$ and for each $I, J \in \mathcal{I}$:

$$\begin{aligned} (\mathbb{A}_{kl})_{IJ} &= \lambda \int_\Omega \text{div}\, \mathbf{w}_k^I \,\text{div}\, \mathbf{w}_l^J \, d\Omega + 2\mu \sum_{i,j=1,2} \int_\Omega \varepsilon_{ij}(\mathbf{w}_k^I)\, \varepsilon_{ij}(\mathbf{w}_l^J)\, d\Omega \\ (\vec{B}_k)_I &= \int_{\Gamma_c} \mathbf{h}\cdot \mathbf{w}_k^I \, d\Gamma + \int_\Omega \mathbf{f}\cdot \mathbf{w}_k^I \, d\Omega. \end{aligned}$$

Assuming that the data $\mathbf{h}$ and $\mathbf{f}$ are sufficiently regular, that is to say, $\mathbf{h} \in (\mathcal{C}^0(\overline{\Gamma}_c))^2$, $\mathbf{f} \in (\mathcal{C}^0(\overline{\Omega}))^2$, then we can use their interpolates:

$$\left|\begin{array}{l} \pi_h \mathbf{h} = \displaystyle\sum_{J=1,ns,\, M_J\in\Gamma_c} h_1(M_J)\mathbf{w}_1^J + h_2(M_J)\mathbf{w}_2^J \\ \pi_h \mathbf{f} = \displaystyle\sum_{J=1,ns} f_1(M_J)\mathbf{w}_1^J + f_2(M_J)\mathbf{w}_2^J \end{array}\right.$$

to construct an approximation $\overrightarrow{B}^\star$ of $\overrightarrow{B}$. For this, consider $\overrightarrow{F}_k$ with components $(f_k(M_J))_J$ and $\overrightarrow{H}_k$ with components $(h_k(M_J))_J$ for $k = 1, 2$:

$$\overrightarrow{B}^\star = \begin{pmatrix} \mathbb{M}\overrightarrow{F}_1 \\ \mathbb{M}\overrightarrow{F}_2 \end{pmatrix} + \begin{pmatrix} \mathbb{M}_c\overrightarrow{H}_1 \\ \mathbb{M}_c\overrightarrow{H}_2 \end{pmatrix}$$

with $\mathbb{M}$ and $\mathbb{M}_C$ the volume and surface masses:

$$\mathbb{M}_{IJ}=\int_\Omega w_J w_I \, d\Omega, \; (\mathbb{M}_c)_{IJ}=\int_{\Gamma_c} w_J w_I \, d\Gamma.$$

To construct the system [5.17], it suffices to evaluate the integrals:

$$\int_\Omega \operatorname{div} \mathbf{w}_k^I \operatorname{div} \mathbf{w}_l^J \, d\Omega, \; \int_\Omega \varepsilon_{ij}(\mathbf{w}_k^I)\, \varepsilon_{ij}(\mathbf{w}_l^J) \, d\Omega, \; \int_\Omega w_I w_J \, d\Omega \text{ and } \int_{\Gamma_c} w_I w_J \, d\Gamma.$$

These integrals are very nearly those that we have already calculated in the Laplace equation with Dirichlet conditions. In practice, they are calculated as well as the matrix $\mathbb{A}$ without worrying about any embedding conditions (i.e. for each $I, J = 1, ns$). Using a pseudo-elimination technique similar to the one previously used, we come back to a system that is equivalent to the one in [5.17].

In what follows, we will provide a Matlab script that allows us to solve the problem of an ideal loaded shelf of length 40 cm, height 20 cm and with a thickness of 2 cm, and that is made from an aluminum-type material, via a P^2 finite element approximation, component by component.

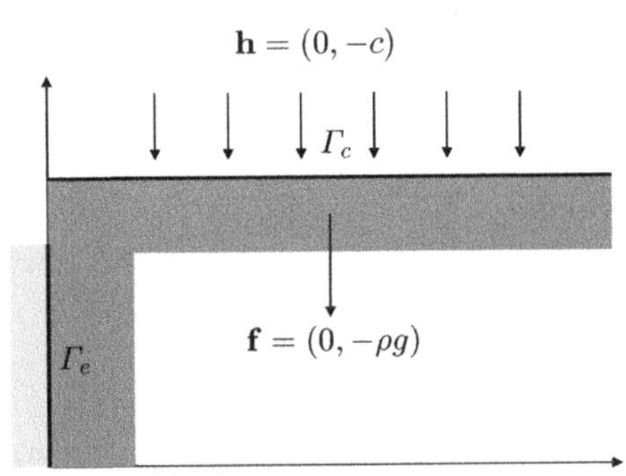

Figure 5.8. *A partially clamped shelf subject to a uniform load*

```
%data
lambda0=10^10;c=100;                                              %dimensionless consts.
rho=2800;lambda=1;mu=1;                                                   %parameters
g=9.81/lambda0;charge=100/lambda0;
%P2 mesh of the shelf (P1 shelf beam, P1 support, P1 junction, P2)
[S1,T1,BR1,RT1]=triangle_rectangle([0 0.4 0 0.02],200,10,0,[1121]);
[S2,T2,BR2,RT2]=triangle_rectangle([0 0.02 -0.1 0],10,50,0,[1103]);
[SF,TF,BRF,RTF]=junction_triangulation(S1,T1,BR1,RT1,S2,T2,BR2,RT2);
[SF,TF,BRF,RTF]=meshP2(SF,TF,BRF,RTF);[TF,SF]=renum(TF,SF);
%calculate the finite element matrices
[M,DivDiv,EpsEps,Dx,Dy]=femat(SF,TF);                                 %volume matrices
[MB2]=FE_calc_1D(SF,TF,BRF,2);}                                       %boundary matrix
K=lambda*DivDiv+2*mu*EpsEps;                                        %elasticity matrix
%calculate the loads
ns=size(SF,1);un=ones(ns,1);ze=zeros(ns,1);
B=[ze;-c*MB2*un-rho*g*M*un];
%eliminating the clamped nodes
Node_dir=node_bound(SF,TF,BRF,[3]);                                     %clamped nodes
ND=[Node_dir;Node_dir];
[K,B]=cd_Dirichlet(K,B,ND);                                         %pseudo-elimination
%solve and calculate the constraints
U=K\B                                                               %sol. of the system
Def=SF+5000*reshape(U,ns,2);                                          %deformed shelf
[Sigma]=cal_sigma(U,M,Dx,Dy,...
        lambda*lambda0,mu*lambda0);                                       %constraints
tr=tresca(Sigma);                                                   %Tresca's criterion
%visualization
TF21=isop2(TF);figure;trisurf(TF21,Def(:,1),Def(:,2),tr);
shadinginterp;view(2);axis image;
```

This script is essentially based on the **femat.m** function that constructs the finite element matrices. The details that we provide here are to highlight the vectorial aspects of the problem. It should be noted that the principles when it comes to calculating the integrals are the same as those for a scalar problem (looping over the triangles, change in variable and numerical quadrature methods):

```
function [M,DivDiv,EpsEps,Dx,Dy]=femat(S,T)
ns=size(S,1);nt=size(T,1);                                                     %sizes
DivDiv=sparse(2*ns,2*ns);                                             %initialization
EpsEps=sparse(2*ns,2*ns);M=sparse(ns,ns);                            %sparse matrices
Dx=sparse(ns,ns);Dy=sparse(ns,ns);
for t=1:nt,                                                %loop over each triangle t
    I=T(t,:);J=ns+I;                                           %num. of dof's (ux,uy)
    [Mt,DXX,DYY,DXY,DX,DY]=elmmat(S(I,:));                        %calc. elem. matrices
   M(I,I)=M(I,I)+Mt;                                                 %scalar matrices
    Dx(I,I)=Dx(I,I)+DX;Dy(I,I)=Dy(I,I)+DY;
    DivDiv(I,I)=DivDiv(I,I)+DXX;                                   %assembled matrices
    DivDiv(J,J)=DivDiv(J,J)+DYY;
    DivDiv(I,J)=DivDiv(I,J)+DXY;
    DivDiv(J,I)=DivDiv(J,I)+DXY';
    EpsEps(I,I)=EpsEps(I,I)+DXX+DYY/2;
    EpsEps(J,J)=EpsEps(J,J)+DXX+DYY/2;
    EpsEps(I,J)=EpsEps(I,J)+DXY/2;
    EpsEps(J,I)=EpsEps(J,I)+DXY'/2;
end,
```

The function **elmmat.m** evaluates the elementary (scalars) integrals:

$$\int_{T_\ell} \tau_i^\ell \tau_j^\ell \, d\Omega,\ \int_{T_\ell} \partial_x \tau_i^\ell \partial_x \tau_j^\ell \, d\Omega,\ \int_{T_\ell} \partial_y \tau_i^\ell \partial_y \tau_j^\ell \, d\Omega\ ,\ \int_{T_\ell} \partial_x \tau_i^\ell \partial_y \tau_j^\ell \, d\Omega$$

$$\int_{T_\ell} \tau_i^\ell \partial_x \tau_j^\ell \, d\Omega \;,\; \int_{T_\ell} \tau_i^\ell \partial_y \tau_j^\ell \, d\Omega$$

using the same principle as the calculation presented in section 5.3.1. The **DX** and **DY** matrices are used to calculate the constraints.

```
function [M,DXX,DYY,DXY,DX,DY]=elmmat(Nodes)
%initialization
nd=size(Nodes,1);
DXX=zeros(nd,nd);DYY=zeros(nd,nd);DXY=zeros(nd,nd);
M=zeros(nd,nd);DX=zeros(nd,nd);DY=zeros(nd,nd);
%quadrature formula for 7 points
nbq=7;os=sqrt(15);s3=1./3.;
pp1=(6.-os)/21.;pp2=(6.+os)/21.;
pp3=(9.+2.*os)/21.;pp4=(9.-2.*os)/21.;
quadT_pts=[s3 s3;pp1 pp1;pp1 pp3;pp3 pp1;                    %quad. points
           pp2 pp2;pp2 pp4;pp4 pp2];
pp1=(155.-os)/2400.;pp2=(155.+os)/2400.;}}
quadT_wgts=[9./80.;pp1;pp1;pp1;pp2;pp2;pp2];                 %quad. weights
%Jacobian of the (affine) transformation
S21=Nodes(2,:)-Nodes(1,:);S31=Nodes(3,:)-Nodes(1,:);
delta=S21(1)*S31(2)-S21(2)*S31(1);
Jflmt=[S31(2)-S21(2);-S31(1)S21(1)]/delta;
%loop over the quadrature points
for k=1:nbq,                                                 %k num. of the point
    x=quadT_pts(k,1);y=quadT_pts(k,2);                       %quad. points
    if(nd==3) w=[1-x-yxy];gw=[-1 1 0;-1 0 1];                %P1 functions
    else a=1-x-y;b=2*x-1;c=2*y-1;                            %P2 functions
      w=[a*(1-2*x-2*y),b*x*,c*y*,4*x*a,4*x*y,4*y*a];
      gw=[4*(x+y)-3 4*x-1 0 4*(1-2*x-y) 4*y -4*y;
      4*(x+y)-3 0 4*y-1 -4*x 4*x 4*(1-x-2*y)];
    end,
pk=quadT_pts(k)*abs(delta);
M=M+pk*w'*w;                                                 %wi.wj
jx=Jflmt(1,:)*gw;jy=Jflmt(2,:)*gw;
DXX=DXX+pk*jx'*jx;                                           %dx(wi)dx(wj)
DYY=DYY+pk*jy'*jy;                                           %dy(wi)dy(wj)
DXY=DXY+pk*jx'*jy;                                           %dx(wi)dy(wj)
DX=DX+pk*w'*jx;                                              %wi.dx(wj)
DY=DY+pk*w'*jy;                                              %wi.dy(wj)
end,
```

In this example, the boundary mass matrix $\mathbb{M}_c$ also needs to be calculated. This operation is performed using a technique similar to that used to calculate the finite element matrices, apart from the fact that restricting the basis functions, as well as the domain of integration, to the triangle edges located on the boundary concerned by the calculation is advisable. This is done using the **FE_calc_1D.m** function:

```
function [M]=FE_calc_1D(S,T,BR,ref_bound)
s35=sqrt(3./5);quadS_pts=0.5*[1-s35 1 1+s35];                %1D quadrature
os=1/18;quadS_wgts=os*[5 8 5];
nt=size(T,1);ns=size(S,1);q=size(T,2);                       %initialization
nbq=length(quadS_wgts);
M=sparse(ns,ns);
%loop over the triangle vertices
for t=1:nt,
    for a=1:3,
        as=mod(a,3)+1;I=T(t,a);J=T(t,as);                     %edge num.
        if(ismember(BR(t,a),ref_bound))                       %boundary edge
            L=norm(S(I,:)-S(J,:));                            %edge length
```

```
                    in=[I J];if(q==6) in=[in T(t,3+a)];end,                  %global numbering
                    for k=1:nbq,                                              %loop over quad. pts
                        x=quadS_pts(k);c=L*quadS_wgts(k);
                        if(q==3) w=[1-x x];                                   %P1 functions
                        else w=[(1-2*x)*(1-x) (2*x-1)*x 4*x*(1-x)];           %
P2end,M(in,in)=M(in,in)+c*w'*w; \Comment{\%assembly}
                    end,
                end,
            end,
        end,
```

Calculating the constraints and Tresca's criterion requires some knowledge about the derivatives of the approximate displacement field $\mathbf{u}_h$. As part of the Lagrangian finite element approximation, $\mathbf{u}_h$ is not differentiable (it is only continuous) across the interface between any two triangles. To construct an approximation for these derivatives, we have essentially two possibilities: either to be content with discontinuous derivatives or instead construct an approximation from each node of the mesh using the derivatives that come from each triangle that contains the node. To obtain a correctly weighted average from each triangle in the second case, it suffices to consider the following problem, in which we denote $\mathbf{v}_h^x$ the P^1- or P^2-approximation of $\partial_x \mathbf{u}_h$ (*ditto* for $\mathbf{v}_h^y \approx \partial_y \mathbf{u}_h$):

$$\int_\Omega \mathbf{v}_h^x w_I d\Omega = \int_\Omega \partial_x \mathbf{u}_h w_I d\Omega \quad \forall I = 1, ns$$

leading to the two linear systems:

$$\mathbb{M}\overrightarrow{V}_1^x = \mathbb{D}_x \overrightarrow{U}_1, \; \mathbb{M}\overrightarrow{V}_2^x = \mathbb{D}_x \overrightarrow{U}_2$$

where $\mathbb{M}$ is the mass matrix and $\mathbb{D}_x$ is the matrix whose components are $\int_\Omega (\partial_x w_J) w_I \, d\Omega$. Recall that the approximation order of the derivatives is less than the order of their values.

The function **cal_sigma.m** calculates the derivatives according to this principle and forms the stiffness tensor for each node of the mesh:

```
function [Sigma]=cal_sigma(U,M,Dx,Dy,lambda,mu)
ns=size(M,1);Ux=U(1:ns);Uy=U(ns+1:2*ns);                %initialization
DxUx=M\(Dx*Ux);DxUy=M\(Dx*Uy);                         %derivative calc.
DyUx=M\(Dy*Ux);DyUy=M\(Dy*Uy);
Sigma(:,1)=lambda*(DxUx+DyUy)+2*mu*DxUx;               %sigma calc.
Sigma(:,4)=lambda*(DxUx+DyUy)+2*mu*DyUy;
Sigma(:,2)=mu*(DxUy+DyUx);Sigma(:,3)=Sigma(:,2);
```

Tresca's criterion represents the shear stresses being applied, and is given by:

$$tr = |\lambda_1 - \lambda_2|$$

where λ_1 and λ_2 are the eigenvalues of the stiffness tensor.

Given the constraint tensor, the **tresca.m** function calculates Tresca's criterion for any node in the mesh:

```
function [tr]=tresca(Sigma,S,T)
ns=size(Sigma,1);                                    %initialization
for I=1:ns,                                          %nodes loop
    sp=eig([Sigma(I,1),Sigma(I,2);
            Sigma(I,3),Sigma(I,4)]);                 %eigenvalues
    tr(I)=abs(sp(1)-sp(2));                          %Tresca
end,
```

The result for the shelf calculation is presented in Figure 5.9. The (exaggerated) shelf deformation is also presented in the same figure, as well as Tresca's criterion which indicates areas of high shear stress.

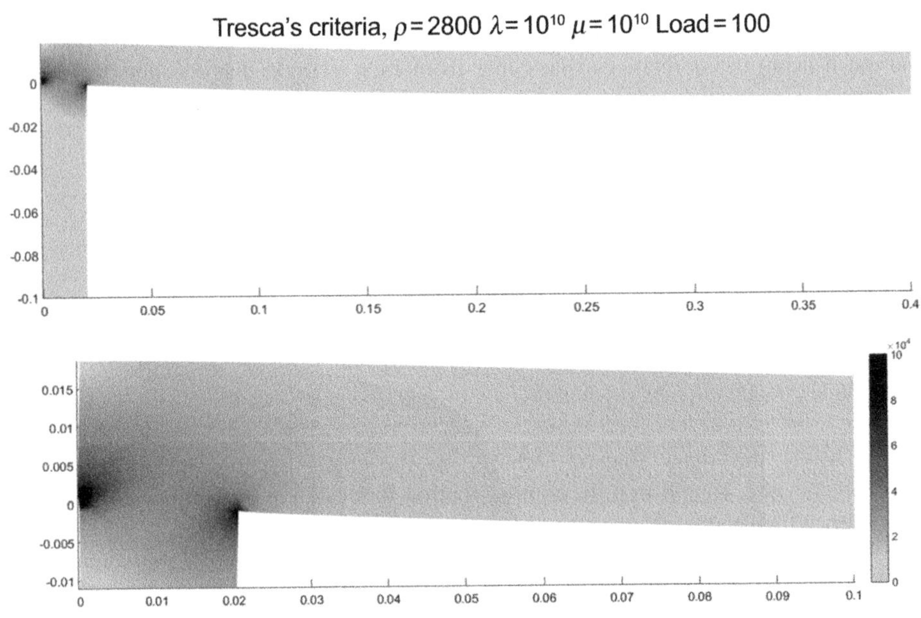

Figure 5.9. *Approximate solution for the elasticity equations using P^2 finite elements*

5.3.4. *Some elementary tools for 2D meshes*

In this last section, we provide Matlab code of some elementary tools which has been used in the previous examples for the 2D meshes. The basic tool (**triangle_square.m**) is the one that realized the mesh "structure" of the unit square, according to a grid containing m (resp. n) intervals along the x (resp. y) direction, a reference material table ref_mat (0 by default), and a boundary reference table ref_bound. Each elementary rectangle of this grid gets cut into two triangles depending on either the first or on the second bisector. This division is carried out alternately so as not to generate too much greater of an asymmetry in the mesh, which would create a systematic bias when performing calculations. This function generates the table S (of dimension $ns\times$ 2) with the coordinates of the vertices in the triangulation, table T (of dimension $nt \times 3$) with the vertex numbers, the table RT (of dimension nt) which indicates a reference for each triangle, and finally the table BR (of dimension $nt \times 3$) that indicates a reference boundary for each triangle edge (which is 0 for each internal edge).

```
function [S,T,BR,RT]=triangle_square(m,n,ref_mat,ref_bound)
%initialization
dx=1./m;dy=1./n;x=0:dx:1.;y=0:dy:1.;
ns=(m+1)*(n+1);nt=2*m*n;
BR=zeros(nt,3);
if(nargin==2) ref_mat=0;ref_bound=[1 2 3 4];end,
if(nargin==3) ref_bound=[1 2 3 4];end,
RT=ref_mat*ones(1,nt);
%definition of the vertices
[X,Y]=meshgrid(x,y);
S=[reshape(X',ns,1) reshape(Y',ns,1)];
%definition of the triangle numbering
N1=[];N2=[];mp=m+1;
%type 1 line cutting
for i=1:m
    if(mod(i,2)==1)                                    %i odd
        N1=[N1 i i+1 mp+i i+1+mp i+mp i+1];
    else                                               %i even
        N1=[N1 i+1 i+1+mp i i+mp i i+1+mp];
    end,
end,
N1=reshape(N1,3,2*m)';
%type 2 line cutting
for i=1:m
    if(mod(i,2)==1)                                    %i odd
        N2=[N2 i+1 i+1+mp i i+mp i i+1+mp];
    else                                               %i even
        N2=[N2 i i+1 mp+i i+1+mp i+mp i+1];
    end,
end,
N2=reshape(N2,3,2*m)';
%mesh from alternating lines 1 and 2 (translation of numbering)
T=[];
for j=1:n
    if(mod(j,2)==1) T=[T;N1+(j-1)*mp];
    else T=[T;N2+(j-1)*mp];
    end,
end,
%boundary numbering
p=2*(n-1)*m+1;
p1=ref_bound(3);p2=0;if(mod(n,2)==0) p2=ref_bound(3);p1=0;end,
for i=1:4:2*m,
    BR(i,:)=[ref_bound(1) 0 0];
    if(i+2<=2*m) BR(i+2,:)=[0 0 ref_bound(1)]; end,
    BR(p+i,:)=[p1 0 p2];
    if(p+i+2<=nt) BR(p+i+2,:)=[p2 0 p1];end,
end,
%left and right boundaries (4 2)
t=2*m+1;r=2*(n-1)*m+2;
s=2*(m-mod(m,2))+1;q=2*m-2+mod(m,2);
```

```
q1=ref_bound(2);q2=0;if(mod(m,2)==0) q2=ref_bound(2);q1=0;end,
for i=1:4*m:2*(n-1)*m+1,
    BR(i,:)=BR(i,:)+[0 0 ref_bound(4)];
    if(i+t<=r) BR(i+t,:)=BR(i+t,:)+[ref_bound(4) 0 0];end,
    BR(q+i,:)=BR(q+i,:)+[q2 0 q1];
    if(q+i+s<=nt) BR(q+i+s,:)=BR(q+i+s,:)+[q1 0 q2];end,
end,
```

For example, Figure 5.10 presents the P^1 mesh obtained by evoking the command:

```
    [S,T,BR,RT]=triangle_square(10,20,0,[1 1 2 1]);
    plot_triangulation(S,T,BR,RT,0);
```

A rectangular mesh is then derived by rescaling the square mesh:

```
function [S,T,BR,RT]=triangle_rectangle(rect,m,n,ref_mat,ref_bound)
if(nargin==3) ref_mat=0;ref_bound=[1 2 3 4];end,
if(nargin==4) ref_bound=[1 2 3 4];end,
[S,T,BR,RT]=triangle_square(m,n,ref_mat,ref_bound);
%transform of the S coordinates
l=rect(2)-rect(1);S(:,1)=rect(1)+l*S(:,1);
l=rect(4)-rect(3);S(:,2)=rect(3)+l*S(:,2);
```

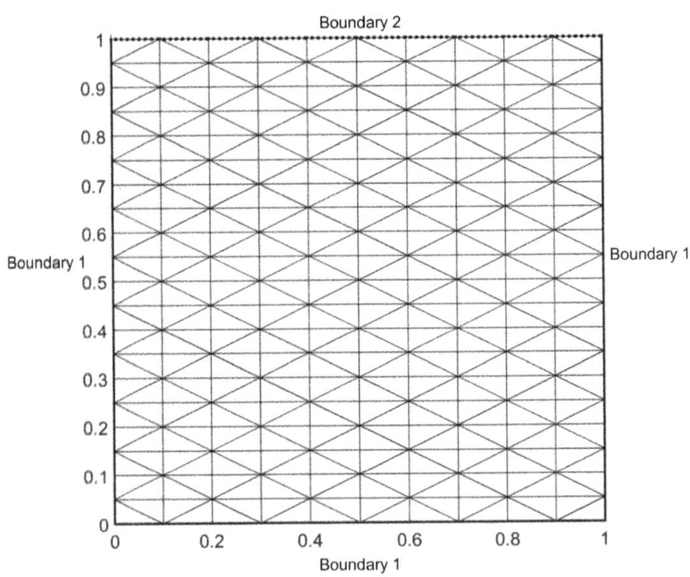

Figure 5.10. *Regular mesh of the unit square*

The following command:

```
[S,T,BR,RT]=triangle_rectangle([-1 1 0 1],20,10,0,[1 1 2 1]);
plot_triangulation(S,T,BR,RT,0);
```

generates the mesh presented in Figure 5.11.

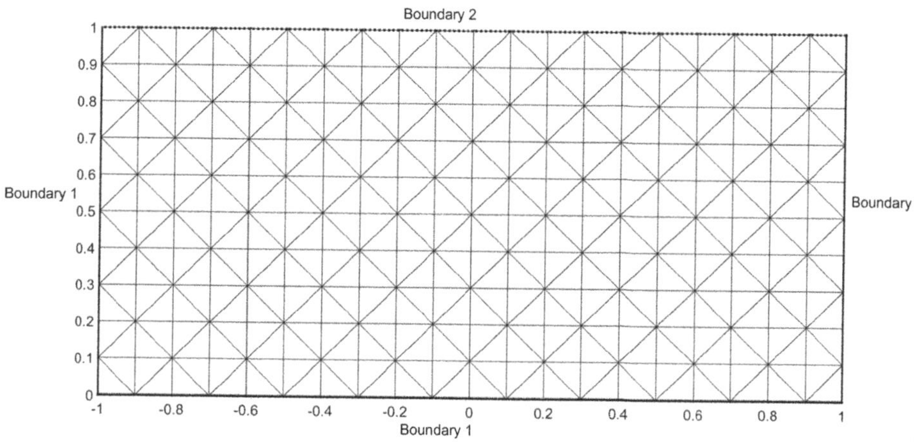

Figure 5.11. *Regular rectangular mesh*

Using the mesh merge tool (**triangulation_merge.m**) allows for meshes of more complex geometric domains to be generated:

```
function [SF,TF,BRF,RTF]=triangulation_merge(S1,T1,BR1,RT1,...
                                              S2,T2,BR2,RT2)
%search for any coinciding nodes within
%an eps precision defined by the domain size /10000
xmax=max(max(S1(:,1)),max(S2(:,1)));
xmin=min(min(S1(:,1)),min(S2(:,1)));
ymax=max(max(S1(:,2)),max(S2(:,2)));
ymin=min(min(S1(:,2)),min(S2(:,2)));
eps=max(xmax-xmin,ymax-ymin)/10000.;
%create the table of edges
A1=edge_graph(S1,T1);
L1=bound_node_detection(A1);
A2=edge_graph(S2,T2);
L2=bound_node_detection(A2);
%coinciding nodes (search by brute-force...)
ns1=size(S1,1);ns2=size(S2,1);
nt1=size(T1,1);nt2=size(T2,1);
C=zeros(ns2,1);D=zeros(ns1,1);
for i=1:size(L1,1),
    I=L1(i);
    for j=1:size(L2,1),
        J=L2(j);
        if(norm(S1(I,:)-S2(J,:))<eps)                        %coincide
            C(J)=I;                          %S2 J num. becomes S1 I num.
            D(I)=J;                          %S2 J num. becomes S1 I num.
```

```
                                                        %break;
            end ,
        end ,
end ,
%merge the triangulations
SF=S1;TF=[T1;T2];
BRF=[BR1;BR2];RTF=[RT1 RT2];
n2=0;
for i=1:ns2 ,                                           %second triangulation numbering
    if(C(i)==0)
        n2=n2+1;C(i)=ns1+n2;
        SF=[SF;S2(i,:)];
    end ,
end ,
for t=1:nt2 ,                                           %update triangle numbering (2nd part)
    for i=1:3,
        I=T2(t,i);
        TF(nt1+t,i)=C(I);
    end ,
end ,
for t=1:nt1;                                            %deleted boundary edges become internal edges
    for a=1:3,
        if(BRF(t,a)~=0)
            I=T1(t,a); if(a==3) J=T1(t,1); else J=T1(t,a+1);end,
            if(D(I)~=0  D(J)~=0) BRF(t,a)=0;end,       %common edge
        end ,
    end ,
end ,
for t=1:nt2;
    for a=1:3,
        if(BRF(nt1+t,a)~=0)
            I=T2(t,a); if(a==3) J=T2(t,1); else J=T2(t,a+1);end,
            if(C(I)<=ns1  C(J)<=ns1) BRF(nt1+t,a)=0;end,   %common edge
        end ,
    end ,
end ,
```

We will illustrate how this function is used by "gluing" together two rectangular meshes which are compatible (see Figure 5.12):

```
%P1 meshes
[S1,T1,BR1,RT1]=triangle_rectangle([0 1 0 2],10,20,0,[1 1 1 1]);
[S2,T2,BR2,RT2]=triangle_rectangle([1 2 0 1],10,10,0,[1 1 1 1]);
%merging the meshes
[SF,TF,BRF,RTF]=triangulation_merge(S1,T1,BR1,RT1,S2,T2,BR2,RT2);
plot_triangulation(SF,TF,BRF,RTF,0);
```

It is quite easy to generate a P^2 mesh from a P^1 one by including the midpoints from each edge of the P^1 triangulation. However, care should be taken when the boundary of a computational domain is curved, since precision is lost when an isoperimetrical P^2 mesh of the computational domain is not created (see section 4.3). The Matlab **P2mesh.m** function realizes this operation:

```
function [S2,T2,BR2,RT2]=P2mesh(S,T,BR,RT)
%Initialization
ns=size(S,1); nt=size(T,1);
ns2=ns+3*nt;                                %overestimate the number of P2 points
S2=zeros(ns2,2); S2(1:ns)=S;
T2=zeros(nt,6); T2(:,1:3)=T;
RT2=RT;BR2=BR;
```

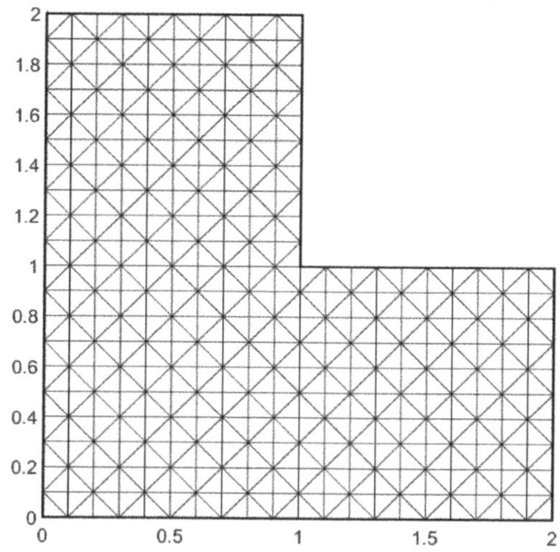

Figure 5.12. *Merging together two (compatible) meshes*

```
%edge graph
Edge=sparse(ns,ns);n=ns;
%loop over the triangles
for l=1:nt,
    %loop over the edges
    for k=1:3,
        I=T(l,k);
        J=T(l,mod(k,3)+1);
        if(J<I) I1=J;I2=I;else I1=I;I2=J;end,
        %update
        t=Edge(I1,I2);
        if(t==0)                                  %add a node
            Edge(I1,I2)=l+k/10.;                  %coding triangle num. + edge num.
            n=n+1;
            S2(n,:)=(S2(I1,:)+S2(I2,:))/2.;
            T2(l,3+k)=n;
        else,                                     %node already added
            q=floor(t);a=round(10*(t-q));
            T2(l,3+k)=T2(q,3+a);
        end,
    end,
end,
%resizing S2
S2=S2(1:n,:);
```

When a P^2 mesh is formed from a P^1 one, the numbering of the nodes is not at all optimal in terms of the matrix profile. It can be interesting to find more effective numberings (not absolutely optimal but nearly). These renumberings are based on the Cuthill–Mackee, or the *reverse* Cuthill–Mackee algorithm, which are available in Matlab. As an example, the following **renum.m** function is based upon such algorithms to optimize a numbering, whether it be for P^1 and P^2 meshes:

```
function [T_s,S_s]=renum(T,S)
%various sizes
nt=size(T,1);ns=size(S,1);q=size(T,2);
%connectivity graph matrix
G=sparse(ns,ns);
for t=1:nt,
    for i=1:q,
        I=T(t,i);
        for j=1:q,
            J=T(t,j);
            G(I,J)=1;
        end,
    end,
end,
%optimization (reverse Cuthill-Mackee)
R = symrcm(G);
%renumbering the nodes
S_s=S(R,:);
T_s=zeros(size(T));
Rm=zeros(ns,1);
%inverse permutation of R
for n=1:ns,
    Rm(R(n))=n;
end,
%renumbering the triangles
for t=1:nt,
    for i=1:q,
        T_s(t,i)=Rm(T(t,i));
    end,
end,
```

To take into account any essential conditions, for example, Dirichlet ones or those clamped in elasticity, it is useful to have tools that allow us to identify the nodes of a boundary edge identified via a reference. To do so, we use the **detect_bound_nodes.m** function that is based on the list of edges, which itself is obtained with the help of the **edge_graph.m** function:

```
function L=detect_bound_nodes(A)
[r,c,v]=find(A==1);                    %indices of the unshared edges
L=unique(r);
```

```
function A=edge_graph(S,T)
ns=size(S,1); nt=size(T,1);A=sparse(ns,ns);
for t=1:nt,
    I=T(t,1);J=T(t,2);K=T(t,3);
    A(I,J)=A(I,J)+1;
    A(J,I)=A(J,I)+1;
    A(I,K)=A(I,K)+1;
    A(K,I)=A(K,I)+1;
    A(J,K)=A(J,K)+1;
    A(K,J)=A(K,J)+1;
end,
```

Finally, note that it can be useful to have mesh visualization tools that are a little more elaborate than that provided by Matlab (*trimesh*). For example, we can use the following function **plot_triangulation.m**, which proposes the display of numberings and edge references:

```
function plot_triangulation(S,T,BR,RT,num)
%create the list of default colors (11 colors)
color=[1 1 1; 0 0 1; 0 1 0; 1 0 0; 1 1 0; 0 1 1; 1 0 1;
       0.5 0.5 0.5; 0.5 0.5 0; 0 0.5 0.5; 0.5 0 0.5];
%definition of the colors
reft=unique(RT);n=length(reft);refc=11*ones(1,n);c=0;m=min(reft)+1;
for i=1:min(n,11) c=c+1;refc(reft(i)+m)=c;end;
%plot of the mesh
figure;hold on;rmax=0;
for l=1:size(T,1)
    X=[S(T(l,1),1) S(T(l,2),1) S(T(l,3),1)];
    Y=[S(T(l,1),2) S(T(l,2),2) S(T(l,3),2)];
    i=refc(RT(l)+m);c=color(i,:);      patch(X,Y,c);
    %color of the boundaries
     for k=1:3,
         ks=mod(k,3)+1;
         r=BR(l,k); if(r>rmax) rmax=r;end,
         if(r>0  r<11)
            line([X(k) X(ks)],[Y(k) Y(ks)],'LineWidth',...
                 2,'Color',color(r+1,:));
         end,
     end,
    if(num>0)                                       %display the numbering
        if(num>=10)                                 %display the triangle numbering
            G=[X;Y]*[1;1;1]/3;                      %triangle barycenter
            text(G(1),G(2),int2str(l),'FontSize',8);
        end,
        if(num~=10)                                 %display the P1 node numbering
               for k=1:3,
                   text(X(k),Y(k),int2str(T(l,k)),...
                        'FontSize',8,'BackgroundColor',[1 1 1]);
               end,
        end,
        if((num==2 || num==12 ) size(T,2)==6)       % P2 numbering
            for k=1:3,
                k2=mod(k,3)+1;
                text((X(k)+X(k2))*.5,(Y(k)+Y(k2))*.5, ...
                     int2str(T(l,k+3)),'BackgroundColor', ...
                     [1 1 1],'FontSize',8);
            end,
        end,
    end,
end
```

Note that, if we wanted to use the primitive *trisurf* Matlab function to plot the isosurfaces, then one should note that the aforementioned function only works with P^1 triangulations, and that it is necessary to convert any P^2 triangulation into a P^1 one. This is the purpose of the **isop2.m** function which follows and is based upon dividing a P^2 triangle into four P^1 triangles:

```
function [T1]=isop2(T2)
nt=size(T2,1);T1=zeros(4*nt,3);
k=0;
for t=1:nt,
    k=k+1; T1(k,1)=T2(t,1);T1(k,2)=T2(t,4);T1(k,3)=T2(t,6);
    k=k+1; T1(k,1)=T2(t,4);T1(k,2)=T2(t,2);T1(k,3)=T2(t,5);
    k=k+1; T1(k,1)=T2(t,4);T1(k,2)=T2(t,5);T1(k,3)=T2(t,6);
    k=k+1; T1(k,1)=T2(t,3);T1(k,2)=T2(t,6);T1(k,3)=T2(t,5);
end;
```

Appendices

Appendix 1

Solving Linear Systems

When finite elements are used to discretize linear elliptic problems, large linear systems arise whose matrix structures are very particular. They are sparse, which is to say that they possess a very large number of zero entries (see theorems 5.2 and 5.3). It is therefore important to understand the main techniques that are used to solve linear systems, so that the most appropriate one can be chosen depending on the nature of the linear system. This appendix is for those who may have questions about how the finite element method can be implemented in Fortran, C/C++, or any other fast programming language. In the programming environments used within scientific computing (Matlab, Scilab, Octave, Python, Julia, etc.), matrix storage issues and system-solving algorithms are obfuscated by the terminology used (*sparse*-type in Matlab), but understanding these techniques will be very informative.

There is no question about presenting the entire theory of linear systems here, as this topic is covered extensively in the classical literature. We will endeavor to recall the principles of classical algorithms (the direct and iterative methods) and, in particular, we will focus on how the sparse matrix storage structures influence the choice of solving algorithms.

In the sequel, we will look at how one may solve the following linear system:

$$\mathbb{A}X = B \qquad \text{[A1.1]}$$

where $\mathbb{A}$ is an invertible matrix of rank n, B is a given vector in $\mathbb{R}^n$, and X denotes the solution vector. For the sake of legibility in this appendix, we will omit the arrows usually written above each vector.

We will begin by recalling some of the theoretical elements that contributes toward numerical error propagation, which can occur when one is solving a linear system. Then, we will briefly present some **direct** solving methods, producing the solution in a finite number of steps, as well as some **iterative** methods that generate a vector sequence that converges toward the solution of the linear system. These methods are first exposed in the case of "dense" matrices, before then discussing how they can be adapted to complex storage structures.

A1.1. Propagation of numerical errors

The question here is to estimate the error inherited by the solution when one solves, not the linear system [A1.1], but rather instead the perturbed system:

$$(\mathbb{A} + \delta\mathbb{A})Y = B + \delta B \qquad \text{[A1.2]}$$

where $\delta\mathbb{A}$ and δB represent the errors within the matrix system and on the right-hand side, respectively. On the one hand, in practice these errors arise from discrepancies of the initial data due to any rounding carried out by the computer; on the other hand, from the rounding errors being propagated throughout the course of solving the linear system.

A1.1.1. *Estimates for errors*

If we assume that there are no rounding errors in our chosen solving method, but rather just an initial error provided by the data B, then the perturbed problem [A1.2] takes on the form:

$$\mathbb{A}Y = B + \delta B$$

for which it is easy to show that:

$$| Y - X | \leq \| \mathbb{A}^{-1} \| | \delta B |,$$

coinciding with the usual notion of stability. In the sequel, $| \, . \, |$ will denote any norm of $\mathbb{R}^n$, and $\|.\|$ will denote the matrix norm subordinate to this norm. Conversely, if rounding errors are taken into account, which in fact always exist when solving using a computer, then the error in the solutions is linked to the **condition number** of the matrix $\mathbb{A}$, which we denote by $\gamma(\mathbb{A})$ and define it by:

$$\gamma(\mathbb{A}) = \| \mathbb{A} \| \| \mathbb{A}^{-1} \| \qquad \text{[A1.3]}$$

which, when the matrix $\mathbb{A}$ is Hermitian, coincides with the ratio of its largest eigenvalue over its smallest eigenvalue (using their moduli), if the Euclidean norm $| \cdot |_2$ is used. We have the next fundamental result:

PROPOSITION A1.1.– *Let $\mathbb{A}$ be a rank n invertible matrix, $\delta\mathbb{A}$ be a rank n matrix, and let B and δB be two vectors in $\mathbb{R}^n$.*

i) If $\|\delta\mathbb{A}\| < \dfrac{1}{\|\mathbb{A}^{-1}\|}$, then $(\mathbb{A}+\delta\mathbb{A})$ is invertible.

ii) If X denotes the solution to the linear system $\mathbb{A}X = B$, and $Y = X + \delta X$ is the solution to the linear system $(\mathbb{A}+\delta\mathbb{A})Y = B + \delta B$, then we have the following error estimate:

$$\frac{|\delta X|}{|X|} \leq \frac{\gamma(\mathbb{A})}{1-\gamma(\mathbb{A})\frac{\|\delta\mathbb{A}\|}{\|\mathbb{A}\|}}\left(\frac{|\delta B|}{|B|} + \frac{\|\delta\mathbb{A}\|}{\|\mathbb{A}\|}\right). \qquad [A1.4]$$

Proof.

As $\mathbb{A}$ is invertible, we have:

$$\mathbb{A} + \delta\mathbb{A} = \mathbb{A}\left(\mathbb{I} + \mathbb{A}^{-1}\delta\mathbb{A}\right)$$

which, from the assumptions on $\delta\mathbb{A}$, shows that:

$$\left(\mathbb{I} + \mathbb{A}^{-1}\delta\mathbb{A}\right) = \mathbb{I} - \mathbb{C} \quad \text{with } \|\mathbb{C}\| < 1.$$

Now let: $\mathbb{S}_k = \sum_{j=0}^{k} \mathbb{C}^j$. The sequence $(\mathbb{S}_k)_{k\geq 0}$ converges to some matrix $\mathbb{S}$ because the series converges normally. Furthermore, we have:

$$\mathbb{S}_k(\mathbb{I}-\mathbb{C}) = \mathbb{I} - \mathbb{C}^{k+1} \quad \text{and} \quad \lim_{k\to\infty} \mathbb{C}^k = 0$$

which proves that:

$$\mathbb{S}(\mathbb{I}-\mathbb{C}) = \mathbb{I}.$$

We also show that: $(\mathbb{I}-\mathbb{C})\mathbb{S} = \mathbb{I}$ and consequently that $(\mathbb{I}-\mathbb{C})$ is invertible, with inverse:

$$\mathbb{S} = \sum_{j=0}^{\infty} \mathbb{C}^j.$$

ii) It is easy to show that:

$$\|\mathbb{S}\| \leq \frac{1}{1-\|\mathbb{C}\|}.$$

By applying this estimate $(\mathbb{C} = -\mathbb{A}^{-1}\delta\mathbb{A})$, we obtain:

$$||(\mathbb{A} + \delta\mathbb{A})^{-1}|| \leq \frac{||\mathbb{A}^{-1}||}{1 - ||\mathbb{A}^{-1}||\ ||\delta\mathbb{A}||}$$

since we have:

$$(\mathbb{A} + \delta\mathbb{A})^{-1} = \left(\mathbb{I} + \mathbb{A}^{-1}\delta\mathbb{A}\right)^{-1}\mathbb{A}^{-1}.$$

However, we have that:

$$(\mathbb{A} + \delta\mathbb{A})\delta X = B + \delta B - \mathbb{A}X - \delta\mathbb{A}X = \delta B - \delta\mathbb{A}X$$

which is to say that:

$$\delta X = (\mathbb{A} + \delta\mathbb{A})^{-1}(\delta B - \delta\mathbb{A}X)$$

from which we get the estimate:

$$|\delta X| \leq \frac{||\mathbb{A}^{-1}||}{1 - ||\mathbb{A}^{-1}||\ ||\delta\mathbb{A}||}(|\delta B| + ||\delta\mathbb{A}|||X|)$$

thus leading to [A1.4]. ■

The condition $||\delta\mathbb{A}|| < \frac{1}{||\mathbb{A}^{-1}||}$ ensures that $1 > 1 - \gamma(\mathbb{A})\frac{||\delta\mathbb{A}||}{||\mathbb{A}||} > 0$. The relative error of the solution X is therefore proportional to the relative errors of the matrix and the right-hand side, with a proportionality coefficient greater than the matrix condition number. Therefore, if the condition number is large, then a significant error may ruin the solution of the linear system. Let us illustrate this phenomenon with the following simple example.

A1.1.2. *Example*

Consider the following example: $\mathbb{A} = \begin{bmatrix} 2.23 & 1.2 \\ 1.3 & 0.7 \end{bmatrix}$ and $B = \begin{pmatrix} -2.06 \\ -1.2 \end{pmatrix}$.

The linear system $\mathbb{A}X = B$ has the solution: $X = \begin{pmatrix} -2 \\ 2 \end{pmatrix}$.

Now, consider the perturbation: $\delta B = \begin{pmatrix} 10^{-3} \\ 10^{-3} \end{pmatrix}$.

Then the linear system $\mathbb{A}Y = B + \delta B$ has the solution: $Y = \begin{pmatrix} -2.5 \\ 2.93 \end{pmatrix}$.

Now let us calculate the condition number for the matrix $\mathbb{A}$ via the matrix norm:

$$\|\mathbb{A}\|_\infty = \max_i \sum_j |\mathbb{A}_{ij}|\,.$$

We have: $\|\mathbb{A}\|_\infty = 3.43$ and $\left\|\mathbb{A}^{-1}\right\|_\infty = 3.53\,10^3$, which gives for the condition number (∞-norm):

$$\gamma_\infty(\mathbb{A}) \simeq 1.2\,10^4.$$

Other examples are straightforward to construct. For example, let us point out that of Rutihauser:

$$\mathbb{A} = \begin{bmatrix} 10 & 1 & 4 & 0 \\ 1 & 10 & 5 & -1 \\ 4 & 5 & 10 & 7 \\ 0 & -1 & 7 & 9 \end{bmatrix} \quad \text{with } B = \begin{pmatrix} 15 \\ 15 \\ 26 \\ 15 \end{pmatrix}$$

for which the solution of $\mathbb{A}X = B$ is $X = (1,1,1,1)^t$, with $\gamma_\infty(\mathbb{A}) = 6.26 \times 10^4$.

The Hilbert matrices $\left(\mathbb{A}_{ij} = \frac{1}{i+j}\right)$ provide examples of invertible matrices whose condition number tends to infinity as their ranks tend to infinity. When the matrix condition number is large, then we say that the matrix is **ill-conditioned**. No numerical method allows us to overcome this situation. One must necessarily increase the computational precision by using double – or perhaps even quadruple – precision number formats, or instead apply correction methods to the rounding errors, so as to obtain a solution with the desired precision for the linear system. On the other hand, one must be careful of the fact that numerical methods can exacerbate the effects of ill-conditioning, since this method is very unstable (amplification of rounding errors).

A1.2. Direct-solving methods

Recall that these methods are supposed to lead us to the solution for the linear system after performing a *finite* number of operations. What factors differentiate one method from another? Of course, it is the number of operations that have to be executed, before one obtains the solution for the linear system. What we mean by an **operation** is any of the following: addition, subtraction, multiplication, division and radicals. On any given computer, the execution time between each operation certainly differs but, since we will be interested in what happens asymptotically relative to n (the matrix rank), we will assume that each of these operations is equivalent.

A1.2.1. *The determinant method*

This well-known method is based on calculating the determinants of the rank $n-1$ cofactors and allows us to calculate the matrix inverse after a finite number of operations. It requires the rank n determinant to be calculated, as well as the n^2 determinants of rank $n-1$. But the calculation of a rank p determinant requires p calculations of rank $p-1$ determinants and $p-1$ operations, which can be summarized in the following formula (N_{op} denotes the number of operations):

$$N_{op}(\text{Det}_p) = p \times N_{op}(\text{Det}_{p-1}) + (p-1)$$

from which we get: $N_{op}(\text{Det}_p) = O(p!)$, by induction. This calculation shows that the number of operations required to invert a rank n matrix via the determinant method is:

$$N_{op}(\mathbb{A}^{-1}) = O((n+1)!).$$

This method is unusable in practice ($n = 50 \Longrightarrow N_{op} \simeq 1.5 \times 10^{66}$).

A1.2.2. *Solving a triangular system*

The determinant method calculates the inverse. But to solve a linear system, calculating it is not required. We are now going to propose various alternative methods that require a finite number of operations, proportional to n^3 (compare this with $n!$). These methods, which are called *direct* ones, are based on the Gaussian elimination method (which involves combining the equations of linear systems); they allow us to reduce the system to a triangular one:

$$\mathbb{L}X = \tilde{B} \text{ or } \mathbb{U}X = \tilde{B}$$

– where $\mathbb{L}$ is a lower-triangular matrix, that is:

$$\mathbb{L}_{ij} = 0 \quad \text{if } i < j$$

– and $\mathbb{U}$ is an upper-triangular matrix, that is:

$$\mathbb{U}_{ij} = 0 \quad \text{if } i > j.$$

Solving a triangular system is a straightforward operation. Indeed, suppose that we have to solve the lower triangular system $\mathbb{L}X = \tilde{B}$. Then we have the following equations:

$$\begin{cases} \mathbb{L}_{11}X_1 & = \tilde{B}_1 \\ \mathbb{L}_{21}X_1 + \mathbb{L}_{22}X_2 & = \tilde{B}_2 \\ \dots & \\ \mathbb{L}_{i1}X_1 + \ldots + \mathbb{L}_{ii}X_i & = \tilde{B}_i \\ \dots & \\ \mathbb{L}_{n1}X_1 + \ldots + \mathbb{L}_{nn}X_i & = \tilde{B}_n. \end{cases}$$

The first equation yields X_1, the second, X_2 (as then X_1 is now known), and so on, until X_n has been determined. We thus obtain algorithm A1.1, which is called **forward substitution** .

```
for i = 1, n
    S = B̃_i
    for j = 1, i − 1
        S = S − L_ij X_j
    if L_ii = 0 then
        STOP: non-invertible matrix
    X_i = S / L_ii
```

Algorithm A1.1. *Forward substitution*

This algorithm makes it easy to identify non-invertible matrices (from a zero term along the diagonal), and we can even reduce the memory occupied by "overwriting" the right-hand side, that is, by replacing X_i and X_j in the algorithm with $\tilde{B}_i$ and $\tilde{B}_j$. In this case, $\tilde{B}$ is the calculated solution once the algorithm has finished.

Counting the number of operations is easy, since we have:

$$N_{op}(\text{Forward substitution}) = O(n^2). \quad \text{[A1.5]}$$

Similarly, the algorithm for solving an upper triangular system, which is called a **backward substitution**, is represented in algorithm A1.2.

```
for i = n, 1, -1
    S = B̃_i
    for j = n, i + 1, -1
        S = S - U_ij X_j
    if U_ii = 0 then
        STOP: non-invertible matrix
    X_i = S / U_ii
```

$$
\begin{array}{l}
\textbf{for } i = n, 1, -1 \\
\quad S = \tilde{B}_i \\
\quad \textbf{for } j = n, i+1, -1 \\
\quad\quad S = S - \mathbb{U}_{ij} X_j \\
\quad \textbf{if } \mathbb{U}_{ii} = 0 \textbf{ then} \\
\quad\quad \text{STOP: non-invertible matrix} \\
\quad X_i = \dfrac{S}{\mathbb{U}_{ii}}
\end{array}
$$

Algorithm A1.2. *Back substitution algorithm*

There are versions of these algorithms where the *loops* over i and j are reversed. They have advantages in certain situations (for the matrix storage structures, or vectorization) of which we will return to. In doing so, forward substitution is shown in algorithm A1.3.

$$
\begin{array}{l}
\textbf{for } i = 1, n \\
\quad \textbf{if } \mathbb{L}_{ii} = 0 \textbf{ then} \\
\quad\quad \text{STOP: non-invertible matrix} \\
\quad X_i = \dfrac{\tilde{B}_i}{\mathbb{L}_{ii}} \\
\quad \textbf{for } j = i+1, n \\
\quad\quad \tilde{B}_j = \tilde{B}_j - \mathbb{L}_{ji} X_i
\end{array}
$$

Algorithm A1.3. *Forward substitution, column format*

Now let us see which algorithms aid us when determining the solution of a triangular system.

A1.2.3. *Factorization in triangular form*

The **Gaussian elimination** method consists of taking linear combinations of the equations, before then reducing them to an upper triangular system. Algorithm A1.4 transforms a matrix $\mathbb{A}$ into an upper triangular one by modifying additionally the right-hand side.

for $k = 1, n-1$
  if $\mathbb{A}_{kk} = 0$ **then**
    STOP: non-invertible matrix
  for $i = k+1, n$
    $p_i = \dfrac{\mathbb{A}_{ik}}{\mathbb{A}_{kk}}$
    for $j = k+1, n$
      $\mathbb{A}_{ij} = \mathbb{A}_{ij} - p_i \mathbb{A}_{kj}$
    $\mathbb{A}_{ik} = 0$
    $B_i = B_i - p_i B_k$

Algorithm A1.4. *Gaussian elimination algorithm*

The fact that a pivot $\mathbb{A}_{kk}$ vanishes does not mean that the matrix $\mathbb{A}$ is non-invertible. To convince oneself, consider the rank 3 matrix:

$$\mathbb{A} = \begin{bmatrix} 1 & 1 & 0 \\ 0 & 1 & 0 \\ 0 & 0 & 1 \end{bmatrix}$$

which, after iterating Gauss's elimination algorithm once, becomes:

$$\mathbb{A} = \begin{bmatrix} 1 & 1 & 0 \\ 0 & \mathbf{0} & -1 \\ 0 & -1 & 0 \end{bmatrix}.$$

The modification that the matrix has undergone after its kth iteration is represented in Figure A1.1.

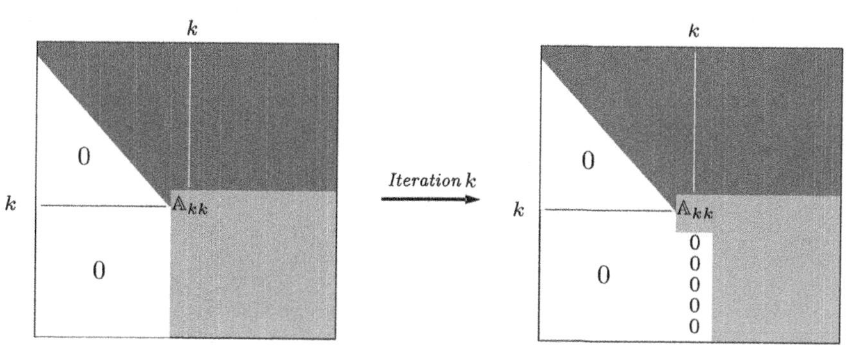

Figure A1.1. *kth iteration of the Gaussian elimination algorithm*

The Gaussian elimination algorithm is actually an $\mathbb{LU}$ factorization algorithm for the matrix $\mathbb{A}$, where $\mathbb{L}$ is a lower triangular matrix and $\mathbb{U}$ is the upper triangular matrix obtained from the elimination. Under certain hypotheses, if the factorization is possible then it is unique; moreover, whenever the matrix $\mathbb{A}$ is symmetric, then we have symmetric factorizations. This gives us the following result:

PROPOSITION A1.2.–

a) Any invertible matrix $\mathbb{A}$, whose principal submatrices[1] *are invertible, can be factored into the form $\mathbb{LU}$.*

b) The factorization $\mathbb{A} = \mathbb{LU}$ is unique, where $\mathbb{L}$ has a unit diagonal.

c) If the matrix $\mathbb{A}$ is symmetric and $\mathbb{LU}$-factorizable, then $\mathbb{A}$ can be factored into the form:

$$\mathbb{A} = \mathbb{LDL}^t$$

where $\mathbb{D}$ is a diagonal matrix.

d) If the matrix $\mathbb{A}$ is symmetric and positive-definite, then $\mathbb{A}$ can be factored into the form:

$$\mathbb{A} = \mathbb{LL}^t.$$

These properties are not too difficult to prove, though doing so is somewhat technical.

The $\mathbb{LU}$-factorization algorithms are obtained by using the following equalities:

$$\left|\begin{array}{lll} \mathbb{A}_{ij} = \sum\limits_{k=1}^{\min(i,j)} \mathbb{L}_{ik}\mathbb{U}_{kj} \ \forall i,j = 1,n & & \\ \mathbb{L}_{ii} = 1 & \forall i = 1,n & \text{(uniqueness condition)} \end{array}\right.$$

which factors a matrix into a product of a lower triangular matrix and an upper triangular one. From these formulae, we obtain the following relations:

$$\left|\begin{array}{ll} \mathbb{A}_{ij} = \sum\limits_{k=1}^{j-1} \mathbb{L}_{ik}\mathbb{U}_{kj} + \mathbb{L}_{ij}\mathbb{U}_{jj} & \forall j < i \\ \mathbb{A}_{ij} = \sum\limits_{k=1}^{i-1} \mathbb{L}_{ik}\mathbb{U}_{kj} + \mathbb{U}_{ij} & \forall j \geq i \end{array}\right.$$

1 Principal submatrix of rank k: $\begin{bmatrix} \mathbb{A}_{11} & \dots & \mathbb{A}_{1k} \\ \vdots & \vdots & \vdots \\ \mathbb{A}_{k1} & \dots & \mathbb{A}_{kk} \end{bmatrix}$.

which allow us to conclude the $\mathbb{LU}$-factorization algorithm in algorithm A1.5, which is sometimes called **Doolittle**'s algorithm.

for $i = 1, n$
  $\mathbb{L}_{ii} = 1$
for $p = 1, n$
  for $j = p, n$

$$\mathbb{U}_{pj} = \mathbb{A}_{pj} - \sum_{k=1}^{p-1} \mathbb{L}_{pk}\mathbb{U}_{kj}$$

  if $\mathbb{U}_{pp} = 0$ **then**
    STOP: non-factorizable matrix
  for $i = p+1, n$

$$\mathbb{L}_{ip} = \left(\mathbb{A}_{ip} - \sum_{k=1}^{p-1} \mathbb{L}_{ik}\mathbb{U}_{kp}\right) / \mathbb{U}_{pp}$$

Algorithm A1.5. *LU-factorization algorithm*

In the case where the matrix $\mathbb{A}$ is symmetric and positive-definite (item (d) of proposition A1.2), then algorithm A1.5 takes on a simpler form that is presented in algorithm A1.6, called **Cholesky**'s algorithm.

for $p = 1, n$

$$\mathbb{L}_{pp} = \left(\mathbb{A}_{pp} - \sum_{k=1}^{p-1} |\mathbb{L}_{pk}|^2\right)^{1/2}$$

  if $\mathbb{L}_{pp} = 0$ or $\mathbb{L}_{pp} \notin \mathbb{R}$ **then**
    STOP: matrix not pos.-def
  for $i = p+1, n$

$$\mathbb{L}_{ip} = \left(\mathbb{A}_{ip} - \sum_{k=1}^{p-1} \mathbb{L}_{ik}\mathbb{L}_{pk}\right) / \mathbb{L}_{pp}$$

Algorithm A1.6. *Cholesky's algorithm*

Note that this algorithm makes it easy to check whether a matrix is positive-definite or not (the latter would involve the square root of zero or of a negative number). Algorithm A1.6 is said to be in **column** form because, during the pth iteration, there are $n - p$ terms $(\mathbb{L}_{ip})_{i=p+1,n}$ calculated along the pth column. There is also a form called the **row** form for which, during the pth iteration, there are $p - 1$ terms $(\mathbb{L}_{pj})_{j=1,p-1}$ calculated along the pth row.

A1.2.3.1. *The partial pivoting method*

In order to remedy the issue of zero pivots appearing in the factorization algorithms (item (a) of proposition A1.2), a technique that permutes the rows of the matrix is used, and it is based on the following property:

PROPOSITION A1.3.– *Let $\mathbb{A}$ be an invertible matrix, then there exists a permutation matrix $\mathbb{P}$ such that $\mathbb{PA}$ is $\mathbb{LU}$-factorizable.*

As an example, we will describe the Gaussian elimination method using partial pivoting. In the Gaussian elimination algorithm, when a zero pivot $\mathbb{A}_{kk}$ appears in the kth column, then we look along this column for a non-zero term $\mathbb{A}_{\ell k}$ with $\ell > k$. If one exists, then the ℓ and k rows are swapped before the elimination is applied. If one does not exist, then this means that the matrix is non-invertible, since then we have found a non-invertible $(n - k + 1)$ submatrix. In practice, an n-dimensional vector q is used to represent the permutation of the n rows. Initialized first to $q(i) = i,\ i = 1, n$, this vector is updated after each row permutation by exchanging the values that it holds.

This technique does not allow us in practice to exchange two rows of the matrix, since the permutation is virtual. In algorithm A1.7, a pivoting technique is used to search for the maximal column pivot, thus reducing the numerical error propagation at just a small additional cost.

```
for i = 1, n
  q(i) = i
for k = 1, n − 1
  K = q(k) and M = 0                              look for max. pivot
  for i = k, n
    if |A_{q(i)K}| > M then
      ℓ = i and M = |A_{q(i)K}|
  if M < ε then
    STOP: non-invertible matrix
  q(k) = q(ℓ) and q(ℓ) = K                        permute the rows
  K = q(k)
  for i = k + 1, n
    I = q(i)
    p_I = A_{IK} / A_{KK}
    for j = k + 1, n
      J = q(j)
      A_{IJ} = A_{IJ} − p_I A_{KJ}
    B_I = B_I − p_I B_K
```

Algorithm A1.7. *Gaussian elimination algorithm with row pivoting*

Note that this partial-pivoting version of the Gaussian elimination algorithm allows us to detect whether a matrix is invertible or not. Indeed, the appearance of a zero pivot corresponds to a full column vanishing in a submatrix, and therefore the matrix is non-invertible. In practice, it is not possible to observe the vanishing of a pivot due to rounding errors, but only to the quasi-vanishing of the pivot relative to some tolerance parameter, $\varepsilon > 0$, which we assign ourselves. Obtaining a very small pivot is evidence of an ill-conditioned matrix, and also therefore of low-quality results. Consequently, the tolerance parameter needs to be chosen according to the accuracy of the calculation and the quality of the expected result.

We can generalize this pivoting method by also allowing for columns to be pivoted, that is, by finding the maximal pivot during the kth iteration among the current $(\mathbb{A}_{ij})_{i\geq k, j\geq k}$ terms, which further alleviates any numerical error propagation.

The partial pivoting technique also applies to $\mathbb{LU}$-factorization algorithms. The maximal pivot during the pth step of the algorithm is then given by:

$$\max_{j=p,n} \left| \mathbb{A}_{pj} - \sum_{k=1}^{p-1} \mathbb{L}_{pk}\mathbb{U}_{kj} \right|$$

and it is attained at the $\ell\mathrm{th}$ index. We then interchange the $p\mathrm{th}$ and $\ell\mathrm{th}$ indices.

A1.2.3.2. *Number of operations*

Each algorithm that we have just presented contains three nested loops. So we have:

$$N_{op}(\text{Factorization}) = O(n^3).$$

A1.2.3.3. *The forward-back substitution phase*

When we have an $\mathbb{LU}$-factorization (respectively, an $\mathbb{LL}^t$-factorization) for a matrix $\mathbb{A}$, then it is easy to solve the linear system $\mathbb{A}X = B$. One just solves the lower triangular system $\mathbb{L}Y = B$ before then solving the upper triangular system $\mathbb{U}X = Y$ (resp. $\mathbb{L}^tX = Y$).

A1.3. Iterative methods

Iterative methods are based upon constructing a vector sequence $\left(X^{(k)}\right)$ that converges to the solution vector $\bar{X}$ for the system $\mathbb{A}\bar{X} = B$. Iterative methods are typically split into two classes:

– splitting methods;

– gradient methods.

However, their similarities reunite them. Before briefly outlining these methods, let us recall some general results regarding convergence.

A1.3.1. *Convergence of iterative methods*

Consider a vector sequence $(X^{(k)})$ defined by the recurrence relation:

$$X^{(k+1)} = \mathbb{C}X^{(k)} + D. \quad \text{[A1.6]}$$

If the sequence $(X^{(k)})_{k\geq 0}$ converges to the vector $\bar{X}$, then we must necessarily have that:

$$\bar{X} = \mathbb{C}\bar{X} + D$$

from which we can deduce:

$$D = (\mathbb{I} - \mathbb{C})\mathbb{A}^{-1}B. \quad \text{[A1.7]}$$

We then say that the iterative method [A1.6] converges, if the sequence $\left(X^{(k)}\right)$ converges for any initial vector $X^{(0)}$. Now, we may state the following classical convergence result:

PROPOSITION A1.4.– *The iterative method [A1.6] converges if and only if:*

$$\rho(\mathbb{C}) < 1. \quad [A1.8]$$

Moreover, if the equality [A1.7] is satisfied, then the sequence $(X^{(k)})$ *converges to* $\bar{X}$.

The number $\rho(\mathbb{C})$ represents the spectral radius of the matrix $\mathbb{C}$, that is, the modulus of the largest (a priori, complex) eigenvalue in $\mathbb{C}$. In order to compare the iterative methods, a tool to characterize the different speeds of convergence is required. Therefore, let us define:

– the **rate of convergence** after m iterations:

$$R_m(\mathbb{C}) = -\log\left(\|\mathbb{C}^m\|^{\frac{1}{m}}\right) \quad [A1.9]$$

– the **asymptotic rate of convergence**:

$$R(\mathbb{C}) = -\log(\rho(\mathbb{C})). \quad [A1.10]$$

The number $N_m = \frac{1}{R_m(\mathbb{C})}$ can then be interpreted as the estimated number of required iterations for the error to be reduced by a factor of e $(2, 718...)$. Note that:

$$\lim_{m\to\infty} R_m(\mathbb{C}) = R(\mathbb{C}) \quad \text{and} \quad \lim_{m\to\infty} N_m = N = \frac{1}{R(\mathbb{C})}.$$

In general, the number N is used to characterize the rate of convergence for an iterative method.

A1.3.2. *Splitting methods*

The first type of method is based upon the matrix splitting:

$$\mathbb{A} = \mathbb{M} - \mathbb{N} \quad [A1.11]$$

where the matrix $\mathbb{M}$ is assumed to be *invertible*. This splitting is associated with the iterative process:

$$\mathbb{M}X^{(k+1)} = \mathbb{N}X^{(k)} + B \quad [A1.12]$$

which is of the general form [A1.6], with:

$$\mathbb{C} = \mathbb{M}^{-1}\mathbb{N} \text{ and } D = \mathbb{M}^{-1}B \quad [A1.13]$$

and which satisfies the property [A1.7].

Now it is a matter of knowing whether $\rho\left(\mathbb{M}^{-1}\mathbb{N}\right)$ is less than 1 or not. In the two following propositions, we will indicate that classes of matrices are valid when proving this result (see, for example, Ciarlet (1989)).

For *Hermitian matrices* (or symmetric in the real case), we have the following:

PROPOSITION A1.5.– *If $\mathbb{A}$ is a self-adjoint matrix, then the matrix $(\mathbb{M}^* + \mathbb{N})$ is self-adjoint. Furthermore, if we assume that the matrix $(\mathbb{M}^* + \mathbb{N})$ is positive-definite, then $\rho\left(\mathbb{M}^{-1}\mathbb{N}\right) < 1$ if, and only if, $\mathbb{A}$ is positive-definite.*

There is a more general result concerning matrices with *non-negative inverses*. A vector Z is said to be non-negative ($Z \geq 0$) if each of its components is either positive or zero. A matrix $\mathbb{H}$ is said to be non-negative if we have that $\mathbb{H}Z \geq 0$ for every $Z \geq 0$.

PROPOSITION A1.6.– *If the matrix $\mathbb{A}$ is invertible and has a non-negative inverse ($\mathbb{A}^{-1} \geq 0$), and if the matrices $\mathbb{M}^{-1}$ and $\mathbb{N}$ are both non-negative ($\mathbb{M}^{-1} \geq 0$, $\mathbb{N} \geq 0$), then we have:*

$$\rho\left(\mathbb{M}^{-1}\mathbb{N}\right) = \frac{\rho\left(\mathbb{A}^{-1}\mathbb{N}\right)}{1 + \rho\left(\mathbb{A}^{-1}\mathbb{N}\right)} < 1.$$

Now let us present some classical splittings of the form $\mathbb{A} = \mathbb{M} - \mathbb{N}$. An important splitting method is based upon using the submatrices $\mathbb{D}$, $-\mathbb{E}$ and $-\mathbb{F}$, which represent the diagonal, the strictly lower triangular and the strictly upper triangular sections of the matrix $\mathbb{A}$.

$\mathbb{A} = \mathbb{D} - \mathbb{E} - \mathbb{F} =$ 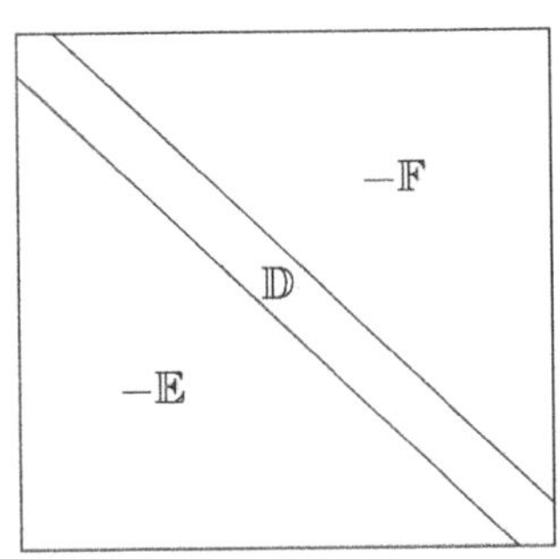

Figure A1.2. *Splitting into the strictly triangular and diagonal parts*

A1.3.2.1. *The Jacobi method*

We will use the following splitting: $\mathbb{M} = \mathbb{D}$ and $\mathbb{N} = \mathbb{E} + \mathbb{F}$. The iteration [A1.12] then takes the form of algorithm A1.8.

Given $X^{(0)}$
for $k = 1, N_{max}$
  for $i = 1, n$

$$X_i^{(k+1)} = \left(B_i - \sum_{j=1}^{i-1} \mathbb{A}_{ij} X_j^{(k)} - \sum_{j=i+1}^{n} \mathbb{A}_{ij} X_j^{(k)} \right) / \mathbb{A}_{ii}$$

  $R = \left| B - \mathbb{A} X^{(k+1)} \right|$ *residual modulus*
  if $R \leq \varepsilon |B|$ **then**
    STOP: converges up to ε

Algorithm A1.8. *The Jacobi algorithm*

This method is *explicit* in the sense that, when calculating the $(k+1)$th iteration, only values produced from the previous iterations are used (here, k of them). Obviously, to apply the Jacobi method, the matrix $\mathbb{A}$ cannot present any vanishing terms along its diagonal.

Here, we apply a convergence test to the sequence $(X^{(k)})$ that uses the **relative residue** $R^{(k)}/|B|$ (where $R^{(k)} = |B - \mathbb{A}X^{(k)}|$) with respect to a given norm and allows us to estimate the quality of the solution obtained for a given linear system. Indeed, we have (for an invertible $\mathbb{A}$ and a solution X):

$$X - X^{(k)} = \mathbb{A}^{-1}(B - \mathbb{A}X^{(k)}) = \mathbb{A}^{-1}R^{(k)} = \mathbb{A}^{-1}|\mathbb{A}X|\frac{R^{(k)}}{|B|}$$

that is:

$$\frac{|X - X^{(k)}|}{|X|} \leq \gamma(\mathbb{A})\frac{|R^{(k)}|}{|B|}$$

which, if the matrix is well conditioned, shows that the relative residue of B provides a good estimate for the relative error of the solution. This stop test is the most classic, though other options do exist:

– $R \leq \varepsilon$, an absolute error test, well-suited provided that $||A^{-1}||$ is not too large;
– $R \leq \varepsilon|B - \mathbb{A}X^{(0)}|$, comparing the relative residue with the initial residue.

For safety reasons, one should always preemptively choose a maximum number of iterations (N_{max}) to run.

A1.3.2.2. *The Gauss-Seidel method*

During the kth iteration in the Jacobi method, when calculating the ith component of $X^{(k+1)}$, the previous $i-1$ components of $X^{(k+1)}$ have already been updated. They make for a better approximation of the solution, and so we should be interested in using them. This leads us to the Gauss–Seidel method, presented in algorithm A1.9.

Given $X^{(0)}$
for $k = 1, N_{max}$
  for $i = 1, n$

$$X_i^{(k+1)} = \left(B_i - \sum_{j=1}^{i-1} \mathbb{A}_{ij} X_j^{(k+1)} - \sum_{j=i+1}^{n} \mathbb{A}_{ij} X_j^{(k)} \right) / \mathbb{A}_{ii}$$

$$R = \left| B - \mathbb{A} X^{(k+1)} \right|$$

  if $R \leq \varepsilon |B|$ **then**
    STOP: converges up to ε

Algorithm A1.9. *The Gauss-Seidel algorithm*

On the one hand, it should be noted that the number of operations is identical to that of the Jacobi method and, on the other hand, that it is possible to reduce the memory space by using a single vector. Although this is an *explicit* method, a partial inversion of the matrix (by solving the lower triangular part by forward substitution) is performed. Therefore, a priori the Gauss–Seidel method makes for an improved Jacobi method. Indeed, it is a splitting method that takes on the form $\mathbb{A} = \mathbb{M} - \mathbb{N}$ with:

$$\mathbb{M} = \mathbb{D} - \mathbb{E} \text{ and } \mathbb{N} = \mathbb{F}$$

and is subject to the same conditions of application as the Jacobi method, namely that the diagonal of the matrix $\mathbb{A}$ must not have any zero terms.

The Gauss–Seidel method is generalized by considering a splitting of the form:

$$\mathbb{M} = \frac{1}{\omega}\mathbb{D} - \mathbb{E} \text{ and } \mathbb{N} = \left(\frac{1}{\omega} - 1\right)\mathbb{D} + \mathbb{F}$$

where ω is a real parameter called the **relaxation** parameter. It leads to a method called the **SOR** (*successive over-relaxation*) method. We have the following convergence results:

PROPOSITION A1.7.– *If the matrix $\mathbb{A}$ is Hermitian and positive-definite, then the Jacobi, Gauss-Seidel and the SOR methods, where $\omega \in (0, 2)$, converge to the solution of the linear system.*

In the case of definite-positive symmetric tridiagonal matrices, by comparing the spectral radii of the different iteration matrices associated with these methods, we can obtain more precise results (Ciarlet 1989):

– the Gauss-Seidel method converges more rapidly than the Jacobi method;

– if $\omega \in (0, 1) \cup (\omega_2, 2)$ (for a given $1 < \omega_2 < 2$, depending on $\mathbb{A}$), the SOR method converges less rapidly than the Gauss–Seidel method;

– if $\omega \in (1, \omega_2)$, the SOR method converges more rapidly than the Gauss–Seidel method, with an optimal value for the relaxation parameter $\omega = \omega_0$.

A1.3.3. *Gradient methods*

Gradient methods, based upon minimizing the quadratic functional associated with a linear system, apply mainly to positive-definite symmetric matrices. However, they can be generalized to non-symmetric cases too. When $\mathbb{A}$ is positive-definite and symmetric, solving the linear system $\mathbb{A}X = B$ is equivalent to finding a minimum of the function from $\mathbb{R}^n$ into $\mathbb{R}$ defined by:

$$\Phi(X) = \frac{1}{2}(\mathbb{A}X \mid X) - (B \mid X) \qquad [A1.14]$$

since we have:

$$\nabla\Phi(X) = \mathbb{A}X - B.$$

To search for the minimum of the function Φ, typically so-called **gradient** methods are used, the principle of which we will now recall.

A1.3.3.1. *Constant-step gradient method*

Denote $W^{(k)} = \mathbb{A}X^{(k)} - B = \nabla\Phi(X^{(k)})$. Let $\theta > 0$, and consider the following iterative process:

$$X^{(k+1)} = X^{(k)} - \theta W^{(k)} \qquad [A1.15]$$

which fits into the splitting method frameworks [A1.6]-[A1.7] with: $\mathbb{C} = \mathbb{I} - \theta\mathbb{A}$ and $D = \theta B$. A necessary and sufficient condition for the process [A1.15] to converge toward the solution of the linear system is that:

$$\rho(\mathbb{I} - \theta\mathbb{A}) < 1.$$

Denote by λ (respectively, Λ) the smallest (respectively, the greatest) eigenvalue of $\mathbb{A}$ ($0 < \lambda < \Lambda$, as we assumed that $\mathbb{A}$ is positive-definite). Then we have the following convergence result:

PROPOSITION A1.8.– *The constant-step gradient method converges if and only if:*

$$0 < \theta < \frac{2}{\Lambda}.$$

Furthermore, the minimum of $\rho(\mathbb{I} - \theta\mathbb{A})$ is attained precisely when:

$$\theta = \theta_0 = \frac{2}{\Lambda + \lambda} \quad \text{and} \quad \rho\left(\mathbb{I} - \theta_0\mathbb{A}\right) = \frac{\Lambda - \lambda}{\Lambda + \lambda}.$$

Its proof is based on the fact that: $\rho(\mathbb{I} - \theta\mathbb{A}) = \text{Max}(|1 - \theta\Lambda|, |1 - \theta\lambda|)$. As a general rule, neither Λ nor λ are known to us. That is why the constant-step gradient descent converges erratically and subsequently has little use in practice. A more reliable strategy consists of adjusting each step θ for each iteration k.

A1.3.3.2. *Optimal-step gradient method*

During each iteration, choose a $\theta^{(k)}$ for which the real-valued function:

$$\theta \mapsto h(\theta) = \Phi\left(X^{(k)} - \theta W^{(k)}\right)$$

realizes its minimum.

The optimal value of $\theta^{(k)}$ is given by the formula (which is characterized by the condition $h'(\theta^{(k)}) = 0$):

$$\theta^{(k)} = \frac{\left(\mathbb{A}X^{(k)} - B \mid W^{(k)}\right)}{\left(\mathbb{A}W^{(k)} \mid W^{(k)}\right)} = \frac{\left|W^{(k)}\right|^2}{\left(\mathbb{A}W^{(k)} \mid W^{(k)}\right)}. \qquad [A1.16]$$

This choice leads to the optimal-step gradient method:

$$X^{(k+1)} = X^{(k)} - \theta^{(k)}W^{(k)} \qquad \theta^{(k)} \text{ given by } [A1.16]$$

and to algorithm A1.10.

$X^{(0)}$ given
for $k = 0, N_{max}$
$W^{(k)} = B - \mathbb{A}X^{(k)}$ *residual*
if $\left|W^{(k)}\right| \leq \varepsilon|B|$ **then**
STOP: converges up to ε
$\theta^{(k)} = \dfrac{\left|W^{(k)}\right|^2}{(\mathbb{A}W^{(k)}|W^{(k)})}$
$X^{(k+1)} = X^{(k)} - \theta^{(k)}W^{(k)}$

Algorithm A1.10. *Optimal-step gradient algorithm*

This method is no longer presented via a splitting, though it can be shown using different techniques that it does indeed converge. More precisely, we have:

PROPOSITION A1.9.– *The optimal-step gradient method converges, and we have the estimate:*

$$\left|X^{(k)} - X\right|_2^2 \leq C\Big(1 - \frac{\lambda}{\Lambda}\Big)^k. \qquad [A1.17]$$

The estimate [A1.17] shows that the closer the matrix condition number $\gamma(\mathbb{A}) = \Lambda/\lambda$ of $\mathbb{A}$ is to 1, the better that the method converges. The aforementioned gradient methods use the descent direction, which has the steepest slope (opposite to the gradient). We will now see a method that uses a descent direction which, although it is less suitable locally, actually turns out to be more efficient overall.

A1.3.3.3. *Conjugate gradient method*

This method consists of choosing the descent directions $W^{(k)}$ that are mutually orthogonal for the scalar product associated with $\mathbb{A}$, when $\mathbb{A}$ is positive-definite and symmetric. Here, these are called **conjugate directions**.

In algorithm A1.11, $G^{(k)}$ represents the gradient of Φ at the point $X^{(k)}$ and $W^{(k)}$ represents the conjugate direction. The algorithm terminates as soon as $G^{(k)} = \nabla\Phi(X^k) = 0$, since then $\mathbb{A}X^{(k)} = B$. In reality, the conjugate gradient algorithm is not an iterative method in the strict sense, because it converges *after at most n iterations*. Indeed, by induction we can prove, though that is tedious enough in itself, that the following orthogonality relations hold:

$$\left|\begin{array}{l} \left(\nabla\Phi\left(X^{(k+1)}\right) \mid W^{(j)}\right) = 0 \\ \left(\nabla\Phi\left(X^{(k+1)}\right) \mid \nabla\Phi\left(X^{(j)}\right)\right) = 0 \ \ \forall j < k+1 \\ \left(\mathbb{A}W^{(k+1)} \mid W^{(j)}\right) = 0 \end{array}\right. \qquad [A1.18]$$

which prove that at the nth iteration, the family $\left(\nabla\Phi\left(X^{(j)}\right)\right)_{j=0,n}$ is an orthogonal family of $(n+1)$ vectors in $\mathbb{R}^n$. Consequently, one of these vectors must be zero, which is to say that there exists an index $\ell \leq n$ such that $\mathbb{A}X^{(\ell)} = B$. This result is theoretical and not always guaranteed in practice due to rounding errors propagating throughout the iterative process. Note that the previous algorithm requires only one computationally expensive operation: the matrix-vector product, $\mathbb{A}W^{(k)}$.

$X^{(0)}$, given $G^{(0)} = \mathbb{A}X^{(0)} - B$ and $W^{(0)} = G^{(0)}$

for $k = 0, N_{max}$

$$\theta^{(k)} = \frac{\left(G^{(k)} \mid W^{(k)}\right)}{\left(\mathbb{A}W^{(k)} \mid W^{(k)}\right)} \qquad \textit{optimal step}$$

$$X^{(k+1)} = X^{(k)} - \theta^{(k)} W^{(k)}$$

$$G^{(k+1)} = \nabla\Phi\left(X^{(k+1)}\right) = \mathbb{A}X^{(k+1)} - B = G^{(k)} - \theta^{(k)}\mathbb{A}W^{(k)}$$

if $\left|G^{(k+1)}\right| \leq \varepsilon |B|$ **then**

STOP: converges up to ε

$$\gamma^{(k)} = \frac{\left|G^{(k+1)}\right|^2}{\left|G^{(k)}\right|^2}$$

$$W^{(k+1)} = G^{(k+1)} + \gamma^{(k)} W^{(k)} \qquad \textit{new descent direction}$$

Algorithm A1.11. *Conjugate gradient algorithm*

When $n = 2$, we can get an idea of the differences between the gradient method and the conjugate gradient one. The functional $X \mapsto \Phi(X)$ represents a paraboloid in $\mathbb{R}^3$, and the isovalue curves $\Phi(X) = cst$ are thus ellipses in $\mathbb{R}^2$.

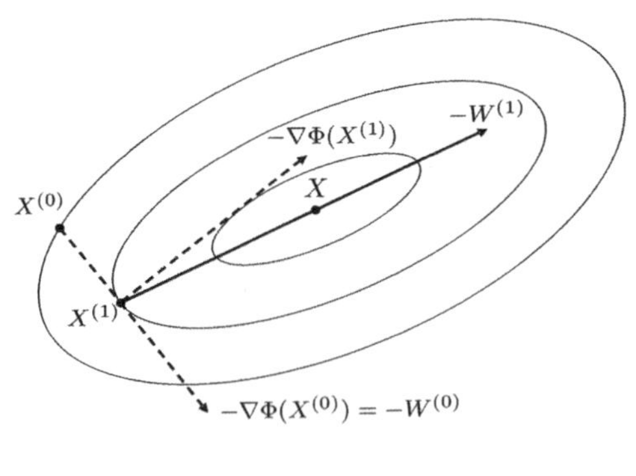

Figure A1.3. *Geometric interpretation of the gradient and of the conjugate gradient methods*

For both methods, the same point $X^{(1)}$ is obtained after the first iteration, which is characterized geometrically by the fact that the line passing through $X^{(0)}$ in the direction of $-\nabla\Phi\left(X^{(0)}\right)$, is tangent to the isovalue curve passing through $X^{(1)}$. The descent direction of the gradient $-\nabla\Phi\left(X^{(1)}\right)$ is orthogonal to the tangent vector of the isovalue curve at the point $X^{(1)}$. The conjugate direction $W^{(1)}$ is built in such a way so that $\left(W^{(1)} \mid \mathbb{A}\nabla\Phi\left(X^{(0)}\right)\right) = 0$ (orthogonality relative to the inner product associated with $\mathbb{A}$). This direction passes through the center of the ellipse $\Phi = cst$, which is a minimum of Φ, and thus a solution of $\mathbb{A}X = B$. After two iterations, the conjugate gradient method provides the solution. However, the descent directions $-\nabla\Phi\left(X^{(k)}\right)$ constructed when using the gradient method alternately goes from one iteration to the other two mutually orthogonal directions. Hence, it is impossible for the gradient method to converge after a finite number of iterations, unless the ellipses are themselves actually circles (with a condition number equal to 1, or rather $\mathbb{A} = \alpha\mathbb{I}$), or if the initial point $X^{(0)}$ is located at one of the extremities for either the major or minor axes of the ellipse.

A1.3.4. *Linear system preconditioning*

The precision of the solution to a linear system is related to the matrix condition number of the system (see A1.3). To circumvent this dilemma, one technique involves changing the linear system (without changing the solution) by using a better conditioned matrix; such methods are called *preconditioning* techniques.

To be precise, the idea involves multiplying the linear system with a matrix $\mathbb{P}$ that approximates the inverse of a matrix $\mathbb{A}$ corresponding to the initial linear system. As the matrix $\mathbb{P}\mathbb{A}$ is then subsequently better conditioned, we expect direct methods to become more stable and iterative methods to converge more quickly. The ideal choice is obviously $\mathbb{P} = \mathbb{A}^{-1}$ because, in this case, a solution to the system is immediate. The quality of a preconditioned iterative method is therefore based upon the choice of preconditioning matrix that achieves the best compromise between the quality of the approximation for the inverse $\mathbb{A}^{-1}$ and the computational cost of determining $\mathbb{P}$. The general preconditioning technique for linear systems consists of choosing a positive-definite symmetric preconditioning matrix $\mathbb{P}$, which is then factorized into the form $\mathbb{P} = \mathbb{L}\mathbb{L}^t$.

Then we set:

$$\tilde{\mathbb{A}} = \mathbb{L}^{-1}\mathbb{A}\mathbb{L}^{-t}, \quad \tilde{B} = \mathbb{L}^{-1}B, \quad \tilde{X} = \mathbb{L}^t X$$

which leads us to the new linear system:

$$\tilde{\mathbb{A}}\tilde{X} = \tilde{B}$$

where $\tilde{\mathbb{A}}$ is a positive-definite symmetric matrix provided that $\mathbb{A}$ is. Calculating $\tilde{B}$ is simple and inexpensive (solving a triangular system), and if we know $\tilde{X}$, then it is clear that we can simply deduce X by solving another triangular system. This approach is particularly suitable for iterative methods. Let us look at the optimal form of the preconditioned conjugate gradient method (see algorithm A1.12).

$X^{(0)}$ given, $G^{(0)} = \mathbb{A}X^{(0)} - B$ and $W^{(0)} = \mathbb{L}^{-1}G^{(0)}$
for $k = 0, N_{max}$

$$\theta^{(k)} = \frac{\left(\mathbb{P}^{-1}G^{(k)} \mid G^{(k)}\right)}{\left(\mathbb{A}W^{(k)} \mid W^{(k)}\right)} \qquad \textit{optimal step}$$

$$X^{(k+1)} = X^{(k)} - \theta^{(k)}W^{(k)}$$

$$G^{(k+1)} = G^{(k)} - \theta^{(k)}\mathbb{A}W^{(k)}$$

if $\left|G^{(k+1)}\right| \leq \varepsilon|B|$ **then**

STOP: converges up to ε

$$\gamma^{(k)} = \frac{\left(\mathbb{P}^{-1}G^{(k+1)} \mid G^{(k+1)}\right)}{\left(\mathbb{P}^{-1}G^{(k)} \mid G^{(k)}\right)}$$

$$W^{(k+1)} = \mathbb{P}^{-1}G^{(k+1)} + \gamma^{(k)}W^{(k)} \qquad \textit{new descent direction}$$

Algorithm A1.12. *Preconditioned conjugate gradient algorithm*

If we know the Cholesky factorization $\mathbb{P} = \mathbb{L}\mathbb{L}^t$, then calculating the product $\mathbb{P}^{-1}Y$ leads to the solution of the triangular system:

$$\mathbb{L}Z = Y \text{ then } \mathbb{L}^tV = Z$$

and provides the vector $V = \mathbb{P}^{-1}Y$. It can be seen that algorithm A1.12 essentially requires just one inversion (forward-back substitution) per iteration, and directly produces $X^{(k)} = \mathbb{L}^{-t}\tilde{X}^{(k)}$.

In practice, several methods (Meurant 1999; Saad 2003), which are essentially heuristic, can be used to determine the preconditioning matrix $\mathbb{P}$. Let us state the two most often used:

– the **SSOR** approach (*Symmetric SOR*): $\mathbb{P} = \left(\frac{\mathbb{D}}{\omega} - \mathbb{E}\right)\left(\frac{\mathbb{D}}{\omega}\right)^{-1}\left(\frac{\mathbb{D}}{\omega} - \mathbb{F}\right)$;

– the incomplete Cholesky factorization for a matrix $\mathbb{A}$, which consists of omitting some terms in $\mathbb{A}$ during the factorization and keeping just a few diagonals – or even just the main diagonal.

The iterative methods presented work essentially for definite-positive symmetric matrices. When faced with non-symmetric matrices, the use of either a direct method

(Gaussian elimination) or a more sophisticated iterative method (Meurant 1999; Saad 2003) is recommended. For example, there are the **GMRES** (*gradient minimal residue*) or the **BICGSTAB** (biconjugate gradient stabilized) methods, with the first method being one with the best performance, *provided that it converges.*

A1.4. Matrix storage structures

The matrices which arise from finite element discretizations (or from finite differences) are "structurally" sparse, that is to say, they possess a significant number of zero terms, the locations of which being known in advance. A reasonable idea is to avoid performing calculations with these zero terms, in order to reduce the required calculation time when inverting a linear system. Moreover, it is not possible to "store" matrices with very high ranks (10^5, for example) in the usual format within the computer memory (using two-entry row-column tables); for a real matrix of rank 10^5, that constitutes up to 10^{10} terms, requiring 40 gigabytes of memory! This is why **compression** techniques, which involves storing the matrix as a vector with the maximal number of structurally null terms having been eliminated, are used and moreover are essential. Let us first present the main techniques that are used to store matrices, which are called compressions, before then discussing how to adapt the aforementioned methods that are used to solve linear systems when compression is involved.

A1.4.1. *Matrix compression*

There are four main types of compression which, in a roughly decreasing order of occupied memory space, are as follows: *diagonal* compression, *band* compression, *profile* compression and *sparse* compression. Clearly, when the matrix is symmetric, the upper triangular part does not have to be stored. We will consider the case of a symmetric matrix explicitly, with the generalization toward non-symmetric matrices being rather straightforward (see, for example, section A1.4.1.4).

A1.4.1.1. *Diagonal compression*

Denote by D_0 the diagonal of a matrix $\mathbb{A}$, and D_k^- the kth subdiagonal of $\mathbb{A}$, which is defined by:

$$D_k^- = \{\mathbb{A}_{ij}, \text{ such that } i - j = k,\ i,j = 1,n\} \qquad \left(D_0^- = D_0\right)$$

where n will always represent the rank of the matrix $\mathbb{A}$. The length of D_k^- is $(n-k)$, and we will denote the ℓth term in the diagonal D_k^- by $\left(D_k^-\right)_\ell$. Then we have the following formulas:

$$\left| \begin{array}{l} \left(D_k^-\right)_\ell = \mathbb{A}_{\ell+k,\ell} \quad \forall k = 0, n-1,\ \ell = 1, n-k \\ \mathbb{A}_{ij} = \left(D_{i-j}^-\right)_j \quad i \geq j,\ i,j = 1,n \end{array} \right.$$

and which lets us pass from the diagonal representation to the matrix representation and vice versa. As soon as the matrix $\mathbb{A}$ presents a vanishing diagonal, that is, $(D_k^-)_\ell = 0, \forall \ell = 1, n-k$, then we do not store it. Therefore, we construct a compressed vector for the matrix $\mathbb{A}$:

$$\mathbb{C}_d = \left(D_{k_1}^-, D_{k_2}^-, \cdots, D_{k_p}^-\right)$$

where $(k_1, k_2, \cdots, k_p)$ are arranged in an increasing order. The vector $\mathbb{C}_d$ has length:

$$\sum_{m=1,p} (n - k_m) = pn - \sum_{m=1,p} k_m.$$

In order to find a (non-zero) term $\mathbb{A}_{ij}$ in the vector $\mathbb{C}_d$, additional information is required, for example, a list of the stored diagonals and their position along the vector $\mathbb{C}_d$. To do so, we construct a mapping ℓ_c defined by:

$$\ell_c(k) = \begin{cases} 0 \text{ if } D_k^- \notin \mathbb{C}_d \\ p \text{ if } D_k^- \in \mathbb{C}_d \end{cases}$$

where p is the position along $\mathbb{C}_d$ of the $(D_k^-)_1$ term.

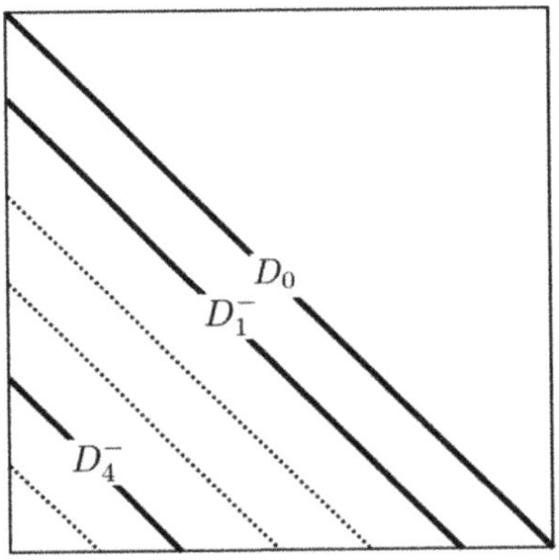

$\mathbb{C}_d = (D_0, D_1^-, D_4^-)$
of size $\mathbb{C}_d = 3n - 5$

Figure A1.4. *Diagonal compression storage*

Using the mapping ℓ_c, it is then easy to locate the position of the $\mathbb{A}_{ij}$ term along the vector $\mathbb{C}_d$, since:

$$\left| \begin{array}{l} \text{if } \ell_c(i-j) = 0 \text{ then } \mathbb{A}_{ij} = 0 \\ \text{if } \ell_c(i-j) = p \text{ then } \mathbb{A}_{ij} = (\mathbb{C}_d)_{p+j-1}\,. \end{array} \right.$$

Note that the description of the compressed structure occupies a vector of length n, which is very small. Conversely, we are able to locate what $\mathbb{A}_{ij}$ term corresponds to which $(\mathbb{C}_d)_\ell$ term of the compressed vector $\mathbb{C}_d$. But this requires the use of an expensive process, since one must locate the index k for which:

$$0 \neq \ell_c(k) \leq \text{ position of } (\mathbb{C}_d)_\ell < \ell_c(k+m) \neq 0.$$

Using such a process should be banned in practice, as it is much better to use an alternative descriptor from ℓ_c. As a general rule, note that an economical choice of descriptor will favor one of the following two directions, $\mathbb{A} \rightarrow \mathbb{C}_d$ or $\mathbb{C}_d \rightarrow \mathbb{A}$. Consequently, one should choose the descriptor best suited for whatever calculations one wishes to carry out. For direct methods, it is preferable therefore to go in the direction $\mathbb{A} \rightarrow \mathbb{C}_d$, whereas, for iterative methods, both directions are possible since only products of matrices with vectors are required.

A1.4.1.2. *Band compression*

Diagonal compression is not compatible with $\mathbb{LU}$- nor $\mathbb{LL}^t$-factorization techniques. Indeed, if there is a subdiagonal that vanishes (and is therefore not stored) between two non-zero diagonals, then the factorization process changes this diagonal. This is why we introduce the so-called **band** compression, in which we keep each diagonal lying between two non-zero diagonals. Clearly, memory space is larger but, as we will see later, this compression is compatible with the $\mathbb{LU}$- and $\mathbb{LL}^t$-factorizations.

Let us set: $\mu = \max\{i-j,\ i \geq j \text{ tel que } \mathbb{A}_{ij} \neq 0\}$, which corresponds to the index of the non-zero diagonal D_μ^- that is the furthest away from the main diagonal, D_0. Here, μ is called the **bandwidth** of the matrix $\mathbb{A}$. The following properties are immediate:

– if the matrix $\mathbb{A}$ is diagonal; we have $\mu = 0$;

– when $\mathbb{A}_{n1} \neq 0$; we have $\mu = n - 1$.

Band compression consists of storing each coefficient $\mathbb{A}_{ij}$ for which $i \geq j$ and $(i-j) \leq \mu$. It occupies more memory than diagonal compression does, since non-zero diagonals D_k^- have to be additionally stored. In practice, either diagonal-by-diagonal storage:

$$\mathbb{C}_b = \left(D_0, D_1^-, \cdots, D_\mu^-\right)$$

or, alternatively, row-by-row storage is adopted:

$$\mathbb{C}_l = \Big(\mathbb{A}_{11}\ \underbrace{\mathbb{A}_{21}\mathbb{A}_{22}}_{\text{row } 2}\ \cdots\ \underbrace{\mathbb{A}_{\mu+1,1}\cdots\mathbb{A}_{\mu+1,\mu+1}}_{\text{row } \mu+1}\ \cdots\ \underbrace{\mathbb{A}_{n,n-\mu}\cdots\mathbb{A}_{n,n}}_{\text{row } n}\Big).$$

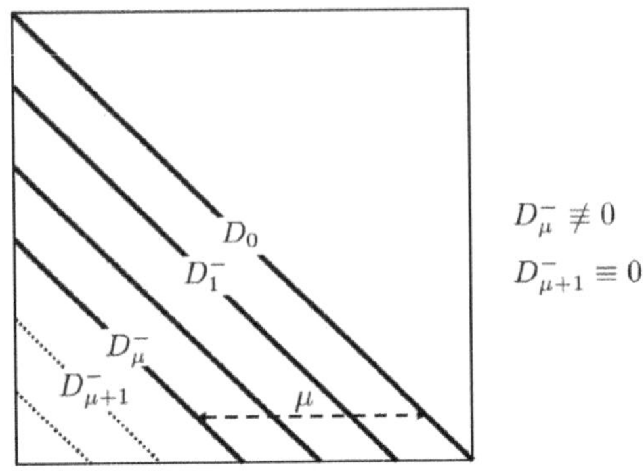

Figure A1.5. *Band matrix*

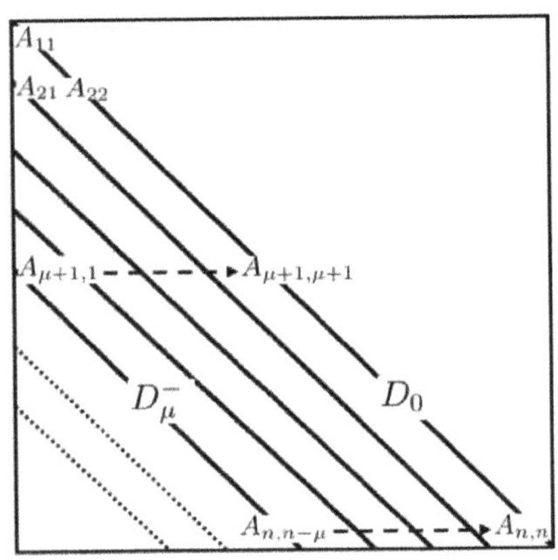

Figure A1.6. *Band (row-by-row) storage*

Whichever storage scheme is used, the length of the compressed vector, $\mathbb{C}_b$ or $\mathbb{C}_l$, is given by:

$$\sum_{m=0,\mu} (n-m) = n(\mu+1) - \frac{\mu(\mu+1)}{2}.$$

The $\mathbb{A}_{ij}$ term is easily located within the compressed vector, $\mathbb{C}_b$ or $\mathbb{C}_l$, via the band number μ. Indeed:

– in the compressed $\mathbb{C}_b$ case, we have:

$$\left| \begin{array}{l} \text{if } i-j>\mu \text{ then } \mathbb{A}_{ij}=0 \\ \text{if } i-j\leq\mu \text{ then } \mathbb{A}_{ij}=(\mathbb{C}_b)_\ell \ \text{ with } \ell=\sum\limits_{k=0}^{i-j-1}(n-k)+j. \end{array} \right.$$

– in the compressed row-by-row $\mathbb{C}_l$ case, we have:

$$\left| \begin{array}{lll} \text{if } i-j>\mu & \text{then } \mathbb{A}_{ij}=0 & \\ \text{if } i-j\leq\mu \text{ and } i\leq\mu+1 & \text{then } \mathbb{A}_{ij}=(\mathbb{C}_l)_\ell & \text{with } \ell=\sum\limits_{k=1}^{i-1}k+j=\frac{1}{2}i(i-1)+j \\ \text{if } i-j\leq\mu \text{ and } i>\mu+1 & \text{then } \mathbb{A}_{ij}=(\mathbb{C}_l)_\ell & \text{with } \ell=\sum\limits_{k=1}^{\mu}k+(\mu+1)(i-\mu)+j. \end{array} \right.$$

A1.4.1.3. *Profile compression*

As soon as μ becomes close to n, band compression nearly fills up all the storage, and is therefore not very helpful in practice. This is why we prefer the so-called **profile** (or even **skyline**) compression, with per-row storage and compatibility with the $\mathbb{LU}$-and $\mathbb{LL}^t$-factorizations.

We begin by constructing the profile mapping P^-, which is defined by:

$$P^-(i)=\begin{cases} \min\limits_{j\in J^-(i)} j & \text{if } J^-(i)\neq\emptyset \\ i & \text{if } J^-(i)=\emptyset \end{cases}$$

where, for the ith row of the matrix, we have denoted:

$$J^-(i)=\{j\in\{1,\cdots,i\} \text{ such that } \mathbb{A}_{ij}\neq 0\}.$$

In other terms, $P^-(i)$ is the column index of the first non-zero term along the ith row. The "–" sign indicates that we consider only the lower triangular part. The upper profile mapping defined with regard to the upper-triangular part can be defined analogously. The definition of the mapping P^- implies trivially that:

$$\mathbb{A}_{ij}=0 \text{ if } j<P^-(i)$$

and allows us to draw a "separation", called the matrix profile, between the zero and the non-zero components (Figure A1.7).

When using row-storage, only terms belonging to the profile (the shaded part) are involved:

$$\mathbb{C}_p = \big(\mathbb{A}_{11} \; \underbrace{\mathbb{A}_{2,P^-(2)}\mathbb{A}_{22}}_{\text{row } 2} \; \cdots \; \underbrace{\mathbb{A}_{i,P^-(i)} \cdots \mathbb{A}_{ii}}_{\text{row } i} \; \cdots \; \underbrace{\mathbb{A}_{n,P^-(n)} \cdots \mathbb{A}_{n,n}}_{\text{row } n}\big).$$

By default, the diagonal terms are always preserved, even if they are zero. One must associate a descriptor to this compression, which will allow us to locate a $\mathbb{A}_{ij}$ term along the compressed vector, $\mathbb{C}_p$. The most commonly used is that which associates the diagonal term $\mathbb{A}_{ii}$ with its position along the vector $\mathbb{C}_p$, which we denote by $\ell_p(i)$; so we have:

$$\mathbb{A}_{ii} = (\mathbb{C}_p)_{\ell_p(i)} \, .$$

For $i \geq 2$, the difference $e_i = \ell_p(i) - \ell_p(i-1)$ indicates the number of terms stored along the ith row, and which allows us to determine the profile. Indeed, we have the equality:

$$P^-(i) = i - e_i + 1 \text{ for } 2 \leq i \leq n.$$

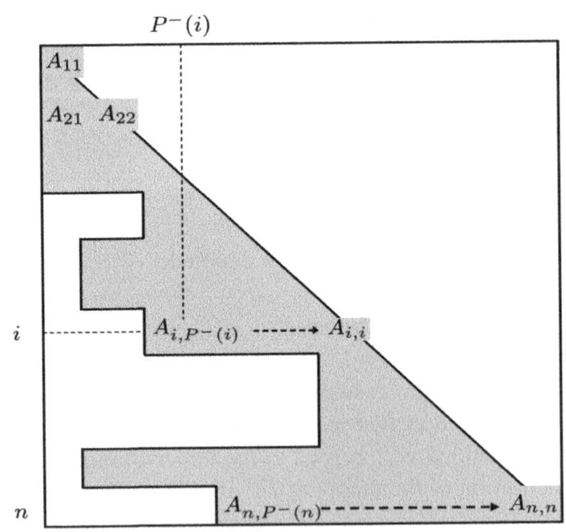

Figure A1.7. *Profile (row) storage*

Consequently, to determine the location of a $\mathbb{A}_{ij}$ term along the compressed vector $\mathbb{C}_p$ quickly, we can use the following formulae:

$$\left| \begin{array}{ll} \text{if } j < P^-(i) & \text{then } \mathbb{A}_{ij} = 0 \\ \text{if } j \geq P^-(i) & \text{then } \mathbb{A}_{ij} = (\mathbb{C}_p)_\ell \text{ with } \ell = \ell_p(i) + j - i. \end{array} \right.$$

A1.4.1.4. *Storing non-symmetric matrices*

We can consider non-symmetric matrices using symmetric profiles (which is the general case for finite element matrices), by using the same profile P^- to store the upper triangular part with regard to the columns. We store the compressed vector, following the lower triangular part, with the columns $\left(\mathbb{A}_{P^-(j),j} \cdots \mathbb{A}_{j-1,j}\right)_{j=1,n}$.

For any non-symmetric matrix, there are two possible methods of compression:

– the row compression in which we define the upper profile mapping P^+ such that:

$$P^+(i) = \begin{cases} \max\limits_{j \in J^+(i)} j & \text{if } J^+(i) \neq \emptyset \\ i & \text{if } J^+(i) = \emptyset \end{cases}$$

where, for the ith row of the matrix, $J^+(i) = \{j \in \{i, \cdots, n\} \text{ s.t. } \mathbb{A}_{ij} \neq 0\}$. Then by construction, we have $\mathbb{A}_{ij} = 0$ for each $j > P^+(i)$. We then arrange the matrix row by row:

$$\left(\mathbb{A}_{i,P^-(i)} \cdots \mathbb{A}_{i,P^+(i)}\right)_{i=1,n}$$

– the column/row compression in which we define the upper profile mapping $\tilde{P}^+$ such that:

$$\tilde{P}^+(j) = \begin{cases} \min\limits_{i \in \tilde{J}^+(j)} i & \text{if } \tilde{J}^+(j) \neq \emptyset \\ j-1 & \text{if } \tilde{J}^+(j) = \emptyset \end{cases}$$

where, for each $j \geq 2$ column of the matrix, we have denoted:

$$\tilde{J}^+(j) = \{i \in \{1, \cdots, j-1\} \text{ such that } \mathbb{A}_{ij} \neq 0\}.$$

By construction, we therefore have $\mathbb{A}_{ij} = 0$ for each $i < \tilde{P}^+(j)$. We arrange the matrix by starting with the lower triangular part, then completing with the upper triangular part (off from the diagonal):

$$\left(\left(\mathbb{A}_{i,P^-(i)} \cdots \mathbb{A}_{i,i}\right)_{i=1,n}, \left(\mathbb{A}_{\tilde{P}^+(j),j} \cdots \mathbb{A}_{j-1,j}\right)_{j=2,n}\right).$$

In order to find the position of the $\mathbb{A}_{ij}$ terms efficiently, one must know the location along the compressed vector for a particular term of the ith row or of the jth column, in general either the first term or the diagonal term. It is left to the reader to find the formulae that allow the position of any term to be found efficiently.

A1.4.1.5. *Sparse compression*

Clearly, profile compression is not optimal either because, within a truncated row, there can be many zeros. In particular, in the event that the first column of the matrix has no zero components, then the profile storage scheme will become dense. Introducing **sparse** storage, which consists of storing only the non-zero terms, makes it possible to achieve a near-optimal storage scheme with regard to memory usage. It is of course no longer compatible with the $\mathbb{LU}$- nor $\mathbb{LL}^t$-factorizations, but it proves to be particularly effective in the context of iterative methods. There are essentially two techniques of sparse compression:

– one based upon the notion of the *range* of non-zero terms along each row;

– the other involves the explicit knowledge of the column (or row) indices, which determine the non-zero terms along a row (or along a column).

In terms of memory management, the first method is generally more economical than the second in terms of memory space, albeit with a detrimentally slower performance. The second method, called *compressed sparse row/compressed sparse column* format, is the most common. In particular, this is the technique used when dealing with sparse matrices (*sparse* type in Matlab). We are going to briefly describe these two techniques.

The **range** descriptor consists of describing the non-zero terms along each row of the matrix, listed successively as the ranges of the non-zero terms with variable lengths, and where the number of ranges varies too (Figure A1.8).

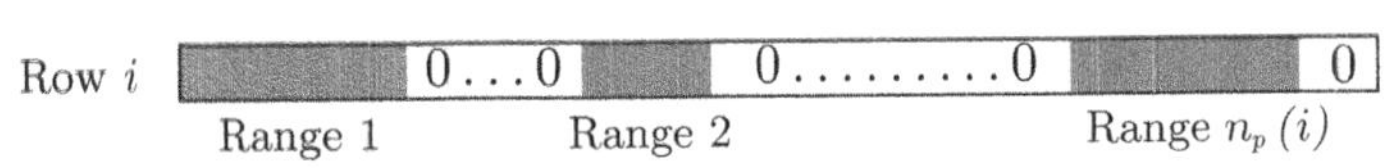

Figure A1.8. *Range storage with one row*

The compressed vector $\mathbb{C}_m$ is made up then from the ranges of the non-zero terms, arranged row-by-row, and where each term has the location (i, j). The following descriptors are used for the ith row of the matrix:

– $n_p(i)$: the number of ranges of non-zero terms in the ith row;

– for each range P_k, $k = 1, n_p(i)$:

- $c(i, k)$: the index of the jth column where the range P_k begins,

- $\ell(i,k)$: the position along the $\mathbb{C}_s$ vector of the last term in the P_k range.

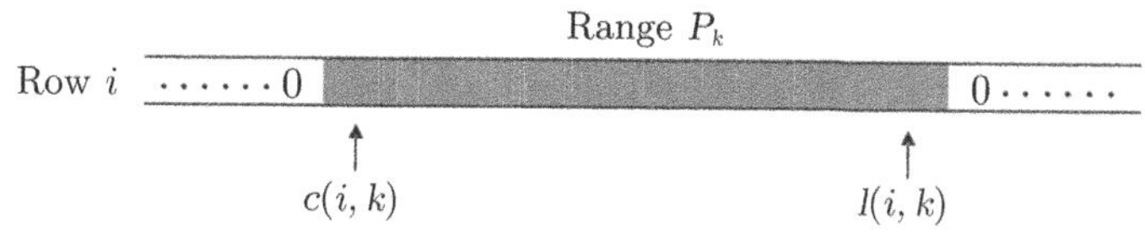

Figure A1.9. *Range descriptors*

Note that the descriptors used do not make it very straightforward to find the position (i,j) of each term along $\mathbb{C}_s$. Nevertheless, it suffices when a *sequential* calculation involving the $\mathbb{C}_s$ vector is performed, for example, a matrix-vector product operation. In fact, one must know how to find what column corresponds to which $\mathbb{C}_s(q)$ term that is located along a range P_k of the ith row, which is easy because $c(i,k)$ indicates the column index j where the kth range begins.

The *range* storage can become prohibitive in certain situations, because the arrays used to describe the sparse structure use up the memory space. If N_p denotes the total number of ranges, then these arrays would occupy: $n + 2N_p$ places. Thus, for a *checkerboard* matrix (alternating 1's and 0's) we would have: $N_p = n^2/2$. The length of the compressed vector $\mathbb{C}_s$ is also $n^2/2$, which gives a total occupancy of $n + \frac{3}{2}n^2$, greater than the dense storage.

The *compressed sparse row* descriptor uses, on the one hand, a vector J that indicates the column index j for each non-zero term stored row-by-row within the compressed vector $\mathbb{C}_s$ and, on the other hand, a vector P that indicates, for the ith row, the position of the first non-zero term in the row along the vector $\mathbb{C}_s$ (or within J) (Figure A1.10).

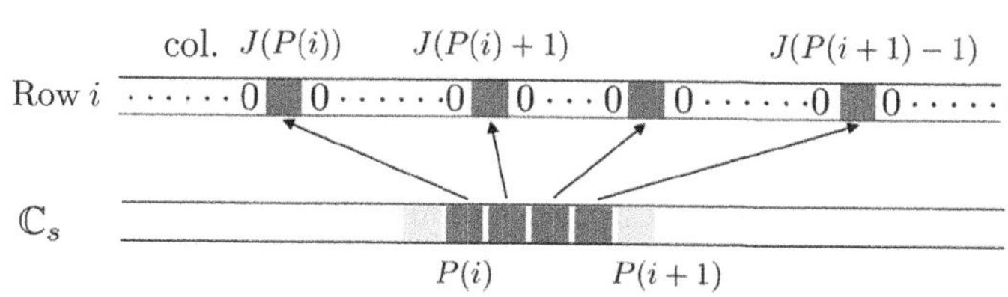

Figure A1.10. *"Compressed sparse row" storage*

The non-zero terms of the ith row are therefore given by:

$\mathbb{C}_s(k)$ for $P(i) \leq k < P(i+1)$ where $J(k)$ is the index of the jth column.

The vector J has the same size as the vector $\mathbb{C}_s$ (the number of non-zero terms within the matrix), and the size of the vector P is given by the number of rows.

For example, given a rank 5 matrix:

$$\begin{bmatrix} 1\,7\,0\,0\,0 \\ 2\,0\,4\,0\,1 \\ 0\,2\,8\,0\,0 \\ 4\,0\,0\,3\,6 \\ 0\,0\,1\,1\,0 \end{bmatrix} \longrightarrow \left| \begin{array}{ll} \mathbb{C}_s = & (1\,7\,2\,4\,1\,2\,8\,4\,3\,6\,1\,1) \\ J = & (1\,2\,1\,3\,5\,2\,3\,1\,4\,5\,3\,4) \\ P = & (1\,3\,6\,8\,11). \end{array} \right.$$

Just like the *range* descriptor, the *compressed sparse row* descriptor does not allow us to ascertain the location (i, j) of a term within the compressed vector directly (it is necessary to perform a search for it). On the other hand, algorithms which iterate through the compressed vector $\mathbb{C}_s$ sequentially will be effective here. If the matrix is not very sparse, then the *compressed sparse row* descriptor occupies more space than the *range* descriptor. Regardless, it is quicker since obtaining the jth column index for a term in the compressed vector $\mathbb{C}_s$ is immediate (vector J), while a calculation is necessary for the *range* descriptor.

Clearly, if the matrix is symmetric then one only needs to store the lower triangular part.

A1.4.2. *Algorithms for compressed matrices*

We have just presented the four main storage types. In the context of finite element discretization (or of finite differences), then the choice of numbering for the unknowns will play a major role in what structure the matrices that we end up using will take on. In particular, memory space optimization algorithms (Cuthill-Mackee, for example) seek the numbering which minimizes the bandwidth μ and, for some, the effective size of the compressed vector. Sparse storage is essentially insensitive to numbering.

For example, in the case of a Lagrangian finite element mesh, taking into account the support properties of the global basis functions, one will seek to minimize the variation in the index $\mid i - j \mid$ of the points (M_i, M_j) that belong to the same element K_ℓ. The bandwidth μ is given in this case by:

$$\mu = \max_{K_\ell \in \mathcal{T}_h} \Big(\max \big\{ \mid i - j \mid \text{ such that } M_i, M_j \in K_\ell \big\} \Big).$$

Let us now indicate some advantages and disadvantages of these storage structures, as well as some selection criteria.

A1.4.2.1. *Memory occupancy*

Denote by m_d, m_b, m_p and m_s the respective lengths of the compressed vector $\mathbb{C}$ that arises from the *diagonal*, *band*, *profile* and *sparse*, storage schemes. We will always have:

$$m_b \geq m_p \geq m_s$$

because, by construction, $P^-(i) \leq \mu$ and also *sparse* storage schemes exclude every zero term, a fortiori those which are not included within the profile. Furthermore, we also have:

$$m_b \geq m_d$$

because the *band* storage scheme includes each diagonal included within the *diagonal* storage scheme. Conversely, we cannot evaluate m_d relative to m_p, because it depends on the matrix structure. Finally, we have that $m_d \geq m_s$.

As an example, let us look at the case of the Q^1-Lagrangian finite element mesh, as shown in Figure A1.11 (with p^2 nodes). A somewhat tedious calculation shows that, for the rank p^2 symmetric stiffness matrix, we obtain the following estimates:

$$m_b = O(p^3), \quad m_p = O(p^3), \quad m_d = O(5p^2), \quad m_s = O(5p^2)$$

which is to be compared to the dense storage at $O(\frac{1}{2}p^4)$. The gain presented by the *diagonal* and *sparse* storage schemes is therefore appreciable. The case we are in is that with a structured mesh, which is why the *diagonal* storage scheme is comparable to the *sparse* one.

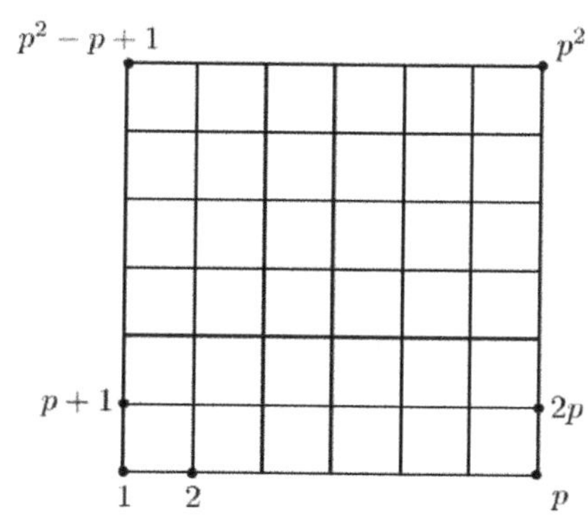

Figure A1.11. Q^1 *mesh*

If memory space is an important criterion, the calculation time of the matrix operations (factorizations and products) is then just as important. Additionally, it is important to know what matrix operations are permitted by which storage structure schemes.

A1.4.2.2. *Compression and matrix operations*

We are only interested in the matrix operations that intervene in the solving methods for linear systems, which were presented previously. These operations come in three flavors:

– factorizations (Gaussian elimination, $\mathbb{LU}$-, or $\mathbb{LL}^t$-factorization);

– matrix-vector products that come up in iterative techniques;

– extractions of the submatrices that appear in iterative Gauss–Seidel techniques.

As the last type of operation is similar to that of the matrix-vector product, it therefore fits into the previous case. In order to simplify the presentation, let us always assume that the matrix $\mathbb{A}$ defining the system is a positive-definite symmetric one.

A1.4.2.3. *Cholesky-type factorizations*

Recall that this is a question of factorizing the matrix $\mathbb{A}$ into the form of: $\mathbb{LL}^t = \mathbb{A}$. We have the following basic result:

PROPOSITION A1.10.– *Let $\mathbb{A}$ be a matrix factorization of the form $\mathbb{LL}^t$, then the lower triangular matrix $\mathbb{L}$ has the same profile structure (respectively, band) as the matrix $\mathbb{A}$.*

This property can be proved using induction. We can take advantage of the band or profile structure of the matrix $\mathbb{A}$ to reduce the calculation time when performing the Cholesky factorization, on the one hand, by not calculating the $\mathbb{L}_{ij}$ terms which do not belong to the profile and, on the other hand, by not taking any products with a zero term, also not belonging to the profile. This leads to a particularly efficient algorithm based upon the "row" form of Cholesky's algorithm (algorithm A1.13).

for $p = 1, n$

  for $j = 1, P^-(p) - 1$

    $m = \max(P^-(p), P^-(j))$

$$\mathbb{L}_{pj} = \Big(\mathbb{A}_{pj} - \sum_{k=m}^{j-1} \mathbb{L}_{pk}\mathbb{L}_{jk}\Big)/\mathbb{L}_{jj}$$

$$\mathbb{L}_{pp} = \Big(\mathbb{A}_{pp} - \sum_{k=P^-(p)}^{p-1} |\mathbb{L}_{pk}|^2\Big)^{1/2}$$

Algorithm A1.13. *Cholesky's algorithm with profile compression*

REMARK A1.1.– *It is easy to write a programmable version of this algorithm using only the compression pointer ℓ_p (exercise). Using Cholesky's algorithm in the "column" form is a less efficient method because one must perform a profile membership test. It is not possible to use pivoting techniques, even in the case of band or profile storage, since then the profile mapping is lost.*

One can similarly adapt the Gaussian elimination algorithm and that of the LU-factorization to the *profile* compressed matrices. The *diagonal* and the *sparse* storage schemes are incompatible with Cholesky's or the LU-factorization algorithms since they change the zeros belonging to the profile. The fact that the profile is conserved in addition to the optimality largely explain why profile compression is quite often used in finite element programs.

A1.4.2.4. *Matrix-vector products*

The matrix-vector product is the main operation used in the iterative solving methods. Since there is no creation or modification of the array during this operation, each storage scheme can support this operation. The *sparse* storage scheme is therefore be the most competitive provided that we know how to write the optimal (for the number of operations) algorithm of the product. Let us indicate which form it is that this algorithm takes on in both the case with a *profile* storage scheme, as well as with and the *sparse* storage scheme, for a symmetric matrix.

Given the profile pointer P^- structure (inline addressing), the algorithm for the matrix-vector product is optimal for the *profile* storage scheme, if it can be broken down into the sum of two products:

$$\mathbb{A}X = \mathbb{L}X + \mathbb{U}X$$

where $\mathbb{L}$ denotes the lower triangular part of the matrix $\mathbb{A}$ (including the main diagonal), and where $\mathbb{U}$ denotes the upper triangular part (excluding the main diagonal). Then the product $\mathbb{L}X$ is realized using the "row" form of the product algorithm, whereas the product $\mathbb{U}X$ is realized using the "column" form, hence avoiding any profile membership tests (algorithm A1.14).

for $i = 1, n$
    $Y_i = 0$
    for $j = 1, P^-(i)$
        $Y_i = Y_i + \mathbb{A}_{ij} X_j$
for $j = 2, n$
    for $i = 1, P^-(j) - 1$
        $Y_i = Y_i + \mathbb{A}_{ij} X_j$

Algorithm A1.14. *Matrix-vector product algorithm with profile compression*

In the case of sparse matrices, writing an efficient matrix-vector product algorithm relies on the fact that the compression descriptors allow for the (i,j)-position of the ℓth element of the compressed vector $\mathbb{C}_s$ to be found rather easily. Then, we can perform the operation:

$$Y_i = Y_i + (\mathbb{C}_s)_\ell \times X_j.$$

In the case of any matrix using the *compressed sparse row* descriptor, algorithm A1.15 is written next, with the convention that $P(n+1) = L = \dim(\mathbb{C}_s) + 1$.

for $i = 1, n$
  $Y_i = 0$
  for $k = P(i), P(i+1) - 1$
    $Y_i = Y_i + \mathbb{C}_s(k) X_{J(k)}$

Algorithm A1.15. *Matrix-vector product algorithm with sparse compression*

The matrix-vector product algorithms that have been presented here are optimal, in the sense that only products involving non-zero terms in the matrix are calculated.

Appendix 2

Solutions

A2.1. Exercises from Chapter 1

Solution 1. Integration by parts formulae

Formulae regarding the various differential operators are recalled in Appendix 3.

a) The two following relations are equivalent:

$$\int_\Omega \left(u\frac{\partial v}{\partial x_i} + \frac{\partial u}{\partial x_i}v\right) d\Omega = \int_{\partial\Omega} uvn_i \, d\Gamma \ (1 \leq i \leq 3), \qquad \forall u, v \in \mathcal{C}^1(\Omega) \cap \mathcal{C}^0(\overline{\Omega}) \qquad [\mathrm{C1}_a]$$

$$\int_\Omega (u \ \mathrm{div}\, \mathbf{v} + \nabla u \cdot \mathbf{v}) \, d\Omega = \int_{\partial\Omega} u(\mathbf{v}\cdot\mathbf{n}) \, d\Gamma, \qquad \begin{array}{l} \forall u \in \mathcal{C}^1(\Omega) \cap \mathcal{C}^0(\overline{\Omega}) \\ \forall \mathbf{v} \in \left(\mathcal{C}^1(\Omega) \cap \mathcal{C}^0(\overline{\Omega})\right)^3 \end{array} \qquad [\mathrm{C1}_b].$$

Suppose that the relation [C1$_a$] is satisfied, so that for each component v_i of $\mathbf{v}$, we have

$$\int_\Omega \left(u\frac{\partial v_i}{\partial x_i} + \frac{\partial u}{\partial x_i}v_i\right) d\Omega = \int_{\partial\Omega} uv_in_i \, d\Gamma, \ \forall u, v_i \quad (1 \leq i \leq 3)$$

that is to say, by summing over the i indices:

$$\begin{aligned}\sum_{i=1}^{3} \int_\Omega \left(u\frac{\partial v_i}{\partial x_i} + \frac{\partial u}{\partial x_i}v_i\right) d\Omega &= \sum_{i=1}^{3} \int_{\partial\Omega} uv_in_i \, d\Gamma, \quad \forall u, v_1, v_2, v_3 \\ &= \int_{\partial\Omega} u \sum_{i=1}^{3} v_in_i \, d\Gamma, \quad \forall u, v_1, v_2, v_3.\end{aligned}$$

From this, we arrive at the relation [C1$_b$] by using definitions of the differential operators. Conversely, if we assume that relation [C1$_b$] is satisfied, then by definition:

$$\int_\Omega \Big(u \sum_{i=1}^3 \frac{\partial v_i}{\partial x_i} + \sum_{i=1}^3 \frac{\partial u}{\partial x_i} v_i \Big)\, d\Omega = \int_{\partial\Omega} u \sum_{i=1}^3 v_i n_i \, d\Gamma, \ \forall u, \ \forall \mathbf{v} = \begin{pmatrix} v_1 \\ v_2 \\ v_3 \end{pmatrix}.$$

In particular, for $\mathbf{v} = v e_i$, $\forall i = 1, 2, 3$, we find that

$$\int_\Omega \Big(u \frac{\partial v}{\partial x_i} + \frac{\partial u}{\partial x_i} v \Big)\, d\Omega = \int_{\partial\Omega} u\, v\, n_i \, d\Gamma,$$

and hence [C1$_a$].

b) Let $\mathbf{u}, \ \mathbf{v} \in \big(\mathcal{C}^1(\Omega) \cap \mathcal{C}^0(\overline{\Omega})\big)^3$:

$$\begin{aligned}
&\int_\Omega \mathbf{u} \cdot \mathbf{curl}\, \mathbf{v}\, d\Omega = \int_\Omega \Big(\sum_{i=1}^3 u_i \, (\mathbf{curl}\, \mathbf{v})_i \Big)\, d\Omega \\
&= \int_\Omega \Big(u_1 \Big(\frac{\partial v_3}{\partial x_2} - \frac{\partial v_2}{\partial x_3} \Big) + u_2 \Big(\frac{\partial v_1}{\partial x_3} - \frac{\partial v_3}{\partial x_1} \Big) + u_3 \Big(\frac{\partial v_2}{\partial x_1} - \frac{\partial v_1}{\partial x_2} \Big) \Big)\, d\Omega \\
&= \int_\Omega \Big(\Big(- \frac{\partial u_1}{\partial x_2} v_3 + \frac{\partial u_1}{\partial x_3} v_2 \Big) + \Big(- \frac{\partial u_2}{\partial x_3} v_1 + \frac{\partial u_2}{\partial x_1} v_3 \Big) \\
&+ \Big(- \frac{\partial u_3}{\partial x_1} v_2 + \frac{\partial u_3}{\partial x_2} v_1 \Big) \Big)\, d\Omega \\
&+ \int_{\partial\Omega} \Big(u_1 (v_3 n_2 - v_2 n_3) + u_2 (v_1 n_3 - v_3 n_1) + u_3 (v_2 n_1 - v_1 n_2) \Big)\, d\Gamma \\
&= \int_\Omega \Big(v_1 \Big(\frac{\partial u_3}{\partial x_2} - \frac{\partial u_2}{\partial x_3} \Big) + v_2 \Big(\frac{\partial u_1}{\partial x_3} - \frac{\partial u_3}{\partial x_1} \Big) + v_3 \Big(\frac{\partial u_2}{\partial x_1} - \frac{\partial u_1}{\partial x_2} \Big) \Big)\, d\Omega \\
&+ \int_{\partial\Omega} \Big(v_1 (u_2 n_3 - u_3 n_2) + v_2 (u_3 n_1 - u_1 n_3) + v_3 (u_1 n_2 - u_2 n_1) \Big)\, d\Gamma \\
&= \int_\Omega (\mathbf{curl}\, \mathbf{u} \cdot \mathbf{v})\, d\Omega + \int_{\partial\Omega} (\mathbf{u} \times \mathbf{n}) \cdot \mathbf{v}\, d\Gamma.
\end{aligned}$$

This shows that:

$$\int_\Omega (\mathbf{u} \cdot \mathbf{curl}\, \mathbf{v} - \mathbf{curl}\, \mathbf{u} \cdot \mathbf{v})\, d\Omega = \int_{\partial\Omega} (\mathbf{u} \times \mathbf{n}) \cdot \mathbf{v}\, d\Gamma, \quad \forall \mathbf{u}, \ \mathbf{v} \in \big(\mathcal{C}^1(\Omega) \cap \mathcal{C}^0(\overline{\Omega})\big)^3.$$

Solution 2. Classical solutions to the Laplacian

a) ($\beta = 0$) Assume that the solution u of [P2] belongs to $\mathcal{C}^2(\Omega) \cap \mathcal{C}^1(\overline{\Omega})$. By multiplying the equation of [P2] on Ω by $v \in \mathcal{C}^1(\overline{\Omega})$, before then integrating it over Ω, we obtain:

$$-\int_\Omega \Delta u\, v\, d\Omega = \int_\Omega f v\, d\Omega, \qquad \forall v \in \mathcal{C}^1(\overline{\Omega}).$$

By integrating the above by parts once, we obtain:

$$\int_\Omega \nabla u \cdot \nabla v\, d\Omega = \int_\Omega f v\, d\Omega + \int_{\partial\Omega} \frac{\partial u}{\partial n} v\, d\Gamma, \qquad \forall v \in \mathcal{C}^1(\overline{\Omega})$$

that is, from the boundary condition [P2] on $\partial\Omega$:

$$\int_\Omega \nabla u \cdot \nabla v\, d\Omega = \int_\Omega f v\, d\Omega + \int_{\partial\Omega} g v\, d\Gamma, \qquad \forall v \in \mathcal{C}^1(\overline{\Omega}).$$

Thus, for $f \in \mathcal{C}^0(\overline{\Omega})$ and $g \in \mathcal{C}^0(\partial\Omega)$, the obtained variational formulation is:

$$\left| \begin{array}{l} \text{Find } u \in \mathcal{C}^1(\overline{\Omega}) \text{ such that} \\ \displaystyle\int_\Omega \nabla u \cdot \nabla v\, d\Omega = \int_\Omega f v\, d\Omega + \int_{\partial\Omega} g v\, d\Gamma, \qquad \forall v \in \mathcal{C}^1(\overline{\Omega}). \end{array} \right. \qquad [\mathrm{F2}_a]$$

To summarize, if $u \in \mathcal{C}^2(\Omega) \cap \mathcal{C}^1(\overline{\Omega})$ satisfies [P2] with $\beta = 0$, then u satisfies [F2$_a$] since then Green's classical formulae can be applied here. Let u_1 and u_2 in $\mathcal{C}^1(\overline{\Omega})$ be two solutions of [F2$_a$], then $w = u_1 - u_2 \in \mathcal{C}^1(\overline{\Omega})$ satisfies:

$$\int_\Omega \nabla w \cdot \nabla v\, d\Omega = 0, \qquad \forall v \in \mathcal{C}^1(\overline{\Omega})$$

and so, in particular when $v = w$:

$$\int_\Omega |\nabla w|^2\, d\Omega = 0.$$

Hence, we see that $w = cst$: the solution of [P2], if it exists, is indeed unique, up to a constant provided that $\beta = 0$. Actually, $u + cst$ is a solution of [P2].

b) ($\beta = 1$) We assume that the solution u of [P2] belongs to $\mathcal{C}^2(\Omega) \cap \mathcal{C}^1(\overline{\Omega})$. In multiplying the equation of [P2] on Ω by $v \in \mathcal{C}^1(\overline{\Omega})$, before then integrating it over Ω, we obtain:

$$\int_\Omega uv\, d\Omega - \int_\Omega \Delta u\, v\, d\Omega = \int_\Omega f v\, d\Omega, \qquad \forall v \in \mathcal{C}^1(\overline{\Omega}).$$

Proceeding as before, in this case the variational formulation is:

Find $u \in \mathcal{C}^1(\overline{\Omega})$ such that

$$\int_\Omega uv\, d\Omega + \int_\Omega \nabla u \cdot \nabla v\, d\Omega = \int_\Omega fv\, d\Omega + \int_{\partial\Omega} gv\, d\Gamma, \qquad \forall v \in \mathcal{C}^1(\overline{\Omega}). \qquad [\mathrm{F2}_b]$$

If $u \in \mathcal{C}^2(\Omega) \cap \mathcal{C}^1(\overline{\Omega})$ satisfies [P2] with $\beta = 1$, then u satisfies [F2$_b$].

Let u_1 and u_2 in $\mathcal{C}^1(\overline{\Omega})$ be two solutions of [F2$_b$], then $w = u_1 - u_2 \in \mathcal{C}^1(\overline{\Omega})$ is such that:

$$\int_\Omega |w|^2\, d\Omega + \int_\Omega |\nabla w|^2\, d\Omega = 0.$$

From this, we see that $w = 0$: the solution of [P2], if it exists, is unique, provided that $\beta = 1$.

c) ($\beta = 1$) With regard to the existence of a "classical" solution, no general abstract result exists and, in particular, it may happen that there are no "classical" solutions whatsoever. Furthermore, $C^1(\bar{\Omega})$ endowed with the norm:

$$\|u\|_{C^1(\overline{\Omega})} = \sup_{x \in \overline{\Omega}} |u(x)| + \sum_{i=1}^{3} \sup_{x \in \overline{\Omega}} \left| \frac{\partial u}{\partial x_i}(x) \right|$$

is not a Hilbert space: the Lax–Milgram theorem cannot be applied here. After examining the variational formulation [F2$_b$], we note that its associated bilinear form is:

$$(u, v) \mapsto \int_\Omega uv\, d\Omega + \int_\Omega \nabla u \cdot \nabla v\, d\Omega$$

i.e., the canonical inner product $(\cdot,\cdot)_{H^1(\Omega)}$. It is natural therefore to place oneself in the space $H^1(\Omega)$. We then obtain a new variational formulation:

Find $u \in H^1(\Omega)$ such that

$$\int_\Omega uv\, d\Omega + \int_\Omega \nabla u \cdot \nabla v\, d\Omega = \int_\Omega fv\, d\Omega + \int_{\partial\Omega} gv\, d\Gamma, \qquad \forall v \in H^1(\Omega). \qquad [\mathrm{F2}_c]$$

Now, the space $H^1(\Omega)$ endowed with the inner product $(\cdot,\cdot)_{H^1(\Omega)}$ becomes a Hilbert space, allowing for Riesz's representation theorem (or the Lax–Milgram theorem) to be applied and establish the existence and uniqueness for the solution of the variational formulation [F2$_c$]. *A posteriori*, the solution is measured using the $H^1(\Omega)$ norm:

$$\|u\|_{H^1(\Omega)} = \left((u, u)_{H^1(\Omega)}\right)^{1/2} = \left(\int_\Omega |u|^2\, d\Omega + \int_\Omega |\nabla u|^2\, d\Omega\right)^{1/2}.$$

Solution 3. Examples of elements of H^1

a) By Proposition 1.7, given a function in $L^2(0,2)$, one can differentiate it in the distribution sense and see that the result belongs to $L^2(0,2)$. Given the function $v \in L^2(0,2)$ and any function $\varphi \in \mathcal{D}(0,2)$, we have:

$$\begin{aligned}\left\langle \frac{\partial v}{\partial x}, \varphi \right\rangle &= -\left\langle v, \frac{\partial \varphi}{\partial x} \right\rangle \\ &= -\int_{(0,2)} v \frac{\partial \varphi}{\partial x}\, dx = -\int_{(0,1)} v \frac{\partial \varphi}{\partial x}\, dx - \int_{(1,2)} v \frac{\partial \varphi}{\partial x}\, dx \\ &= -\int_{(0,2)} \frac{\partial \varphi}{\partial x}\, dx - \int_{(1,2)} (2-x) \frac{\partial \varphi}{\partial x}\, dx = -\varphi(1) + \left[\varphi(1) - \int_{(1,2)} \varphi\, dx\right] \\ &= -\int_{(0,2)} \chi_{(1,2)} \varphi\, dx\end{aligned}$$

where $\chi_{(1,2)}$ is the indicator function on $(1,2)$, which belongs to $L^2(0,2)$, so $v \in H^1(0,2)$. Now given the function $w \in L^2(0,2)$ and any function $\varphi \in \mathcal{D}(0,2)$, we have:

$$\begin{aligned}\left\langle \frac{\partial w}{\partial x}, \varphi \right\rangle &= -\left\langle w, \frac{\partial \varphi}{\partial x} \right\rangle \\ &= -\int_{(0,2)} w \frac{\partial \varphi}{\partial x}\, dx = -\int_{(0,1)} w \frac{\partial \varphi}{\partial x}\, dx - \int_{(1,2)} w \frac{\partial \varphi}{\partial x}\, dx \\ &= -\int_{(0,1)} a \frac{\partial \varphi}{\partial x}\, dx - \int_{(1,2)} b \frac{\partial \varphi}{\partial x}\, dx = -a\varphi(1) + b\varphi(1) \\ &= \left\langle (b-a)\delta_{x=1}, \varphi \right\rangle.\end{aligned}$$

If $b = a$, then evidently $w \in H^1(0,2)$. Otherwise, as the Dirac mass is not measurable, we have $w \notin H^1(0,2)$. Finally, the function z is square-integrable if, and only if, x^{2p} is integrable, which is to say, if, and only if, $p > -1/2$. Likewise, its derivative $\partial_x z = px^{p-1}$ is integrable if, and only if, $p > 1/2$. From this, we see that:

$$z \in H^1(0,2) \quad \Longleftrightarrow \quad p > \frac{1}{2}.$$

b) For $p \leq 0$, then there is no difficulty here. So suppose that $p > 0$. Relative to polar coordinates centered at $(0,0)$: $u(r,\theta) = 2^p |\log r|^p$, for $(r,\theta) \in (0,1/2) \times [0,2\pi)$. If we define $u_\varepsilon = u\,\chi_{r>\varepsilon}$ for $\varepsilon \in (0,1/2)$, we see that $\lim_{\varepsilon\to 0+} u_\varepsilon(x,y) = u(x,y)$ for every $(x,y) \in B \setminus \{(0,0)\}$. Furthermore, $u_\varepsilon \in L^2(B)$, and:

$$\lim_{\varepsilon\to 0+} \int_B u_\varepsilon^2\, dxdy = 2^{2p+1}\pi \lim_{\varepsilon\to 0+} \int_\varepsilon^{1/2} r\,|\log r|^{2p} dr < \infty.$$

By Fatou's lemma, $u \in L^2(B)$. Let $\varphi \in \mathcal{D}(B)$, then we have $u\partial_x\varphi \in L^1(B)$, and:

$$\left\langle \frac{\partial u}{\partial x}, \varphi \right\rangle = -\int_B u \frac{\partial \varphi}{\partial x}\, dxdy = -\lim_{\varepsilon \to 0+} \int_{B_\varepsilon} u \frac{\partial \varphi}{\partial x}\, dxdy$$

with $B_\varepsilon = \{(r,\theta) \in B \text{ such that } r > \varepsilon\}$. After integrating this by parts:

$$\left\langle \frac{\partial u}{\partial x}, \varphi \right\rangle = \lim_{\varepsilon \to 0+} \Big(-\int_{\partial B_\varepsilon} u\varphi n_x d\Gamma + \int_{B_\varepsilon} \frac{\partial u}{\partial x} \varphi\, dxdy \Big).$$

Since $\varphi \in \mathcal{D}(B)$, for the left-hand side we get:

$$\begin{aligned} \Big| \int_{\partial B_\varepsilon} u\varphi n_x d\Gamma \Big| &= \Big| \int_0^{2\pi} u(\varepsilon,\theta)\varphi(\varepsilon,\theta) n_x(\varepsilon,\theta)\varepsilon\, d\theta \Big| \\ &= 2^p \varepsilon |\log\varepsilon|^p \Big| \int_0^{2\pi} \varphi(\varepsilon,\theta) n_x(\varepsilon,\theta)\, d\theta \Big| \\ &\le 2^{p+1}\pi\, \varepsilon |\log\varepsilon|^p \|\varphi\|_{L^\infty(B)} \to 0 \text{ when } \varepsilon \to 0^+. \end{aligned}$$

Although for the right-hand side, we have $\int_{B_\varepsilon} \frac{\partial u}{\partial x}\varphi\, dxdy = \int_B v_\varepsilon \varphi\, dxdy$ with $v_\varepsilon = g_x\, \chi_{r>\varepsilon}$ for $\varepsilon \in (0, 1/2)$ and $g_x(r,\theta) = \partial_x u(r,\theta) = -2^p p r^{-1} |\log r|^{p-1} \cos\theta$, for $r > 0$. When $p < 1/2$, using Fatou's lemma we can check that $g_x \in L^2(B)$. Indeed, $v_\varepsilon(r,\theta)$ converges to $g_x(r,\theta)$ as soon as $r \neq 0$, $v_\varepsilon \in L^2(B)$, and finally:

$$\begin{aligned} \int_B v_\varepsilon^2\, dxdy &\le 2^{2p+1} p^2 \pi \int_\varepsilon^{1/2} r^{-1} |\log r|^{2p-2} dr \\ &= \frac{2^{2p+1} p^2 \pi}{2p-1} \left(|\log\varepsilon|^{2p-1} - |\log 2|^{2p-1} \right). \end{aligned}$$

Therefore, $\lim_{\varepsilon \to 0+} \int_B v_\varepsilon^2\, dxdy < \infty$ since $p < 1/2$, and indeed then we see that $g_x \in L^2(B)$. After passing to the limit, we conclude:

$$\left\langle \frac{\partial u}{\partial x}, \varphi \right\rangle = \int_B g_x \varphi\, dxdy \qquad \forall \varphi \in \mathcal{D}(B).$$

For the other condition $p < 1/2$, one can reason similarly to find that $\partial_y u \in L^2(B)$, and hence that $u \in H^1(B)$ by proposition 1.7.

Solution 4. Continuous extensions

a) We will argue this in several (simple) steps.

- *Definition of a continuous extension $\tilde{\ell}$ (a reminder).* Let $x \in X$. By a density argument, there exists a sequence of elements in V, denoted $(v_k)_k$, such that $\lim_{k\to\infty} \|v_k - x\|_X = 0$. Let $(z_k)_k$ be the sequence of reals defined by $z_k = \ell(v_k)$ for each k. Let us look into the $(z_k)_k$:

$$|z_k - z_m| = |\ell(v_k) - \ell(v_m)| \overset{\ell \text{ lin.}}{=} |\ell(v_k - v_m)| \leq C_\ell \|v_k - v_m\|_X.$$

As $(v_k)_k$ converges (to x) in X, it is necessarily a Cauchy sequence, and so $\lim_{k,m\to\infty} \|v_k - v_m\|_X = 0$. Therefore, $(z_k)_k$ is a Cauchy sequence in $\mathbb{R}$. However, $\mathbb{R}$ is complete: thus $(z_k)_k$ is convergent. From here, one defines the continuous extension of ℓ at x as being given by the limit of $(z_k)_k$:

$$\tilde{\ell}(x) = \lim_k \ell(v_k).$$

Indeed, performing the same reasoning for every element x of X allows us to define $\tilde{\ell} : X \to \mathbb{R}$.

- *Uniqueness of the extension $\tilde{\ell}$.* Let us revisit the method used above just now. Given $x \in X$, let $(v_k)_k$ and $(v'_k)_k$ be two sequences of elements in V which converge to x. Now let us check that the definition of $\tilde{\ell}(x)$ is independent of either sequence, and consequently proving uniqueness:

$$|\lim_k \ell(v_k) - \lim_k \ell(v'_k)| = |\lim_k \{\ell(v_k) - \ell(v'_k)\}| \overset{\ell \text{ lin.}}{=} |\lim_k \ell(v_k - v'_k)|.$$

However, $|\ell(v_k - v'_k)| \leq C_\ell \|v_k - v'_k\|_X$, and $\lim_k \|v_k - v'_k\|_X = 0$ by the triangular inequality ($\|v_k - v'_k\|_X \leq \|v_k - x\|_X + \|x - v'_k\|_X$). From this, we conclude that $\lim_k \ell(v_k) = \lim_k \ell(v'_k)$.

- *Linearity of the extension $\tilde{\ell}$.* We obtain linearity by passing to the limit. Let $x^1, x^2 \in X$, $\alpha^1, \alpha^2 \in \mathbb{R}$: by density, there exist two sequences, $(v^1_k)_k$ and $(v^2_k)_k$, of elements in V which converge to x^1 and x^2, respectively, in X, and:

$$\begin{aligned}\tilde{\ell}(\alpha^1 x^1 + \alpha^2 x^2) &= \lim_k \ell(\alpha^1 v^1_k + \alpha^2 v^2_k) \\ &\overset{\ell \text{ lin.}}{=} \lim_k \{\alpha^1 \ell(v^1_k) + \alpha^2 \ell(v^2_k)\} = \alpha^1 \tilde{\ell}(x^1) + \alpha^2 \tilde{\ell}(x^2).\end{aligned}$$

- *Continuity of the extension $\tilde{\ell}$ from X into $\mathbb{R}$.* We want to show that $\exists C > 0$, $\forall x \in X$ implies that $|\tilde{\ell}(x)| \leq C\|x\|_X$. To do so, for a given $x \in X$, let $(v_k)_k$ be a sequence of elements in V converging to x in X. We write:

$$|\tilde{\ell}(x)| = |\lim_k \ell(v_k)| = \lim_k |\ell(v_k)|$$

where, for each k, we have $|\ell(v_k)| \leq C_\ell \|v_k\|_X$. As $\lim_k \|v_k\|_X = \|x\|_X$, from this we see that:

$$|\tilde{\ell}(x)| \leq C_\ell \|x\|_X.$$

REMARK A2.1.– *The continuity modulus is the same for both ℓ and its extension $\tilde{\ell}$.*

b) More generally, if one considers a linear mapping ℓ from V into a Banach space Y, then the steps are logically identical. Let us briefly cover the first one.

- *Definition of the continuous extension $\tilde{\ell}$.* Let $x \in X$: there exists a sequence of elements in V, denoted $(v_k)_k$, such that $\lim_{k\to\infty} \|v_k - x\|_X = 0$. Introduce the sequence $(z_k)_k$ of elements in Y defined by $z_k = \ell(v_k)$ for each k. By linearity, we see that $\|z_k - z_m\|_Y \leq C_\ell \|v_k - v_m\|_X$. As $(v_k)_k$ is Cauchy in V, then $(z_k)_k$ is Cauchy in Y. Given that Y is complete, then $(z_k)_k$ must converge, and we define the continuous extension of ℓ at x to be equal to the limit of $(z_k)_k$, that is to say, $\tilde{\ell}(x) = \lim_k \ell(v_k)$. Therefore, we can construct $\tilde{\ell} : X \to Y$.

- *Uniqueness, linearity and continuity of the extension $\tilde{\ell}$.* These are obtained in exactly the same way as before, by replacing the absolute value with the norm of Y.

REMARK A2.2.– *The same results hold for Banach spaces defined over $\mathbb{C}$.*

Solution 5. Trace theorem

a) Given $v \in \mathcal{C}^1[0,1]$: we write:

$$v(0) = v(x) - \int_0^x v'(t)\,dt \implies |v(0)| \leq |v(x)| + \left|\int_0^x v'(t)\,dt\right|.$$

Next, we begin by bounding from above the right-hand side via the Cauchy–Schwarz inequality:

$$\begin{aligned}\left|\int_0^x v'(t)\,dt\right| &\leq \int_0^x |v'(t)|\,dt \leq \Big(\int_0^x 1\,dt\Big)^{1/2}\Big(\int_0^x |v'(t)|^2\,dt\Big)^{1/2} \\ &\leq \sqrt{x}\Big(\int_0^x |v'(t)|^2\,dt\Big)^{1/2} \leq 1 \times \Big(\int_0^1 |v'(t)|^2\,dt\Big)^{1/2}.\end{aligned}$$

To do the same for the left-hand side ($|v(x)|$), we will use a "clever trick" to pass over to an integral: for any constant c, the identity $c = \int_0^1 c\,dx$ holds. From which:

$$|v(0)| = \int_0^1 |v(0)|\,dx \leq \int_0^1 |v(x)|\,dx + \Big(\int_0^1 |v'(t)|^2\,dt\Big)^{1/2}.$$

By bounding the "newly integrated" left-hand side then, again by using the Cauchy–Schwarz inequality, we get:

$$\int_0^1 |v(x)|\,dx \leq 1 \times \Big(\int_0^1 |v(x)|^2\,dx\Big)^{1/2}.$$

From this we conclude that, for every element v of $\mathcal{C}^1([0,1])$, there exists the following upper bound:

$$|v(0)| \leq \Big(\int_0^1 |v(x)|^2\,dx\Big)^{1/2} + \Big(\int_0^1 |v'(t)|^2\,dt\Big)^{1/2} \leq \sqrt{2}\|v\|_{H^1(]0,1[)}.$$

REMARK A2.3.– *Above, we have used the fact that* $(a+b)^2 \leq 2a^2 + 2b^2$, *which applies to every* $a, b \in \mathbb{R}$.

b) Let us apply exercise 4 (Question (a)) by choosing:

$H = H^1(0,1)$, endowed with the norm $\|v\|_{H^1(0,1)} = \big(\int_0^1 v^2\,dx + \int_0^1 (v')^2\,dx\big)^{1/2}$ and $V = \mathcal{C}^1[0,1]$.

That is to say, $\mathcal{C}^1[0,1]$ is dense in $H^1(0,1)$ from proposition 1.10. In conclusion, we can extend the trace mapping $\gamma_\star : v \mapsto v(0)$ uniquely to a continuous linear mapping (still denoted by $\gamma_\star$) from $H^1(0,1)$ into $\mathbb{R}$.

c) The answer is no. To convince oneself of this, let $(v_m)_{m\geq 1}$ be the sequence of elements in $L^2(0,1)$ defined by $v_m(x) = e^{-mx}$. By a direct calculation:

$$\int_0^1 v_m^2\,dx = \int_0^1 e^{-2mx}\,dx = \Big[\frac{1}{-2m}e^{-2mx}\Big]_0^1 = \frac{1}{2m}(1 - e^{-2m}).$$

Therefore, $\lim_{m\to\infty} \|v_m\|_{L^2(0,1)} = 0$, whereas $v_m(0) = 1$ for each $m \geq 1$. Therefore, we cannot find a constant $C > 0$ such that, for each $m \geq 1$, we have $|v_m(0)| \leq C\,\|v_m\|_{L^2(]0,1[)}$.

d) Let $v \in \mathcal{C}_c^\infty(\overline{\Omega_2})$ (see definition 1.9), then, for every $y \in \mathbb{R}$, the function $x \mapsto v(x,y)$ belongs to $\mathcal{C}^1[0,1]$. From Question (a):

$$|v(0,y)|^2 \leq 2\int_{x=0}^{x=1} \Big(|v(x,y)|^2 + \Big|\frac{\partial v}{\partial x}(x,y)\Big|^2\Big)\,dx.$$

As v has compact support in $\overline{\Omega_2}$ by definition, $y \mapsto v(0,y)$ belongs to $L^2(\mathbb{R}_y)$. Using Fubini's theorem:

$$\|v(0,\cdot)\|^2_{L^2(\mathbb{R}_y)} \leq 2\int_{\Omega_2} \Big(|v|^2 + \Big|\frac{\partial v}{\partial x}\Big|^2\Big)\,dxdy \leq 2\|v\|^2_{H^1(\Omega_2)}$$

the result follows.

e) Similarly, one can prove that $\|v(1,\cdot)\|^2_{L^2(\mathbb{R}_y)} \leq 2\,\|v\|^2_{H^1(\Omega_2)}$ for every $v \in \mathcal{C}_c^\infty(\overline{\Omega_2})$. So, if we define $\gamma_0 : \mathcal{C}_c^\infty(\overline{\Omega_2}) \to L^2(\partial\Omega_2)$ that associates v with $v_{|\partial\Omega_2}$, we see that:

$$\|\gamma_0 v\|_{L^2(\partial\Omega_2)} \leq 2\|v\|_{H^1(\Omega_2)}, \qquad \forall v \in \mathcal{C}_c^\infty(\overline{\Omega_2}).$$

From proposition 1.10 and exercise 4 (Question (b)), we can therefore extend the trace mapping to a continuous linear mapping from $H^1(\Omega_2)$ into $L^2(\partial\Omega_2)$.

Solution 6. Sobolev spaces of fractional orders and the trace theorem

a) The pair (u,μ) solves the eigenvalue problem in $(0,L)$ if, and only if:

$$\begin{cases} -u''(x) = \mu\, u(x) \text{ for } x \in (0,L) \\ u'(0) = u'(L) = 0 \end{cases} \quad \text{and } u \neq 0.$$

We can check painlessly that the (normalized) eigenfunctions and their corresponding eigenvalues are given by:

$$u_k(x) = \sqrt{\frac{2}{L}}\cos\left(\frac{k\pi x}{L}\right), \; \mu_k = \frac{k^2\pi^2}{L^2}, \; \text{for } k \in \mathbb{N}.$$

In this case, for $s \in [0,1]$:

$$H^s(0,L) = \{v \in L^2(0,L) \text{ such that } v = \sum_{k\geq 0} \alpha_k u_k \text{ with } \sum_{k\geq 0} k^{2s}|\alpha_k|^2 < \infty\}.$$

When $v = \chi_{(0,L/2)}$, we have for each $k \geq 0$:

$$\alpha_k = (v, u_k)_{L^2(0,L)} = \int_0^{L/2} \sqrt{\frac{2}{L}}\cos\left(\frac{k\pi x}{L}\right)dx = \frac{\sqrt{2L}}{k\pi}\sin\left(\frac{k\pi}{2}\right).$$

When k is even, then $\alpha_k = 0$; when k is odd, then $|\alpha_k| = \dfrac{\sqrt{2L}}{k\pi}$. From this, we observe that $\chi_{(0,L/2)}$ is an element of $H^s(0,L)$ if and only if:

$$\sum_{k \text{odd}} k^{2s-2} < \infty \quad \Longleftrightarrow \quad s < 1/2.$$

More generally, we can check easily that a (discontinuous) staircase function on $(0,L)$ belongs to $H^s(0,L)$ if and only if $s < 1/2$.

b) In the open subset $\Omega_2 = (0, L_1) \times (0, L_2)$, by a separation of variables, we find that the (normalized) eigenfunctions and their associated eigenvalues are given by:

$$u_{k,l}(x,y) = \frac{2}{\sqrt{L_1 L_2}} \cos\left(\frac{k\pi x}{L_1}\right) \cos\left(\frac{l\pi y}{L_2}\right), \; \mu_{k,l} = \frac{k^2\pi^2}{L_1^2} + \frac{l^2\pi^2}{L_2^2}, \text{ for } k, l \in \mathbb{N}.$$

For $s \in [0,1]$:

$$H^s(\Omega_2) = \Big\{ v \in L^2(\Omega_2) \text{ such that } v = \sum_{k,l \geq 0} \alpha_{k,l} u_{k,l} \text{ with } \sum_{k,l \geq 0} (k^2+l^2)^s |\alpha_{k,l}|^2 < \infty \Big\}.$$

Let $v \in H^1(\Omega_2)$, then note first that the norms are equivalent:

$$\|v\|^2_{H^1(\Omega_2)} \approx \|v\|^2_{L^2(\Omega_2)} + \sum_{k,l \geq 0} (k^2 + l^2)|\alpha_{k,l}|^2. \qquad [C6_a]$$

For $K, L \geq 0$, let us denote:

$$v_{K,L} = \sum_{0 \leq k \leq K, 0 \leq l \leq L} \alpha_{k,l} u_{k,l}.$$

Then, on the one hand, note that $v_{K,L} \in \mathcal{C}^1(\overline{\Omega_2})$ and, on the other hand, that $(v_{K,L})_{K,L}$ converges to v in $H^1(\Omega_2)$ when K and L both tend toward infinity. Moreover, the trace of $v_{K,L}$ on $\Gamma = \{0\} \times (0, L_2)$, denoted $\gamma_\Gamma v_{K,L}$, is:

$$\begin{aligned} \gamma_\Gamma v_{K,L}(y) &= \frac{2}{\sqrt{L_1 L_2}} \sum_{0 \leq k \leq K, 0 \leq l \leq L} \alpha_{k,l} \cos\left(\frac{l\pi y}{L_2}\right) \\ &= \sum_{0 \leq l \leq L} \left(\sqrt{\frac{2}{L_1}} \sum_{0 \leq k \leq K} \alpha_{k,l}\right) \sqrt{\frac{2}{L_2}} \cos\left(\frac{l\pi y}{L_2}\right) \qquad [C6_b] \\ &= \sum_{0 \leq l \leq L} \beta_l u_l(y), \quad \text{with } \beta_l = \sqrt{\frac{2}{L_1}} \sum_{0 \leq k \leq K} \alpha_{k,l}. \end{aligned}$$

In particular, $\|\gamma_\Gamma v_{K,L}\|^2_{L^2(\Gamma)} = \sum_{0 \leq l \leq L} |\beta_l|^2 = \frac{2}{L_1} \sum_{0 \leq l \leq L} \big| \sum_{0 \leq k \leq K} \alpha_{k,l} \big|^2$. For $z \geq 0$, by the Cauchy–Schwarz inequality we have that:

$$\begin{aligned} \Big| \sum_{0 \leq k \leq K} \alpha_{k,l} \Big|^2 &= \Big| \sum_{0 \leq k \leq K} \frac{1}{(1+k)^z} (1+k)^z \alpha_{k,l} \Big|^2 \\ &\leq \Big(\sum_{0 \leq k \leq K} \frac{1}{(1+k)^{2z}} \Big) \Big(\sum_{0 \leq k \leq K} (1+k)^{2z} |\alpha_{k,l}|^2 \Big). \end{aligned} \qquad [C6_c]$$

Furthermore, if we choose $z \in (1/2, 1]$, then there exists a $C_z > 0$ such that, for each $K \geq 0$, we have $\sum_{0 \leq k \leq K} (1+k)^{-2z} < C_z$, which implies:

$$\|\gamma_\Gamma v_{K,L}\|^2_{L^2(\Gamma)} \leq \frac{2C_z}{L_1} \sum_{0 \leq k \leq K, 0 \leq l \leq L} (1+k)^{2z} |\alpha_{k,l}|^2.$$

Moreover, from [C6$_a$] we see that there exists a $C > 0$ (independent of v) such that:

$$\|\gamma_\Gamma v_{K,L}\|_{L^2(\Gamma)} \leq C \, \|v_{K,L}\|_{H^1(\Omega_2)}, \qquad \forall K, L \in \mathbb{N}.$$

By passing to the limit and recalling exercise 4 (Question (b)), from this we see that:

$$\|\gamma_\Gamma v\|_{L^2(\Gamma)} \leq C \, \|v\|_{H^1(\Omega_2)}$$

which is to say that the trace mapping γ_Γ is indeed continuous from $H^1(\Omega_2)$ into $L^2(\Gamma)$.

c) By revisiting [C6$_b$], we have $\|\gamma_\Gamma v_{K,L}\|^2_{H^s(\Gamma)} = \sum_{0 \leq l \leq L} (1 + (\frac{l^2\pi^2}{L_2^2})^s) |\beta_l|^2$ for every $s \in (0, 1/2)$. By adapting [C6$_c$], we can choose $z = 1 - s \in (1/2, 1)$ to find that there exists a $C_z > 0$ such that:

$$\|\gamma_\Gamma v_{K,L}\|^2_{H^s(\Gamma)} \leq \frac{2C_z}{L_1} \sum_{0 \leq k \leq K, 0 \leq l \leq L} \left(1 + \left(\frac{l^2\pi^2}{L_2^2}\right)^s\right)(1+k)^{2-2s} |\alpha_{k,l}|^2.$$

However, the function $x \mapsto \log x$ is concave, and consequently $a^s b^{1-s} \leq sa + (1-s)b$ for $a, b > 0$. Therefore, for each factor, we have the upper bound:

$$\left(1+\left(\frac{l^2\pi^2}{L_2^2}\right)^s\right)(1+k)^{2-2s} \leq 1+k^2+\left(\frac{l^2\pi^2}{L_2^2}\right)^s \left((1+k)^2\right)^{1-s} \leq s\frac{l^2\pi^2}{L_2^2}+(2-s)(1+k)^2.$$

By reviewing [C6$_a$] again, there exists a $C_s > 0$ (independent of v) such that:

$$\|\gamma_\Gamma v_{K,L}\|_{H^s(\Gamma)} \leq C_s \, \|v_{K,L}\|_{H^1(\Omega_2)}, \text{ for } K, L \in \mathbb{N}.$$

Again in passing to the limit (see exercise 4, Question (b)), we see that the trace mapping γ_Γ is indeed continuous from $H^1(\Omega_2)$ into $H^s(\Gamma)$.

Solution 7. Equivalence of norms: Poincaré inequalities in one dimension

a) First of all, recall that in exercise 5, we proved that the trace mapping:

$$\gamma_\star \ : \ \left| \begin{array}{l} C^1[0,1] \to \mathbb{R} \\ v \mapsto v(0) \end{array} \right.$$

can be uniquely extended to a continuous linear mapping from $H^1(0,1)$ into $\mathbb{R}$.

Let v be a function in $C^1[0,1]$, and write:

$$v(x) = v(0) + \int_0^x v'(y)\, dy \qquad [\mathrm{C7}_a].$$

Applying the triangular inequality to the right-hand side (with respect to the $L^2(0,1;dx)$ norm), we deduce:

$$\begin{aligned} \Big(\int_0^1 |v(x)|^2 dx\Big)^{\frac{1}{2}} &\leq \Big(\int_0^1 \Big(\int_0^x v'(y)\, dy\Big)^2 dx\Big)^{\frac{1}{2}} + \Big(\int_0^1 |v(0)|^2 dx\Big)^{\frac{1}{2}} \\ &\leq \Big(\int_0^1 \Big(\int_0^1 v'(y)\, dy\Big)^2 dx\Big)^{\frac{1}{2}} + |v(0)| \\ &\leq \Big(\int_0^1 |v'(y)|^2 dy\Big)^{\frac{1}{2}} + |v(0)| \end{aligned}$$

where we have used the Cauchy–Schwarz inequality to obtain the last inequality. Using the same density and continuity arguments as we did for the trace mapping at 0, observe that $C_1 = 1$ is such that:

$$\|v\|_{L^2(0,1)} \leq C_1 \left[\|v'\|_{L^2(0,1)} + |v(0)|\right] \qquad \forall v \in H^1(0,1).$$

b) From the definition of the Sobolev space $H_0^1(0,1)$, we will consider $v \in \mathcal{D}(0,1)$ first before reasoning by a density argument. As $v(0) = 0$ for every $v \in \mathcal{D}(0,1)$, after using the same calculation as we did in Question (a), we conclude that:

$$\|v\|_{L^2(0,1)} \leq C_1 \|v'\|_{L^2(0,1)} \qquad \forall v \in H_0^1(0,1).$$

This last inequality is also true (evidently) for elements belonging to the closed vector subspace of $H^1(0,1)$:

$$\{v \in H^1(0,1) \text{ such that } v(0) = 0\}.$$

REMARK A2.4.– *In the same way, using the relation $v(x) = v(1) - \int_x^1 v'(y)\, dy$, one may show that the inequality also holds for elements of:*

$$\{v \in H^1(0,1) \textit{ such that } v(1) = 0\}.$$

c) First, by using the Cauchy–Schwarz inequality in addition to the same arguments as those in exercise 5, the mean of a function:

$$m : \left| \begin{array}{ccc} \mathcal{C}^1([0,1]) & \to & \mathbb{R} \\ v & \mapsto & \displaystyle\int_0^1 v(y)\,dy \end{array} \right.$$

extends uniquely to a continuous linear mapping from $H^1(0,1)$ into $\mathbb{R}$. For $v \in \mathcal{C}^1[0,1]$, by integrating the identity:

$$v(x) = \int_y^x v'(z)\,dz + v(y)$$

over $[0,1]$ using the mean of a function with respect to the variable y, we obtain:

$$v(x) = \int_0^1 \int_y^x v'(z)\,dz\,dy + m(v).$$

We then bound it from above using the absolute value:

$$|v(x)| \leq \int_0^1 \left| \int_y^x v'(z)\,dz \right| dy + |m(v)| \leq \int_0^1 \int_0^1 |v'(z)|\,dzdy + |m(v)|$$
$$|v(x)| \leq \int_0^1 |v'(z)|\,dz + |m(v)|.$$

Integrating their squares over $(0,1)$:

$$\|v\|_{L^2(0,1)} \leq \| \int_0^1 |v'(z)|\,dz + |m(v)|\, \|_{L^2(0,1)}.$$

Then by applying the triangle inequality (see the calculations that follow [C7$_a$]) again, for $v \in \mathcal{C}^1[0,1]$, we see therefore that:

$$\|v\|_{L^2(0,1)} \leq \|v'\|_{L^2(0,1)} + |m(v)|$$

and, by a density argument (with $C_2 = 1$):

$$\|v\|_{L^2(0,1)} \leq C_2 \left\{ \|v'\|_{L^2(0,1)} + |m(v)| \right\} \qquad \forall v \in H^1(0,1).$$

d) We can consequently produce several subspaces that can serve as examples in $H^1(0,1)$, for which the semi-norm $|\cdot|_{1,(0,1)}$ is actually a norm that is furthermore equivalent to the $\|\cdot\|_{H^1(0,1)}$ norm:

$H_0^1(0,1)$, $\{v \in H^1(0,1)$ such that $v(0) = 0\}$, $\{v \in H^1(0,1)$ such that $v(1) = 0\}$

$\left\{v \in H^1(0,1) \text{ such that } \displaystyle\int_0^1 v(x)\,dx = 0\right\}$.

Solution 8. Equivalence of norms: Poincaré inequalities in two dimensions or higher

Recall the statement of Rellich's theorem from theorem 1.6: *let Ω be a bounded open subset of $\mathbb{R}^\mathrm{d}$ with a Lipschitzian boundary. Then, for every bounded sequence in $H^1(\Omega)$, we can extract a subsequence that converges in $L^2(\Omega)$*. In practice, what is the use of this result?

a) Reasoning by contradiction, we can obtain the result:

$$\forall C,\ \exists v \in H^1(\Omega),\ \|v\|_{L^2(\Omega)} > C\left\{\|\nabla v\|_{(L^2(\Omega))^\mathrm{d}} + \|\gamma_\Gamma v\|_{L^2(\Gamma)}\right\}.$$

Let us set $C = k$, for each $k \in \mathbb{N} \setminus \{0\}$. Then there exists a $v_k \in H^1(\Omega)$, such that:

$$\|v_k\|_{L^2(\Omega)} = 1, \quad \|\nabla v_k\|_{(L^2(\Omega))^\mathrm{d}} + \|\gamma_\Gamma v_k\|_{L^2(\Gamma)} < \frac{1}{k}. \qquad [\mathrm{C}8_a]$$

By construction, observe that the sequence $(\nabla v_k)_k$ tends towards 0 in $(L^2(\Omega))^\mathrm{d}$. Furthermore, the sequence $(v_k)_k$ is bounded in $H^1(\Omega)$, as:

$$\|v_k\|^2_{H^1(\Omega)} = \|v_k\|^2_{L^2(\Omega)} + \|\nabla v_k\|^2_{(L^2(\Omega))^\mathrm{d}} < 1 + \frac{1}{k^2} < 2.$$

By Rellich's theorem, there exists a subsequence which can be extracted from $(v_k)_k$, *that we continue to denote by* $(v_k)_k$, and which strongly converges in $L^2(\Omega)$ to a limit $v \in L^2(\Omega)$. In particular, we have that $\|v\|_{L^2(\Omega)} = 1$.

What does ∇v equal in the distribution sense? Let $\mathbf{z} \in (\mathcal{D}(\Omega))^\mathrm{d}$. We have:

$$\begin{aligned}
\langle \nabla v, \mathbf{z}\rangle &= -\langle v, \operatorname{div}\mathbf{z}\rangle \qquad \text{(distribution-sense derivative)}\\
&= -\int_\Omega v \operatorname{div}\mathbf{z}\, d\Omega \ \ (v \in L^2(\Omega),\ \text{identity [1.22]}).\\
&= -\lim_{k\to\infty}\int_\Omega v_k \operatorname{div}\mathbf{z}\, d\Omega \ \ (v = \lim_{k\to\infty} v_k \ \text{in } L^2(\Omega))\\
&= +\lim_{k\to\infty}\int_\Omega \nabla v_k \cdot \mathbf{z}\, d\Omega \ \ \text{(integrate by parts)}\\
&= 0 \qquad (\lim_{k\to\infty} \nabla v_k = 0 \text{ in } (L^2(\Omega))^\mathrm{d}).
\end{aligned}$$

Therefore, $\nabla v = 0$. As $0 \in (L^2(\Omega))^\mathrm{d}$, we see that v belongs to $H^1(\Omega)$. And since its gradient vanishes on the *connected* set Ω, we must have that $v = cst$. On the one hand, as $\lim_{k\to\infty} v_k = v$ in $L^2(\Omega)$ and, on the other hand, $\lim_{k\to\infty} \nabla v_k = 0 = \nabla v$ in $(L^2(\Omega))^\mathrm{d}$, finally we have:

$$\lim_{k\to\infty} v_k = v \text{ in } H^1(\Omega).$$

What benefits do we gain from knowing the convergent nature of $(v_k)_k$ in $L^2(\Omega)$? *The trace mapping is continuous from $H^1(\Omega)$ into $L^2(\partial\Omega)$, but it is not continuous from $L^2(\Omega)$ in $L^2(\partial\Omega)$*. From the fact that $(v_k)_k$ converges to v in $H^1(\Omega)$, we see that the traces $(\gamma_\Gamma v_k)_k$ tend towards $\gamma_\Gamma v$ in $L^2(\Gamma)$ by continuity. From [C8$_a$], we have $\gamma_\Gamma v = 0$, and we know that v is a constant and finally that $v = 0$ in Ω. This contradicts the fact that $\|v\|_{L^2(\Omega)} = 1$.

b) The procedure is essentially identical to the one used in Question (a). By a contradiction, we obtain the result:

$$\forall C,\ \exists v \in H^1(\Omega),\ \|v\|_{L^2(\Omega)} > C(\|\nabla v\|_{(L^2(\Omega))^{\mathrm{d}}} + |m(v)|).$$

Again, we can pick $C = k$ for each $k \in \mathbb{N} \setminus \{0\}$: there exists a $v_k \in H^1(\Omega)$ such that:

$$\|v_k\|_{L^2(\Omega)} = 1, \quad \|\nabla v_k\|_{(L^2(\Omega))^{\mathrm{d}}} + |m(v_k)| < \frac{1}{k}.$$

By construction, observe that the sequence $(\nabla v_k)_k$ tends towards 0 in $(L^2(\Omega))^{\mathrm{d}}$. Moreover, the sequence $(v_k)_k$ is bounded in $H^1(\Omega)$, since for each $k \in \mathbb{N} \setminus \{0\}$, we have $\|v_k\|^2_{H^1(\Omega)} < 2$. From Rellich's theorem, we can extract a subsequence from $(v_k)_k$, *which we again denote by* $(v_k)_k$, and that strongly converges in $L^2(\Omega)$ towards a limit $v \in L^2(\Omega)$. We have then that $\|v\|_{L^2(\Omega)} = 1$.

What does ∇v equal in the distribution sense? By using exactly the same arguments as in Question (a), one can show that $\nabla v = 0$. As $0 \in (L^2(\Omega))^{\mathrm{d}}$, v therefore belongs to $H^1(\Omega)$. And since its gradient vanishes on the connected set Ω, we must have that $v = cst$.

However in this case, we have:

$$v = \frac{\int_\Omega v\, d\Omega}{\int_\Omega d\Omega} = \lim_{k\to\infty} m(v_k) = 0$$

which contradicts the equality $\|v\|_{L^2(\Omega)} = 1$.

c) If the linear form L satisfies $L(1) = 0$, then there is no constant $C_3 > 0$ such that:

$$\|1\|_{L^2(\Omega)} \le C_3 \left(\|\nabla 1\|_{(L^2(\Omega))^{\mathrm{d}}} + |L(1)|\right)$$

since $\|1\|_{L^2(\Omega)} > 0$, whereas the right-hand side equals 0.

Suppose now that $L(1) \neq 0$ and let us continue as in Questions (a) and (b) with a proof by contradiction: to begin with, for each $k \in \mathbb{N} \setminus \{0\}$, there exists a $v_k \in H^1(\Omega)$ such that:

$$\|v_k\|_{L^2(\Omega)} = 1, \quad \|\nabla v_k\|_{(L^2(\Omega))^{\mathrm{d}}} + |L(v_k)| < \frac{1}{k}$$

until we arrive at the limit v in $L^2(\Omega)$, which is constant with $\|v\|_{L^2(\Omega)} = 1$.

On the one hand, since $\lim_{k\to\infty} v_k = v$ in $L^2(\Omega)$ and, on the other hand, $\lim_{k\to\infty} \nabla v_k = 0 = \nabla v$ in $(L^2(\Omega))^{\mathrm{d}}$, we have again that $\lim_{k\to\infty} v_k = v$ in $H^1(\Omega)$. Since the form L is continuous on $H^1(\Omega)$, this implies that:

$$L(v) = \lim_{k\to\infty} L(v_k) = 0.$$

By linearity and since v is a constant, $L(v) = L(v \times 1) = v \times L(1)$ and, as $L(1) \neq 0$, we must necessarily have $v = 0$, contradicting finally that $\|v\|_{L^2(\Omega)} = 1$.

d) Therefore, we can give a class of vector subspaces of $H^1(\Omega)$ for which the semi-norm of $H^1(\Omega)$ is actually a norm, namely from $\forall \Gamma \subset \partial\Omega$ with a non-zero measure:

$$\mathcal{V}_\Gamma = \left\{ v \in H^1(\Omega) \text{ such that } v_{|\Gamma} = 0 \right\}$$

and notably $H_0^1(\Omega) = \mathcal{V}_{\partial\Omega}$. The following space is also a suitable example:

$$\mathcal{V} = \left\{ v \in H^1(\Omega) \text{ such that } \int_\Omega v\, d\Omega = 0 \right\}.$$

The first result is known under the name of the Poincaré–Friedrichs inequality [2.24], and the second under the name of the Poincaré–Wirtinger inequality [2.23].

Solution 9. Equivalence of norms and semi-norms

To begin with, we note that $v \in H^{m+1}(\Omega)$ if, and only if, $v \in L^2(\Omega)$ and $\partial_i v \in H^m(\Omega)$, for $i = 1, \mathrm{d}$. Moreover:

$$\|v\|_{H^{m+1}(\Omega)} = \left(\|v\|^2_{L^2(\Omega)} + \sum_{i=1}^{\mathrm{d}} \|\partial_i v\|^2_{H^m(\Omega)} \right)^{1/2}.$$

a) The function J_v is continuous and strictly convex on the finite-dimensional vector space, $\mathbb{P}^m(\Omega)$. According to a classical optimization result (see Ciarlet 1989), it possesses a unique minimum, denoted from now on by p_v.

b) Let $m = 0$: we know that by Rellich's theorem (theorem 1.6) from every bounded sequence in $H^1(\Omega)$, we can extract a subsequence that converges in $L^2(\Omega) = H^0(\Omega)$, and which is the desired result.

For $M \geq 0$, suppose that the assertion is true for $m \leq M$. To establish the statement for $M + 1$, we must show that, for every bounded sequence $(v_k)_k$ in

$H^{M+2}(\Omega)$, that there exists a subsequence that converges in $H^{M+1}(\Omega)$. On the one hand, $(v_k)_k$ is a bounded sequence in $H^1(\Omega)$ and, on the other hand, $(\partial_i v_k)_k$, where $i = 1, \mathrm{d}$ are bounded sequences in $H^{M+1}(\Omega)$. By the induction hypothesis (respectively, for $m = M$ and $m = 0$), there exists the elements $g_1, \cdots, g_{\mathrm{d}}$ in $H^M(\Omega)$, an element v in $L^2(\Omega)$, and a subsequence of $(v_k)_k$ which, by an abuse of notation, we denote by $(v_k)_k$ again, such that:

$$\lim_{k\to\infty} \|\partial_i v_k - g_i\|_{H^M(\Omega)} = 0 \quad \text{and} \quad \lim_{k\to\infty} \|v_k - v\|_{L^2(\Omega)} = 0.$$

Then we check that $\partial_i v = g_i$ for $i = 1, \mathrm{d}$ in the distribution sense. Indeed, let $\varphi \in \mathcal{D}(\Omega)$:

$$\begin{aligned}\langle \partial_i v, \varphi\rangle &= -\langle v, \partial_i \varphi\rangle = -\int_\Omega v\, \partial_i \varphi\, d\Omega \\ &= -\lim_{k\to\infty} \int_\Omega v_k\, \partial_i \varphi\, d\Omega = +\lim_{k\to\infty} \int_\Omega \partial_i v_k\, \varphi\, d\Omega \quad \text{(integrate by parts)} \\ &= \int_\Omega g_i\, \varphi\, d\Omega = \langle g_i, \varphi\rangle.\end{aligned}$$

Therefore, each first derivative of $v \in L^2(\Omega)$ belongs to $H^M(\Omega)$, and so $v \in H^{M+1}(\Omega)$. Moreover:

$$\begin{aligned}\lim_{k\to\infty} \|v_k - v\|_{H^{M+1}(\Omega)} &= \Big(\lim_{k\to\infty} \|v_k - v\|^2_{L^2(\Omega)} + \sum_{i=1}^{\mathrm{d}} \lim_{k\to\infty} \|\partial_i v_k - \partial_i v\|^2_{H^M(\Omega)}\Big)^{1/2} \\ &= 0.\end{aligned}$$

The statement is therefore true for $M + 1$.

c) Let $m = 0$. If $v \in H^1(\Omega)$ is such that $|v|_{1,\Omega} = 0$, then $\nabla v = 0$ on Ω, which is connected and hence it follows that $v = cst$ everywhere on Ω. Therefore, $v \in \mathbb{P}^0(\Omega)$.

For $M \geq 0$, suppose that the statement is true for $m \leq M$. To establish the statement for the $M + 1$ step, we must prove that if $v \in H^{M+2}(\Omega)$ is such that $|v|_{M+2,\Omega} = 0$, then $v \in \mathbb{P}^{M+1}(\Omega)$. For each multi-index $\alpha' \in \mathbb{N}^{\mathrm{d}}$ of order $M + 1$, define $y_{\alpha'} = \partial_{\alpha'} v \in H^1(\Omega)$: by hypothesis, $\partial_i y_{\alpha'} = 0$ for $i = 1, \mathrm{d}$. Indeed, we have the equalities $\partial_i y_{\alpha'} = \partial_i \partial_{\alpha'} v = \partial_\alpha v$, where α is a multi-index of order $M + 2$. By the induction hypothesis ($m = 0$), we have $y_{\alpha'} = cst$ on Ω. Set ($x^{\alpha'} = x_1^{\alpha'_1} \cdots x_{\mathrm{d}}^{\alpha'_{\mathrm{d}}}$, $\alpha'! = \alpha'_1! \cdots \alpha'_{\mathrm{d}}!$):

$$y = v - \sum_{\alpha' \in \mathbb{N}^{\mathrm{d}} \text{ s.t. } |\alpha'| = M+1} y_{\alpha'} \frac{x^{\alpha'}}{\alpha'!}.$$

We get $y \in H^{M+2}(\Omega)$ and moreover, for each $\alpha'' \in \mathbb{N}^d$ of order $M+1$:

$$\partial_{\alpha''} y = \partial_{\alpha''} v - \sum_{\substack{\alpha' \in \mathbb{N}^d \text{ s.t.} \\ |\alpha'| = M+1}} \frac{y_{\alpha'}}{\alpha'!} \partial_{\alpha''} x^{\alpha'} = y_{\alpha''} - \sum_{\substack{\alpha' \in \mathbb{N}^d \text{ s.t.} \\ |\alpha'| = M+1}} \frac{y_{\alpha'}}{\alpha'!} \partial_{\alpha''} x^{\alpha'}.$$

Now, notice that $|\alpha''| = M+1$ and consequently, if in the sum we have $\alpha' \neq \alpha''$, then $\partial_{\alpha''} x^{\alpha'} = 0$, while $\partial_{\alpha''} x^{\alpha''} = \alpha''!$, and therefore:

$$\partial_{\alpha''} y = y_{\alpha''} - \frac{y_{\alpha''}}{\alpha''!} \alpha''! = 0.$$

From this, we see that $y \in H^{M+1}(\Omega)$ satisfies $|y|_{M+1,\Omega} = 0$ and, by the induction hypothesis ($m = M$), we have $y \in \mathbb{P}^M(\Omega)$. Finally, it follows that:

$$v = y + \sum_{\alpha' \in \mathbb{N}^d \text{ s.t. } |\alpha'| = M+1} \frac{y_{\alpha'}}{\alpha'!} x^{\alpha'} \in \mathbb{P}^{M+1}(\Omega).$$

The statement is therefore true for $M+1$.

d) Let $m \in \mathbb{N}$ be given. Reasoning by contradiction, suppose that [I9] does not hold: then for each $k \in \mathbb{N} \setminus \{0\}$, there exists $v_k \in H^{m+1}(\Omega)$ such that $\inf_{p \in \mathbb{P}^m(\Omega)} \|v_k + p\|_{H^{m+1}(\Omega)} > k\, |v_k|_{m+1,\Omega}$. In other words:

$$|v_k|_{m+1,\Omega} < \frac{1}{k} \|v_k + p_{v_k}\|_{H^{m+1}(\Omega)}, \qquad \forall k \geq 1$$

from Question (a). Now set $v'_k = v_k + p_{v_k}$. Since $p_{v_k} \in \mathbb{P}^m(\Omega)$, we have $|v'_k|_{m+1,\Omega} = |v_k|_{m+1,\Omega}$. There exists therefore a sequence $(v'_k)_{k \geq 1}$ of elements in $H^{m+1}(\Omega)$ satisfying:

$$|v'_k|_{m+1,\Omega} < \frac{1}{k} \|v'_k\|_{H^{m+1}(\Omega)}, \qquad \forall k \geq 1.$$

We have $v'_k \neq 0$ for each $k \geq 1$. Define $w_k = v'_k / \|v'_k\|_{H^{m+1}(\Omega)}$ to get a bounded sequence $(w_k)_{k \geq 1}$ of elements in $H^{m+1}(\Omega)$, with $\lim_{k \to \infty} |w_k|_{m+1,\Omega} = 0$. As $(w_k)_{k \geq 1}$ is bounded in $H^{m+1}(\Omega)$, from Question (b), we can extract a subsequence, again denoted by $(w_k)_{k \geq 1}$, that converges in $H^m(\Omega)$ to w ($\in H^m(\Omega)$).

Now, in the sense of distributions, let us calculate the partial derivatives of w of order $m+1$. Let $\varphi \in \mathcal{D}(\Omega)$ and consider $\alpha \in \mathbb{N}^d$ of order $m+1$. There exists

$i \in \{1, \cdots, \mathrm{d}\}$ such that $\alpha_i > 0$: denote by α' the multi-index of order m satisfying $\alpha'_i = \alpha_i - 1$, and $\alpha'_j = \alpha_j$ if $j \neq i$. It follows that:

$$\begin{aligned}
\langle \partial_\alpha w, \varphi \rangle &= \langle \partial_i \partial_{\alpha'} w, \varphi \rangle = -\langle \partial_{\alpha'} w, \partial_i \varphi \rangle = -\int_\Omega \partial_{\alpha'} w \, \partial_i \varphi \, d\Omega \quad (\partial_{\alpha'} w \in L^2(\Omega)) \\
&= -\lim_{k \to \infty} \int_\Omega \partial_{\alpha'} w_k \, \partial_i \varphi \, d\Omega \qquad (w = \lim_{k \to \infty} w_k \text{in } H^m(\Omega)) \\
&= +\lim_{k \to \infty} \int_\Omega \partial_\alpha w_k \, \varphi \, d\Omega \qquad \text{(integrate by parts)} \\
&= 0 \qquad (\lim_{k \to \infty} |w_k|_{m+1,\Omega} = 0).
\end{aligned}$$

Therefore, $\partial_\alpha w = 0$ for each multi-index of order $m+1$, implying that $w \in H^{m+1}(\Omega)$ and $|w|_{m+1,\Omega} = 0$. From this, first we observe that:

$$\begin{aligned}
\lim_{k \to \infty} \|w_k - w\|_{H^{m+1}(\Omega)} &= \lim_{k \to \infty} \left(\|w_k - w\|^2_{H^m(\Omega)} + |w_k - w|^2_{m+1,\Omega} \right)^{1/2} \\
&= 0.
\end{aligned} \qquad [C9_a]$$

From Question (c), we also observe that $w \in \mathbb{P}^m(\Omega)$. By revisiting the definition of the p_{v_k} from the solution to the optimization problem from Question (a), we have:

$$\|v_k + p_{v_k}\|_{H^{m+1}(\Omega)} \leq \|v_k + p_{v_k} + \lambda w\|_{H^{m+1}(\Omega)}, \qquad \forall k \geq 1, \ \forall \lambda \in \mathbb{R}$$

since $p_{v_k} + \lambda w \in \mathbb{P}^m(\Omega)$. By the definition of $(v'_k)_{k \geq 1}$ $(v'_k = v_k + p_{v_k})$ and $(w_k)_{k \geq 1}$ $(w_k = v'_k / \|v'_k\|_{H^{m+1}(\Omega)})$, we end up with:

$$\|v'_k\|_{H^{m+1}(\Omega)} \leq \left\| \|v'_k\|_{H^{m+1}(\Omega)} w_k + \lambda w \right\|_{H^{m+1}(\Omega)}, \qquad \forall k \geq 1, \ \forall \lambda \in \mathbb{R}.$$

For each k, choose $\lambda = -\|v'_k\|_{H^{m+1}(\Omega)}$ and, by factorizing the $\|v'_k\|_{H^{m+1}(\Omega)} \neq 0$, it follows finally that $1 \leq \|w_k - w\|_{H^{m+1}(\Omega)}$ for each $k \geq 1$, contradicting the convergence result in [C9$_a$].

Solution 10. Differentiation in the distribution sense

a) Let $\Omega = I$ be an interval in $\mathbb{R}$, and consider $v \in L^2(I)$. To prove that v' belongs to $(H^1_0(I))'$, we are going to use the continuous extension (exercise 4). By definition, $\mathcal{D}(I)$ is dense in $H^1_0(I)$ when endowed with the norm $\|\cdot\|_{H^1(I)}$. With the help of the Cauchy-Schwarz inequality, for every $\varphi \in \mathcal{D}(I)$, we have:

$$|\langle v', \varphi \rangle| = |\langle v, \varphi' \rangle| = \left| \int_I v \varphi' \, dx \right| \leq \|v\|_{L^2(I)} \|\varphi'\|_{L^2(I)} \leq \|v\|_{L^2(I)} \|\varphi\|_{H^1(I)}.$$

Therefore, $\ell : \varphi \mapsto \langle v', \varphi \rangle$ is a linear form on $\mathcal{D}(I)$, continuous with respect to the $\|\cdot\|_{H^1(I)}$ norm. From exercise 4, we can extend ℓ via density to a continuous linear form on $H_0^1(I)$: we deduce that $v' \in (H_0^1(I))'$. Then equally, we have shown that $\|v'\|_{(H_0^1(I))'} \leq \|v\|_{L^2(I)}$. From this we can conclude that the derivative operator $v \mapsto v'$, which is a linear mapping, is also a continuous one from $L^2(I)$ into $(H_0^1(I))'$.

REMARK A2.5.– *Why can we not say that* $v' \in \big(H^1(I)\big)'$ *(when* $v \in L^2(I)$*)? The problem is that we have reasoned using a* density argument, *starting from* $\mathcal{D}(I)$. *However,* $I = \mathbb{R}$, $\mathcal{D}(I)$ *is not dense in* $H^1(I)$.

b) Let Ω be an open subset of $\mathbb{R}^d$, and consider $v \in L^2(\Omega)$. Reasoning as in the previous question, for every $\varphi \in \mathcal{D}(\Omega)$, we can easily check that:

$$|\langle \partial_i v, \varphi \rangle| \leq \|v\|_{L^2(\Omega)} \|\partial_i \varphi\|_{L^2(\Omega)} \leq \|v\|_{L^2(\Omega)} \|\varphi\|_{H^1(\Omega)}.$$

From this, we conclude therefore that $\partial_i v \in (H_0^1(\Omega))'$, and that the ith partial derivative operator is continuous from $L^2(\Omega)$ into $(H_0^1(\Omega))'$.

c) Let Ω be an open subset of $\mathbb{R}^3$, and consider $\mathbf{v} \in (L^2(\Omega))^3$. In the sense of distributions, we have:

$$\operatorname{div} \mathbf{v} = \frac{\partial v_1}{\partial x_1} + \frac{\partial v_2}{\partial x_2} + \frac{\partial v_3}{\partial x_3}.$$

As v_1, v_2, and v_3 belong to $L^2(\Omega)$, we see from Question (b) that $\operatorname{div} \mathbf{v} \in (H_0^1(\Omega))'$.

REMARK A2.6.– *To construct the divergence in the distribution sense, for every* $\varphi \in \mathcal{D}(\Omega)$, *one has the sequence of equalities:*

$$\langle \mathbf{v}, \nabla\varphi \rangle = \sum_{i=1,3} \langle v_i, \partial_i \varphi \rangle = -\sum_{i=1,3} \langle \partial_i v_i, \varphi \rangle = -\langle \sum_{i=1,3} \partial_i v_i, \varphi \rangle = -\langle \operatorname{div} \mathbf{v}, \varphi \rangle.$$

Similarly, let us check that each of the three components that make up the curl of $\mathbf{v}$ belongs to $(H_0^1(\Omega))'$: this time we can write $\mathbf{curl}\, \mathbf{v} \in \big((H_0^1(\Omega))'\big)^3$.

REMARK A2.7.– *To construct the curl in the distribution sense, for every* $\mathbf{z} \in (\mathcal{D}(\Omega))^3$, *we write:* $\langle \mathbf{v}, \mathbf{curl}\, \mathbf{z} \rangle = \cdots = \langle \mathbf{curl}\, \mathbf{v}, \mathbf{z} \rangle$.

Solution 11. Extension and restriction

a) To establish that:

$$v_j \in H^1(\Omega_j),\ j = 1, 2 \text{ and } \gamma_1 v_1 = \gamma_2 v_2 \text{ on } F \quad \Longrightarrow \quad v \in H^1(\Omega)$$

we revisit the proof of Proposition 1.1 with some minor modifications in its hypotheses. The converse is proven as in Proposition 1.2.

b) This is again a variant of the previous question. Indeed, if we denote by $\tilde{v}$ the extension of v by 0 to $\mathbb{R}^d \setminus \overline{\Omega}$ then, by construction, we have that $\tilde{v} \in L^2(\mathbb{R}^d)$. Otherwise, the two open subsets are firstly, $\Omega_1 = \Omega$, and secondly, $\Omega_2 = \mathbb{R}^d \setminus \overline{\Omega}$, along with $\mathbb{R}^d = \overline{\Omega_1} \cup \overline{\Omega_2}$. However, the restriction of $\tilde{v}$ to Ω_1, which equals v, belongs to $H^1(\Omega_1)$, and likewise the restriction of $\tilde{v}$ to Ω_2, which equals 0, belongs to $H^1(\Omega_2)$. Finally, the interface between Ω_1 and Ω_2 equals $\partial\Omega$, and the traces of the two restrictions coincide along $\partial\Omega$: they both equal to 0, as the trace of an element of $H_0^1(\Omega_1)$, and naturally as the trace of the zero function on Ω_2.

c) As the boundary is Lipschitzian, it suffices to prove that $\gamma_0 v = 0$ on $\partial\Omega$ (see Theorem 1.8).

- By hypothesis, $\tilde{v} \in H^1(\mathbb{R}^d)$ so, for every $\mathbf{z} \in (\mathcal{D}(\mathbb{R}^d))^d$:

$$\int_{\mathbb{R}^d} \tilde{v} \operatorname{div} \mathbf{z} \, d\Omega = -\int_{\mathbb{R}^d} \nabla\tilde{v} \cdot \mathbf{z} \, d\Omega.$$

- Let us show that the restriction of $\nabla\tilde{v}$ to Ω, which we denote by $(\nabla\tilde{v})_{|\Omega}$, equals ∇v. We reason as in Proposition 1.2. We want to obtain this result in $(L^2(\Omega))^d$, and we can justify this via a density argument. Let $\boldsymbol{q} \in (\mathcal{D}(\Omega))^d$. Denote by $\mathbf{z}$ the extension of $\boldsymbol{q}$ by 0 on $\mathbb{R}^d \setminus \overline{\Omega}$: by construction, $\mathbf{z} \in (\mathcal{D}(\mathbb{R}^d))^d$. With this remark, we write successively:

$$\begin{aligned}\int_{\Omega} (\nabla\tilde{v})_{|\Omega} \cdot \boldsymbol{q} \, d\Omega &= \int_{\mathbb{R}^d} \nabla\tilde{v} \cdot \mathbf{z} \, d\Omega = -\int_{\mathbb{R}^d} \tilde{v} \operatorname{div} \mathbf{z} \, d\Omega = -\int_{\Omega} v \operatorname{div} \boldsymbol{q} \, d\Omega \\ &= \int_{\Omega} \nabla v \cdot \boldsymbol{q} \, d\Omega.\end{aligned}$$

The last equality uses the fact that $v \in H^1(\Omega)$. Indeed, we have $(\nabla\tilde{v})_{|\Omega} = \nabla v$. In a similar fashion, the restriction of $\nabla\tilde{v}$ to $\mathbb{R}^d \setminus \overline{\Omega}$, which we denote by $(\nabla\tilde{v})_{|\mathbb{R}^d \setminus \overline{\Omega}}$, vanishes.

- From the above, for every $\mathbf{z} \in (\mathcal{D}(\mathbb{R}^d))^d$, we find:

$$\int_{\Omega} v \, (\operatorname{div} \mathbf{z}_{|\Omega}) \, d\Omega = \int_{\mathbb{R}^d} \tilde{v} \operatorname{div} \mathbf{z} \, d\Omega = -\int_{\mathbb{R}^d} \nabla\tilde{v} \cdot \mathbf{z} \, d\Omega = -\int_{\Omega} \nabla v \cdot \mathbf{z}_{|\Omega} \, d\Omega.$$

Otherwise, from Green's formula [1.34] on Ω, we know that:

$$\begin{aligned}\int_{\Omega} \left(v \, (\operatorname{div} \mathbf{z}_{|\Omega}) + \nabla v \cdot \mathbf{z}_{|\Omega}\right) d\Omega &= \langle \gamma_n \mathbf{z}, \gamma_0 v \rangle_{(H^{1/2}(\partial\Omega))', H^{1/2}(\partial\Omega)} \\ &= \int_{\partial\Omega} \gamma_0 v (\mathbf{z} \cdot \mathbf{n})_{|\partial\Omega} \, d\Gamma\end{aligned}$$

since the trace of an element in $H^1(\Omega)$ automatically belongs to $L^2(\partial\Omega)$. Hence, for every $\mathbf{z} \in (\mathcal{D}(\mathbb{R}^{\mathrm{d}}))^{\mathrm{d}}$, we have established:

$$\int_{\partial\Omega} \gamma_0 v (\mathbf{z} \cdot \mathbf{n})_{|\partial\Omega}\, d\Gamma = 0.$$

Therefore we have established that $\gamma_0 v = 0$, and hence that $v \in H_0^1(\Omega)$ (Theorem 1.8).

Solution 12. The function space $H(\mathrm{div}, \Omega)$

We equip $H(\mathrm{div}; \Omega)$ with the inner product:

$$(\cdot,\cdot)_{H(\mathrm{div};\Omega)} \;:\; (\mathbf{v}, \mathbf{w}) \mapsto \int_\Omega \mathbf{v} \cdot \mathbf{w}\, d\Omega + \int_\Omega \mathrm{div}\, \mathbf{v}\ \mathrm{div}\, \mathbf{w}\, d\Omega.$$

a) Suppose $\ell \in (H(\mathrm{div}; \Omega))'$ vanishes on $(\mathcal{C}_c^\infty(\overline{\Omega}))^{\mathrm{d}}$. From Riesz's representation theorem, there exists an element $\mathbf{l} \in H(\mathrm{div}; \Omega)$ such that:

$$\ell(\mathbf{w}) = (\mathbf{l}, \mathbf{w})_{H(\mathrm{div};\Omega)} = \int_\Omega \mathbf{l} \cdot \mathbf{w}\, d\Omega + \int_\Omega D\ \mathrm{div}\, \mathbf{w}\, d\Omega, \quad \forall \mathbf{w} \in H(\mathrm{div}; \Omega) \quad [\mathrm{C}12_a]$$

where we have denoted $D = \mathrm{div}\, \mathbf{l} \in L^2(\Omega)$. We extend $\mathbf{l}$, resp. D, by 0 to $\mathbb{R}^{\mathrm{d}} \setminus \overline{\Omega}$: denote by $\tilde{\mathbf{l}}$, resp. $\tilde{D}$, the respective extension. Now, let $\mathbf{z} \in (\mathcal{D}(\mathbb{R}^{\mathrm{d}}))^{\mathrm{d}}$, then:

$$\int_{\mathbb{R}^{\mathrm{d}}} (\tilde{\mathbf{l}} \cdot \mathbf{z} + \tilde{D}\ \mathrm{div}\, \mathbf{z})\, d\Omega = \int_\Omega (\mathbf{l} \cdot \mathbf{z}_{|\Omega} + D\ \mathrm{div}(\mathbf{z}_{|\Omega}))\, d\Omega = \ell(\mathbf{z}_{|\Omega}) = 0$$

since, from Definition 1.9, we have $\mathbf{z}_{|\Omega} \in (\mathcal{C}_c^\infty(\overline{\Omega}))^{\mathrm{d}}$. Therefore:

$$\int_{\mathbb{R}^{\mathrm{d}}} \tilde{\mathbf{l}} \cdot \mathbf{z}\, d\Omega = -\int_{\mathbb{R}^{\mathrm{d}}} \tilde{D}\ \mathrm{div}\, \mathbf{z}\, d\Omega = \langle \nabla \tilde{D}, \mathbf{z} \rangle, \qquad \forall \mathbf{z} \in (\mathcal{D}(\mathbb{R}^{\mathrm{d}}))^{\mathrm{d}}.$$

In other words, $\nabla \tilde{D} = \tilde{\mathbf{l}}$ in $(\mathcal{D}'(\mathbb{R}^{\mathrm{d}}))^{\mathrm{d}}$, implying that $\tilde{D} \in H^1(\mathbb{R}^{\mathrm{d}})$. Since additionally, $\tilde{D}_{|\mathbb{R}^{\mathrm{d}} \setminus \overline{\Omega}} = 0$, we know from exercise 11 that $D \in H_0^1(\Omega)$ and $\mathbf{l} = (\nabla \tilde{D})_{|\Omega} = \nabla D$. By the definition of $H_0^1(\Omega)$, there exists $(D_k)_k \in (\mathcal{D}(\Omega))^{\mathbb{N}}$ such that $\lim_{k\to\infty} \|D - D_k\|_{H^1(\Omega)} = 0$.

Now let $\mathbf{w} \in H(\mathrm{div}; \Omega)$ then, from $[\mathrm{C}12_a]$, we find:

$$\begin{aligned} \ell(\mathbf{w}) &= \int_\Omega \nabla D \cdot \mathbf{w}\, d\Omega + \int_\Omega D\ \mathrm{div}\, \mathbf{w}\, d\Omega \\ &= \lim_{k\to\infty} \Big(\int_\Omega \nabla D_k \cdot \mathbf{w}\, d\Omega + \int_\Omega D_k\ \mathrm{div}\, \mathbf{w}\, d\Omega \Big) = 0 \end{aligned}$$

where we have integrated by parts for each k. Therefore, the linear form ℓ vanishes and, following the hint, we may conclude that $(\mathcal{C}_c^\infty(\overline{\Omega}))^{\mathrm{d}}$ is indeed dense in $H(\mathrm{div};\Omega)$.

b) Let $\mathbf{w} \in (\mathcal{C}_c^\infty(\overline{\Omega}))^{\mathrm{d}}$. We are interested in how $(\mathbf{w}\cdot\mathbf{n})_{|\partial\Omega}$ operates on $H^{1/2}(\partial\Omega)$. As the space $H^{1/2}(\partial\Omega)$ coincides with the space of traces of $H^1(\Omega)$, consider a $v \in H^1(\Omega)$ and, by reasoning as we did to prove Proposition 1.13, we get:

$$\int_{\partial\Omega} (\mathbf{w}\cdot\mathbf{n})_{|\partial\Omega}\gamma_0 v\, d\Gamma = \int_\Omega (\mathbf{w}\cdot\nabla v + (\mathrm{div}\,\mathbf{w})v)\, d\Omega. \qquad [\mathrm{C}12_b]$$

By applying the Cauchy-Schwarz inequality twice successively, we observe that:

$$\begin{aligned}\left|\int_{\partial\Omega} (\mathbf{w}\cdot\mathbf{n})_{|\partial\Omega}\gamma_0 v\, d\Gamma\right| &\le \|\mathbf{w}\|_{(L^2(\Omega))^{\mathrm{d}}}\|\nabla v\|_{(L^2(\Omega))^{\mathrm{d}}} + \|\,\mathrm{div}\,\mathbf{w}\|_{L^2(\Omega)}\|v\|_{L^2(\Omega)}\\ &\le \|\mathbf{w}\|_{H(\mathrm{div};\Omega)}\|v\|_{H^1(\Omega)}.\end{aligned}$$

By the definition of $H^{1/2}(\partial\Omega)$, the trace mapping γ_0 is surjective, continuous, and linear from $H^1(\Omega)$ into $H^{1/2}(\partial\Omega)$: for every $\lambda \in H^{1/2}(\partial\Omega)$, there exists a $v \in H^1(\Omega)$ such that $\gamma_0 v = \lambda$. From the above, it follows that:

$$\left|\int_{\partial\Omega} (\mathbf{w}\cdot\mathbf{n})_{|\partial\Omega}\lambda\, d\Gamma\right| \le \|\mathbf{w}\|_{H(\mathrm{div};\Omega)} \inf_{v\in H^1(\Omega)\ s.t.\ \gamma_0 v=\lambda} \|v\|_{H^1(\Omega)}.$$

By using Definition [1.32] of the $H^{1/2}(\partial\Omega)$ norm, finally we find that:

$$\left|\int_{\partial\Omega} (\mathbf{w}\cdot\mathbf{n})_{|\partial\Omega}\lambda\, d\Gamma\right| \le \|\mathbf{w}\|_{H(\mathrm{div};\Omega)}\|\lambda\|_{H^{1/2}(\partial\Omega)},\ \forall\lambda \in H^{1/2}(\partial\Omega). \qquad [\mathrm{C}12_c]$$

Therefore, $\lambda \mapsto \int_{\partial\Omega}(\mathbf{w}\cdot\mathbf{n})_{|\partial\Omega}\lambda\, d\Gamma$ is a continuous linear form on $H^{1/2}(\partial\Omega)$, and we have indeed that $(\mathbf{w}\cdot\mathbf{n})_{|\partial\Omega} \in (H^{1/2}(\partial\Omega))'$, where:

$$\langle(\mathbf{w}\cdot\mathbf{n})_{|\partial\Omega},\lambda\rangle_{(H^{1/2}(\partial\Omega))',H^{1/2}(\partial\Omega)} = \int_{\partial\Omega} (\mathbf{w}\cdot\mathbf{n})_{|\partial\Omega}\lambda\, d\Gamma.$$

c) Introduce the normal trace mapping γ_n, defined from $(\mathcal{C}_c^\infty(\overline{\Omega}))^{\mathrm{d}}$ into $(H^{1/2}(\partial\Omega))'$ by $\gamma_n\mathbf{w} = (\mathbf{w}\cdot\mathbf{n})_{|\partial\Omega}$. From [C12$_c$], for every $\mathbf{w} \in (\mathcal{C}_c^\infty(\overline{\Omega}))^{\mathrm{d}}$ we have:

$$\|\gamma_n\mathbf{w}\|_{(H^{1/2}(\partial\Omega))'} = \sup_{\lambda\in H^{1/2}(\partial\Omega)\setminus\{0\}} \frac{\left|\int_{\partial\Omega}(\mathbf{w}\cdot\mathbf{n})_{|\partial\Omega}\lambda\, d\Gamma\right|}{\|\lambda\|_{H^{1/2}(\partial\Omega)}} \le \|\mathbf{w}\|_{H(\mathrm{div};\Omega)}. \qquad [\mathrm{C}12_d]$$

The mapping γ_n is therefore continuous from $(\mathcal{C}_c^\infty(\overline{\Omega}))^{\mathrm{d}}$ into $(H^{1/2}(\partial\Omega))'$ *with regards to* the norm of $H(\mathrm{div};\Omega)$. By exercise 4 (Question (b)) and the density result from Question (a), we can extend γ_n to a continuous linear mapping from $H(\mathrm{div};\Omega)$ into $(H^{1/2}(\partial\Omega))'$, where $\|\gamma_n \mathbf{w}\|_{(H^{1/2}(\partial\Omega))'} \leq \|\mathbf{w}\|_{H(\mathrm{div};\Omega)}$ for every $\mathbf{w} \in H(\mathrm{div};\Omega)$ using the upper bound of [C12$_d$].

d) We reason simply by using the density of $(\mathcal{C}_c^\infty(\overline{\Omega}))^{\mathrm{d}}$ in $H(\mathrm{div};\Omega)$ from [C12$_b$], which holds for every $\mathbf{w} \in (\mathcal{C}_c^\infty(\overline{\Omega}))^{\mathrm{d}}$ and every $v \in H^1(\Omega)$.

e) Let $g \in (H^{1/2}(\partial\Omega))'$ be given. Since $v \mapsto \langle g, \gamma_0 v\rangle_{(H^{1/2}(\partial\Omega))',H^{1/2}(\partial\Omega)}$ defines a continuous linear form on $H^1(\Omega)$, by Riesz's representation theorem, we know that the variation formulation [F12] admits a solution, which is necessarily unique.

If we choose a test function $v \in \mathcal{D}(\Omega)$ in [F12], we have that $\gamma_0 v = 0$, from which it follows:

$$0 = (u,v)_{H^1(\Omega)} = \int_\Omega \nabla u \cdot \nabla v \, d\Omega + \int_\Omega uv \, d\Omega$$
$$= \sum_{i=1}^{\mathrm{d}} \langle \frac{\partial u}{\partial x_i}, \frac{\partial v}{\partial x_i}\rangle + \langle u,v\rangle = -\sum_{i=1}^{\mathrm{d}} \langle \frac{\partial^2 u}{\partial x_i^2}, v\rangle + \langle u,v\rangle = \langle -\Delta u + u, v\rangle.$$

From which we see that $-\Delta u + u = 0$ in $\mathcal{D}'(\Omega)$. But since $u \in L^2(\Omega)$, we have shown therefore that $\Delta u = u$ in $L^2(\Omega)$. In particular, $u \in H^1(\Delta;\Omega)$, and we can use the integration by parts formula [1.39] provided that we choose a $v \in H^1(\Omega)$ from [F12]. We have (with $\gamma_1 u = \left(\frac{\partial u}{\partial n}\right)_{|\partial\Omega}$):

$$\langle g, \gamma_0 v\rangle_{(H^{1/2}(\partial\Omega))',H^{1/2}(\partial\Omega)} = (u,v)_{H^1(\Omega)} = \int_\Omega \nabla u \cdot \nabla v \, d\Omega + \int_\Omega uv \, d\Omega$$
$$= \int_\Omega (-\Delta u + u)\, v \, d\Omega + \langle \gamma_1 u, \gamma_0 v\rangle_{(H^{1/2}(\partial\Omega))',H^{1/2}(\partial\Omega)}$$
$$= \langle \gamma_1 u, \gamma_0 v\rangle_{(H^{1/2}(\partial\Omega))',H^{1/2}(\partial\Omega)}$$

where we have used the fact that $\Delta u = u$ in $L^2(\Omega)$, to go into the last equality. As this equality holds for every $v \in H^1(\Omega)$, and the trace mapping γ_0 is surjective from $H^1(\Omega)$ onto $H^{1/2}(\partial\Omega)$, we have proven that $\gamma_1 u = g$ in $(H^{1/2}(\partial\Omega))'$. This establishes the surjectivity of $\gamma_1 : H^1(\Delta;\Omega) \to (H^{1/2}(\partial\Omega))'$.

Finally, if we introduce $\mathbf{w} = \nabla v$ then, on the one hand, we have $H(\mathrm{div};\Omega)$ since $\mathrm{div}\, \mathbf{w} = \Delta v \in L^2(\Omega)$, whereas, on the other hand, $\gamma_n \mathbf{w} = \gamma_1 u$ in $(H^{1/2}(\partial\Omega))'$.

Indeed, for every $v \in H^1(\Omega)$, with the help of the integration by parts formulae [1.34] and [1.39], we get:

$$\begin{aligned}\langle \gamma_n \mathbf{w}, \gamma_0 v \rangle_{(H^{1/2}(\partial\Omega))', H^{1/2}(\partial\Omega)} &= \int_\Omega (\operatorname{div} \mathbf{w})\, v \, d\Omega + \int_\Omega \mathbf{w} \cdot \nabla v \, d\Omega \\ &= \int_\Omega \Delta u \, v \, d\Omega + \int_\Omega \nabla u \cdot \nabla v \, d\Omega \\ &= \langle \gamma_1 u, \gamma_0 v \rangle_{(H^{1/2}(\partial\Omega))', H^{1/2}(\partial\Omega)}.\end{aligned}$$

Therefore, $\gamma_n \mathbf{w} = g$ in $(H^{1/2}(\partial\Omega))'$, and hence the surjectivity of $\gamma_n : H(\mathrm{div};\Omega) \to (H^{1/2}(\partial\Omega))'$ follows from it.

f) For every $\mathbf{w} \in (\mathcal{D}(\Omega))^{\mathrm{d}}$, we have that $(\mathbf{w}\cdot\mathbf{n})_{|\partial\Omega} = 0$, which is to say $\gamma_n \mathbf{w} = 0$. As γ_n is continuous on $H(\mathrm{div};\Omega)$, we see that:

$$H_0(\mathrm{div};\Omega) \subset \{\mathbf{w} \in H(\mathrm{div};\Omega) \text{ such that } \gamma_n \mathbf{w} = 0\}.$$

It is intuitive to think of these two spaces as being *equal*. This result can be proven with the help of the hint and, more precisely, by proving that $(\mathcal{D}(\Omega))^{\mathrm{d}}$ is dense in $\{\mathbf{w} \in H(\mathrm{div};\Omega) \text{ such that } \gamma_n \mathbf{w} = 0\}$. We refer to Girault and Raviart (1986) for the details.

A2.2. Exercises from Chapter 2

Solution 13. Variational formulation for Fourier boundary condition problems

a) Let $u, f \in L^2(\Omega)$ with $\Delta u \in L^2(\Omega)$. After multiplying the first equation from [P13] by $v \in H^1(\Omega)$ and integrating over Ω, we obtain:

$$\int_\Omega uv \, d\Omega - \int_\Omega \Delta u \, v \, d\Omega = \int_\Omega fv \, d\Omega, \qquad \forall v \in H^1(\Omega).$$

By applying Green's formula [1.39] to the right-hand side, we have:

$$\int_\Omega uv \, d\Omega + \int_\Omega \nabla u \cdot \nabla v \, d\Omega - \langle \frac{\partial u}{\partial n}, v \rangle_{(H^{1/2}(\partial\Omega))', H^{1/2}(\partial\Omega)} = \int_\Omega fv \, d\Omega, \quad \forall v \in H^1(\Omega).$$

Finally, it suffices to use the second equation from [P13] to find the associated variational formulation (noting that $g, u_{|\partial\Omega} \in L^2(\partial\Omega)$):

$$\left| \begin{array}{l} \text{Find } u \in H^1(\Omega) \text{ such that} \\ \displaystyle\int_\Omega uv\, d\Omega + \int_\Omega \nabla u \cdot \nabla v\, d\Omega + \lambda \int_{\partial\Omega} uv\, d\Gamma \\ \displaystyle\qquad\qquad = \int_\Omega fv\, d\Omega + \int_{\partial\Omega} gv\, d\Gamma, \quad \forall v \in H^1(\Omega). \end{array} \right. \qquad \text{[F13]}$$

b) If $\lambda > 0$, then it is obvious that the bilinear form:

$$a_\lambda(u, v) = \int_\Omega uv\, d\Omega + \int_\Omega \nabla u \cdot \nabla v\, d\Omega + \lambda \int_{\partial\Omega} uv\, d\Gamma$$

defines an inner product on $H^1(\Omega)$ that is equivalent to the standard one. Indeed, it is a symmetric bilinear form that satisfies (from the trace Theorem 1.7):

$$||v||^2_{H^1(\Omega)} \le a_\lambda(v, v) \le ||v||^2_{H^1(\Omega)} + \lambda ||v||^2_{L^2(\partial\Omega)} \le (1 + C^2\lambda)||v||^2_{H^1(\Omega)}.$$

Besides, the form:

$$\ell(v) = \int_\Omega fv\, d\Omega + \int_{\partial\Omega} gv\, d\Gamma$$

is linear and continuous on $H^1(\Omega)$ because, by the virtue of the Cauchy-Schwarz inequality and the trace Theorem 1.7, we have the following upper bound:

$$|\ell(v)| \le \left(||f||_{L^2(\Omega)} + C||g||_{L^2(\partial\Omega)}\right) ||v||_{H^1(\Omega)}.$$

The Lax–Milgram theorem establishes the existence of a unique $u \in H^1(\Omega)$ for us, with $\ell(v) = a_\lambda(u, v)$, $\forall v \in H^1(\Omega)$.

c) According to the above, if u is a solution of [P13], then u satisfies [F13]. Let us investigate the converse. In [F13], if we choose $v \in \mathcal{D}(\Omega)(\subset H^1(\Omega))$, then we have:

$$\int_\Omega (uv + \nabla u \cdot \nabla v)\, d\Omega = \int_\Omega fv\, d\Omega$$

since $v = 0$ on $\partial\Omega$. We can then replace the integrals $\int_\Omega \cdot\, \partial_i v\, d\Omega$ with the duality brackets $\langle \cdot, \partial_i v\rangle$, before differentiating in the distribution sense:

$$\langle f, v\rangle = \langle u, v\rangle + \sum_{i=1,d} \langle \frac{\partial u}{\partial x_i}, \frac{\partial v}{\partial x_i}\rangle = \langle u, v\rangle - \sum_{i=1,d} \langle \frac{\partial^2 u}{\partial x_i^2}, v\rangle = \langle u, v\rangle - \langle \Delta u, v\rangle.$$

From this, we deduce that:

$$\langle u - \Delta u, v \rangle = \langle f, v \rangle \,, \; \forall v \in \mathcal{D}(\Omega)$$

which is to say, $u - \Delta u = f$ in the distribution sense. Since u and f belong to $L^2(\Omega)$, $\Delta u \in L^2(\Omega)$, and therefore $u - \Delta u = f$ almost everywhere in Ω. In particular, we have:

$$\int_\Omega (u - \Delta u)\, v \, d\Omega = \int_\Omega f v \, d\Omega, \; \forall v \in H^1(\Omega). \qquad [\mathrm{F13}_a]$$

Now, returning to [F13] and after applying Green's formula [1.39], which is valid since $u \in H^1(\Delta; \Omega)$, we find:

$$\begin{aligned}\int_\Omega (u - \Delta u)\, v \, d\Omega + \langle \frac{\partial u}{\partial n}, v \rangle_{(H^{1/2}(\partial\Omega))', H^{1/2}(\partial\Omega)} + \lambda \langle u, v \rangle_{(H^{1/2}(\partial\Omega))', H^{1/2}(\partial\Omega)} \\ = \int_\Omega f v \, d\Omega + \langle g, v \rangle_{(H^{1/2}(\partial\Omega))', H^{1/2}(\partial\Omega)}, \; \forall v \in H^1(\Omega).\end{aligned}$$

Comparing this with [F13$_a$], we get:

$$\langle \frac{\partial u}{\partial n} + \lambda u - g, v \rangle_{(H^{1/2}(\partial\Omega))', H^{1/2}(\partial\Omega)} = 0, \; \forall v \in H^1(\Omega).$$

Since, by definition, $(\frac{\partial u}{\partial n}_{|\partial\Omega} + \lambda u_{|\partial\Omega} - g)$ is a continuous linear form on $H^{1/2}(\partial\Omega)$, and the trace mapping γ_0 is surjective from $H^1(\Omega)$ onto $H^{1/2}(\partial\Omega)$, the equality above that holds for every $v \in H^1(\Omega)$ proves that $\frac{\partial u}{\partial n}_{|\partial\Omega} + \lambda u_{|\partial\Omega} = g$. To conclude, as g and $u_{|\partial\Omega}$ belong to $L^2(\partial\Omega)$, we observe from this that $\frac{\partial u}{\partial n} + \lambda u = g$ almost everywhere on $\partial\Omega$.

d) If $\lambda < 0$, we can no longer apply the Lax–Milgram theorem because the bilinear form a_λ is no longer coercive on $H^1(\Omega)$. One must adapt the Fredholm alternative (see section 2.4.6) by using:

$$a_+(u, v) = (u, v)_{H^1(\Omega)} \text{ and } c(u, v) = \lambda \int_{\partial\Omega} uv \, d\Gamma.$$

Solution 14. Variational formulation for heat diffusion

a) Let us set $u = u^\star - \tau_0$. Since $u \in H^1(\Omega)$ and $u_{|\partial\Omega} = 0$, we must look for a u in $H_0^1(\Omega)$. It is clear that u satisfies:

$$\begin{cases} \text{Find } u \in H_0^1(\Omega) \text{ such that} \\ -\operatorname{div}(k\nabla u) = f & \text{in } \Omega \\ u = 0 & \text{on } \partial\Omega. \end{cases} \qquad [\text{P}14_b]$$

b) After multiplying the first equation of [P14$_b$] by $v \in H_0^1(\Omega)$ and integrating over Ω, we obtain:

$$-\int_\Omega \operatorname{div}(k\nabla u)\, v\, d\Omega = \int_\Omega f v\, d\Omega, \qquad \forall v \in H_0^1(\Omega).$$

From here, there are two possibilities. Either, we remember that, according to Stokes' formula (proved in exercise 1 when $\mathrm{d} = 3$):

$$\forall \mathbf{u} \in (\mathcal{C}^\infty(\overline{\Omega}))^{\mathrm{d}},\ \forall v \in \mathcal{C}^\infty(\overline{\Omega}), \quad \int_\Omega (\operatorname{div} \mathbf{u}\, v + \mathbf{u} \cdot \nabla v)\, d\Omega = \int_{\partial\Omega} v(\mathbf{u}\cdot\mathbf{n})\, d\Gamma.$$

If we set $\mathbf{u} = k\nabla u$ and take into account the fact that $v_{|\partial\Omega} = 0$, then by "analogy" we obtain:

$$-\int_\Omega \operatorname{div}(k\nabla u)\, v\, d\Omega = \int_\Omega k\nabla u \cdot \nabla v\, d\Omega.$$

Or, by using the fact that $\mathcal{D}(\Omega)$ is dense in $H_0^1(\Omega)$, we can replace the test function v with a sequence $(v_\ell)_\ell$ of elements in $\mathcal{D}(\Omega)$ that satisfy $\lim_{\ell\to\infty} \|v_\ell - v\|_{H^1(\Omega)} = 0$, $\lim_{\ell\to\infty} \|v_\ell - v\|_{L^2(\Omega)} = 0$ and $\lim_{\ell\to\infty} \|\partial_i v_\ell - \partial_i v\|_{L^2(\Omega)} = 0$, for each $i = 1, \cdots, \mathrm{d}$.

Then, by differentiating in the distribution sense:

$$\begin{aligned} -\int_\Omega \operatorname{div}(k\nabla u)\, v\, d\Omega &= -\lim_\ell \int_\Omega \operatorname{div}(k\nabla u)\, v_\ell\, d\Omega = -\lim_\ell \langle \operatorname{div}(k\nabla u), v_\ell\rangle \\ &= -\lim_\ell \langle \sum_{i=1}^{\mathrm{d}} \frac{\partial}{\partial x_i}\left(k\frac{\partial u}{\partial x_i}\right), v_\ell\rangle = \lim_\ell \sum_{i=1}^{\mathrm{d}} \langle k\frac{\partial u}{\partial x_i}, \frac{\partial v_\ell}{\partial x_i}\rangle \\ &= \lim_\ell \int_\Omega \sum_{i=1}^{\mathrm{d}} k\frac{\partial u}{\partial x_i}\frac{\partial v_\ell}{\partial x_i}\, d\Omega = \sum_{i=1}^{\mathrm{d}} \int_\Omega k\frac{\partial u}{\partial x_i}\frac{\partial v}{\partial x_i}\, d\Omega \\ &= \int_\Omega k\nabla u.\nabla v\, d\Omega. \end{aligned}$$

In either case, the variational formulation associated to [P14$_b$] becomes:

$$\left| \begin{array}{l} \text{Find } u \in H_0^1(\Omega) \text{ such that} \\ \displaystyle\int_\Omega k\nabla u \cdot \nabla v \, d\Omega = \int_\Omega f v \, d\Omega, \quad \forall v \in H_0^1(\Omega). \end{array} \right. \qquad [\text{F}14_b]$$

c) Let u_1 and u_2 be two solutions of [F14$_b$], then:

$$\int_\Omega k\nabla(u_1 - u_2) \cdot \nabla v \, d\Omega = 0, \quad \forall v \in H_0^1(\Omega).$$

In choosing $v = u_1 - u_2$ for the test function, we get:

$$\int_\Omega k|\nabla(u_1 - u_2)|^2 \, d\Omega = 0.$$

By the definition of k, we know that:

$$k_{min} \int_\Omega |\nabla(u_1 - u_2)|^2 \, d\Omega \leq \int_\Omega k|\nabla(u_1 - u_2)|^2 \, d\Omega$$

with $k_{min} > 0$, and hence we conclude that $\nabla(u_1 - u_2) = 0$ on Ω. Thus, $u_1 - u_2$ is a constant function over each connected component of the open subset Ω. Moreover, as we know that $(u_1 - u_2)_{|\partial\Omega} = 0$, the constant vanishes and therefore $u_1 = u_2$.

d) From the above, if u is a solution of [P14$_b$], then u satisfies [F14$_b$]. The converse is very straightforward to establish. Indeed, since u is a solution of [F14$_b$] that belongs to $H_0^1(\Omega)$, we have $u_{|\partial\Omega} = 0$. Then, we choose a test function v of $\mathcal{D}(\Omega)$ in [F14$_b$] and, by reasoning conversely to the first question, we find that $-\operatorname{div}(k\nabla u) = f$ in the distribution sense. Since f belongs to $L^2(\Omega)$, $\operatorname{div}(k\nabla u) \in L^2(\Omega)$ and finally $-\operatorname{div}(k\nabla u) = f$ almost everywhere on Ω. To conclude, note that $u^\star = u + \tau_0$ satisfies [P14].

Solution 15. Problem with a discontinuous positive coefficient

Denote $\ell(v) = \displaystyle\int_\Sigma g v \, d\Sigma$ for $v \in H_0^1(\Omega)$.

a) First, since $u \in H_0^1(\Omega)$, we know that $u \in H^1(\Omega)$ and $u = 0$ on $\partial\Omega$.

b) Let $\phi_i \in \mathcal{D}(\Omega_i)$ where $i = 1, 2$, and extend it by 0 to Ω: its extension, denoted by ϕ, belongs to $\mathcal{D}(\Omega)$ from exercise 11. As $\mathcal{D}(\Omega) \subset H_0^1(\Omega)$, we can use it as a test function in [F15], and we have that $\ell(\phi) = 0$, that is:

$$0 = \int_\Omega \kappa \nabla u \cdot \nabla \phi \, d\Omega = \kappa_i \int_{\Omega_i} (\nabla u)_{|\Omega_i} \cdot \nabla \phi_i \, d\Omega_i = \kappa_i \langle \nabla(u_{|\Omega_i}), \nabla \phi_i \rangle$$
$$= -\kappa_i \langle \Delta(u_{|\Omega_i}), \phi_i \rangle.$$

REMARK A2.8.– *To justify the equality* $(\nabla u)_{|\Omega_i} = \nabla(u_{|\Omega_i})$*, see exercise 11 (Question (c)).*

In what follows, denote $u_i = u_{|\Omega_i}$ for $i = 1, 2$. As $u \in H_0^1(\Omega)$, we have $u_1 \in H^1(\Omega_1)$ and $u_2 \in H^1(\Omega_2)$; moreover, we get $u_{2|\partial\Omega} = 0$ since $\partial\Omega \subset \partial\Omega_2$, as well as $u_{1|\Sigma} = u_{2|\Sigma}$ (once again, see exercise 11). Finally, from the above, $\Delta u_i = 0$ in the distribution sense, namely in $\mathcal{D}'(\Omega_i)$; in particular, $\Delta u_i = 0$ almost everywhere.

c) To conclude, take $v \in H_0^1(\Omega)$ and integrate [1.39] by parts over Ω_1 and over Ω_2 ($v_i = v_{|\Omega_i}$):

$$\ell(v) = \int_\Omega \kappa \nabla u \cdot \nabla v \, d\Omega = \kappa_1 \int_{\Omega_1} \nabla u_1 \cdot \nabla v_1 \, d\Omega_1 + \kappa_2 \int_{\Omega_2} \nabla u_2 \cdot \nabla v_2 \, d\Omega_2$$
$$= \kappa_1 \Big(\langle \nabla u_1 \cdot \mathbf{n}_1, v_1 \rangle_{(H^{1/2}(\partial\Omega_1))', H^{1/2}(\partial\Omega_1)} - \int_{\Omega_1} \Delta u_1 \, v_1 \, d\Omega_1 \Big)$$
$$+ \kappa_2 \Big(\langle \nabla u_2 \cdot \mathbf{n}_2, v_2 \rangle_{(H^{1/2}(\partial\Omega_2))', H^{1/2}(\partial\Omega_2)} - \int_{\Omega_2} \Delta u_2 \, v_2 \, d\Omega_2 \Big)$$
$$= \kappa_1 \langle \nabla u_1 . \mathbf{n}_1, v_1 \rangle_{(H^{1/2}(\partial\Omega_1))', H^{1/2}(\partial\Omega_1)} + \kappa_2 \langle \nabla u_2 . \mathbf{n}_2, v_2 \rangle_{(H^{1/2}(\partial\Omega_2))', H^{1/2}(\partial\Omega_2)}.$$

By definition, v vanishes along the boundary $\partial\Omega$. Thus, the support of $v_{|\partial\Omega_i}$ is included in the interface, Σ, for $i = 1, 2$. In this case, we can therefore replace the duality brackets (between the spaces defined) on $\partial\Omega_i$ with the duality brackets (between the spaces defined) on Σ. The similar is followed for the left-hand side. Finally, note that $\mathbf{n}_1 = -\mathbf{n}_2$ on Σ. It therefore becomes:

$$\langle g, v_{|\Sigma} \rangle_{(H_\Sigma)', H_\Sigma} = \langle (\kappa_1 \nabla u_1 - \kappa_2 \nabla u_2) \cdot \mathbf{n}_1, v_{|\Sigma} \rangle_{(H_\Sigma)', H_\Sigma}, \quad \forall v \in H_0^1(\Omega)$$

where we denote by H_Σ the space of traces on Σ of the elements in $H_0^1(\Omega)$. Since $\Sigma = \partial\Omega_1$, we have $H^{1/2}(\Sigma) = H^{1/2}(\partial\Omega_1)$ by definition, and we observe that H_Σ is equal to $H^{1/2}(\Sigma)$. By the surjectivity of the trace mapping, we find:

$$(\kappa_1 \nabla u_1 - \kappa_2 \nabla u_2) \cdot \mathbf{n}_1 = g \text{ in } \left(H^{1/2}(\Sigma) \right)'.$$

As $g \in L^2(\Sigma)$, we see finally that:

$$(\kappa_1 \nabla u_1 - \kappa_2 \nabla u_2) \cdot \mathbf{n}_1 = g \text{ almost everywhere on } \Sigma.$$

Problem [F15] can therefore be reinterpreted in the terms of a partial differential equation with the following boundary and transmission conditions:

$$\begin{cases} \text{Find } (u_1, u_2) \in H^1(\Omega_1) \times H^1(\Omega_2) & \text{such that} \\ \Delta u_1 = 0 & \text{in } \Omega_1 \\ \Delta u_2 = 0 & \text{in } \Omega_2 \\ u_2 = 0 & \text{on } \partial\Omega \\ u_1 = u_2 \text{ and } (\kappa_1 \nabla u_1 - \kappa_2 \nabla u_2) \cdot \mathbf{n}_1 = g & \text{on } \Sigma. \end{cases}$$

Solution 16. An elementary study of coercivity

a) Let $H = \mathbb{R}^p$, and let $\mathbb{S}$ be the unit sphere in H: $\mathbb{S} = \{u \in H \text{ such that } \|u\|_H = 1\}$. Since $\mathbb{S}$ is closed and bounded in $\mathbb{R}^p$ (which is a finite-dimensional vector space), we see that $\mathbb{S}$ is compact. Denote by ℓ_a the following function:

$$\ell_a : \left| \begin{array}{l} \mathbb{S} \to \mathbb{R} \\ u \mapsto a(u, u). \end{array} \right.$$

The function ℓ_a is continuous on $\mathbb{S}$, which is compact: ℓ_a therefore attains both its maximum and its minimum on $\mathbb{S}$. By reasoning as in the proof of lemma 2.1, we see that they both have the same sign. Without loss of generality, we will assume then that they are both strictly positive. Let α denote the minimum: $\exists u_{min} \in \mathbb{S}$ such that $\ell_a(u_{min}) = \alpha$.

Pick an arbitrary $v \in H \setminus \{0\}$. Set $u = v/\|v\|_H$: then $u \in \mathbb{S}$ and we have $\ell_a(u) \geq \alpha$ which, using bilinearity, we can rewrite as:

$$a(v, v) \geq \alpha \|v\|_H^2.$$

From this, we observe that a is coercive.

b) In infinite dimensions, a closed and bounded subset is no longer necessarily compact. We cannot therefore reason as we did previously. Moreover, we have the following counterexample: let $(e_k)_{k \geq 1}$ be a Hilbert basis for H (which we assume to be separable, guaranteeing the existence of such a basis). We may write any element v in H in the form:

$$v = \sum_1^\infty v_k e_k, \; v_k = (v, e_k)_H \; \forall k \geq 1, \text{ and } \|v\|_H = \Big(\sum_1^\infty v_k^2\Big)^{1/2}.$$

To define a continuous bilinear form, it suffices to describe how it operates on any pair of basis elements. We set:

$$a_\star(e_i, e_j) = \delta_{i,j}\frac{1}{j}, \ \forall i, j \geq 1.$$

From this, we see that:

$$a_\star(v, v) = \sum_1^\infty \frac{v_k^2}{k}.$$

By construction, $a_\star$ is therefore positive-definite. On the other hand, it is not coercive due to the fact that $\lim_{k\to+\infty} a_\star(e_k, e_k) = 0$.

REMARK A2.9.– *In an infinite-dimensional separable Hilbert space, the unit sphere* $\mathbb{S}$ *is never compact. Otherwise, the reasoning used in Question (a) could be applied and this would imply that the positive-definite form* $a_\star$ *is coercive.*

Solution 17. Variational formulation with a symmetric form

a) Consider the functional:

$$J : \left| \begin{array}{l} H \rightarrow \mathbb{R} \\ v \ \mapsto \dfrac{1}{2}a(v, v) - \ell(v). \end{array} \right.$$

Using the fact that a is a symmetric bilinear form, for $v \in H$ we have:

$$J(u+v) = \frac{1}{2}a(u+v, u+v) - \ell(u+v) = \frac{1}{2}a(u, u) - \ell(u) + a(u, v) - \ell(v) + \frac{1}{2}a(v, v).$$

If u is a solution of [F17], then:

$$J(u + v) = J(u) + \frac{1}{2}a(v, v). \qquad \text{[J17]}$$

As a is positive-definite, we have that $J(u + v) > J(u)$, $\forall v \in H \setminus \{0\}$, that is u is a (strict) minimum of J. Conversely, if u is a minimum of J, then $J(u) \leq J(u + \alpha v)$, $\forall v \in H$, $\forall \alpha \in \mathbb{R}$, that is:

$$0 \leq J(u + \alpha v) - J(u) = \alpha \left(a(u, v) - \ell(v) + \frac{\alpha}{2}a(v, v) \right), \ \forall v \in H, \ \forall \alpha \in \mathbb{R}$$

which shows that:

$$a(u,v) - \ell(v) + \frac{\alpha}{2}a(v,v) \geq 0 \quad \forall\alpha > 0.$$

By letting α tend toward 0, we see that $a(u,v) - \ell(v) \geq 0$, and this result holds for every $v \in H$. Reasoning now with $-v$ in the place of v, we get $a(u,v) - \ell(v) \leq 0$, thus proving that u is a solution of [F17].

b) Let u_1 and u_2 be two minima. According to [J17], we have:

$$J(u_1) = J(u_2) = J(u_1 + u_2 - u_1) = J(u_1) + \frac{1}{2}a(u_1 - u_2, u_2 - u_1)$$

and so $a(u_1 - u_2, u_2 - u_1) = 0$; this implies that $u_1 = u_2$.

c) Now let us assume that a is continuous and coercive on $H \times H$, and ℓ is continuous on H. From the Lax–Milgram theorem, the problem [F17] admits a unique solution which, according to the above, is a minimum of J.

Solution 18. Spectral corollary: an application of the spectral theorem

Recall that we say the injection from H into L is:

– *continuous*, which means that: $\exists C_{H\to L} > 0, \forall v \in H, \|v\|_L \leq C_{H\to L} \|v\|_H$;

– *compact*, which means that: for every bounded sequence $(v_k)_k$ in H, we can extract a subsequence that converges in L, that is, $\exists (v_{k'})_{k'}$ a subsequence, $\exists v_L \in L$, with $\lim_{k'\to\infty} \|v_{k'} - v_L\|_L = 0$.

a) Let $u^* \in L$, then the linear form $v \mapsto (u^*, v)_L$ is continuous on H. From the Lax–Milgram theorem, since a_+ is coercive there exists a unique solution to [F18$_b$]. Furthermore, $\|T_+u^*\|_H \leq C\|u^*\|_L$, where $C > 0$ is independent of u^* (it suffices to use the coercivity of a_+, in conjunction with the continuous injection $H \subset L$). Therefore, we can define the linear mapping $T_+ \in \mathcal{L}(L, H)$. Set $T = i_{H\to L} \circ T_+ \in \mathcal{L}(L)$. As $i_{H\to L}$ is compact, T is also *compact*.

T is *positive* because, for every $u^* \in L$, from [F18$_b$] we have:

$$(u^*, Tu^*)_L = (u^*, T_+u^*)_L = a_+(T_+u^*, T_+u^*) \geq 0.$$

Let $u^* \in L$ be such that $Tu^* = 0$. By the definition of T, we have $T_+u^* = 0$ and, from [F18$_b$], we get that:

$$(u^*, v)_L = 0 \quad \forall v \in H.$$

From the density of H in L, the same equality holds for every $v \in L$, and therefore $u^* = 0$: T is *injective*.

T is *self-adjoint* since, for every $u^*, v^* \in L$, by the symmetry of a_+ and using the definitions of T_+u^* and of T_+v^*, we get:

$$\begin{aligned}(u^*, Tv^*)_L &= (u^*, T_+v^*)_L = a_+(T_+u^*, T_+v^*) = a_+(T_+v^*, T_+u^*) \\ &= (v^*, T_+u^*)_L = (u^*, Tv^*)_L.\end{aligned}$$

Every hypothesis for the spectral theorem 1.5 is satisfied: there exists a Hilbert basis $(e_n)_{n\geq 1}$ for L that is generated by the eigenvectors of T. Let us denote their corresponding eigenvalues by $(\mu_n)_{n\geq 1}$, which are all strictly positive and have finite multiplicity. Additionally, $\lim_{n\to+\infty} \mu_n = 0$.

To say that $(e_n)_{n\geq 1}$ is a Hilbert basis of L means that:

– for each $n, m \geq 1$, we have $(e_n, e_m)_L = \delta_{nm}$;

– we can represent every element u of L as a sum with respect to the $(e_n)_{n\geq 1}$. More precisely, in denoting $a_n = (u, e_n)_L$ for $n \geq 1$, then we have $u = \sum_{n\geq 1} a_n e_n$ in L, which means that if we introduce $\tilde{u}_N = \sum_{n=1,N} a_n e_n$ for $N \geq 1$, we get:

$$\lim_{N\to\infty} \|u - \tilde{u}_N\|_L = 0 \text{ and } \|u\|_L = \Big(\sum_{n\geq 1} a_n^2\Big)^{1/2}.$$

Finally, if $(a_n)_{n\geq 1} \in \ell^2$, then $u = \sum_{n\geq 1} a_n\, e_n$ belongs to L and $\|u\|_L^2 = \sum_{n\geq 1} a_n^2$.

b) By the construction of T, note that for each $n \geq 1$, we have $e_n \in H$; indeed, we see that:

$$\begin{aligned}Te_n = \mu_n e_n \text{ in } L &\Longrightarrow i_{H\to L}(T_+e_n) = \mu_n e_n \text{ in } L \\ &\Longrightarrow T_+e_n = \mu_n e_n \text{ in } H \\ &\Longrightarrow e_n = (\mu_n)^{-1} T_+ e_n \text{ in } H.\end{aligned}$$

Now let us set $\lambda_n = (\mu_n)^{-1}$. By definition, the sequence $(\lambda_n)_{n\geq 1}$ consists of strictly positive real numbers with $\lim_{n\to\infty} \lambda_n = \infty$. And, for each $n \geq 1$, we have $e_n = \lambda_n T_+ e_n$. From the definition of T_+, this means that:

$$a_+(e_n, v) = \lambda_n (e_n, v)_L \quad \forall n \geq 1, \quad \forall v \in H.$$

Thus, $(e_n)_{n\geq 1}$ does indeed consist of the eigenfunctions from the problem [F18], with corresponding eigenvalues $(\lambda_n)_{n\geq 1}$. Conversely, if $(e, \lambda) \in H \times \mathbb{R}$ is a solution of [F18], then $\lambda > 0$ and we easily can check that (e, λ^{-1}) is an eigenvector-eigenvalue pair of T.

c) For each $n, m \geq 1$, we have $a_+(e_n, e_m) = \lambda_n (e_n, e_m)_L = \delta_{nm} \lambda_n$. Consequently, if we denote by $\alpha > 0$ the coercivity constant of the form a_+, then for $N < M$ we get:

$$\alpha \, \|\tilde{u}_N - \tilde{u}_M\|_H^2 \leq a_+(\tilde{u}_N - \tilde{u}_M, \tilde{u}_N - \tilde{u}_M) = \sum_{n=N+1,M} \lambda_n \, a_n^2.$$

Therefore, $(\tilde{u}_N)_{N \geq 1}$ is *Cauchy in H* provided that $\sum \lambda_n \, a_n^2$ converges. However, we have:

$$\sum_{n=1,N} \lambda_n \, a_n^2 = a_+(\tilde{u}_N, \tilde{u}_N), \quad \forall N \geq 1. \qquad \text{[E18]}$$

It is therefore necessary to show that the sequence $(a_+(\tilde{u}_N, \tilde{u}_N))_{N \geq 1}$, which from the above is increasing, is also bounded, so we can conclude that it converges. Since $u \in H$, $a_+(u, u)$ exists, and we have:

$$\begin{aligned} a_+(u,u) - a_+(\tilde{u}_N, \tilde{u}_N) &= a_+(u - \tilde{u}_N, u) + a_+(\tilde{u}_N, u) - a_+(\tilde{u}_N, \tilde{u}_N) \\ &= a_+(u - \tilde{u}_N, u - \tilde{u}_N) + a_+(u - \tilde{u}_N, \tilde{u}_N) + a_+(\tilde{u}_N, u) - a_+(\tilde{u}_N, \tilde{u}_N) \\ &= a_+(u - \tilde{u}_N, u - \tilde{u}_N) + 2a_+(\tilde{u}_N, u - \tilde{u}_N) \end{aligned}.$$

where the last line follows from the symmetry of the form a_+.

Moreover, $\tilde{u}_N = \sum_{n=1,N} a_n e_n$ and we have $a_+(e_n, u - \tilde{u}_N) = \lambda_n (e_n, u - \tilde{u}_N)_L = 0$ for each $1 \leq n \leq N$, since $u - \tilde{u}_N = \sum_{m>N} (u, e_m)_L e_m$ and also that the $(e_n)_{1 \leq n \leq N}$ are orthogonal in L with respect to the $(e_m)_{m>N}$. So $a_+(\tilde{u}_N, u - \tilde{u}_N) = 0$.

As a_+ is coercive, ($a_+(u - \tilde{u}_N, u - \tilde{u}_N) \geq 0$), and therefore we have $a_+(u, u) \geq a_+(\tilde{u}_N, \tilde{u}_N)$; from this, we may conclude that the sequence $(a_+(\tilde{u}_N, \tilde{u}_N))_{N \geq 1}$ converges. By the identity in [E18], the same can be said for the series $\sum \lambda_n \, a_n^2$, and hence for the sequence $(\tilde{u}_N)_{N \geq 1}$ is indeed Cauchy in H. As H is a Hilbert space, it converges. Let us denote its limit by $v \in H$: $\lim_{N \to \infty} \|\tilde{u}_N - v\|_H = 0$. For each $N \geq 1$, we have:

$$\|u - v\|_L \leq \|u - \tilde{u}_N\|_L + \|\tilde{u}_N - v\|_L \leq \|u - \tilde{u}_N\|_L + C_{H \to L} \, \|\tilde{u}_N - v\|_H.$$

If we pass to the limit ($N \to \infty$), we end up with $u = v$, and so $(\tilde{u}_N)_{N \geq 1}$ indeed converges to u in H. Finally, as $a_+(\tilde{u}_N, \tilde{u}_N) = a_+(u, u) - a_+(u - \tilde{u}_N, u - \tilde{u}_N)$ and also as $\lim_{N\to\infty} a_+(u - \tilde{u}_N, u - \tilde{u}_N) = 0$, from [E18] and by the continuity of the form a_+, we have:

$$a_+(u,u) = \lim_{N\to\infty} a_+(\tilde{u}_N, \tilde{u}_N) = \sum_{n\geq 1} \lambda_n\, a_n^2. \qquad \text{[E18}_b\text{]}$$

d) First, it is easy to check that $(u,v) \mapsto a_+(u,v)$ is an inner product on H. Next, define $f_n = (\lambda_n)^{-1/2} e_n$ for each $n \geq 1$. Then, for each $n, m \geq 1$:

$$a_+(f_n, f_m) = \frac{1}{\sqrt{\lambda_n\,\lambda_m}}\, a_+(e_n, e_m) = \frac{\lambda_n}{\sqrt{\lambda_n\,\lambda_m}}\, (e_n, e_m)_L = \frac{\lambda_n}{\sqrt{\lambda_n\,\lambda_m}}\, \delta_{nm} = \delta_{nm}.$$

Let $u \in H$, and now define $u_N^\star = \sum_{n=1,N} a_+(u, f_n) f_n$ for each $N \geq 1$. By the definition of $(f_n)_{n\geq 1}$ and since a_+ is symmetric, we have:

$$u_N^\star = \sum_{n=1,N} \frac{1}{\lambda_n} a_+(u, e_n) e_n = \sum_{n=1,N} a_n e_n = \tilde{u}_N, \ \forall N \geq 1.$$

From the previous question, we have $\lim_{N\to\infty} \|u_N^\star - u\|_H = 0$: $(f_n)_{n\geq 1}$ is indeed a Hilbert basis of H. The result for the corresponding norm to a_+ is an immediate consequence of [E18$_b$].

Solution 19. Solving the Laplacian with non-homogeneous conditions

a) Consider the problem with the inhomogeneous Fourier condition:

$$\begin{cases} \text{Find } u \in H^1(\Omega) & \text{such that} \\ -\Delta u = f & \text{in } \Omega \\ \dfrac{\partial u}{\partial n} + \lambda\, u = g & \text{on } \partial\Omega \end{cases} \qquad \text{[P19}_a\text{]}$$

with $\lambda > 0$, $f \in L^2(\Omega)$, and $g \in L^2(\partial\Omega)$. From Proposition 2.3, this problem is equivalent to the following variational formulation:

$$\left|\begin{array}{l} \text{Find } u \in H^1(\Omega) \text{ such that} \\ \displaystyle\int_\Omega \nabla u \cdot \nabla v \, d\Omega + \lambda \int_{\partial\Omega} uv \, d\Gamma = \int_\Omega fv \, d\Omega + \int_{\partial\Omega} gv \, d\Gamma, \quad \forall v \in H^1(\Omega). \end{array}\right. \qquad \text{[F19}_a\text{]}$$

The form $\ell(v) = \displaystyle\int_\Omega fv\,d\Omega + \int_{\partial\Omega} gv\,d\Gamma$ is linear. For the continuity, note that:

$$\begin{aligned} \forall v \in H^1(\Omega), \quad |\ell(v)| &\leq \left|\int_\Omega fv\,d\Omega\right| + \left|\int_{\partial\Omega} gv\,d\Gamma\right| \\ &\leq \|f\|_{L^2(\Omega)} \|v\|_{H^1(\Omega)} + \|g\|_{L^2(\partial\Omega)} \|v_{|\partial\Omega}\|_{L^2(\partial\Omega)}. \end{aligned}$$

From the continuity of the trace mapping from theorem 1.7: there exists a $C > 0$ such that:

$$\|v_{|\partial\Omega}\|_{L^2(\partial\Omega)} \leq C\, \|v\|_{H^1(\Omega)} \quad \forall v \in H^1(\Omega)$$

from which we deduce the continuity of the linear form:

$$|\ell(v)| \leq \left(\|f\|_{L^2(\Omega)} + C\,\|g\|_{L^2(\partial\Omega)}\right) \|v\|_{H^1(\Omega)} \quad \forall v \in H^1(\Omega).$$

The form $a(u,v) = \int_\Omega \nabla u \cdot \nabla v \, d\Omega + \lambda \int_{\partial\Omega} uv \, d\Gamma$ is bilinear. Continuity arises from the following estimate:

$$|a(u,v)| \leq \|u\|_{H^1(\Omega)}\|v\|_{H^1(\Omega)} + \lambda\,\|u_{|\partial\Omega}\|_{L^2(\partial\Omega)}\|v_{|\partial\Omega}\|_{L^2(\partial\Omega)} \quad \forall u, v \in H^1(\Omega).$$

Again, the continuity of the trace mapping is used to finally obtain:

$$|a(u,v)| \leq (1 + \lambda\, C^2)\|u\|_{H^1(\Omega)}\|v\|_{H^1(\Omega)} \quad \forall u, v \in H^1(\Omega).$$

Only the coercivity of a remains to be shown. Note that:

$$\begin{aligned} a(u,u) &= \int_\Omega |\nabla u|^2 \, d\Omega + \lambda \int_{\partial\Omega} u^2 \, d\Gamma \\ &= \|\nabla u\|^2_{(L^2(\Omega))^d} + \lambda\,\|u_{|\partial\Omega}\|^2_{L^2(\partial\Omega)} \quad \forall u \in H^1(\Omega). \end{aligned}$$

From exercise 8 (Question (a), with $\Gamma = \partial\Omega$), we have:

$$\frac{1}{C_1}\|u\|^2_{L^2(\Omega)} \leq \left(\|\nabla u\|^2_{(L^2(\Omega))^d} + \|u_{|\partial\Omega}\|^2_{L^2(\partial\Omega)}\right)$$

where $C_1 > 0$ is independent of u, that is, for every $u \in H^1(\Omega)$:

$$\begin{aligned} 2a(u,u) &\geq \|\nabla u\|^2_{(L^2(\Omega))^d} + \min(\lambda, 1)\left(\|\nabla u\|^2_{(L^2(\Omega))^d} + \|u_{|\partial\Omega}\|^2_{L^2(\partial\Omega)}\right) \\ &\geq \|\nabla u\|^2_{(L^2(\Omega))^d} + \frac{\min(\lambda,1)}{C_1}\|u\|^2_{L^2(\Omega)} \geq \min(1, \frac{\min(\lambda,1)}{C_1})\|u\|^2_{H^1(\Omega)}. \end{aligned}$$

Here, the Lax–Milgram theorem can be applied, asserting the existence and uniqueness for the solution of [F19$_a$] and hence of [P19$_a$].

b) Consider the following problem with the inhomogeneous Neumann condition ($f \in L^2(\Omega)$ and $g \in L^2(\partial\Omega)$):

$$\begin{cases} \text{Find } u \in H^1(\Omega) & \text{such that} \\ -\Delta u = f & \text{in } \Omega \\ \dfrac{\partial u}{\partial n} = g & \text{on } \partial\Omega. \end{cases} \qquad [\text{P19}_b]$$

Note that, if a solution exists, then it cannot be unique. Indeed, we can easily see that, if u is a solution, then $u + c$ (with $c \in \mathbb{R}$) is also a solution.

Let us introduce the vanishing mean condition to the space:

$$V = \{v \in H^1(\Omega) \text{ such that } \int_\Omega v \, d\Omega = 0\}$$

so that we can eliminate any constant. The problem then becomes:

$$\begin{cases} \text{Find } u \in V & \text{such that} \\ -\Delta u = f & \text{in } \Omega \\ \dfrac{\partial u}{\partial n} = g & \text{on } \partial\Omega. \end{cases} \qquad [\text{P19}_c]$$

Moreover, by integrating the first equation over Ω before then integrating by parts [1.39]:

$$\begin{aligned} \int_\Omega f \, d\Omega &= - \int_\Omega \Delta u \, 1 \, d\Omega \\ &= \int_\Omega \nabla u \cdot \nabla 1 \, d\Omega - \langle \frac{\partial u}{\partial n}, 1 \rangle_{(H^{1/2}(\partial\Omega))', H^{1/2}(\partial\Omega)} = - \int_{\partial\Omega} g \, d\Gamma \end{aligned}$$

which highlights the following *necessary condition* for the existence of a solution:

$$\int_\Omega f \, d\Omega + \int_{\partial\Omega} g \, d\Gamma = 0. \qquad [\text{C19}_c]$$

We will assume that this condition is satisfied in the following. Let us check (as usual) that if u is a solution of the problem [P19$_c$], then it is also a solution of the variational formulation:

$$\left| \begin{array}{l} \text{Find } u \in V \text{ such that} \\ \displaystyle\int_\Omega \nabla u \cdot \nabla v \, d\Omega = \int_\Omega f v \, d\Omega + \int_{\partial\Omega} g v \, d\Gamma \quad \forall v \in V. \end{array} \right. \qquad [\text{F19}_c]$$

Conversely, if u is a solution of [F19$_c$], we will want to reason in the sense of distributions, and return to the original problem. But *one has to be cautious*. Indeed, a test function $v \in \mathcal{D}(\Omega)$ does not necessarily belong to V, since its mean may not necessarily equal 0: the idea is to use $v - m(v) \in V$ instead, where $m(v) = \int_\Omega v\, d\Omega \,/\text{meas}\,\Omega$ is the mean of v. For every $v \in \mathcal{D}(\Omega)$, we then write:

$$\begin{aligned}
\langle f, v\rangle &= \int_\Omega f v\, d\Omega + \int_{\partial\Omega} g v\, d\Gamma \quad \text{since } v_{|\partial\Omega} = 0 \\
&= \int_\Omega f\,(v - m(v))\; d\Omega + m(v) \int_\Omega f\, d\Omega + \int_{\partial\Omega} g v\, d\Gamma \\
&= \int_\Omega f\,(v - m(v))\; d\Omega - m(v) \int_{\partial\Omega} g\, d\Gamma + \int_{\partial\Omega} g v\, d\Gamma \quad \text{from [C19}_c\text{]} \\
&= \int_\Omega f\,(v - m(v))\; d\Omega + \int_{\partial\Omega} g\,(v - m(v))\; d\Gamma \\
&= \int_\Omega \nabla u \cdot \nabla\,(v - m(v))\; d\Omega \quad \text{as } u \text{ is a solution of [F19}_c\text{], and } v - m(v) \in V \\
&= \int_\Omega \nabla u \cdot \nabla v\, d\Omega = \langle -\Delta u, v\rangle.
\end{aligned}$$

We have therefore that $-\Delta u = f$ in the distribution sense and, since f is in $L^2(\Omega)$, this equality holds true almost everywhere.

Now let us recover the boundary conditions; for $v \in H^1(\Omega)$ (and in noting that $v - m(v) \in V$), we write:

$$\begin{aligned}
\int_{\partial\Omega} g v\, d\Gamma &= \int_{\partial\Omega} g\,(v - m(v))\; d\Gamma + m(v) \int_{\partial\Omega} g\, d\Gamma \\
&= \int_{\partial\Omega} g\,(v - m(v))\; d\Gamma - m(v) \int_\Omega f\, d\Omega \quad \text{from [C19}_c\text{]} \\
&= \int_\Omega \nabla u \cdot \nabla\,(v - m(v))\; d\Omega - \int_\Omega f\,(v - m(v))\; d\Omega \\
&\quad - m(v) \int_\Omega f\, d\Omega \qquad \text{as } u \text{ is a solution of [F19}_c\text{] and } v - m(v) \in V \\
&= \int_\Omega \nabla u \cdot \nabla v\, d\Omega - \int_\Omega f v\, d\Omega \\
&= \int_\Omega (-\Delta u - f) v\, d\Omega + \langle \frac{\partial u}{\partial n}, v\rangle_{(H^{1/2}(\partial\Omega))', H^{1/2}(\partial\Omega)} \\
&= \langle \frac{\partial u}{\partial n}, v\rangle_{(H^{1/2}(\partial\Omega))', H^{1/2}(\partial\Omega)}.
\end{aligned}$$

Above, we have integrated [1.39] by parts to obtain the penultimate equality. With this being true for all $v \in H^1(\Omega)$, we obtain the Neumann boundary condition on $\partial\Omega$, as usual.

It remains to be seen whether [F19$_c$] is well posed. Our goal is to see whether the hypotheses of the Lax–Milgram theorem are satisfied or not. Endow the space V with the $H^1(\Omega)$ norm. As a closed vector subspace of $H^1(\Omega)$, V is indeed a Hilbert space. Using the same arguments as in the case with the Fourier boundary condition, we will show that the linear form $\ell(v) = \int_\Omega fv\, d\Omega + \int_{\partial\Omega} gv\, d\Gamma$ is continuous. Using the same arguments as in the Dirichlet boundary condition case, $a(u,v) = \int_\Omega \nabla u \cdot \nabla v\, d\Omega$ is bilinear and continuous. As for coercivity, it suffices to use the equivalence between the $|\cdot|_{1,\Omega}$ semi-norm and the $H^1(\Omega)$ norm in the space V, which had been proven in exercise 8 (Question (b)). Finally, by the virtue of the Lax–Milgram theorem and by assuming that [C19$_c$] is satisfied, the variational formulation [F19$_c$] and hence the problem [P19$_c$] are well posed.

Solution 20. Dirichlet–Neumann or Dirichlet–Fourier mixed boundary condition problems

a) To get the variational formulation, we place ourselves in the following vector subspace of $H^1(\Omega)$:

$$H^1_{0,\Gamma_1}(\Omega) = \{v \in H^1(\Omega) \text{ such that } v_{|\Gamma_1} = 0\}$$

which includes the homogeneous Dirichlet boundary condition on Γ_1, because $u_{|\Gamma_1} = 0$ is an essential boundary condition. Let $v \in H^1_{0,\Gamma_1}(\Omega)$, and then multiply the first equation of [P20] by v, and then integrate over Ω before then integrating by parts [1.39] over Ω:

$$\begin{aligned}
\int_\Omega fv\, d\Omega &= -\int_\Omega \Delta u\, v\, d\Omega = \int_\Omega \nabla u \cdot \nabla v\, d\Omega - \langle \frac{\partial u}{\partial n}, v\rangle_{(H^{1/2}(\partial\Omega))', H^{1/2}(\partial\Omega)} \\
&= \int_\Omega \nabla u \cdot \nabla v\, d\Omega - \int_{\Gamma_2} \frac{\partial u}{\partial n} v\, d\Gamma \qquad (\text{ since } \frac{\partial u}{\partial n}_{|\Gamma_2} \in L^2(\Gamma_2)) \\
&= \int_\Omega \nabla u \cdot \nabla v\, d\Omega + \lambda \int_{\Gamma_2} uv\, d\Gamma - \int_{\Gamma_2} gv\, d\Gamma.
\end{aligned}$$

We end up with the following variational formulation:

Find $u \in H^1_{0,\Gamma_1}(\Omega)$ such that

$$\int_\Omega \nabla u \cdot \nabla v\, d\Omega + \lambda \int_{\Gamma_2} uv\, d\Gamma = \int_\Omega fv\, d\Omega + \int_{\Gamma_2} gv\, d\Gamma \quad \forall v \in H^1_{0,\Gamma_1}(\Omega). \qquad \text{[F20]}$$

b) In analyzing [F20], we see that:

- $V = H^1_{0,\Gamma_1}(\Omega)$ is a closed vector subspace of $H^1(\Omega)$, since the mapping $\gamma_{\Gamma_1} : H^1(\Omega) \to L^2(\Gamma_1)$ is continuous: it is therefore a Hilbert space. We can endow it with the inner product along with the norm of $H^1(\Omega)$;

- $a(u,v) = \int_\Omega \nabla u \cdot \nabla v \, d\Omega$ is a continuous bilinear form on V. Moreover, according to exercise 8 (Question (a)), it is coercive on V:

$$\exists C_1 > 0\,, \forall v \in H^1_{0,\Gamma_1}(\Omega), \int_\Omega |\nabla v|^2 \, d\Omega \geq C_1 \, \|v\|^2_{H^1(\Omega)}.$$

- $\ell(v) = \int_\Omega f v \, d\Omega + \int_{\Gamma_2} g v \, d\Gamma$ is a continuous linear form on V (see exercise 19, Question (a)).

According to the Lax–Milgram theorem, the variational formulation [F20] is well posed.

c) Let u be the solution of the variational formulation. As $u \in H^1_{0,\Gamma_1}(\Omega)$, we have that $u_{|\Gamma_1} = 0$. Then, given v in $\mathcal{D}(\Omega)$ ($\subset H^1_{0,\Gamma_1}(\Omega)$) and differentiating in the distribution sense, we get:

$$\begin{aligned} \ell(v) &= \langle f, v \rangle\,; \\ a(u,v) &= \int_\Omega \nabla u \cdot \nabla v \, d\Omega = \sum_{i=1}^{\mathrm{d}} \int_\Omega \frac{\partial u}{\partial x_i} \frac{\partial v}{\partial x_i} \, d\Omega = \sum_{i=1}^{\mathrm{d}} \langle \frac{\partial u}{\partial x_i}, \frac{\partial v}{\partial x_i} \rangle \\ &= -\sum_{i=1}^{\mathrm{d}} \langle \frac{\partial^2 u}{\partial x_i^2}, v \rangle = -\langle \Delta u, v \rangle.. \end{aligned}$$

Thus, $-\Delta u = f$ in $\mathcal{D}'(\Omega)$. And, since $f \in L^2(\Omega)$, we have $-\Delta u = f$ almost everywhere in Ω. To conclude, pick v in $H^1_{0,\Gamma_1}(\Omega)$ and then integrate by parts. We get:

$$\int_{\Gamma_2} g v \, d\Gamma = \lambda \int_{\Gamma_2} u v \, d\Gamma + \langle \frac{\partial u}{\partial n}, v_{|\Gamma_2} \rangle_{(H_{\Gamma_2})', H_{\Gamma_2}}$$

where we denote by H_{Γ_2} the space of traces on Γ_2 of elements in $H^1_{0,\Gamma_1}(\Omega)$. As H_{Γ_2} is a dense subspace of $L^2(\Gamma_2)$, the last relation follows:

$$\left(\lambda u + \frac{\partial u}{\partial n} \right)_{|\Gamma_2} = g.$$

It holds almost everywhere on Γ_2 since $g \in L^2(\Gamma_2)$. Thus, we have found all the equations of [P20].

Solution 21. Regularity of the solutions to Laplace's equation

a) We have that u_s is measurable and, moreover:

$$\int_\Omega u_s^2\,d\Omega = \int_{r=0}^{r=1}\int_{\theta=0}^{\theta=\frac{3}{2}\pi} r^{-4/3}\sin^2(\frac{2}{3}\theta)\,rdrd\theta$$
$$= \int_0^1 r^{-1/3}\,dr \times \int_0^{\frac{3}{2}\pi}\sin^2(\frac{2}{3}\theta)\,d\theta.$$

We find then $\int_0^1 r^{-1/3}\,dr = \frac{3}{2}$, and $\int_0^{\frac{3}{2}\pi}\sin^2(\frac{2}{3}\theta)\,d\theta = \frac{3}{4}\pi$, then finally:

$$\int_\Omega u_s^2\,d\Omega = \frac{9}{8}\pi, \text{ that is to say } u_s \in L^2(\Omega) \text{ with } \|u_s\|_{L^2(\Omega)} = \sqrt{\frac{9}{8}\pi}.$$

b) When $r > 0$, we have $\frac{\partial u_s}{\partial r} = -\frac{2}{3}r^{-5/3}\sin(\frac{2}{3}\theta)$, from which:

$$\frac{\partial^2 u_s}{\partial r^2} = \frac{10}{9}r^{-8/3}\sin(\frac{2}{3}\theta), \text{ and } \frac{1}{r}\frac{\partial u_s}{\partial r} = -\frac{2}{3}r^{-8/3}\sin(\frac{2}{3}\theta).$$

On the other hand:

$$\frac{1}{r^2}\frac{\partial^2 u_s}{\partial \theta^2} = -\frac{4}{9}r^{-8/3}\sin(\frac{2}{3}\theta).$$

We conclude that:

$$\Delta u_s = \left(\frac{10}{9} - \frac{2}{3} - \frac{4}{9}\right) r^{-8/3}\sin(\frac{2}{3}\theta) = 0.$$

Thus, u_s is a solution of the Laplace equation belonging to $L^2(\Omega)$.

c) On Γ_D^1, we have $\theta = 0$: $u_{s|\Gamma_D^1} = 0$. On Γ_D^2, we have $\theta = \frac{3}{2}\pi$: $u_{s|\Gamma_D^2} = 0$. Finally, on Γ_N, we have $\frac{\partial u_s}{\partial n} = \frac{\partial u_s}{\partial r}_{|r=1} = -\frac{2}{3}\sin(\frac{2}{3}\theta)$.

d) Hence we have $u_s \in L^2(\Omega)$, $\Delta u_s = 0$ in Ω, and $u_{s|\Gamma_D^1} = u_{s|\Gamma_D^2} = 0$. Moreover, on Γ_N, observe that $u_s = \sin(\frac{2}{3}\theta)$, and hence we have the relation:

$$\left(\frac{2}{3}u_s + \frac{\partial u_s}{\partial n}\right)_{|\Gamma_N} = 0.$$

In the end, u_s satisfies the equations of exercise 20, with $\Gamma_1 = \Gamma_D$, $\Gamma_2 = \Gamma_N$, $\lambda = \frac{2}{3}$ and with the data $f = 0$ and $g = 0$. We can draw two conclusions from this:

- $u_s \notin H^1(\Omega)$; indeed, if we had $u_s \in H^1(\Omega)$, we would see from exercise 20 that $u_s = 0$, which is clearly not the case. Therefore, u_s is not a *weak solution* of the mixed Dirichlet–Fourier problem;

- if we solve the mixed Dirichlet–Fourier problem with a (more general) condition for the solution, namely that of membership in $L^2(\Omega)$, we will no longer have uniqueness necessarily.

Solution 22. Problem with a discontinuous coefficient that changes sign

a) The equivalent variational formulation of the problem [P22] in $V = H_0^1(\Omega)$ can be written as:

$$\left| \begin{array}{l} \text{Find } u \in V \text{ such that} \\ a(u,v) = \displaystyle\int_\Omega f v \, d\Omega \quad \forall v \in V \end{array} \right. \qquad \text{[F22]}$$

with $a(u,v) = \displaystyle\int_\Omega \sigma \nabla u \cdot \nabla v \, d\Omega$. In the following, we endow V with the norm $v \mapsto \|\nabla v\|_{(L^2(\Omega))^d}$, which is equivalent to the $\|\cdot\|_{H^1(\Omega)}$ norm (see exercise 8). Let us show that $a(\cdot,\cdot)$ is not coercive.

- Let $v_1 \in H_0^1(\Omega_1)$, with $v_1 \neq 0$: extend v_1 by 0 to the entirety of Ω to define $\widetilde{v_1} \in H_0^1(\Omega)$ (see exercise 11). Since $\sigma_1 > 0$, we have $a(\widetilde{v_1}, \widetilde{v_1}) > 0$.

- Let $v_2 \in H_0^1(\Omega_2)$, with $v_2 \neq 0$: extend v_2 by 0 to the entirety of Ω to define $\widetilde{v_2} \in H_0^1(\Omega)$. As $\sigma_2 < 0$, we have $a(\widetilde{v_2}, \widetilde{v_2}) < 0$.

- If we reason by using the continuity (see the proof of lemma 2.1), we get $v_{\gamma^\star} \neq 0$ such that $a(v_{\gamma^\star}, v_{\gamma^\star}) = 0$.

b) For v to belong to $H^1(\Omega)$, it is necessary and sufficient (see exercise 11) that the restrictions match along the interface, namely that $\gamma_\Sigma v_1 = \gamma_\Sigma v_2$, so that we have the equivalence:

$$v \in H_0^1(\Omega) \iff v_j \in H_{0,\Gamma_j}^1(\Omega_j) \text{ for } j = 1,2 \text{ and } \gamma_\Sigma v_1 = \gamma_\Sigma v_2.$$

c) For $v, w \in V$, we simply write:

$$a(v,w) = \int_{\Omega_1} \sigma_1 \nabla v_1 \cdot \nabla w_1 \, d\Omega + \int_{\Omega_2} \sigma_2 \nabla v_2 \cdot \nabla w_2 \, d\Omega,$$

$$a_{\Omega_j}(v_j, w_j) = \int_{\Omega_j} \sigma_j \nabla v_j \cdot \nabla w_j \, d\Omega, \text{ for } j = 1,2.$$

d) As Γ_j has a non-zero measure, for $j = 1, 2$, we can apply the Poincaré–Friedrichs inequality [2.24]:

$$\exists C_j > 0\,, \forall v_j \in H^1_{0,\Gamma_j}(\Omega_j), \int_{\Omega_j} |\nabla v_j|^2 \, d\Omega \geq C_j \, \|v_j\|^2_{H^1(\Omega_j)}.$$

Therefore, $\|\cdot\|_{V_j} : v_j \mapsto \left(\int_{\Omega_j} |\nabla v_j|^2 \, d\Omega\right)^{1/2}$ is a norm equivalent to the $\|\cdot\|_{H^1(\Omega_j)}$ norm in $H^1_{0,\Gamma_j}(\Omega_j)$. From this, we see that the forms a_{Ω_1} and a_{Ω_2} are coercive.

e) The inf-sup condition [2.43] can be written as:

$$\exists \alpha',\ \forall v \in V,\quad \sup_{w \in V \setminus \{0\}} \frac{|a(v,w)|}{\|w\|_V} \geq \alpha' \, \|v\|_V. \qquad [C22_a]$$

For $v \neq 0$, if we choose $w = T_0 v$, then, by using the operator T_0, we can isolate the contribution coming from Ω_1 and Ω_2, which have the same sign. On the other hand, $T_0 \notin \mathcal{L}(V)$: indeed, if $v \in V$ is such that $\gamma_\Sigma v \neq 0$, then the restrictions $(T_0 v)_{|\Omega_1}$ and $(T_0 v)_{|\Omega_2}$ *do not match along the interface*. In short, the operator T_0 does not allow us to obtain the inf-sup condition [$C22_a$].

f) Note that Ω_2 is the reflection of Ω_1 along the plane defined by the equation $\{x = 0\}$. From here, we can estimate $\|R_1 v_1\|_{V_2}$ by the change in variables, $(x, y, z) \mapsto (-x, y, z)$. Then we find:

$$\begin{aligned}
\|R_1 v_1\|^2_{V_2} &= \int_{\Omega_2} |\nabla (R_1 v_1)|^2 d\Omega \\
&= \int_{\Omega_2} \left(\left(\frac{\partial v_1}{\partial x}\right)^2 + \left(\frac{\partial v_1}{\partial y}\right)^2 + \left(\frac{\partial v_1}{\partial z}\right)^2\right)(-x, y, z)\, d\Omega \\
&= \int_{\Omega_1} \left(\left(-\frac{\partial v_1}{\partial x}\right)^2 + \left(\frac{\partial v_1}{\partial y}\right)^2 + \left(\frac{\partial v_1}{\partial z}\right)^2\right)(x, y, z)\, d\Omega = \|v_1\|^2_{V_1}.
\end{aligned}$$

Therefore, we have $|||R_1||| = 1$.

g) The idea here is to adjust T_0 (which has the correct "sign") to obtain a continuous linear mapping from V into V. Depending on the choice of R_1, we note that $\gamma_\Sigma (R_1 v_1) = \gamma_\Sigma v_1$ for all $v_1 \in V_1$. We can then consider the linear mapping:

$$Tv = \begin{cases} v_1 & \text{in } \Omega_1 \\ -v_2 + 2R_1 v_1 & \text{in } \Omega_2. \end{cases}$$

Let us check that $Tv \in V$. Indeed, along the interface, the trace of $(Tv)_1 = (Tv)_{|\Omega_1}$ equals $\gamma_\Sigma v_1 = \gamma_\Sigma v$, while the trace of $(Tv)_2 = (Tv)_{|\Omega_2}$ equals $\gamma_\Sigma(-v_2 + 2 + R_1 v_1) = -\gamma_\Sigma v + 2\gamma_\Sigma(R_1 v_1) = \gamma_\Sigma v$. We have that $Tv \in V$. On the other hand, T is, of course, continuous on V, that is $|||T||| < \infty$ (see the previous question).

Now, let us study the inf-sup condition [C22$_a$] by applying Young's inequality with $\eta > 0$, and recalling that $\|R_1 v_1\|_{V_2} = \|v_1\|_{V_1}$ from Question (f):

$$\begin{aligned}
a(v, Tv) &= a_{\Omega_1}(v_1, (Tv)_1) + a_{\Omega_2}(v_2, (Tv)_2) \\
&= a_{\Omega_1}(v_1, v_1) + a_{\Omega_2}(v_2, -v_2 + 2R_1 v_1) \\
&= \int_{\Omega_1} \sigma_1 |\nabla v_1|^2 \, d\Omega + \int_{\Omega_2} |\sigma_2||\nabla v_2|^2 \, d\Omega - 2\int_{\Omega_2} |\sigma_2| \nabla v_2 \cdot \nabla(R_1 v_1) \, d\Omega \\
&\geq \int_{\Omega_1} \sigma_1 |\nabla v_1|^2 \, d\Omega + \int_{\Omega_2} |\sigma_2||\nabla v_2|^2 \, d\Omega \\
&\quad - \int_{\Omega_2} |\sigma_2|(\eta^2 |\nabla v_2|^2 + \frac{1}{\eta^2}|\nabla(R_1 v_1)|^2) \, d\Omega \qquad \text{(Young's inequality)} \\
&= \int_{\Omega_1} (\sigma_1 - \frac{1}{\eta^2}|\sigma_2|)|\nabla v_1|^2 \, d\Omega + \int_{\Omega_2} (1 - \eta^2)|\sigma_2||\nabla v_2|^2 \, d\Omega \\
&\geq \min(\sigma_1 - \tfrac{1}{\eta^2}|\sigma_2|, (1 - \eta^2)|\sigma_2|)\|\nabla v\|^2_{(L^2(\Omega))^d}.
\end{aligned}$$

Then we can obtain [C22$_a$] with $\alpha' > 0$ provided, on the one hand, that $\eta^2 > |\sigma_2|/\sigma_1$ and, on the other hand, that $1 > \eta^2$. Indeed, if these conditions are met then, since $\|Tv\|_V \leq |||T||| \, \|v\|_V$, we get:

$$\frac{|a(v, Tv)|}{\|Tv\|_V} \geq \frac{1}{|||T|||} \frac{|a(v, Tv)|}{\|v\|_V} \geq \frac{1}{|||T|||} \min(\sigma_1 - \frac{1}{\eta^2}|\sigma_2|, (1 - \eta^2)|\sigma_2|)\|v\|_V$$

with $|||T|||^{-1} \min(\sigma_1 - \frac{1}{\eta^2}|\sigma_2|, (1 - \eta^2)|\sigma_2|) > 0$. Since Young's inequality holds for every $\eta > 0$ (and given the signs of σ_1 and σ_2), this is equivalent to the condition:

$$1 > \frac{|\sigma_2|}{\sigma_1} \iff \frac{\sigma_2}{\sigma_1} > -1.$$

If this condition holds true, then the Banach–Nečas–Babuška corollary 2.1 applies, as the form a is symmetric. Hence, the problem with discontinuous coefficient whose sign changes is well posed.

h) Reasoning as before, this time let us introduce:

$$Tv = \begin{cases} v_1 - 2R_2 v_2 & \text{in } \Omega_1 \\ -v_2 & \text{in } \Omega_2. \end{cases}$$

By calculations which are very similar to those used in the previous question, we can check that an inf-sup condition [C22$_a$] exists with $\alpha' > 0$, provided that the following (new) condition holds:

$$\frac{\sigma_2}{\sigma_1} < -1.$$

Thus, when the open subset Ω is symmetric, the problem with discontinuous coefficient that changes sign is well-posed as soon as $\sigma_2/\sigma_1 \neq -1$ holds. Finally, one can prove that this problem is ill posed when $\sigma_2/\sigma_1 = -1$, though this is somewhat complicated.

Solution 23. Linear elasticity

a) This is a Dirichlet problem; we therefore seek a solution in $\mathbf{V} = (H_0^1(\Omega))^3$. The displacement vector $\mathbf{u}$ is an element of $\mathbf{V}$; multiply the equations on Ω with the components v_i of a test field $\mathbf{v} \in \mathbf{V}$, before then integrating by parts [1.39]. For each $i = 1, 2, 3$, we get:

$$\int_\Omega \sum_j \sigma_{ij}(\mathbf{u}) \frac{\partial v_i}{\partial x_j}\, d\Omega = \int_\Omega f_i v_i\, d\Omega.$$

One then adds these equalities together, before reexpressing the Cauchy stress tensor in terms of the deformation tensor. We get:

$$\sum_{i=1}^{3} \left(\lambda \int_\Omega \operatorname{div} \mathbf{u}\, \frac{\partial v_i}{\partial x_i}\, d\Omega + 2\mu \sum_{j=1}^{3} \int_\Omega \varepsilon_{ij}(\mathbf{u}) \frac{\partial v_i}{\partial x_j}\, d\Omega \right) = \int_\Omega \mathbf{f} \cdot \mathbf{v}\, d\Omega.$$

By the symmetry of the indices i, j, we note that:

$$\sum_{i,j=1}^{3} \int_\Omega \varepsilon_{ij}(\mathbf{u}) \frac{\partial v_i}{\partial x_j}\, d\Omega = \sum_{i,j=1}^{3} \int_\Omega \varepsilon_{ij}(\mathbf{u}) \varepsilon_{ij}(\mathbf{v})\, d\Omega.$$

The variational formulation of the problem [P23] is therefore:

$$\left\{ \begin{array}{l} \text{Find } \mathbf{u} \in \mathbf{V} \text{ such that} \\ a(\mathbf{u}, \mathbf{v}) = \ell(\mathbf{v}), \quad \forall \mathbf{v} \in \mathbf{V} \end{array} \right. \qquad \text{[F23]}$$

where a is the bilinear form on $\mathbf{V} \times \mathbf{V}$ defined by:

$$a(\mathbf{u}, \mathbf{v}) = \lambda \int_\Omega \operatorname{div} \mathbf{u} \operatorname{div} \mathbf{v} \, d\Omega + 2\mu \sum_{i,j=1}^{3} \int_\Omega \varepsilon_{ij}(\mathbf{u}) \, \varepsilon_{ij}(\mathbf{v}) \, d\Omega$$

and ℓ is the linear form on $\mathbf{V}$ defined by:

$$\ell(\mathbf{v}) = \int_\Omega \mathbf{f} \cdot \mathbf{v} \, d\Omega.$$

We have therefore established that, if $\mathbf{u}$ is a solution of [P23], then it is also a solution of [F23].

Conversely, suppose that $\mathbf{u}$ is a solution of [F23]. Consider $\mathbf{v} \in (\mathcal{D}(\Omega))^3 \subset \mathbf{V}$. In the distribution sense, we obtain:

$$\lambda \sum_{i=1}^{3} \left\langle \operatorname{div} \mathbf{u}, \frac{\partial v_i}{\partial x_i} \right\rangle + 2\mu \sum_{i,j=1}^{3} \langle \varepsilon_{ij}(\mathbf{u}), \varepsilon_{ij}(\mathbf{v}) \rangle = \sum_{i=1}^{3} \langle f_i, v_i \rangle .$$

Once again, by the symmetry of the indices i, j, we have:

$$\sum_{i,j=1}^{3} \langle \varepsilon_{ij}(\mathbf{u}), \varepsilon_{ij}(\mathbf{v}) \rangle = \sum_{i,j=1}^{3} \left\langle \varepsilon_{ij}(\mathbf{u}), \frac{\partial v_i}{\partial x_j} \right\rangle.$$

Therefore, we arrive at:

$$\sum_{i,j=1}^{3} \left\langle \sigma_{ij}(\mathbf{u}), \frac{\partial v_i}{\partial x_j} \right\rangle = \sum_{i=1}^{3} \langle f_i, v_i \rangle .$$

Then, we pick $\mathbf{v} \in (\mathcal{D}(\Omega))^3$ with only one non-zero component, which shows, for each $i = 1, 2, 3$, that $\sum_{j=1}^{3} \frac{\partial}{\partial x_j} \sigma_{ij}(\mathbf{u}) = f_i$ in the distribution sense, and in fact almost everywhere on Ω because $\mathbf{f}$ belongs to $(L^2(\Omega))^3$. Since $\mathbf{u} \in \mathbf{V}$, $\mathbf{u}$ is indeed a solution of [P23].

b) Let $\mathbf{v} \in (\mathcal{D}(\Omega))^3$, by differentiating in the distribution sense, we have:

$$\left\langle \frac{\partial v_i}{\partial x_j}, \frac{\partial v_j}{\partial x_i} \right\rangle = -\left\langle v_i, \frac{\partial^2 v_j}{\partial x_i \partial x_j} \right\rangle = \left\langle \frac{\partial v_i}{\partial x_i}, \frac{\partial v_j}{\partial x_j} \right\rangle.$$

From this, we deduce:

$$\int_\Omega (\varepsilon_{ij}(\mathbf{v}))^2 \, d\Omega = \frac{1}{4}\int_\Omega \left(\frac{\partial v_i}{\partial x_j}\right)^2 d\Omega + \frac{1}{4}\int_\Omega \left(\frac{\partial v_j}{\partial x_i}\right)^2 d\Omega + \frac{1}{2}\int_\Omega \frac{\partial v_i}{\partial x_i}\frac{\partial v_j}{\partial x_j} \, d\Omega.$$

After summing over i and j, one finds:

$$\sum_{i,j=1}^{3} \int_\Omega (\varepsilon_{ij}(\mathbf{v})))^2 \, d\Omega = \frac{1}{2}\sum_{i,j=1}^{3} \int_\Omega \left(\frac{\partial v_i}{\partial x_j}\right)^2 d\Omega + \frac{1}{2}\int_\Omega (\operatorname{div} \mathbf{v})^2 \, d\Omega.$$

Using Korn's first inequality for functions in $(\mathcal{D}(\Omega))^3$, it follows that:

$$\sum_{i,j=1}^{3} \int_\Omega (\varepsilon_{ij}(\mathbf{v}))^2 \, d\Omega \geq \frac{1}{2}\sum_{i=1}^{3} \int_\Omega |\nabla v_i|^2 \, d\Omega \quad \forall \mathbf{v} \in (\mathcal{D}(\Omega))^3.$$

This result easily extends to functions on $\mathbf{V}$ by using density.

c) The Lax–Milgram theorem guarantees the existence and uniqueness for the solution of the problem [F23], provided that is hypotheses are satisfied. We will work with the gradient norm of $\mathbf{V}$ (see the Poincaré inequality [2.22]), that is to say, $\mathbf{v} \mapsto \sum_{i=1}^{3} \left(\int_\Omega |\nabla v_i|^2\right)^{1/2}$.

In order to see that the bilinear form a is continuous, it suffices to use the Cauchy–Schwarz inequality. The positivity hypothesis on the Lamé coefficients, in addition to Korn's first inequality, allows us to prove that a is coercive. The continuity of the linear form ℓ then follows from the Cauchy–Schwarz inequality.

d) Apply the results from exercise 17, with:

$$\begin{aligned} J(\mathbf{v}) &= \frac{1}{2}a(\mathbf{v}, \mathbf{v}) - \ell(\mathbf{v}) \\ &= \frac{\lambda}{2}\int_\Omega (\operatorname{div} \mathbf{v})^2 \, d\Omega + \mu \sum_{i,j=1}^{3} \int_\Omega (\varepsilon_{ij}(\mathbf{v}))^2 \, d\Omega - \int_\Omega \mathbf{f} \cdot \mathbf{v} \, d\Omega. \end{aligned}$$

Solution 24. Solution to the plate equation

a) Recall that the space $H^2(\Omega)$, which is defined by:

$$H^2(\Omega) = \left\{ v \in H^1(\Omega) \text{ such that } \frac{\partial^2 v}{\partial x_i \partial x_j} \in L^2(\Omega) \quad \forall i, j = 1, 2 \right\}$$

endowed with the norm:

$$\|v\|_{H^2(\Omega)} = \left(\|v\|^2_{L^2(\Omega)} + \|\nabla v\|^2_{(L^2(\Omega))^2} + \sum_{i,j=1,2} \left\| \frac{\partial^2 v}{\partial x_i x_j} \right\|^2_{L^2(\Omega)} \right)^{1/2}$$

defines a Hilbert space.

b) Note that by definition, $H_0^2(\Omega)$ coincides with the closure of $\mathcal{D}(\Omega)$ in $H^2(\Omega)$, and therefore it is a closed vector subspace of $H^2(\Omega)$: it is therefore a Hilbert space endowed with the norm of $H^2(\Omega)$. Moreover, via the second definition, we observe that $H_0^2(\Omega)$ equals the kernel of the linear mapping $v \mapsto (\gamma_0 v, \gamma_1 v)$, which is continuous from $H^2(\Omega)$ into $L^2(\partial\Omega) \times L^2(\partial\Omega)$ (see theorem 1.7 and proposition 1.15). Hence the fact that $H_0^2(\Omega)$ is a closed vector subspace of $H^2(\Omega)$ is recovered: the same conclusion then follows.

c) By definition, $\mathcal{D}(\Omega)$ is dense in $H_0^2(\Omega)$. Let $v_k \in \mathcal{D}(\Omega)$:

$$\begin{aligned}\int_\Omega f\, v_k \, d\Omega &= \int_\Omega \Delta^2 u\, v_k\, d\Omega = \langle \Delta^2 u, v_k \rangle = -\sum_{i=1}^{2} \left\langle \frac{\partial}{\partial x_i}(\Delta u), \frac{\partial v_k}{\partial x_i} \right\rangle \\ &= \sum_{i=1}^{2} \left\langle \Delta u, \frac{\partial^2 v_k}{\partial x_i^2} \right\rangle = \langle \Delta u, \Delta v_k \rangle = \int_\Omega \Delta u\, \Delta v_k \, d\Omega.\end{aligned}$$

For $v \in H_0^2(\Omega)$, let $(v_k)_k \in (\mathcal{D}(\Omega))^{\mathbb{N}}$ be such that $\lim_{k\to\infty} \|v - v_k\|_{H^2(\Omega)} = 0$. In particular, we have $\lim_{k\to\infty} \|v - v_k\|_{L^2(\Omega)} = \lim_{k\to\infty} \|\Delta v - \Delta v_k\|_{L^2(\Omega)} = 0$.

Therefore, u is a solution of:

$$\left| \begin{array}{l} \text{Find } u \in H_0^2(\Omega) \text{ such that} \\ \displaystyle\int_\Omega \Delta u\, \Delta v \, d\Omega = \int_\Omega f\, v\, d\Omega \quad \forall v \in H_0^2(\Omega). \end{array} \right. \qquad \text{[F24]}$$

Subsequently, we set:

$$a(u, v) = \int_\Omega \Delta u\, \Delta v\, d\Omega \text{ and } \ell(v) = \int_\Omega f\, v\, d\Omega.$$

Conversely, by reasoning in the distribution sense, it is easy to prove that the solution u of [F24] is also a solution of [P24].

d) The form a is symmetric, bilinear and continuous on $H_0^2(\Omega)$, since $\forall u, v \in H_0^2(\Omega)$:

$$|a(u,v)| = \left| \int_\Omega \Delta u \, \Delta v \, d\Omega \right| \le \|\Delta u\|_{L^2(\Omega)} \, \|\Delta v\|_{L^2(\Omega)} \le 2\|u\|_{H^2(\Omega)} \|v\|_{H^2(\Omega)}.$$

Moreover, the linear form ℓ is continuous on $H_0^2(\Omega)$, since $f \in L^2(\Omega)$:

$$|\ell(v)| = \left| \int_\Omega f \, v d\Omega \right| \le \|f\|_{L^2(\Omega)} \, \|v\|_{L^2(\Omega)} \le \|f\|_{L^2(\Omega)} \, \|v\|_{H^2(\Omega)} \quad \forall v \in H_0^2(\Omega).$$

e) Let $v \in H_0^2(\Omega)$: there exists $(v_k)_k \in (\mathcal{D}(\Omega))^{\mathbb{N}}$ such that $\lim_{k\to\infty} \|v - v_k\|_{H^2(\Omega)} = 0$. In particular, $\lim_{k\to\infty} \|\frac{\partial v}{\partial x_i} - \frac{\partial v_k}{\partial x_i}\|_{L^2(\Omega)} = 0$ for $i = 1, 2$. Since $(\frac{\partial v_k}{\partial x_i})_k \in (\mathcal{D}(\Omega))^{\mathbb{N}}$ for $i = 1, 2$, we get that $\frac{\partial v}{\partial x_i} \in H_0^1(\Omega)$ for $i = 1, 2$ (see definition 1.29).

Now, after applying the Poincaré inequality [2.22] to $\frac{\partial v}{\partial x_i}$, we get:

$$\left\| \frac{\partial v}{\partial x_i} \right\|_{L^2(\Omega)}^2 \le C_p \left\| \nabla \left(\frac{\partial v}{\partial x_i} \right) \right\|_{L^2(\Omega)}^2 \le C_p \sum_{j=1,2} \int_\Omega \left(\frac{\partial^2 v}{\partial x_i \partial x_j} \right)^2 d\Omega \le C_p \, |v|_{2,\Omega}^2 .$$

Then again applying the Poincaré inequality, this time to v, using the previous estimate we get:

$$\|v\|_{L^2(\Omega)}^2 \le C_p \, \|\nabla v\|_{L^2(\Omega)}^2 \le 2C_p^2 \, |v|_{2,\Omega}^2$$

which finally proves:

$$\|v\|_{H^2(\Omega)} \le \left(1 + 2C_p + 2C_p^2\right)^{1/2} |v|_{2,\Omega} \quad \forall v \in H_0^2(\Omega).$$

f) For any $v \in \mathcal{D}(\Omega)$, by differentiating in the distribution sense, we get the following identities:

$$\begin{aligned} \int_\Omega \frac{\partial^2 v}{\partial x_1^2} \frac{\partial^2 v}{\partial x_2^2} d\Omega &= \left\langle \frac{\partial^2 v}{\partial x_1^2}, \frac{\partial^2 v}{\partial x_2^2} \right\rangle = -\left\langle \frac{\partial}{\partial x_2} \left(\frac{\partial^2 v}{\partial x_1^2} \right), \frac{\partial v}{\partial x_2} \right\rangle \\ &= -\left\langle \frac{\partial}{\partial x_1} \left(\frac{\partial^2 v}{\partial x_1 \partial x_2} \right), \frac{\partial v}{\partial x_2} \right\rangle = \left\langle \frac{\partial^2 v}{\partial x_1 \partial x_2}, \frac{\partial^2 v}{\partial x_1 \partial x_2} \right\rangle \\ &= \int_\Omega \left(\frac{\partial^2 v}{\partial x_1 \partial x_2} \right)^2 d\Omega \end{aligned}$$

which, by density, remain true for every function $v \in H_0^2(\Omega)$. Then we obtain the identity:

$$a(v,v) = \int_\Omega \left(\left(\frac{\partial^2 v}{\partial x_1^2}\right)^2 + \left(\frac{\partial^2 v}{\partial x_2^2}\right)^2 + 2\,\frac{\partial^2 v}{\partial x_1^2}\,\frac{\partial^2 v}{\partial x_2^2} \right) d\Omega = |v|_{2,\Omega}^2 \quad \forall v \in H_0^2(\Omega).$$

These previous results establish the coercivity for the form a on $H_0^2(\Omega)$:

$$a(v,v) = |v|_{2,\Omega}^2 \geq C\,\|v\|_{H^2(\Omega)}^2 \quad \forall v \in H_0^2(\Omega)$$

where $C = (1 + 2C_p + 2C_p^2)^{-1/2} > 0$.

Each hypothesis of the Lax–Milgram theorem is satisfied, and consequently the variational formulation [F24], and hence the problem [P24], admits a unique solution u satisfying the continuity estimate for the data:

$$\|u\|_{H^2(\Omega)} \leq C\|f\|_{L^2(\Omega)}$$

where $C > 0$ is independent of both u and f.

A2.3. Exercises from Chapter 3

Solution 25. Discretization in one dimension

a) As we have a Dirichlet condition, consider the Hilbert space:

$$H_0^1(0,1) = \left\{ v \in H^1(0,1) \text{ such that } v(0) = v(1) = 0 \right\}.$$

For every $v \in H_0^1(0,1)$, by integrating by parts [1.42], we have:

$$-\int_0^1 u''v\,dx = \int_0^1 u'\,v'\,dx$$

leading to the variational formulation:

$$\left| \begin{array}{l} \text{Find } u \in H_0^1(0,1) \text{ such that} \\ \displaystyle\int_0^1 u'\,v'\,dx = \int_0^1 f\,v\,dx \quad \forall v \in H_0^1(0,1). \end{array} \right. \qquad \text{[F25]}$$

Since the variational problem [F25] is a special case of the homogeneous Dirichlet problem in dimension 1, it admits a unique solution u depending continuously on the data f.

b) Introduce a partition of the line segment $[0,1] = \bigcup_{\ell=1,n+1} S_\ell$, where $S_\ell = [x_{\ell-1}, x_\ell]$ with the endpoints $0 = x_0 < x_1 < \ldots < x_n < x_{n+1} = 1$. Introduce the approximation space:

$$V_h^0 = \{v_h \in C^0[0,1] \text{ such that } v_h(0) = v_h(1) = 0 \text{ and } v_{h|S_\ell} \in P^1(S_\ell)\, \forall \ell = 1, n+1\}$$

where we recall that P^1 denotes the space of polynomials with degree at most one with respect to x. By construction, $V_h^0 \subset H_0^1(0,1)$. For $1 \leq \ell \leq n$, let w_ℓ be the piecewise affine function defined by:

$$w_\ell(x) = \begin{cases} \dfrac{x - x_{\ell-1}}{x_\ell - x_{\ell-1}} & \text{if } x \in [x_{\ell-1}, x_\ell] \\ \dfrac{x_{\ell+1} - x}{x_{\ell+1} - x_\ell} & \text{if } x \in [x_\ell, x_{\ell+1}] \\ 0 & \text{otherwise.} \end{cases}$$

For each $1 \leq \ell \leq n$, it is easy to check that $w_\ell \in V_h^0$ (with $w_\ell(0) = w_\ell(1) = 0$). Moreover, $(w_\ell)_{1 \leq \ell \leq n}$ is a linearly independent set because $w_\ell(x_i) = \delta_{i\ell}$ and also generates V_h^0 because, for all $v_h \in V_h^0$:

$$v_h(x) = \sum_{1 \leq \ell \leq n} v_h(x_\ell) w_\ell(x).$$

The discretization of the variational formulation [F25] in the space V_h^0 is:

$$\left| \begin{array}{l} \text{Find } u_h \in V_h^0 \text{ such that} \\ \displaystyle\int_0^1 u_h'\, v_h'\, dx = \int_0^1 f\, v_h\, dx \quad \forall v \in V_h^0. \end{array} \right. \qquad [\text{F25}_h]$$

or rather, by decomposing $u_h = \sum_{1 \leq \ell \leq n} U_\ell\, w_\ell$:

$$\left| \begin{array}{l} \text{Find } \vec{U} \in \mathbb{R}^n \text{ such that} \\ \displaystyle\sum_{1 \leq \ell \leq n} \left(\int_0^1 w_\ell'\, w_i'\, dx \right) U_\ell = \int_0^1 f\, w_i\, dx \quad \forall\, 1 \leq i \leq n. \end{array} \right.$$

This problem is equivalent to solving the following linear system:

$$\mathbb{A}_1 \vec{U} = \vec{L}$$

where the matrix $\mathbb{A}_1 \in \mathbb{R}^{n\times n}$ and the vector $\overrightarrow{L} \in \mathbb{R}^n$ are defined by:

$$(\mathbb{A}_1)_{i\ell} = \int_0^1 w'_\ell \, w'_i \, dx \text{ and } L_i = \int_0^1 f \, w_i \, dx \text{ for } 1 \leq i, \ell \leq n.$$

c) It is easy to calculate the terms of matrix $\mathbb{A}_1$. Indeed, if we set $h_\ell = x_\ell - x_{\ell-1}$ for each $1 \leq \ell \leq n$, we have:

$$w'_\ell(x) = \begin{cases} \dfrac{1}{h_\ell} & \text{if } x \in [x_{\ell-1}, x_\ell] \\ -\dfrac{1}{h_{\ell+1}} & \text{if } x \in [x_\ell, x_{\ell+1}] \\ 0 & \text{otherwise} \end{cases}$$

which leads to:

$$(\mathbb{A}_1)_{i\ell} = \int_0^1 w'_\ell \, w'_i \, dx = \begin{cases} -\dfrac{1}{h_\ell} & \text{if } i = \ell - 1 \\ \dfrac{1}{h_\ell} + \dfrac{1}{h_{\ell+1}} & \text{if } i = \ell \\ -\dfrac{1}{h_{\ell+1}} & \text{if } i = \ell + 1 \\ 0 & \text{otherwise.} \end{cases} \quad \text{for } 1 \leq i, \ell \leq n$$

The matrix $\mathbb{A}_1$ is tridiagonal and, in the case of a uniform mesh ($h_\ell = h$ for all ℓ), the linear system $\mathbb{A}_1 \overrightarrow{U} = \overrightarrow{L}$ can be written (by setting $U_0 = U_{n+1} = 0$) as:

$$\frac{-U_{\ell-1} + 2U_\ell - U_{\ell+1}}{h} = L_\ell \quad \forall 1 \leq \ell \leq n.$$

We find therefore the classical three-point finite difference scheme to approximate the operator $-u''$, up to a factor of h. Recall that the finite difference three-point scheme leads to the following relations:

$$\frac{-U_{\ell-1} + 2U_\ell - U_{\ell+1}}{h^2} = f(x_\ell) \quad \forall 1 \leq \ell \leq n.$$

The P^1-finite element approximation will therefore be strictly equivalent to the finite difference approximation if we approach $L_\ell = \int_0^1 f \, w_\ell \, dx$ using $hf(x_\ell)$. This is possible if a quadrature formula is used that integrates the constants exactly.

d) The matrices $\mathbb{A}_1$ and $\mathbb{B}_1 = h^{-1}\mathbb{A}_1$ are symmetric and tridiagonal. As the bilinear form $a(u,v) = \int_0^1 u'\,v'\,dx$ is coercive on $V_h^0 \subset H_0^1(0,1)$, we observe that the matrix $\mathbb{A}_1$ is definite-positive as well as the matrix $\mathbb{B}_1$.

Solution 26. P^1*-finite elements in two dimensions on a structured mesh*

a) For $0 \leq i_1 \leq n-1$ and $0 \leq i_2 \leq n-1$, define the triangles ($\ell = 2\,i_2 n + 2\,i_1 + 1$):

$$T_\ell = (M_{i_1+1,i_2}, M_{i_1+1,i_2+1}, M_{i_1,i_2}) \quad \text{and} \quad T_{\ell+1} = (M_{i_1,i_2+1}, M_{i_1,i_2}, M_{i_1+1,i_2+1}).$$

Consider the finite element approximation space:

$$V_h^0 = \left\{ v \in H_0^1(\Omega) \text{ such that } v_{|T_\ell} \in P^1(T_\ell)\ \forall\, 1 \leq \ell \leq 2\,n^2 \right\}$$

where $P^1(T_\ell)$ denotes the space of polynomials of degree at most one with respect to (x_1, x_2). Define the local basis functions on each triangle by:

$$T_\ell : \begin{cases} \tau_1^\ell = \dfrac{x_1 - x_2 + (i_2 - i_1)h}{h} \\ \tau_2^\ell = \dfrac{x_2 - i_2 h}{h} \\ \tau_3^\ell = \dfrac{(i_1+1)h - x_1}{h} \end{cases} \qquad T_{\ell+1} : \begin{cases} \tau_1^\ell = \dfrac{x_2 - x_1 + (i_1 - i_2)h}{h} \\ \tau_2^\ell = \dfrac{(i_2+1)h - x_2}{h} \\ \tau_3^\ell = \dfrac{x_1 - i_1 h}{h} \end{cases} . \quad \text{[E26]}$$

For $0 \leq i_1 \leq n$, $0 \leq i_2 \leq n$, introduce the global basis functions w_{i_1,i_2} that are defined on the triangles T_ℓ by:

$$w_{i_1,i_2|T_\ell} = \begin{cases} \tau_k^\ell & \text{if } M_{i_1,i_2} \text{ is the } k\text{th vertex of the triangle } T_\ell \\ 0 & \text{otherwise.} \end{cases}$$

By construction, we have the following properties:

- if $i_1 = 0$ or $i_2 = 0$ or $i_1 = n$ or $i_2 = n$, then $w_{i_1,i_2|\partial\Omega} \neq 0$ and $w_{i_1,i_2} \notin V_h^0$;
- $\forall\, 0 \leq i_1 \leq n$ and $0 \leq i_2 \leq n$, we have $w_{i_1,i_2}(M_{j_1,j_2}) = \delta_{(i_1,i_2),(j_1,j_2)}$;
- $\forall\, 1 \leq i_1 \leq n-1$ and $1 \leq i_2 \leq n-1$, we have $w_{i_1,i_2} \in V_h^0$;
- $\forall v_h \in V_h^0$, we have $v_h = \sum_{1\leq i_1 \leq n-1, 1 \leq i_2 \leq n-1} v_h(M_{i_1,i_2})\, w_{i_1,i_2}$.

From these properties, we see that $V_h^0 = \text{vect}(w_{i_1,i_2})_{1\leq i_1,i_2 \leq n-1}$ and also that $\dim V_h^0 = (n-1)^2$. The discretized problem [F26] in V_h^0 leads us to the approximate problem:

$$\left|\begin{array}{l} \text{Find } u_h \in V_h^0 \text{ such that} \\ \displaystyle\int_\Omega \nabla u_h \cdot \nabla v_h \, d\Omega = \int_\Omega f\, v_h d\Omega \quad \forall v_h \in V_h^0 \end{array}\right. \qquad [\text{F26}_h]$$

which is equivalent to solving the linear system of order $(n-1)^2$:

$$\mathbb{K}\overrightarrow{U} = \overrightarrow{L}$$

where, having used the *double subscripts* (i_1, i_2) or (j_1, j_2):

$$\mathbb{K}_{i_1,i_2;j_1,j_2} = \int_\Omega \nabla w_{j_1,j_2} \cdot \nabla w_{i_1,i_2}\, d\Omega \text{ and } L_{i_1,i_2} = \int_\Omega f\, w_{i_1,i_2} d\Omega.$$

b) From the expression [E26] for the local basis functions, we have:

$$\begin{aligned}
&\text{on } T_\ell: \quad \nabla\tau_1^\ell = \frac{1}{h}\begin{pmatrix}1\\-1\end{pmatrix} \quad \nabla\tau_2^\ell = \frac{1}{h}\begin{pmatrix}0\\1\end{pmatrix} \quad \nabla\tau_3^\ell = \frac{1}{h}\begin{pmatrix}-1\\0\end{pmatrix}\\
&\text{on } T_{\ell+1}: \nabla\tau_1^\ell = \frac{1}{h}\begin{pmatrix}-1\\1\end{pmatrix} \quad \nabla\tau_2^\ell = \frac{1}{h}\begin{pmatrix}0\\-1\end{pmatrix} \quad \nabla\tau_3^\ell = \frac{1}{h}\begin{pmatrix}1\\0\end{pmatrix}
\end{aligned}.$$

It is straightforward to obtain the elementary matrices $\mathbb{K}_{ij}^\ell = \int_{T_\ell} \nabla\tau_j^\ell \cdot \nabla\tau_i^\ell\, d\Omega$:

$$\mathbb{K}^\ell = \mathbb{K}^{\ell+1} = \frac{1}{2}\begin{bmatrix} 2 & -1 & -1\\ -1 & 1 & 0\\ -1 & 0 & 1\end{bmatrix}.$$

Note that, by the *orthogonality* of the gradient, two elements vanish since the triangles $T_\ell, T_{\ell+1}$ are right-angled. In order to determine the total stiffness matrix $\mathbb{K}$, it is a matter of obtaining the elementary matrices. Consider the row with index (i_1, i_2) of the matrix. Then the vertex M_{i_1,i_2} belongs to the following triangles:

$$T1 = (M_{i_1-1,i_2}, M_{i_1-1,i_2-1}, M_{i_1,i_2}) \; T2 = (M_{i_1,i_2-1}, M_{i_1,i_2}, M_{i_1-1,i_2-1})$$

$$T3 = (M_{i_1,i_2}, M_{i_1,i_2-1}, M_{i_1+1,i_2}) \quad T4 = (M_{i_1,i_2}, M_{i_1,i_2+1}, M_{i_1-1,i_2})$$

$$T5 = (M_{i_1,i_2+1}, M_{i_1,i_2}, M_{i_1+1,i_2+1}) \; T6 = (M_{i_1+1,i_2}, M_{i_1+1,i_2+1}, M_{i_1,i_2}).$$

Note that, a priori, the only contributions come from the columns with indices:

$$(i_1{-}1, i_2{-}1),\ (i_1, i_2{-}1),\ (i_1{-}1, i_2),\ (i_1, i_2),\ (i_1{+}1, i_2),\ (i_1, i_2{+}1),\ (i_1{+}1, i_2{+}1).$$

To calculate the diagonal coefficient $\mathbb{K}_{i_1,i_2;i_1,i_2}$, note that the point M_{i_1,i_2} appears in position 1, 2 or 3 in two triangles. Hence:

$$\mathbb{K}_{i_1,i_2;i_1,i_2} = \frac{1}{2}(2 \times 2 + 2 \times 1 + 2 \times 1) = 4.$$

We now will calculate the coefficient $\mathbb{K}_{i_1,i_2;i_1+1,i_2}$. The pair M_{i_1,i_2} and M_{i_1+1,i_2} of points belong to the triangle $T3$ with position $(1,3)$, and to the triangle $T6$ with position $(3,1)$. Let:

$$\mathbb{K}_{i_1,i_2;i_1+1,i_2} = \frac{1}{2}(-1 + -1) = -1.$$

Let us now calculate the coefficient $\mathbb{K}_{i_1,i_2;i_1+1,i_2+1}$. The pair M_{i_1,i_2} and M_{i_1+1,i_2+1} of points belongs to the triangle $T5$ with position $(2,3)$, and to the triangle $T6$ with position $(3,2)$. Hence:

$$\mathbb{K}_{i_1,i_2;i_1+1,i_2+1} = \frac{1}{2}(0 + 0) = 0.$$

The other coefficients can be obtained via symmetry. By adopting a row-by-row arrangement in terms of the *double indices*, we finally get:

$$\text{row } \mathbb{K}_{i_1,i_2;\cdot,\cdot} : (\ldots \overset{(i_1,i_2-1)}{-1} \ldots \overset{(i_1-1,i_2)}{-1} \overset{(i_1,i_2)}{4} \overset{(i_1+1,i_2)}{-1} \ldots \overset{(i_1,i_2+1)}{-1} \ldots).$$

When $i_1 = 1$ or $i_2 = 1$ or $i_1 = n-1$ or $i_2 = n-1$, the row gets truncated by the entries whose indices correspond to points on the boundary (due to the Dirichlet condition).

c) The entries for the elementary mass matrix are given by:

$$\mathbb{M}^{\ell}_{ij} = \int_{T_\ell} \tau^{\ell}_i \tau^{\ell}_i \, d\Omega.$$

By using the expressions [E26] for the local basis functions, one can also calculate explicitly (though it is somewhat more tedious) the elementary mass matrices:

$$\mathbb{M}^{\ell} = \mathbb{M}^{\ell+1} = \frac{h^2}{24}\begin{bmatrix} 2 & 1 & 1 \\ 1 & 2 & 1 \\ 1 & 1 & 2 \end{bmatrix}.$$

Using the same assembly principle, for the following expression for the (i_1, i_2)th row of the matrix $\mathbb{M}$, we get:

$$\frac{h^2}{12}(\ldots \overset{(i_1-1,i_2-1)}{1} \overset{(i_1,i_2-1)}{1} \ldots \overset{(i_1-1,i_2)}{1} \overset{(i_1,i_2)}{6} \overset{(i_1+1,i_2)}{1} \ldots \overset{(i_1,i_2+1)}{1} \overset{(i_1+1,i_2+1)}{1} \ldots).$$

d) To calculate the right-hand side, we will deal with f using its interpolation:

$$f_h = \sum_{1\leq i_1,i_2\leq n-1} F_{i_1,i_2} w_{i_1,i_2} \quad (F_{i_1,i_2} = f(M_{i_1,i_2})).$$

Thus, each component of the right-hand side $\overrightarrow{L}$ is given by:

$$L_{i_1,i_2} = \int_\Omega f_h \, w_{i_1,i_2} \, d\Omega = \sum_{1\leq j_1,j_2\leq n-1} F_{j_1,j_2} \int_\Omega w_{i_1,i_2} w_{j_1,j_2} \, d\Omega = \left(\mathbb{M}\overrightarrow{F}\right)_{i_1,i_2}.$$

We can reinterpret each row within the system $\mathbb{K}\overrightarrow{U} = \overrightarrow{L}$ as a finite difference:

$$\frac{1}{h^2}\left(-U_{i_1,i_2-1} - U_{i_1-1,i_2} + 4U_{i_1,i_2} - U_{i_1+1,i_2} - U_{i_1,i_2+1}\right) =$$

$$\frac{1}{2} f_{i_1,i_2} + \frac{1}{12}\left(f_{i_1-1,i_2-1} + f_{i_1,i_2-1} + f_{i_1-1,i_2} + f_{i_1+1,i_2} + f_{i_1,i_2+1} + f_{i_1+1,i_2+1}\right).$$

Thus, we obtain the five-point finite difference scheme to approximate the Laplacian operator. The right-hand side is not quite the same as that in the standard approximation, namely $f(M_{i_1,i_2})$. Here, it is actually a weighted average of the function f in the neighborhood of the point M_{i_1,i_2}.

Solution 27. Q^2-Serendipity finite elements

a) Let $p \in P^2(\hat{C})$. Then $D^\alpha p = 0$ for every $|\alpha| \geq 3$. Consequently, its second-order Taylor series is accurate. Expand $p(\hat{M}_i)$ about $p(\hat{M}_9)$ to get:

$$p(\hat{M}_i) = p(\hat{M}_9) + \nabla p(\hat{M}_9).(\hat{M}_i - \hat{M}_9) + 1/2(\hat{M}_i - \hat{M}_9)^T\mathbb{H}(\hat{M}_i - \hat{M}_9)$$

where $\mathbb{H}$ denotes the Hessian matrix (here a constant, since $P \in P^2(\hat{C})$).

As:

$$\sum_{i=1}^{4}(\hat{M}_i - \hat{M}_9) = \sum_{i=5}^{8}(\hat{M}_i - \hat{M}_9) = 0$$

we obtain:

$$\sum_{i=1}^{4} p(\hat{M}_i) = 4p(\hat{M}_9) + 1/2\sum_{i=1}^{4}(\hat{M}_i - \hat{M}_9)^T\mathbb{H}(\hat{M}_i - \hat{M}_9),$$

$$\sum_{i=5}^{8} p(\hat{M}_i) = 4p(\hat{M}_9) + 1/2\sum_{i=5}^{8}(\hat{M}_i - \hat{M}_9)^T\mathbb{H}(\hat{M}_i - \hat{M}_9).$$

In expressing $\hat{M}_5, \hat{M}_6, \hat{M}_7, \hat{M}_8$ as functions of the vertices $\hat{M}_1, \hat{M}_2, \hat{M}_3, \hat{M}_4$, that is as $\hat{M}_5 = 1/2(\hat{M}_1 + \hat{M}_2), \ldots$, we obtain:

$$\sum_{i=5}^{8} p(\hat{M}_i) = 4p(\hat{M}_9) + 1/4 \sum_{i=1}^{4} (\hat{M}_i - \hat{M}_9)^T \mathbb{H} (\hat{M}_i - \hat{M}_9)$$

and so, finally:

$$2 \sum_{i=5}^{8} p(\hat{M}_i) - \sum_{i=1}^{4} p(\hat{M}_i) = 4p(\hat{M}_9). \qquad \text{[E27]}$$

b) Let $(\hat{\tau}_i)_{i=1,9}$ be the standard basis functions for the $Q^2(\hat{C})$ element. Set:

$$\widetilde{\tau}_i = \hat{\tau}_i - \frac{\hat{\tau}_9}{4} \quad i = 1,4 \quad \text{and} \quad \widetilde{\tau}_i = \hat{\tau}_i + \frac{\hat{\tau}_9}{2} \quad i = 5,8.$$

$(\widetilde{\tau}_i)_{i=1,8}$ is a linearly independent set, since:

$$\sum_{i=1}^{8} \lambda_i \widetilde{\tau}_i = 0 \Leftrightarrow \sum_{i=1}^{9} \mu_i \hat{\tau}_i = 0 \text{ with } \mu_i = \lambda_i, i = 1,8 \text{ and } \mu_9 = -1/4 \sum_{i=1}^{4} \lambda_i + 1/2 \sum_{i=5}^{8} \lambda_i.$$

As $(\hat{\tau}_i)_{i=1,9}$ is a basis, we have $\mu_i = 0$, $i = 1,8$, which implies $\lambda_i = 0$, $i = 1,8$. To check that $(\widetilde{\tau}_i)_{i=1,8}$ satisfies the relation [E27], it suffices to replace the $(\widetilde{\tau}_i)_{i=1,8}$ with their expressions as functions of the $(\hat{\tau}_i)_{i=1,9}$, before using $\hat{\tau}_i(M_j) = \delta_{ij}$ to establish the conclusion.

c) The Q^2-Serendipity element is a second-order Lagrangian finite element. Indeed, $\hat{P}$ is eight-dimensional as a hyperplane of $Q^2(\hat{C})$, and since $(\widetilde{\tau}_i)_{i=1,8}$ forms a basis of $\hat{P}$, we have unisolvency. Therefore, by construction, we have $P^2(\hat{C}) \subset \hat{P}$.

d) In the case of a mesh with a square structure, that is, with (n rows, n columns), we can easily count the number of non-zero terms in the stiffness matrix for each case.

In $Q^2(\hat{C})$:

- corner nodes: 25 connections for $(n+1)^2$ term;
- midpoint nodes: 15 connections for $2n(n+1)$ terms;
- central nodes: 9 connections for n^2 terms;
- giving in total: $N_1 = 25(n+1)^2 + 30n(n+1) + 9n^2 = 64n^2 + O(n) \simeq 64n^2$.

In Q^2-Serendipity:

- corner nodes: 21 connection for $(n+1)^2$ terms;

- midpoint nodes: 13 connections for $2n(n+1)$ terms;
- giving in total: $N_2 = 21(n+1)^2 + 26n(n+1) = 47n^2 + O(n) \simeq 47n^2$.

Using the Q^2-Serendipity finite element therefore provides a gain of approximately 27% (to store the rigidity matrix), compared to using the classical Q^2 finite element.

Solution 28. Finite element for the $P^1 - P^2$ coupling

The purpose of this exercise is the practical realization of a coupling between a P^1 finite element and a P^2 finite element.

a) To show that (T_1, Σ, P) is a first-order Lagrangian finite element (see definition 3.5): we note that T_1 is a bounded and non-empty closed subset of $\mathbb{R}^2$, with the set of the degrees of freedom $\Sigma = \{M_1, M_2, M_3, M_4\}$ and $P = Q^1(T_1) = \{p(x,y) = axy + bx + cy + d\}$. If suffices therefore to show that $Q^1(T_1)$ is unisolvent on Σ, which is to say that every element $p \in P$ can be uniquely determined by its values at the vertices of Σ, or equivalently that the mapping:

$$S : \left| \begin{array}{l} P \to \mathbb{R}^4 \\ p \mapsto (p(M_i))_{i=1,4} \end{array} \right.$$

is bijective. In finite-dimensional settings, as dim(P) = card Σ, a simple way to prove unisolvency is by showing that the mapping S is injective (or surjective): $S(p) = 0$ is equivalent to $p(M_i) = 0$ for $1 \leq i \leq 4$, which implies that $d = 0$, and then $b = c = 0$, and also that $a = 0$, from which we obtain $p = 0$.

To determine the local basis functions, we look for the functions $\tau_i(x,y)$, where $1 \leq i \leq 4$, that satisfy $\tau_i(M_j) = \delta_{ij}$. We can simply write down this system of four equations to obtain the coefficients a_i, b_i, c_i, d_i for the basis functions. Another way of doing this is to write down the lines equations for the points where the desired function must vanish. For instance, τ_1 must necessarily vanish at M_2, M_3 and M_4, each being located along the line $1 - x - y = 0$. From this, we get:

$$\tau_1(x,y) = d_1(1 - x - y) \quad \text{and } \tau_1(M_1) = 1 \Rightarrow d_1 = 1.$$

By proceeding in the same way for each τ_i, we obtain:

$$\begin{array}{l} \tau_1(x,y) = 1 - x - y, \ \tau_2(x,y) = -2xy + y \\ \tau_3(x,y) = -2xy + x, \ \tau_4(x,y) = 4xy \end{array}. \qquad \text{[E28]}$$

b) Denote by $\ell_g(\ell, i) = I$ the mapping that allows us to go from a local numbering to a global one, where ℓ is the element number, i is the local degree of freedom number and I is the global degree of freedom number. Here, we have:

$$\ell_g(1,1) = 1,\ \ell_g(1,2) = 2,\ \ell_g(1,3) = 3,\ \ell_g(1,4) = 4,$$
$$\ell_g(3,1) = 1,\ \ell_g(3,2) = 3,\ \ldots .$$

The global basis functions w_I, where $I = 1, ..., 8$, can be written as:

$$w_I = \begin{cases} \tau_i^\ell \text{ on } T_\ell, \text{ if } \ell_g(\ell, i) = I \\ 0 \qquad \text{otherwise} \end{cases}$$

and the space $V_h = \text{vect}(w_I)_{I=1,8}$ therefore equals:

$$V_h = \{v_h \in L^2(\Omega) \text{ such that } v_{h|T_1} \in Q^1(T_1),\ v_{h|T_2} \in P^2(T_2),\ v_{h|T_3} \in P^1(T_3)\}.$$

c) Proving that $V_h \subset \mathcal{C}^0(\overline{\Omega})$ amounts to showing that $w_I \in \mathcal{C}^0(\overline{\Omega})$ for each $1 \leq I \leq 8$. One can use proposition 3.7, but it is quicker to reason directly using the functions. The mesh presents two common edges: the edge $[M_2, M_3]$ shared between the triangles T_1 and T_2, and the edge $[M_1, M_3]$ shared between the triangles T_1 and T_3. We are going to check that the functions are continuous along these two edges.

- Along the edge $[M_2, M_3]$, the restriction of a basis function defined on T_1 (and hence belongs to $Q^1(T_1)$) is a second-order polynomial. Similarly, the restriction of a basis function defined on T_2 (and hence belongs to $P^2(T_2)$) is a second-order polynomial. As these restrictions agree at three distinct points, they therefore agree at every point along the edge $[M_2, M_3]$.

- The situation is not as trivial along the edge $[M_1, M_3]$. Indeed, the restriction to this edge of a basis function defined on T_3 (so belonging to $P^1(T_3)$) is a first-order polynomial. On the other hand, the restriction of a basis function defined on T_1 (and hence belongs to $Q^1(T_1)$) is a priori a second-order polynomial. They do not necessarily coincide along this edge, since we can only impose the relation at two distinct points. This problem concerns the global basis functions $(w_I)_{I=1,4}$. However, from the expressions [E28], the restrictions of these four functions along the edge $[M_1, M_3]$ (with the equation $x = 0$) are first-degree polynomials.

So we have $V_h \subset \mathcal{C}^0(\overline{\Omega})$, and proposition 1.1 states that $V_h \subset H^1(\Omega)$.

Solution 29. Three-dimensional prismatic finite element

a) To show that $(\hat{K}, \hat{\Sigma}, \hat{P})$ defines a finite element, it suffices to find expressions for the basis functions. Denote by $\hat{\phi}_j$, where $j = 1, 3$, the local basis functions for the P^1 element defined on the unit triangle $\hat{T}$ in the $O\hat{x}_1\hat{x}_2$ plane, and by $\hat{\psi}_k$, where $k = 1, 2$, for the P^1 element defined on the unit line interval $\hat{S}$ along the $O\hat{x}_3$ axis, as:

$$\hat{\phi}_1(\hat{x}_1, \hat{x}_2) = 1 - \hat{x}_1 - \hat{x}_2, \quad \hat{\phi}_2(\hat{x}_1, \hat{x}_2) = \hat{x}_1, \quad \hat{\phi}_3(\hat{x}_1, \hat{x}_2) = \hat{x}_2$$
$$\hat{\psi}_1(\hat{x}_3) = 1 - \hat{x}_3, \quad \hat{\psi}_2(\hat{x}_3) = \hat{x}_3.$$

It is then clear that the functions $\hat{\tau}_i$, $i = 1, 6$, defined by:

$$\begin{aligned} \hat{\tau}_i(\hat{x}_1, \hat{x}_2, \hat{x}_3) &= \hat{\phi}_i(\hat{x}_1, \hat{x}_2)\hat{\psi}_1(\hat{x}_3) \quad &\text{for } i = 1, 3 \\ \hat{\tau}_i(\hat{x}_1, \hat{x}_2, \hat{x}_3) &= \hat{\phi}_{i-3}(\hat{x}_1, \hat{x}_2)\hat{\psi}_2(\hat{x}_3) \quad &\text{for } i = 4, 6 \end{aligned}$$

satisfy the fundamental property $\hat{\tau}_i(\hat{M}_j) = \delta_{i,j}$ for each $i, j = 1, 6$. These functions belong to $\hat{P}$, which is six dimensional. They therefore form a basis of $\hat{P}$: thus the unisolvency property follows.

b) The elementary mass and stiffness matrices are written, respectively, as:

$$\mathbb{M}_{ij} = \int_{\hat{T}\times\hat{S}} \hat{\tau}_j \hat{\tau}_i \, d\hat{\Omega}_{12} d\hat{\Omega}_3 \quad \text{and} \quad \mathbb{K}_{ij} = \int_{\hat{T}\times\hat{S}} \nabla\hat{\tau}_j \cdot \nabla\hat{\tau}_i \, d\hat{\Omega}_{12} d\hat{\Omega}_3.$$

In setting $\hat{\tau}_i = \hat{\phi}_k \hat{\psi}_l$ and $\hat{\tau}_j = \hat{\phi}_m \hat{\psi}_n$, we have:

$$\begin{aligned} \mathbb{M}_{ij} &= \int_{\hat{T}} \hat{\phi}_k \, \hat{\phi}_m \, d\hat{\Omega}_{12} \int_{\hat{S}} \hat{\psi}_l \, \hat{\psi}_n \, d\hat{\Omega}_3, \\ \mathbb{K}_{ij} &= \int_{\hat{T}} \nabla\hat{\phi}_k \cdot \nabla\hat{\phi}_m \, d\hat{\Omega}_{12} \int_{\hat{S}} \hat{\psi}_l \, \hat{\psi}_n \, d\hat{\Omega}_3 + \int_{\hat{T}} \hat{\phi}_k \, \hat{\phi}_m \, d\hat{\Omega}_{12} \int_{\hat{S}} \hat{\psi}_l' \, \hat{\psi}_n' \, d\hat{\Omega}_3. \end{aligned}$$

Solution 30. Mean-type essential boundary condition

a) Denote $\Omega = (0, 1)$. If u is a solution of [P30], then $u \in H^1(\Omega)$ and $u'' \in L^2(\Omega)$ (since $f \in L^2(\Omega)$). This shows that u belongs to $H^2(\Omega)$. From proposition 1.11, we have $u \in \mathcal{C}^1(\overline{\Omega})$. Multiplying the first equation of [P30] by v in $H^1(\Omega)$, and then integrating over $\Omega = (0, 1)$ by parts:

$$\int_0^1 u'v' \, dx - [u'v]_0^1 = \int_0^1 fv \, dx.$$

Taking into account the boundary condition, we obtain:

$$\int_0^1 u'v'\,dx = \int_0^1 f\,v dx - \alpha v(0) + \beta v(1) \quad \forall v \in H^1(\Omega). \qquad \text{[E30]}$$

From proposition 1.11, we know that $v \in \mathcal{C}^0(\overline{\Omega})$, so $v(0)$ and $v(1)$ are thus well defined.

b) The constant function $v = 1$ belongs to $H^1(\Omega)$ which, after injecting it into the relation [E30], we see that, in order for the problem [F30] to admit a solution, it is necessary for the data f, α and β to satisfy the compatibility condition:

$$\int_0^1 f dx - \alpha + \beta = 0. \qquad \text{[C30]}$$

c) The space $V = \left\{v \in H^1(\Omega) \text{ such that } \int_0^1 v\,dx = 0\right\}$ is a closed vector subspace of $H^1(\Omega)$, since it equals the preimage of $\{0\}$ under the mapping:

$$\left|\begin{array}{ll} H^1(\Omega) & \to \mathbb{R} \\ v & \mapsto \displaystyle\int_0^1 v(x)\,dx \end{array}\right.$$

which is a continuous linear mapping from the Cauchy–Schwarz inequality:

$$\left|\int_0^1 v(x)\,dx\right| \leq ||v||_{L^2(\Omega)} \leq ||v||_{H^1(\Omega)}$$

endowed with the norm $\|\cdot\|_{H^1(\Omega)}$, V is a Hilbert space.

d) From the above, if u is a solution of [P30], then $u \in V$ and, moreover, u is a solution of the variational problem:

$$\left|\begin{array}{l} \text{Find } u \in V \text{ such that} \\ \displaystyle\int_0^1 u'v'\,dx = \int_0^1 fv\,dx - \alpha v(0) + \beta v(1) \quad \forall v \in V. \end{array}\right. \qquad \text{[F30]}$$

By the Poincaré–Wirtinger inequality [2.23], we know that there exists a $C_{PW} > 0$ such that, for every $v \in H^1(\Omega)$:

$$\int_\Omega v^2 dx \leq C_{PW}\Big(\int_\Omega (v')^2 dx + \frac{1}{\text{meas}\Omega}\Big(\int_\Omega v\,dx\Big)^2\Big).$$

For every $v \in V$, we have thus:

$$\int_\Omega (v')^2 dx \geq \frac{1}{C_{PW}} \int_\Omega v^2 dx.$$

From this, we get:

$$\int_\Omega (v')^2 dx \geq \frac{1}{1 + C_{PW}} ||v||^2_{H^1(\Omega)}$$

which proves the coercivity of the continuous symmetric bilinear form $(u, v) \mapsto \int_0^1 u'v' \, dx$ on $V \times V$. By the virtue of the trace theorem 1.7 and the Cauchy–Schwarz inequality, we also have:

$$\Big| \int_0^1 f \, v dx - \alpha v(0) + \beta v(1) \Big| \leq \big(||f||_{L^2(\Omega)} + C \max(|\alpha|, |\beta|) \big) \, ||v||_{H^1(\Omega)}.$$

From the Lax–Milgram theorem, [F30$_b$] is well posed.

e) Let u be the solution of [F30$_b$], and let $v \in H^1(\Omega)$. Note that $w = v - \int_0^1 v \, dx$ belongs to V. We have:

$$\begin{aligned}
\int_0^1 u'v' dx &= \int_0^1 u'w' dx \\
\text{from [F30}_b\text{]} &= \int_0^1 fw \, dx - \alpha w(0) + \beta w(1) \\
&= \int_0^1 fv \, dx - \alpha v(0) + \beta v(1) - \Big(\int_0^1 v \, dx \Big) \Big(\int_0^1 f \, dx - \alpha + \beta \Big) \\
\text{from [C30]} &= \int_0^1 fv \, dx - \alpha v(0) + \beta v(1)
\end{aligned}$$

which shows that u is a solution of [F30], and hence of [P30].

f) We know that V_h is included in V by construction: from the above, the bilinear form $(u_h, v_h) \mapsto \int_0^1 u_h' v_h' \, dx$ is therefore continuous and coercive on $V_h \times V_h$. By applying the Lax–Milgram theorem to the variational formulation [F30$_h$], we see that it is well posed.

g) We have the following results:

- For each $1 \leq i \leq N$, the function $\hat{w}_i$ belongs to V_h since their means vanish (and each is continuous and piecewise regular). Moreover, we can check that the

$(\hat{w}_i)_{i=1,N}$ are linearly independent: if $\sum_{i=1,N} \alpha_i \hat{w}_i = 0$, then we have successively: $\sum_{i=1,N} \alpha_i \hat{w}_i(x_1) = 0$, which implies $\alpha_1 = 0$, and then $\sum_{i=2,N} \alpha_i \hat{w}_i(x_2) = 0$, which implies $\alpha_2 = 0$, and so forth, until $\alpha_N \hat{w}_N(x_N) = 0$ which implies finally that $\alpha_N = 0$. And, since $\dim(V_h) = N$, then $(\hat{w}_i)_{i=1,N}$ is indeed a basis of V_h.

- If we decompose $u_h \in V_h$ as $u_h = \sum_{j=1,N} U_j \hat{w}_j$, we see that the vector $\overrightarrow{U}$, whose components are $(U_i)_{i=1,N}$, is a solution of $\mathbb{A}\overrightarrow{U} = \overrightarrow{L}$, where:

$$\mathbb{A}_{ij} = \int_0^1 \hat{w}_j' \hat{w}_i' dx, \quad L_i = \int_0^1 f \hat{w}_i dx - \alpha \hat{w}_i(0) + \beta \hat{w}_i(1), \quad 1 \leq i, j \leq N.$$

- For $\mathbb{A}_{ij} \neq 0$ to hold, it is necessary for $\text{supp}(\hat{w}_j') \cap \text{supp}(\hat{w}_i')$ to be a line segment (i.e. not reduced to a point). However, for $3 \leq i, j \leq N-2$, we have:

$$\text{supp}(\hat{w}_j') = [x_{j-1}, x_{j+2}] \text{ and } \text{supp}(\hat{w}_i') = [x_{i-1}, x_{i+2}].$$

To obtain the condition above, it is therefore necessary that $|i - j| \leq 2$. Hence, there are *at most* five non-zero elements per row. After an easy calculation, we find that:

$$\mathbb{A}_{i,i-2} = \frac{1}{h}, \ \mathbb{A}_{i,i-1} = -\frac{4}{h}, \ \mathbb{A}_{i,i} = \frac{6}{h}, \ \mathbb{A}_{i,i+1} = -\frac{4}{h}, \ \mathbb{A}_{i,i+2} = \frac{1}{h}, \ 3 \leq i \leq N-2.$$

For the rows with indices 1 and N, we see that there are three. For example:

$$\mathbb{A}_{1,1} = \frac{10}{h}, \quad \mathbb{A}_{1,2} = -\frac{5}{h}, \quad \mathbb{A}_{1,3} = \frac{1}{h}.$$

For the rows with indices 2 and $N-1$, we see that there are four. For example:

$$\mathbb{A}_{2,1} = -\frac{5}{h}, \quad \mathbb{A}_{2,2} = \frac{6}{h}, \quad \mathbb{A}_{2,3} = -\frac{4}{h}, \quad \mathbb{A}_{2,4} = \frac{1}{h}.$$

- When $f = cst$, as the means of each $\hat{w}_i$ vanish, we get $L_i = -\alpha \hat{w}_i(0) + \beta \hat{w}_i(1)$ for each $1 \leq i \leq N$, so that:

$$L_1 = -\alpha, L_N = \beta \text{ and } L_i = 0, \text{ for } 2 \leq i \leq N-1.$$

A2.4. Exercises from Chapter 4

Solution 31. Integrating barycentric coordinates

a) Given a reference triangle $\hat{T}$ with vertices $(0,0),\ (1,0),\ (0,1)$, the barycentric coordinates are given by:

$$\hat{\lambda}_1(\hat{x}_1,\hat{x}_2) = 1 - \hat{x}_1 - \hat{x}_2,\quad \hat{\lambda}_2(\hat{x}_1,\hat{x}_2) = \hat{x}_1,\quad \hat{\lambda}_3(\hat{x}_1,\hat{x}_2) = \hat{x}_2.$$

Denote by M_{ℓ_1}, M_{ℓ_2}, and M_{ℓ_3} the vertices of the triangle T_ℓ. The affine transformation [3.34]:

$$F_\ell(\hat{x}_1,\hat{x}_2) = \sum_{i=1,3} \hat{\lambda}_i(\hat{x}_1,\hat{x}_2) M_{\ell_i}$$

sends the triangle $\hat{T}$ into the triangle T_ℓ with $F_\ell(0,0) = M_{\ell_1}$, $F_\ell(1,0) = M_{\ell_2}$, and $F_\ell(0,1) = M_{\ell_3}$. Provided that the triangle T_ℓ has a non-zero volume, then the transformation F_ℓ is differentiable and invertible, with

$$|\det J_{F_\ell}| = 2\,\text{Area}(T_\ell).$$

After taking the definition of barycentric coordinates into account, we have, for each $1 \le i \le 3$:

$$\begin{array}{l} \hat{\lambda}_i(\hat{x}_1,\hat{x}_2) = \lambda_{\ell_i}(F_\ell(\hat{x}_1,\hat{x}_2)) \\ \lambda_{\ell_i}(x_1,x_2) = \hat{\lambda}_i(F_\ell^{-1}(x_1,x_2)) \end{array}.$$

The change of variable $(x_1,x_2) \mapsto (\hat{x}_1,\hat{x}_2)$ leads to:

$$\int_{T_\ell} \lambda_{\ell_1}^{k_1}\lambda_{\ell_2}^{k_2}\lambda_{\ell_3}^{k_3}\,dx_1 dx_2 = 2\,\text{Area}(T_\ell)\int_{\hat{T}} \hat{\lambda}_1^{k_1}\hat{\lambda}_2^{k_2}\hat{\lambda}_3^{k_3}\,d\hat{x}_1\,d\hat{x}_2.$$

b) Since $\hat{T} = \{(\hat{x}_1,\hat{x}_2)$ such that $\hat{x}_1 \in [0,1],\ \hat{x}_2 \in [0,1-\hat{x}_1]\}$, by using the expression for the barycentric coordinates $\hat{\lambda}_i$, we are led to the calculation:

$$I_{k_1,k_2,k_3} = \int_0^1 \hat{x}_1^{k_2}\int_0^{1-\hat{x}_1} \hat{x}_2^{k_3}(1-\hat{x}_1-\hat{x}_2)^{k_1}\,d\hat{x}_2\,d\hat{x}_1.$$

By integrating successively by parts, we can show easily that:

$$\int_0^a s^m(a-s)^n ds = \frac{m!\,n!}{(m+n+1)!}a^{m+n+1}. \qquad \text{[R31]}$$

Applying [R31] with $a = 1 - \hat{x}_1$, $m = k_3$ and $n = k_1$, we obtain:

$$\int_0^{1-\hat{x}_1} \hat{x}_2^{k_3}(1-\hat{x}_1-\hat{x}_2)^{k_1} d\hat{x}_2 = \frac{k_3!\,k_1!}{(k_3+k_1+1)!}\,(1-\hat{x}_1)^{k_3+k_1+1}$$

which shows that:

$$I_{k_1,k_2,k_3} = \frac{k_3!\,k_1!}{(k_3+k_1+1)!}\int_0^1 \hat{x}_1^{k_2}(1-\hat{x}_1)^{k_3+k_1+1} d\hat{x}_1.$$

Again, by applying [R31] with $a = 1$, $m = k_2$, and $n = k_3 + k_1 + 1$, we obtain:

$$I_{k_1,k_2,k_3} = \frac{k_3!\,k_1!}{(k_3+k_1+1)!}\,\frac{k_2!\,(k_3+k_1+1)!}{(k_3+k_1+1+k_2+1)!} = \frac{k_1!\,k_2!\,k_3!}{(2+k_1+k_2+k_3)!}$$

which finally leads us to the result:

$$\int_{T_\ell} \lambda_{\ell_1}^{k_1}\lambda_{\ell_2}^{k_2}\lambda_{\ell_3}^{k_3}\,dx\,dy = \frac{2\,k_1!\,k_2!\,k_3!}{(2+k_1+k_2+k_3)!}\text{Area}(T_\ell). \qquad \text{[E31]}$$

c) Consider a simplex K_ℓ in d dimensions, made up from the $(\mathrm{d}+1)$ vertices $(M_{\ell_i})_{i=1,\mathrm{d}+1}$, which is the image of the reference d-simplex for $\hat{K}$ with the vertices $(0,...,0,...,0)$, $(1,0,...,0)$ ... $(0,...,0,1)$:

$$F_\ell : \left| \begin{array}{ccc} \hat{K} & \to & K_\ell \\ (\hat{x}_1,\ldots,\hat{x}_\mathrm{d}) & \mapsto & \displaystyle\sum_{i=1,\mathrm{d}+1} \hat{\lambda}_i(\hat{x}_1,\ldots,\hat{x}_\mathrm{d})M_{\ell_i} \end{array} \right.$$

where $(\hat{\lambda}_i)_{i=1,\mathrm{d}+1}$ denotes the barycentric coordinates of the d-simplex:

$$\hat{\lambda}_1 = 1 - \sum_{i=1,\mathrm{d}} \hat{x}_i \text{ and } \hat{\lambda}_i = \hat{x}_{i-1},\ \forall i = 2,\cdots,\mathrm{d}+1.$$

We proceed as in two dimensions. We have:

$$\int_{K_\ell}\Big(\prod_{i=1,\mathrm{d}+1}\lambda_{\ell_i}^{k_i}\Big)dx_1\ldots dx_\mathrm{d} = \mathrm{d}!\,\text{meas}(K_\ell)\int_{\hat{K}}\Big(\prod_{i=1,\mathrm{d}+1}\hat{\lambda}_i^{k_i}\Big)d\hat{x}_1\ldots d\hat{x}_\mathrm{d}$$

that is, in denoting:

$$I_{k_1,\dots,k_{\mathrm{d}+1}} = \int_{\hat{K}} \Big(\prod_{i=1,\mathrm{d}+1} \hat{\lambda}_i^{k_i} \Big) d\hat{x}_1 \dots d\hat{x}_{\mathrm{d}}$$

and $a_{\mathrm{d}-p} = 1 - \sum_{i=1,\mathrm{d}-p} \hat{x}_i$ for $p = 1, \cdots, \mathrm{d}$:

$$I_{k_1,\dots,k_{\mathrm{d}+1}} = \int_0^1 \hat{x}_1^{k_2} \int_0^{1-\hat{x}_1} \hat{x}_2^{k_3} \dots \int_0^{a_{\mathrm{d}-1}} \hat{x}_{\mathrm{d}}^{k_{\mathrm{d}+1}} (a_{\mathrm{d}-1} - \hat{x}_{\mathrm{d}})^{k_1} d\hat{x}_{\mathrm{d}} \dots d\hat{x}_1.$$

Since $a_{\mathrm{d}-1} = a_{\mathrm{d}-2} - \hat{x}_{\mathrm{d}-1}$, by applying [R31] with $s = \hat{x}_{\mathrm{d}}$, $a = a_{\mathrm{d}-1}$, $m = k_{\mathrm{d}+1}$ and $n = k_1$, we find:

$$\int_0^{a_{\mathrm{d}-1}} \hat{x}_{\mathrm{d}}^{k_{\mathrm{d}+1}} (a_{\mathrm{d}-1} - \hat{x}_{\mathrm{d}})^{k_1} d\hat{x}_{\mathrm{d}} = \frac{k_{\mathrm{d}+1}!\, k_1!}{(k_{\mathrm{d}+1} + k_1 + 1)!} (a_{\mathrm{d}-2} - \hat{x}_{\mathrm{d}-1})^{k_{\mathrm{d}+1}+k_1+1}.$$

Which gives:

$$\begin{aligned} I_{k_1,\dots,k_{\mathrm{d}+1}} = \frac{k_{\mathrm{d}+1}!\, k_1!}{(k_{\mathrm{d}+1} + k_1 + 1)!} \int_0^1 \hat{x}_1^{k_2} \int_0^{1-\hat{x}_1} \hat{x}_2^{k_3} \\ \dots \int_0^{a_{\mathrm{d}-2}} \hat{x}_{\mathrm{d}-1}^{k_{\mathrm{d}}} (a_{\mathrm{d}-2} - \hat{x}_{\mathrm{d}-1})^{k_{\mathrm{d}+1}+k_1+1} d\hat{x}_{\mathrm{d}-1} \dots d\hat{x}_1 \end{aligned}.$$

Since $a_{\mathrm{d}-2} = a_{\mathrm{d}-3} - \hat{x}_{\mathrm{d}-2}$, in applying [R31] again with $s = \hat{x}_{\mathrm{d}-1}$, $a = a_{\mathrm{d}-2}$, $m = k_{\mathrm{d}}$ and $n = k_{\mathrm{d}+1} + k_1 + 1$, we find:

$$\begin{aligned} I_{k_1,\dots,k_{\mathrm{d}+1}} = \frac{k_{\mathrm{d}+1}!\, k_{\mathrm{d}}!\, k_1!}{(k_{\mathrm{d}+1} + k_{\mathrm{d}} + k_1 + 2)!} \int_0^1 \hat{x}_1^{k_2} \int_0^{1-\hat{x}_1} \hat{x}_2^{k_3} \\ \dots \int_0^{a_{\mathrm{d}-3}} \hat{x}_{\mathrm{d}-2}^{k_{\mathrm{d}-1}} (a_{\mathrm{d}-3} - \hat{x}_{\mathrm{d}-2})^{k_{\mathrm{d}+1}+k_{\mathrm{d}}+k_1+1} d\hat{x}_{\mathrm{d}-2} \dots d\hat{x}_1 \end{aligned}.$$

Reiterating this, we arrive at $I_{k_1,\dots,k_{\mathrm{d}+1}} = \dfrac{k_1!\, k_2! \dots k_{\mathrm{d}+1}!}{(\mathrm{d} + k_1 + k_2 + \dots + k_{\mathrm{d}+1})!}$ and, consequently:

$$\int_{K_\ell} \Big(\prod_{i=1,\mathrm{d}+1} \lambda_{\ell_i}^{k_i} \Big) dx_1 \dots dx_{\mathrm{d}} = \frac{\mathrm{d}!\, k_1!\, k_2! \dots k_{\mathrm{d}+1}!}{(\mathrm{d} + k_1 + k_2 + \dots + k_{\mathrm{d}+1})!} \mathrm{meas}(K_\ell).$$

Solution 32. Electrostatics: potentials and capacitance

Consider a connected and bounded open subset Ω of $\mathbb{R}^3$ with a "sufficiently regular" boundary $\partial\Omega$ which decomposes into $K+1$ disjoint connected components:

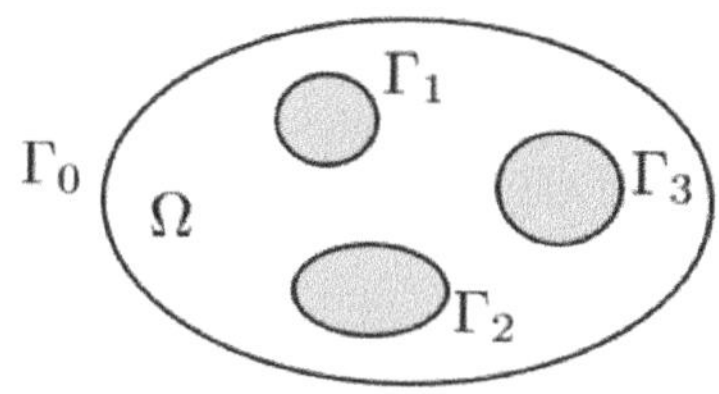

a) Let us show that there exists a function $u_R \in H^1(\Omega)$ such that $u_{R|\Gamma_k} = V_k$, with $0 \leq k \leq K$. For each $1 \leq k \leq K$, we can construct a function $u_{Rk} \in H^1(\Omega)$ such that $u_{Rk|\Gamma_j} = 0$ with $j \in \{0, \cdots, K\} \setminus \{k\}$; $u_{Rk|\Gamma_k} = 1$. Indeed, take a constant function that equals 1 in the neighborhood of Γ_k, that gets "regularly" truncated if we move far enough away from Γ_k, and so that it equals 0 in a neighborhood of each $(\Gamma_j)_{j \neq k}$: we then have a function $u_{Rk} \in \mathcal{C}^2(\overline{\Omega})$ such that $u_{Rk|\Gamma_j} = 0$ for each $j \in \{0, \cdots, K\} \setminus \{k\}$, and $u_{Rk|\Gamma_k} = 1$. The function $u_R = \sum_{k=1,K} V_k u_{Rk}$ belongs to $H^1(\Omega)$ and $u_{R|\Gamma_0} = 0$. By construction, $u_0 = u - u_R$ belongs to $H_0^1(\Omega)$.

Now consider the inner product on $L^2(\Omega)$ given by the first equation of [P32], along with a function v in $H_0^1(\Omega)$, before then integrating it by parts:

$$\int_\Omega \nabla u \cdot \nabla v \, d\Omega = \int_\Omega f \, v \, d\Omega.$$

Letting u_0 be an unknown function, then the variational formulation we have to solve is:

$$\left| \begin{array}{l} \text{Find } u_0 \in H_0^1(\Omega) \text{ such that} \\ \displaystyle\int_\Omega \nabla u_0 \cdot \nabla v \, d\Omega = \int_\Omega f \, v \, d\Omega - \int_\Omega \nabla u_R \cdot \nabla v \, d\Omega \quad \forall v \in H_0^1(\Omega). \end{array} \right. \quad \text{[F32]}$$

As $u_R \in \mathcal{C}^2(\overline{\Omega})$, we can replace the last term of [F32] with $\displaystyle\int_\Omega (\Delta u_R) v \, d\Omega$.

Finally, from theorem 2.3, we know that [P32] and [F32] are equivalent and hence [F32] is well posed.

b) The space W is a K-dimensional vector space. Indeed, for each $1 \leq k \leq K$, let χ_l be the solution of [P32], associated with $f = 0$ and to the values $(V_k)_{1 \leq k \leq K}$, such that $V_k = \delta_{kl}$. The set of functions $(\chi_k)_{1\leq k\leq K}$ is linearly independent since $\sum_{1\leq l\leq K} \alpha_l \chi_l = 0$ implies:

$$0 = \sum_{1\leq l\leq K} \alpha_l \chi_{l|\Gamma_k} = \alpha_k \text{ for each } 1 \leq k \leq K.$$

The set of functions $(\chi_k)_{1\leq k\leq K}$ generates W because every solution u of [P32], associated with $f = 0$ and to the values $(V_k)_{1\leq k\leq K}$, decomposes as:

$$u = \sum_{1\leq k\leq K} V_k \chi_k.$$

Indeed, the difference $u - \sum_{1\leq k\leq K} V_k \chi_k$ satisfies [P32] with the zero data: from the uniqueness of the solution, $u - \sum_{1\leq k\leq K} V_k \chi_k = 0$.

The capacitance matrix $\mathbb{C} \in \mathbb{R}^{K\times K}$ defined by:

$$\mathbb{C}_{kl} = \int_\Omega \nabla\chi_k \cdot \nabla\chi_l \, d\Omega$$

is symmetric. Moreover, it is positive-definite because, for every $\vec{V} \in \mathbb{R}^K$, we have:

$$\begin{aligned}(\mathbb{C}\vec{V}|\vec{V}) &= \sum_{1\leq k\leq K} \Big(\sum_{1\leq l\leq K} V_l \int_\Omega \nabla\chi_k \cdot \nabla\chi_l \, d\Omega \Big) V_k \\ &= \int_\Omega \nabla\Big(\sum_{1\leq k\leq K} V_k\chi_k \Big) \cdot \nabla\Big(\sum_{1\leq l\leq K} V_l\chi_l \Big) \, d\Omega \, . \\ &= \int_\Omega \Big| \nabla\Big(\sum_{1\leq k\leq K} V_k\chi_k \Big) \Big|^2 \, d\Omega \geq 0\end{aligned}$$

If $(\mathbb{C}\vec{V}|\vec{V}) = 0$, then $\sum_{1\leq k\leq K} V_k\chi_k$ is a constant function. The only possible constant is 0 since $\chi_{k|\Gamma_0} = 0$ for each $1 \leq k \leq K$. Consequently, $\vec{V} = 0$.

Let u be a solution of [P32] with $f = 0$ and $Q_k = \int_{\Gamma_k} \frac{\partial u}{\partial n} \, d\Gamma$ for each $0 \leq k \leq K$. We have:

$$0 = \int_\Omega \Delta u \, d\Omega = -\int_\Omega \nabla u \cdot \nabla 1 \, d\Omega + \int_{\partial\Omega} \frac{\partial u}{\partial n} \, d\Gamma = \sum_{k=0,K} Q_k.$$

Otherwise, for $k > 0$, we have:

$$\begin{aligned} Q_k &= \int_{\Gamma_k} \frac{\partial u}{\partial n}\, d\Gamma = \int_{\Gamma_k} \frac{\partial u}{\partial n}\, \chi_k d\Gamma = \int_{\partial\Omega} \frac{\partial u}{\partial n}\, \chi_k d\Gamma \qquad \text{(since } \chi_{k|\Gamma_l} = \delta_{kl}\text{)} \\ &= \int_\Omega \Delta u\, \chi_k\, d\Omega + \int_\Omega \nabla u \cdot \nabla \chi_k\, d\Omega = \int_\Omega \nabla u \cdot \nabla \chi_k\, d\Omega \quad \text{(since } \Delta u = 0\text{)} \\ &= \int_\Omega \nabla\Big(\sum_{l=1,K} V_l \chi_l\Big) \cdot \nabla \chi_k\, d\Omega = \sum_{l=1,K} \mathbb{C}_{kl} V_l. \end{aligned}$$

Hence, in denoting by $\overrightarrow{Q} \in \mathbb{R}^K$ the vector with the components $(Q_k)_{1 \le k \le K}$, then we have just shown that:

$$\overrightarrow{Q} = \mathbb{C}\overrightarrow{V} \text{ and } Q_0 = -\sum_{l=1,K} Q_l.$$

c) Introduce the mesh $\mathcal{T}_h$ of Ω made up from the tetrahedra $(T_\ell)_{l=1,L}$. Assume that the vertices that do not belong to $\partial\Omega$ are numbered from 1 to N: $(M_I)_{I=1,N}$, denote by w_I the global basis function associated with the vertex M_I, and set $V_h = \text{vect}\,(w_I)_{I=1,N} \subset H_0^1(\Omega)$.

Consider the approximate problem:

$$\left| \begin{array}{l} \text{Find } u_{0h} \in V_h \text{ such that} \\ \displaystyle\int_\Omega \nabla u_{0h} \cdot \nabla v_h\, d\Omega = \int_\Omega f\, v_h\, d\Omega + \int_\Omega (\Delta u_R)\, v_h\, d\Omega \quad \forall v_h \in V_h. \end{array} \right. \qquad [\text{F}32_h]$$

This is a well-posed problem whose solution $u_{0h} = \sum_{J=1,N} U_{0J} w_J$ is obtained by solving the linear system:

$$\mathbb{A}\overrightarrow{U} = \overrightarrow{L} + \overrightarrow{G}$$

where $\overrightarrow{U} \in \mathbb{R}^N$ is the vector with components $(U_{0J})_{1 \le J \le N}$ and, for $1 \le I, J \le N$:

$$\mathbb{A}_{IJ} = \int_\Omega \nabla w_I \cdot \nabla w_J\, d\Omega, \quad L_I = \int_\Omega f\, w_I\, d\Omega \text{ and } G_I = \int_\Omega (\Delta u_R)\, w_I\, d\Omega.$$

Since $f + \Delta u_R \in L^2(\Omega)$, we have the following error estimate with $s = 1$, if $\sigma_D = 1$, and $s \in [0, \sigma_D)$, if $\sigma_D < 1$ (see theorem 4.3):

$$\exists C_s > 0 \quad \forall f \in L^2(\Omega) \quad \forall h \quad \|u_0 - u_{0h}\|_{H^1(\Omega)} \le C_s h^s (\|f\|_{L^2(\Omega)} + \|\Delta u_R\|_{L^2(\Omega)}).$$

In the above, we know that $\sigma_D = 1$ if Ω is convex, and that $\sigma_D \in (1/2, 1)$ otherwise (see section 2.5.2 on the regularity of the solutions). Moreover, u_R depends only on Ω.

We have an estimate of the same type for the error $\|\chi_k - \chi_{hk}\|_{H^1(\Omega)}$, namely:

$$\exists D_s > 0 \quad \forall k \quad \forall h \quad \|\chi_k - \chi_{hk}\|_{H^1(\Omega)} \leq D_s h^s$$

with s defined as before. The only difference is that there is no longer any dependence on the data. The error of the capacitance matrix terms can be estimated as:

$$\begin{aligned}
|\mathbb{C}_{kl} - (\mathbb{C}_h)_{kl}| &= \Big| \int_\Omega \nabla\chi_k \cdot \nabla\chi_l \, d\Omega - \int_\Omega \nabla\chi_{hk} \cdot \nabla\chi_{hl} \, d\Omega \Big| \\
&= \Big| \int_\Omega \nabla(\chi_k - \chi_{hk}) \cdot \nabla\chi_l \, d\Omega - \int_\Omega \nabla\chi_{hk} \cdot \nabla(\chi_{hl} - \chi_l) \, d\Omega \Big| \\
&\leq \|\chi_k - \chi_{hk}\|_{H^1(\Omega)} \|\chi_l\|_{H^1(\Omega)} + \|\chi_l - \chi_{hl}\|_{H^1(\Omega)} \|\chi_{hk}\|_{H^1(\Omega)} \\
\text{(Triangle ineq.)} &\leq D_s h^s \|\chi_l\|_{H^1(\Omega)} + D_s h^s \left(\|\chi_k\|_{H^1(\Omega)} + D_s h^s \right) \\
&\leq D_s h^s (2 \max\nolimits_{1 \leq l \leq K} \|\chi_l\|_{H^1(\Omega)} + D_s h^s) \\
&\leq \tilde{C}_s h^s
\end{aligned}$$

where $\tilde{C}_s = D_s \sup_h(2 \max_{1 \leq l \leq K} \|\chi_l\|_{H^1(\Omega)} + D_s h^s) \in (0, +\infty)$. Finally, we have obtained the estimate $\|\mathbb{C} - \mathbb{C}_h\| \leq \tilde{D}_s h^s$ with $\tilde{D}_s > 0$, which we get from $\tilde{C}_s$ depending on the chosen matrix norm.

Solution 33. Bending of a beam

a) Lagrangian finite elements cannot be used to carry out an **internal** approximation of the space $H^2(a, b)$, nor a fortiori of the space $H_0^2(a, b)$, since the functions generated by the Lagrangian finite elements are continuous, yet are not C^1 at the interfaces between the elements. Consequently, these functions cannot belong to the $H^2(a, b)$ space, since of course their second derivatives are Dirac masses supported along certain interfaces.

b) In one dimension, as a third-order polynomial can be uniquely determined by its values and those of its derivatives at two distinct points, we deduce the existence of a unique polynomial $p \in \hat{P}$ such that, for every $(\alpha_0, \alpha_1, \tilde{\alpha}_0, \tilde{\alpha}_1) \in \mathbb{R}^4$:

$$\delta_0(p) = p(0) = \alpha_0, \; \delta_1(p) = p(1) = \alpha_1, \; \delta_0'(p) = p'(0) = \tilde{\alpha}_0, \; \delta_1'(p) = p'(1) = \tilde{\alpha}_1.$$

One easily obtains the expressions for the basis functions on the reference element, as:

$$\tau_1(\hat{x}) = (\hat{x}-1)^2(2\hat{x}+1) \ ; \ \tau_2(\hat{x}) = \hat{x}^2(3-2\hat{x})$$
$$\widetilde{\tau}_1(\hat{x}) = \hat{x}(\hat{x}-1)^2 \ ; \ \widetilde{\tau}_2(\hat{x}) = \hat{x}^2(\hat{x}-1).$$

c) The fact that the family of functions $\{w_i, \, \widetilde{w}_i, \, i = 1, L-1\}$ is linearly independent comes from the following properties, inherited from those of the local basis functions:

$$w_i(x_\ell) = \delta_{i\ell} \ \ w_i'(x_\ell) = 0 \ \ \widetilde{w}_i(x_\ell) = 0 \ \ \widetilde{w}_i'(x_\ell) = \delta_{i\ell} \ \ \forall i, \ell = 1, L-1 \qquad [E33]$$

and the following functions, valid for $1 \leq \ell \leq L-1$:

$$\sum_{i=1,L-1} \left(\alpha_i w_i + \widetilde{\alpha}_i \widetilde{w}_i\right)(x_\ell) = \alpha_\ell \ ; \quad \Big(\sum_{i=1,L-1} \alpha_i w_i + \widetilde{\alpha}_i \widetilde{w}_i \Big)' (x_\ell) = \widetilde{\alpha}_\ell.$$

For each $i = 1, L-1$, let us show that the functions w_i, $\widetilde{w}_i$ belong to $H^2(\Omega)$. It suffices to show that each of the functions, w_i and $\widetilde{w}_i$, as well as their derivatives, coincide at the points x_ℓ since, by construction, they are polynomials on each interval $\overset{\circ}{K}_\ell$, and also that they are piecewise regular. We consider the following three situations.

- At a point x_i of the interface between the elements K_{i-1} and K_i, we have:

$$w_i = \begin{cases} \tau_2^{i-1} & \text{on } K_{i-1} \\ \tau_1^i & \text{on } K_i \end{cases} \quad \text{and} \quad \widetilde{w}_i = \begin{cases} \widetilde{\tau}_2^{i-1} & \text{on } K_{i-1} \\ \widetilde{\tau}_1^i & \text{on } K_i \end{cases}$$

where, by the definition of the local basis functions:

$$\tau_2^{i-1}(x_i) = \tau_1^i(x_i) = 1, \ \left(\tau_2^{i-1}\right)'(x_i) = \left(\tau_1^i\right)'(x_i) = 0,$$
$$\widetilde{\tau}_2^{i-1}(x_i) = \widetilde{\tau}_1^i(x_i) = 0, \ \left(\widetilde{\tau}_2^{i-1}\right)'(x_i) = \left(\widetilde{\tau}_1^i\right)'(x_i) = 1$$

which proves that the functions and their derivatives coincide.

- At either of the two points, x_{i-1} or x_{i+1}, we have, for example with x_{i-1}, and where $2 \leq i \leq L-1$:

$$w_i = \begin{cases} 0 & \text{on } K_{i-2} \\ \tau_2^{i-1} & \text{on } K_{i-1} \end{cases} \quad \text{and} \quad \widetilde{w}_i = \begin{cases} 0 & \text{on } K_{i-2} \\ \widetilde{\tau}_2^{i-1} & \text{on } K_{i-1} \end{cases}$$

and as $\tau_2^{i-1}(x_{i-1}) = \left(\tau_2^{i-1}\right)'(x_{i-1}) = \widetilde{\tau}_2^{i-1}(x_{i-1}) = \left(\widetilde{\tau}_2^{i-1}\right)'(x_{i-1}) = 0$, the functions and their derivatives also coincide.

- At one point x_ℓ with $\ell \notin \{i-1, i, i+1\}$, we have:

$$w_i = \begin{cases} 0 \text{ on } K_{\ell-1} \\ 0 \text{ on } K_\ell \end{cases} \quad \text{and } \widetilde{w}_i = \begin{cases} 0 \text{ on } K_{\ell-1} \\ 0 \text{ on } K_\ell \end{cases}$$

from which we see that the function and its derivative clearly coincide.

It remains to show that, for each $i = 1, L-1$, we have $w_{i|\Gamma} = \widetilde{w}_{i|\Gamma} = w'_{i|\Gamma} = \widetilde{w}'_{i|\Gamma} = 0$, where $\Gamma = \{a, b\}$. To do so, it suffices to examine the previous formulae based on the two points, $x_0 = a$ and $x_L = b$, to convince oneself of this.

d) Taking in account the property [E33], we obtain the following decompositions, valid for $1 \leq \ell \leq L-1$:

$$v_h(x_\ell) = \sum_{i=1,L-1} \left(\alpha_i w_i(x_\ell) + \widetilde{\alpha}_i \widetilde{w}_i(x_\ell)\right) = \alpha_\ell$$
$$v'_h(x_\ell) = \sum_{i=1,L-1} \left(\alpha_i w'_i(x_\ell) + \widetilde{\alpha}_i \widetilde{w}'_i(x_\ell)\right) = \widetilde{\alpha}_\ell$$

and which finally lead to the decomposition:

$$v_h = \sum_{i=1,L-1} \left(v_h(x_i) w_i + v'_h(x_i) \widetilde{w}_i\right)$$

hence showing that the basis coefficients equal the function values, as well as the derivative values, when evaluated at (internal) vertices of the mesh.

e) The discretized problem with Hermitian finite elements is:

| Find $u_h \in V_h$ such that $a(u_h, v_h) = l(v_h) \;\; \forall v_h \in V_h$ [F33$_h$]

with $a(u_h, v_h) = \displaystyle\int_a^b u''_h v''_h \, dx$ and $l(v_h) = \displaystyle\int_a^b f v_h \, dx$.

We write:

$$u_h = \sum_{j=1,L-1} \left(u_h(x_j) w_j + u'_h(x_j) \widetilde{w}_j\right)$$

and we choose $v_h = w_i$ and $v_h = \widetilde{w}_i$ in the discrete formulation, for $i = 1, L-1$:

$$\begin{cases} \displaystyle\sum_{j=1,L-1} u_h(x_j) a(w_j, w_i) + \sum_{j=1,L-1} u'_h(x_j) a(\widetilde{w}_j, w_i) = l(w_i) \;\; \forall i = 1, L-1 \\ \displaystyle\sum_{j=1,L-1} u_h(x_j) a(w_j, \widetilde{w}_i) + \sum_{j=1,L-1} u'_h(x_j) a(\widetilde{w}_j, \widetilde{w}_i) = l(\widetilde{w}_i) \;\; \forall i = 1, L-1 \end{cases}$$

which makes up a square linear system of order $2(L-1)$. There are two options when it comes to storing the unknowns. The first is to separate the unknown "functions" $u_h(x_i)$ from their unknown "derivatives" $u'_h(x_i)$: introduce $\overrightarrow{U} \in \mathbb{R}^{L-1}$ with components $(u_h(x_i))_{1\leq i\leq L-1}$, and $\overrightarrow{U}' \in \mathbb{R}^{L-1}$ with components $(u'_h(x_i))_{1\leq i\leq L-1}$, which leads to a linear system of the form:

$$\begin{bmatrix} \mathbb{K} & \mathbb{C} \\ \mathbb{C}^t & \mathbb{K}' \end{bmatrix} \begin{pmatrix} \overrightarrow{U} \\ \overrightarrow{U}' \end{pmatrix} = \begin{pmatrix} \overrightarrow{L} \\ \overrightarrow{L}' \end{pmatrix}$$

where, for each $1 \leq i, j \leq L-1$:

$$\mathbb{K}_{ij} = \int_\Omega w''_i\, w''_j d\Omega,\ \mathbb{K}'_{ij} = \int_\Omega \widetilde{w}''_i\, \widetilde{w}''_j d\Omega,\ \mathbb{C}_{ij} = \int_\Omega \widetilde{w}''_j\, w''_i\, d\Omega,$$
$$L_i = \int_\Omega f\, w_i\, d\Omega,\ L'_i = \int_\Omega f\, \widetilde{w}_i\, d\Omega.$$

The other possibility consists of successively ordering $u_h(x_i)$ and $u'_h(x_i)$, that is, introduce the $L-1$ vectors $\overrightarrow{U}_i \in \mathbb{R}^2$ with components $(u_h(x_i), u'_h(x_i))$, which leads to a system made up of 2×2 blocks, formally written as:

$$\begin{bmatrix} \mathbb{B}_{1,1} & \mathbb{B}_{1,2} & \dots & \mathbb{B}_{1,L-1} \\ \mathbb{B}_{2,1} & \mathbb{B}_{2,2} & \dots & \mathbb{B}_{2,L-1} \\ \dots & \dots & \dots & \dots \\ \mathbb{B}_{L-1,1} & \mathbb{B}_{L-1,2} & \dots & \mathbb{B}_{L-1,L-1} \end{bmatrix} \begin{pmatrix} \overrightarrow{U}_1 \\ \overrightarrow{U}_2 \\ \vdots \\ \overrightarrow{U}_{L-1} \end{pmatrix} = \begin{pmatrix} \overrightarrow{S}_1 \\ \overrightarrow{S}_2 \\ \vdots \\ \overrightarrow{S}_{L-1} \end{pmatrix}$$

where the $L-1$ vectors $\overrightarrow{S}_i \in \mathbb{R}^2$ are:

$$\overrightarrow{S}_i = \begin{pmatrix} \int_\Omega f\, w_i\, d\Omega \\ \int_\Omega f\, \widetilde{w}_i\, d\Omega \end{pmatrix}.$$

f) The matrix structure depends on the storage choice for the unknowns. In the first case, each matrix $\mathbb{K}$, $\mathbb{K}'$ and $\mathbb{C}$, is tridiagonal, because the support of the basis functions is given by: support(w_i) = support$(\widetilde{w}_i) = [x_{i-1}, x_{i+1}]$.

Due to the support properties, in the second case the only blocks that can be identically non-zero are $\mathbb{B}_{i,i-1}$, $\mathbb{B}_{i,i}$ and $\mathbb{B}_{i,i+1}$. The matrix is then a tridiagonal matrix made up of 2×2 blocks:

$$\begin{bmatrix} \mathbb{B}_{1,1} & \mathbb{B}_{1,2} & 0 & \ddots & \ddots \\ \mathbb{B}_{2,1} & \mathbb{B}_{2,2} & \mathbb{B}_{2,3} & \ddots & \ddots \\ 0 & \mathbb{B}_{3,2} & \mathbb{B}_{3,3} & \ddots & 0 \\ \ddots & \ddots & \ddots & \ddots & \mathbb{B}_{L-2,L-1} \\ \ddots & \ddots & 0 & \mathbb{B}_{L-1,L-2} & \mathbb{B}_{L-1,L-1} \end{bmatrix}.$$

REMARK A2.10.– *In the second case, we can write very efficient algorithms that solve the linear system (algorithms with $O(L)$ operations). This is not the case with the first numbering choice (algorithms with $O(L^2)$ operations). We refer to Appendix A1.4.2.*

Solution 34. Helmholtz problem

Endow $V = H_0^1(\Omega)$ with the inner product $(\cdot,\cdot)_V = (\nabla\cdot,\nabla\cdot)_{(L^2(\Omega))^d}$.

a) The equivalent variational formulation is written as:

$$\left| \begin{array}{ll} \text{Find } u \in V \text{ such that} & \\ a(u,v) = (f,v)_{L^2(\Omega)} & \forall v \in V \end{array} \right. \qquad \text{[F34]}$$

where $a(v,w) = (\nabla v, \nabla w)_{(L^2(\Omega))^d} - \omega^2 (v,w)_{L^2(\Omega)}$ for every $v, w \in V$.

From the spectral corollary (exercise 18), we can decompose v and w into the following:

$$v = \sum_{j \geq 1} \alpha_j e_j, \; w = \sum_{k \geq 1} \beta_k e_k$$

where $\alpha_j = (v, e_j)_{L^2(\Omega)}$ for $j \geq 1$, $\|v\|_V = (\sum_{j \geq 1} \lambda_j \alpha_j^2)^{1/2}$, and $\beta_k = (w, e_k)_{L^2(\Omega)}$ for $k \geq 1$, $\|w\|_V = (\sum_{k \geq 1} \lambda_k \beta_k^2)^{1/2}$. Then we find:

$$\begin{aligned} a(v,w) &= \lim_{J,K \to \infty} a(\sum_{j \leq J} \alpha_j e_j, \sum_{k \leq K} \beta_k e_k) \\ &= \lim_{J,K \to \infty} (\sum_{j \leq J, k \leq K} \alpha_j \beta_k \{(\nabla e_j, \nabla e_k)_{(L^2(\Omega))^d} - \omega^2 (e_j, e_k)_{L^2(\Omega)}\} \\ &= \lim_{J \to \infty} (\sum_{j \leq J} \alpha_j \beta_j (\lambda_j - \omega^2) = \sum_{j \geq 1} \alpha_j \beta_j (\lambda_j - \omega^2) \quad \text{(double orthogonality)} \end{aligned}$$

Above, we have that $\lambda_j > 0$ for each $j \geq 1$.

b) By definition: $Pv = \sum_{j \leq J_0} \alpha_j e_j = \sum_{j \leq J_0} (v, e_j)_{L^2(\Omega)} e_j$. Note that, for every $v \in V$, we have $v = Pv + (I - P)v$, as well as the *orthogonal relations* $(Pv, (I - P)v)_{L^2(\Omega)} = 0$ and $(\nabla(Pv), \nabla((I - P)v))_{(L^2(\Omega))^d} = 0$. It follows that $a(Pv, (I - P)v) = 0$, and:

$$a(v,v) = a(Pv + (I - P)v, Pv + (I - P)v) = a(Pv, Pv) + a((I - P)v, (I - P)v).$$

However, for each $j \leq J_0$, we have $\lambda_j - \omega^2 < 0$, and for each $j > J_0$, we have $\lambda_j - \omega^2 > 0$, and so:

$$a(Pv, Pv) = \sum_{j \leq J_0} \alpha_j^2(\lambda_j - \omega^2) \leq 0 \text{ and } a((I - P)v, (I - P)v) = \sum_{j > J_0} \alpha_j^2(\lambda_j - \omega^2) \geq 0.$$

c) The idea here is to place a minus sign in front of the negative term so that it becomes positive, that is to replace:

$$+ \sum_{j \leq J_0} \alpha_j^2(\lambda_j - \omega^2) + \sum_{j > J_0} \alpha_j^2(\lambda_j - \omega^2)$$

with:

$$- \sum_{j \leq J_0} \alpha_j^2(\lambda_j - \omega^2) + \sum_{j > J_0} \alpha_j^2(\lambda_j - \omega^2).$$

To do this, introduce the linear mapping $T \in \mathcal{L}(V)$:

$$v \mapsto Tv = -Pv + (I - P)v.$$

For every $v \in V$, with the orthogonality relations for P, it now becomes:

$$\begin{aligned} a(v, Tv) &= a(Pv + (I - P)v, -Pv + (I - P)v) \\ &= -a(Pv, Pv) + a((I - P)v, (I - P)v) \geq 0. \end{aligned}$$

Moreover, if $v \neq 0$, then one of the two terms must be strictly positive: $a(v, Tv) > 0$.

d) Again, using the orthogonality property $(Pv, (I-P)v)_V = 0$, we obtain:

$$\begin{aligned}\|Tv\|_V^2 &= \|Pv\|_V^2 + \|(I-P)v\|_V^2 - 2(Pv,(I-P)v)_V \\ &= \|Pv\|_V^2 + \|(I-P)v\|_V^2 + 2(Pv,(I-P)v)_V = \|v\|_V^2\end{aligned}.$$

Hence T is indeed an isometry.

To establish the well-posedness of [F34], we can apply corollary 2.1 of the Banach–Nečas–Babuška theorem since a is symmetric. To establish the inf-sup condition [2.43], the idea is, for $v \neq 0$, to *bound* the following ratio *from below*:

$$\frac{a(v,Tv)}{\|Tv\|_V} \geq \alpha' \|v\|_V$$

where $\alpha' > 0$ is independent of v. Recall that $\|v\|_V = (\sum_{j\geq 1} \lambda_j \alpha_j^2)^{1/2}$. Moreover:

$$\begin{aligned}a(v,Tv) &= \sum_{j\geq 1} \alpha_j^2 |\lambda_j - \omega^2| = \sum_{j\geq 1} \alpha_j^2 \lambda_j \frac{|\lambda_j - \omega^2|}{\lambda_j} \\ &\geq \min_{k\geq 1}\left(\frac{|\lambda_k - \omega^2|}{\lambda_k}\right) \sum_{j\geq 1} \alpha_j^2 \lambda_j = \min_{k\geq 1}\left(\frac{|\lambda_k - \omega^2|}{\lambda_k}\right) \|v\|_V^2\end{aligned}.$$

If we set $\alpha' = \min_{k\geq 1}\left(\frac{|\lambda_k - \omega^2|}{\lambda_k}\right) \in (0,+\infty)$ (recall that $\omega^2 \notin \{\lambda_k, k \geq 1\}$ and $\lim_{k\to\infty} \lambda_k = +\infty$) then, since T is an isometry, we obtain:

$$\frac{a(v,Tv)}{\|Tv\|_V} \geq \alpha' \frac{\|v\|_V^2}{\|Tv\|_V} \geq \alpha' \|v\|_V.$$

The inf-sup condition [2.43] follows from this, and so we may conclude that, if ω^2 does not coincide with any eigenvalue, then the Helmholtz problem is well posed.

Let us base ourselves in the finite-dimensional vector subspaces $(V_h)_h$ of V.

e) The discrete variational formulation can be written as:

$$\left| \begin{array}{ll} \text{Find } u_h \in V_h \text{ such that} & \\ a(u_h, v_h) = (f, v_h)_{L^2(\Omega)} & \forall v_h \in V_h. \end{array} \right. \qquad [\text{F34}_h]$$

f) From the approximability property (see proposition 3.2), for each $j \leq J_0$, we know that $\lim_{h\to 0} \inf_{v_h \in V_h} \|e_j - v_h\|_V = 0$. As there is a *finite number* of eigenfunctions (there are J_0 of them), we see that there exists a postive function $\delta(h)$ with $\lim_{h\to 0} \delta(h) = 0$, and a total of J_0 sequences made up of the elements $(e_j^h)_{j\leq J_0, h}$ such that:

$$\forall j \leq J_0 \quad \forall h \quad \|e_j - e_j^h\|_V \leq \delta(h).$$

In particular, there exists a $h_0 > 0$ such that, for every $h \leq h_0$, we have $\max_{1\leq j\leq J_0} \|e_j - e_j^h\|_V \leq 2$. We restrict our attention to this configuration from now on, that is when $h \leq h_0$.

- From above, a "natural" choice for a linear mapping that approximates P is $P_h \in \mathcal{L}(V, \mathrm{vect}(e_j^h)_{j=1,J_0})$, and is defined by $P_h v = \sum_{j\leq J_0} (v, e_j^h)_{L^2(\Omega)} e_j^h$.

- By definition, for every $v \in V$, we have:

$$Pv = \sum_{j=1}^{J_0} (v, e_j)_{L^2(\Omega)} e_j \ ; \quad P_h v = \sum_{j=1}^{J_0} (v, e_j^h)_{L^2(\Omega)} e_j^h \,.$$

Consequently:

$$\begin{aligned}(P - P_h)v &= \sum_{j=1}^{J_0} \left((v, e_j)_{L^2(\Omega)} e_j - (v, e_j^h)_{L^2(\Omega)} e_j^h\right) \\ &= \sum_{j=1}^{J_0} \left((v, e_j)_{L^2(\Omega)} e_j - (v, e_j)_{L^2(\Omega)} e_j^h + (v, e_j)_{L^2(\Omega)} e_j^h - (v, e_j^h)_{L^2(\Omega)} e_j^h\right). \\ &= \sum_{j=1}^{J_0} \left((v, e_j)_{L^2(\Omega)} (e_j - e_j^h) + (v, e_j - e_j^h)_{L^2(\Omega)} e_j^h\right)\end{aligned}$$

From which we see that, for every $h \leq h_0$:

$$\begin{aligned}\|(P - P_h)v\|_V &\leq \sum_{j=1}^{J_0} \left(\left|(v, e_j)_{L^2(\Omega)}\right| \|e_j - e_j^h\|_V + \left|(v, e_j - e_j^h)_{L^2(\Omega)}\right| \|e_j^h\|_V\right) \\ &\leq \sum_{j=1}^{J_0} \left(\|v\|_{L^2(\Omega)} \|e_j\|_{L^2(\Omega)} \|e_j - e_j^h\|_V + \|v\|_{L^2(\Omega)} \|e_j - e_j^h\|_{L^2(\Omega)} \|e_j^h\|_V\right) \\ &\leq C\delta(h) \|v\|_V\end{aligned}$$

where, to get to the last line, we have used the Poincaré inequality. The constant C depends on J_0, λ_{J_0} (since $\max_{1\leq j\leq J_0} \|e_j\|_V = \sqrt{\lambda_{J_0}}$) and the Poincaré constant, but is independent of h. So we have:

$$\lim_{h\to 0}\left(\sup_{v_h\in V_h\setminus\{0\}} \frac{\|(P-P_h)v_h\|_V}{\|v_h\|_V}\right) \leq \lim_{h\to 0}(C\delta(h)) = 0.$$

- Introduce the discrete linear mapping $T_h \in \mathcal{L}(V_h)$:

$$v_h \mapsto T_h v_h = -P_h v_h + (I - P_h)v_h.$$

- In choosing $v_h \in V_h$, we have:

$$(T - T_h)v_h = -Pv_h + (I - P)v_h + P_h v_h - (I - P_h)v_h = 2(P_h - P)v_h.$$

- From above, we then have:

$$\sup_{v_h\in V_h\setminus\{0\}} \frac{\|(T-T_h)v_h\|_V}{\|v_h\|_V} = 2 \sup_{v_h\in V_h\setminus\{0\}} \frac{\|(P-P_h)v_h\|_V}{\|v_h\|_V}.$$

From this, we see that:

$$\forall h \leq h_0 \qquad \sup_{v_h\in V_h\setminus\{0\}} \frac{\|(T-T_h)v_h\|_V}{\|v_h\|_V} \leq 2C\delta(h).$$

Denote by M the continuity modulus of the form a. For every $h \leq h_0$ and $v_h \in V_h$, we have the lower bound:

$$\begin{aligned} a(v_h, T_h v_h) &= a(v_h, T v_h) + a(v_h, (T_h - T)v_h) \\ &\geq \alpha' \|v_h\|_V^2 - M\,\|v_h\|_V\,\|(T_h - T)v_h\|_V \\ &\geq (\alpha' - 2CM\,\delta(h))\|v_h\|_V^2. \end{aligned}$$

As $\lim_{h\to 0}\delta(h) = 0$, there exists $\alpha^\star > 0$ and $h_1 \in (0, h_0)$ such that, for every $h \leq h_1$, $\alpha' - 2CM\,\delta(h) > \alpha^\star$, and $2C\delta(h) < 1$. The discrete uniform inf-sup condition [3.12] is therefore satisfied with $\alpha'' = \alpha^\star/2$ where $h \leq h_1$. Indeed, after choosing $v_h' = T_h v_h$ in [3.12] and recalling that:

$$\|T_h v_h\|_V \leq \|T v_h\|_V + \|(T_h - T)v_h\|_V \leq (1 + 2C\delta(h))\|v_h\|_V$$

from which we have:

$$\sup_{v'_h \in V_h \setminus \{0\}} \frac{|a(v_h, v'_h)|}{\|v'_h\|_V} \geq \frac{a(v_h, T_h v_h)}{\|T_h v_h\|_V} \geq \frac{\alpha^\star \|v_h\|_V^2}{(1 + 2C\delta(h))\|v_h\|_V} \geq \frac{\alpha^\star}{2} \|v_h\|_V.$$

From this, we observe that Céa's lemma 3.3 can now be applied when h is "sufficiently small":

$$\exists C > 0 \quad \forall h < h_1 \quad \|u - u_h\|_V \leq C \inf_{v_h \in V_h} \|u - v_h\|_V.$$

g) If $u \in \mathcal{P}H^2(\Omega)$ then, by reasoning as in section 4.1.3, we get the following error estimate:

$$\exists C > 0 \quad \forall h < h_1 \quad \|u - u_h\|_V \leq Ch\, |u|_{2,\mathcal{P}}.$$

Solution 35. Regularization of the Dirichlet condition

To solve the Dirichlet problem [P35] (numerically), we are interested in a so-called "penalized" formulation, which is different from the typical one obtained, and which can be stated in the variational form [F35$_\varepsilon$], where $0 < \varepsilon < \varepsilon_{max}$ is a "small" parameter.

a) Let us see whether we fall within the scope of the Lax–Milgram theorem or not. The symmetric bilinear form:

$$a_\varepsilon(u, v) = \int_\Omega (\nabla u \cdot \nabla v + uv)\, d\Omega + \frac{1}{\varepsilon} \int_{\partial\Omega} uv\, d\Gamma$$

is continuous on $H^1(\Omega)$ since, for every $u, v \in H^1(\Omega)$:

$$|a_\varepsilon(u, v)| = \left| (u, v)_{H^1(\Omega)} + \frac{1}{\varepsilon} \int_{\partial\Omega} u\, v\, d\Gamma \right|$$

$$\text{(Cauchy–Schwarz)} \leq \|u\|_{H^1(\Omega)} \|v\|_{H^1(\Omega)} + \varepsilon^{-1} \|u\|_{L^2(\partial\Omega)} \|v\|_{L^2(\partial\Omega)} \cdot$$

$$\text{(trace theorem 1.7)} \leq (1 + C^2\varepsilon^{-1}) \|u\|_{H^1(\Omega)} \|v\|_{H^1(\Omega)}$$

Denoting by M_ε the continuity modulus of the form a_ε, we have $M_\varepsilon \leq C\varepsilon^{-1}$, where $C > 0$ is independent of $\varepsilon \in (0, \varepsilon_{max})$. It is coercive on $H^1(\Omega)$ since, for every $v \in H^1(\Omega)$:

$$a_\varepsilon(v,v) = \|v\|^2_{H^1(\Omega)} + \frac{1}{\varepsilon}\|v\|^2_{L^2(\partial\Omega)} \geq \|v\|^2_{H^1(\Omega)}.$$

Moreover, the linear form:

$$l(v) = \int_\Omega fv\, d\Omega + \frac{1}{\varepsilon}\int_{\partial\Omega} gv\, d\Gamma$$

is continuous on $H^1(\Omega)$, since for every $v \in H^1(\Omega)$:

$$\begin{aligned} |l(v)| &\leq \|f\|_{L^2(\Omega)}\|v\|_{L^2(\Omega)} + \varepsilon^{-1}\|g\|_{L^2(\partial\Omega)}\|v\|_{L^2(\partial\Omega)} \\ \text{(trace theorem)} &\leq \left(\|f\|_{L^2(\Omega)} + C\varepsilon^{-1}\|g\|_{L^2(\partial\Omega)}\right)\|v\|_{H^1(\Omega)} \end{aligned}.$$

Thus, the Lax–Milgram theorem does assert the existence and uniqueness of a solution for the variational formulation [F35$_\varepsilon$].

By choosing $v \in \mathcal{D}(\Omega)$ in [F35$_\varepsilon$], first we see that $-\Delta u_\varepsilon + u_\varepsilon = f$ in the distribution sense. In particular, $\Delta u_\varepsilon \in L^2(\Omega)$, and we can apply Green's formula [1.39]. Then, for every $v \in H^1(\Omega)$, we get:

$$\langle \frac{\partial u_\varepsilon}{\partial n} + \frac{1}{\varepsilon}u_\varepsilon, v\rangle_{(H^{1/2}(\partial\Omega))', H^{1/2}(\partial\Omega)} = \frac{1}{\varepsilon}\int_{\partial\Omega} gv\, d\Gamma.$$

From the surjectivity of the trace mapping, we can see that:

$$\varepsilon\frac{\partial u_\varepsilon}{\partial n} + u_\varepsilon = g_{|\partial\Omega} \text{ on } \partial\Omega$$

which therefore shows that the Dirichlet condition can be approximated (ε is a "small" parameter) by using a Fourier condition.

b) As a reminder, we set $p = \frac{\partial u}{\partial n}_{|\partial\Omega} \in L^2(\partial\Omega)$. By multiplying the equation $-\Delta u + u = f$ by $v \in H^1(\Omega)$, and then integrating it by parts over the open subset Ω, before then using the fact that $u = g$ on $\partial\Omega$, we get the identity $\forall v \in H^1(\Omega)$:

$$\int_\Omega (\nabla u \cdot \nabla v + u\, v)\, d\Omega + \frac{1}{\varepsilon}\int_{\partial\Omega} uv\, d\Gamma = \int_\Omega fv\, d\Omega + \int_{\partial\Omega} pv\, d\Gamma + \frac{1}{\varepsilon}\int_{\partial\Omega} gv\, d\Gamma.$$

c) In taking the difference between the previous relation and [F35$_\varepsilon$], we obtain:

$$\int_\Omega (\nabla v_\varepsilon \cdot \nabla v + v_\varepsilon v) d\Omega + \frac{1}{\varepsilon}\int_{\partial\Omega} v_\varepsilon v \, d\Gamma = \int_{\partial\Omega} pv \, d\Gamma \;\; \forall v \in H^1(\Omega)$$

and which, if we set $v = v_\varepsilon$, leads to the identity:

$$\int_\Omega \left(|\nabla v_\varepsilon|^2 + (v_\varepsilon)^2\right) d\Omega + \frac{1}{\varepsilon}\int_{\partial\Omega} (v_\varepsilon)^2 d\Gamma = \int_{\partial\Omega} pv_\varepsilon d\Gamma.$$

We get the following two inequalities immediately, by applying the Cauchy–Schwarz inequality in $L^2(\partial\Omega)$:

$$\|v_\varepsilon\|^2_{H^1(\Omega)} \leq \|p\|_{L^2(\partial\Omega)} \|v_\varepsilon\|_{L^2(\partial\Omega)} \text{ and } \|v_\varepsilon\|_{L^2(\partial\Omega)} \leq \varepsilon \|p\|_{L^2(\partial\Omega)}.$$

d) By combining the two previous inequalities, we obtain at the following error estimate:

$$\|u - u_\varepsilon\|_{H^1(\Omega)} \leq \sqrt{\varepsilon}\|p\|_{L^2(\partial\Omega)} \qquad [I35]$$

which shows that the solution u_ε of the variational formulation [F35$_\varepsilon$] converges to the solution u of the Dirichlet problem [P35], when ε tends toward 0. However, we will see that this convergence is not very fast (in $\sqrt{\varepsilon}$) and that, from this point of view, it is not a very efficient method when it comes to dealing with the Dirichlet condition.

Furthermore, depending on the data, we can estimate $\|p\|_{L^2(\partial\Omega)}$. Indeed, we have:

$$\|p\|_{L^2(\partial\Omega)} = \|\frac{\partial u}{\partial n}_{|\partial\Omega}\|_{L^2(\partial\Omega)} \leq \|(\nabla u)_{|\partial\Omega}\|_{(L^2(\partial\Omega))^{\mathrm{d}}}$$

since $\frac{\partial u}{\partial n} = \nabla u \cdot \mathbf{n}$ on $\partial\Omega$, and $\mathbf{n} \in (L^\infty(\partial\Omega))^{\mathrm{d}}$ with $|\mathbf{n}| = 1$ almost everywhere.

By applying Theorem 1.7 regarding the trace mapping to $(\partial_i u)_{i=1,\mathrm{d}} \in (H^1(\Omega))^{\mathrm{d}}$, we have $\|(\nabla u)_{|\partial\Omega}\|_{(L^2(\partial\Omega))^{\mathrm{d}}} \leq C\|\nabla u\|_{(H^1(\Omega))^{\mathrm{d}}} \leq C\|u\|_{H^2(\Omega)}$. Moreover, we know that $\|u\|_{H^2(\Omega)}$ depends continuously on the data, f and g (see section 2.5.2):

$$\|u\|_{H^2(\Omega)} \leq C'(\|f\|_{L^2(\Omega)} + \|g\|_{H^2(\Omega)})$$

from which we see that $\|p\|_{L^2(\partial\Omega)} \leq C''(\|f\|_{L^2(\Omega)} + \|g\|_{H^2(\Omega)})$, where $C'' = CC'$, and lastly that:

$$\|u - u_\varepsilon\|_{H^1(\Omega)} \leq C''\sqrt{\varepsilon}(\|f\|_{L^2(\Omega)} + \|g\|_{H^2(\Omega)}). \qquad [I35_a]$$

e) On the approximation space for the P^1-Lagrangian finite element, which we denote by V_h, introduce the following discrete variational formulation with the form a_ε:

$$\left| \begin{array}{l} \text{Find } u_\varepsilon^h \in V_h \text{ such that} \\ a_\varepsilon(u_\varepsilon^h, v_h) = \displaystyle\int_\Omega f v_h \, d\Omega + \frac{1}{\varepsilon} \int_{\partial\Omega} g v_h \, d\Gamma \quad \forall v_h \in V_h. \end{array} \right.$$

Since V_h is included in $H^1(\Omega)$, this problem is well posed because it can be endowed with the induced norm from $H^1(\Omega)$. The form a_ε is a coercive and continuous bilinear one on V_h, and the form l is a continuous linear one on V_h. We therefore find ourselves in the setting where the Lax–Milgram theorem can be applied again.

f) The error estimate is obtained by observing that:

$$\|u - u_\varepsilon^h\|_{H^1(\Omega)} \leq \|u - u_\varepsilon\|_{H^1(\Omega)} + \|u_\varepsilon - u_\varepsilon^h\|_{H^1(\Omega)}.$$

The left-hand side had been bounded in Question (d), see [I35] or [I35$_a$]. Then, as a_ε is symmetric, from Céa's lemma ([3.8] bound) we know that:

$$\|u_\varepsilon - u_\varepsilon^h\|_{H^1(\Omega)} \leq (M_\varepsilon)^{1/2} \inf_{v_h \in V_h} \|u_\varepsilon - v_h\|_{H^1(\Omega)}$$

since the coercivity constant of a_ε is bounded from below by 1. Returning to the proof of theorem 4.1, we find:

$$\|u_\varepsilon - u_\varepsilon^h\|_{H^1(\Omega)} \leq C \frac{h}{\sqrt{\varepsilon}} |u_\varepsilon|_{2,\Omega}$$

where $|u_\varepsilon|_{2,\Omega}$ is uniformly bounded for all ε. Putting the two terms together, we finally end up with:

$$\|u - u_\varepsilon^h\|_{H^1(\Omega)} \leq C\left(\frac{h}{\sqrt{\varepsilon}} + \sqrt{\varepsilon}\right) \qquad \text{[I35}_b\text{]}$$

where $C > 0$ depends only on the data, f and g. In minimizing the function $\varepsilon \mapsto h(\varepsilon) = \frac{h}{\sqrt{\varepsilon}} + \sqrt{\varepsilon}$, we see that $\varepsilon = h$ is optimal. Global convergence is slow, as it is in the order of $\sqrt{h}$.

g) In denoting the global basis function associated with the node M_j by w_j, we have $u_\varepsilon^h = \sum_j U_j w_j$ with $U_j = u_\varepsilon^h(M_j)$. If we choose $v = w_i$ in the discrete variational formulation, then we obtain:

$$\sum_j U_j \int_\Omega (\nabla w_j \cdot \nabla w_i + w_j w_i)\, d\Omega + \frac{1}{\varepsilon} \sum_j U_j \int_{\partial\Omega} w_j w_i \, d\Gamma$$
$$= \int_\Omega f w_i \, d\Omega + \frac{1}{\varepsilon} \int_{\partial\Omega} g w_i \, d\Gamma$$

that is, in matrix form:

$$\left(\mathbb{K} + \mathbb{M} + \frac{1}{\varepsilon}\mathbb{M}^\Gamma\right) \vec{U} = \vec{F} + \frac{1}{\varepsilon}\vec{G}$$

with:

$$\mathbb{K}_{ij} = \int_\Omega \nabla w_j \cdot \nabla w_i d\Omega, \ \ \mathbb{M}_{ij} = \int_\Omega w_j w_i d\Omega, \ \ \mathbb{M}^\Gamma_{ij} = \int_{\partial\Omega} w_j w_i d\Gamma$$

and:

$$F_i = \int_\Omega f w_i d\Omega, \ \ G_i = \int_{\partial\Omega} g w_i d\Omega.$$

We will solve this problem in a subspace of $H^1(\Omega)$, rather than in a subspace of $H_0^1(\Omega)$. This approach allows us to bypass any algorithmic processing that is required to "eliminate" the constraints induced by Dirichlet condition (see, e.g., pseudo-elimination presented in Chapter 5). However, the regularization technique has some major drawbacks: its global convergence is in the order of $\sqrt{h}$, as opposed to h as for the elimination methods, in addition to the deterioration of the conditioning of the linear system that we are trying to solve (which is all the more important as ε is small), and which requires the use of more specialized, and more computationally expensive, solvers.

Solution 36. Stokes' problem

a) Given a solution $(\mathbf{u}, p)$ of [P36], then in particular we have $\mathbf{u} \in (H^1(\Omega))^2$ and, from the boundary condition, $\mathbf{u} \in (H_0^1(\Omega))^2$. We know also that $-\nu\, \Delta u_i + \partial_i p = f_i$

for each $i = 1, 2$. We will proceed by differentiating in the sense of distributions. Let $\mathbf{v} \in (\mathcal{D}(\Omega))^2$, then we have:

$$\begin{aligned}\int_\Omega \mathbf{f} \cdot \mathbf{v}\, d\Omega &= \langle \mathbf{f}, \mathbf{v} \rangle = \sum_{i=1,2} \langle f_i, v_i \rangle \overset{[P36]}{=} \sum_{i=1,2} \langle -\nu \Delta u_i + \frac{\partial p}{\partial x_i}, v_i \rangle \\ &= \nu \sum_{i=1,2} \langle \nabla u_i, \nabla v_i \rangle - \sum_{i=1,2} \langle p, \frac{\partial v_i}{\partial x_i} \rangle \\ &= \nu \sum_{i=1,2} \int_\Omega \nabla u_i \cdot \nabla v_i \, d\Omega - \int_\Omega p \operatorname{div} \mathbf{v}\, d\Omega\end{aligned}.$$

Since $(\mathcal{D}(\Omega))^2$ is dense in $(H_0^1(\Omega))^2$: for every $\mathbf{v} \in (H_0^1(\Omega))^2$, there exists a sequence $(\mathbf{v}_k)_k \in \left((\mathcal{D}(\Omega))^2\right)^{\mathbb{N}}$ such that $\lim_{k\to\infty} |\mathbf{v} - \mathbf{v}_k|_1 = 0$. As a result, we have $\lim_{k\to\infty} \|\mathbf{v} - \mathbf{v}_k\|_{(L^2(\Omega))^2} = 0$ (Poincaré's inequality), $\lim_{k\to\infty} \|\nabla v_i - \nabla(v_k)_i\|_{(L^2(\Omega))^2} = 0$ $(i = 1, 2)$ and $\lim_{k\to\infty} \|\operatorname{div} \mathbf{v} - \operatorname{div} \mathbf{v}_k\|_{L^2(\Omega)} = \lim_{k\to\infty} \| \sum_{i=1,2} (\frac{\partial v_i}{\partial x_i} - \frac{\partial v_i^k}{\partial x_i})\|_{L^2(\Omega)} = 0$. Then we get:

$$\nu \sum_{i=1,2} \int_\Omega \nabla u_i \cdot \nabla v_i \, d\Omega - \int_\Omega p \operatorname{div} \mathbf{v}\, d\Omega = \int_\Omega \mathbf{f} \cdot \mathbf{v}\, d\Omega \quad \forall \mathbf{v} \in (H_0^1(\Omega))^2.$$

Now, if we use the fact that $\operatorname{div} \mathbf{u} = 0$ holds in $L^2(\Omega)$ ($\frac{\partial u_i}{\partial x_i} \in L^2(\Omega)$, for $1 \leq i \leq 2$), we have:

$$\int_\Omega q \operatorname{div} \mathbf{u}\, d\Omega = 0 \quad \forall q \in L^2(\Omega).$$

Therefore, $(\mathbf{u}, p)$ satisfies [F36].

The converse is almost immediate: the boundary condition on $\mathbf{u}$ comes from the fact that $\mathbf{u} \in \left(H_0^1(\Omega)\right)^2$, the relations involving the test function q show that $\operatorname{div} \mathbf{u} = 0$ in $L^2(\Omega)$ and, to find the first equation of [P36], one can reason in the sense of distributions by considering the test function $\mathbf{v}$ in $(\mathcal{D}(\Omega))^2$.

b) It is straightforward to see that [F36] implies $\displaystyle\int_\Omega q \operatorname{div} \mathbf{u}\, d\Omega = 0 \quad \forall q \in L^2_{zmv}(\Omega)$.

For the converse, note that any element $q \in L^2(\Omega)$ can be decomposed as $q = q_0 + c$, where c is the mean value of q on Ω, and $q_0 \in L^2_{zmv}(\Omega)$. Moreover, this decomposition is orthogonal since $(q_0, c)_{L^2(\Omega)} = c \int_\Omega q_0 \, d\Omega = 0$. We can then see that:

$$L^2(\Omega) = L^2_{zmv}(\Omega) \overset{\perp}{\oplus} \mathbb{R}.$$

In particular, as $\mathbf{u}_{|\partial\Omega} = 0$ implies $\mathbf{u} \cdot \mathbf{n}_{|\partial\Omega} = 0$ in $L^2(\Omega)$, we have for $c \in \mathbb{R}$:

$$\int_\Omega c \operatorname{div} \mathbf{u} \, d\Omega = c \int_\Omega \operatorname{div} \mathbf{u} \, d\Omega = c \int_{\partial\Omega} \mathbf{u} \cdot \mathbf{n} \, d\Gamma = 0.$$

c) From Poincaré's inequality, the space:

$$\mathbf{V} = \left\{ \mathbf{v} \in \left(H_0^1(\Omega)\right)^2 \text{ such that } \operatorname{div} \mathbf{v} = 0 \right\}$$

is a Hilbert space endowed with the norm $|\cdot|_1$: $\mathbf{v} \mapsto (|v_1|^2_{1,\Omega} + |v_2|^2_{1,\Omega})^{1/2}$, since it is a closed vector subspace of $(H_0^1(\Omega))^2$. Indeed, $\mathbf{V}$ is the pre-image of $\{0\}$ (a closed subset of $L^2(\Omega)$) under the continuous mapping:

$$\operatorname{div} : \left| \begin{array}{ccc} \left(H_0^1(\Omega)\right)^2 & \to & L^2(\Omega) \\ \mathbf{v} & \mapsto & \operatorname{div} \mathbf{v} \end{array} \right. .$$

Where continuity follows from the inequality:

$$\begin{aligned} \|\operatorname{div} \mathbf{v}\|_{L^2(\Omega)} &= \left\| \frac{\partial v_1}{\partial x_1} + \frac{\partial v_2}{\partial x_2} \right\|_{L^2(\Omega)} \\ &\leq \sqrt{2} \left(\left\| \frac{\partial v_1}{\partial x_1} \right\|^2_{L^2(\Omega)} + \left\| \frac{\partial v_2}{\partial x_2} \right\|^2_{L^2(\Omega)} \right)^{1/2} \leq \sqrt{2}\, |\mathbf{v}|_1 \end{aligned} .$$

d) Using the fact that $\operatorname{div} \mathbf{v} = 0$ for every $\mathbf{v} \in \mathbf{V}$, we see from [F36] that $\mathbf{u}$ is a solution of:

$$\left| \begin{array}{l} \text{Find } \mathbf{u} \in \mathbf{V} \text{ such that} \\ \nu \displaystyle\sum_{i=1,2} \int_\Omega \nabla u_i \cdot \nabla v_i \, d\Omega = \int_\Omega \mathbf{f} \cdot \mathbf{v} \, d\Omega \quad \forall \mathbf{v} \in \mathbf{V}. \end{array} \right. \qquad [F36_a]$$

e) The continuity of the linear form $\mathbf{v} \mapsto \int_\Omega \mathbf{f} \cdot \mathbf{v} \, d\Omega$ on $\mathbf{V}$ follows from the Cauchy–Schwarz inequality. Moreover, as soon as $\nu > 0$, the symmetric bilinear form $(\mathbf{u}, \mathbf{v}) \mapsto \nu \sum_{i=1,2} \int_\Omega \nabla u_i \cdot \nabla v_i \, d\Omega$ coincides with the inner product on $\left(H_0^1(\Omega)\right)^2$. From Riesz's representation theorem, the variational formulation [F36$_a$] admits a unique solution.

f) Even when Ω is polygonal, using Lagrangian finite elements would not give us an internal approximation of $\mathbf{V}$, since the basis functions $(\mathbf{w}_I)_I$ corresponding to the Lagrangian finite elements do not satisfy $\operatorname{div} \mathbf{w}_I = 0$ in Ω.

g) For a given $\varepsilon > 0$, we introduce the following "penalized" problem:

$$\left| \begin{array}{l} \text{Find } \mathbf{u}^{\varepsilon} \in \left(H_0^1(\Omega)\right)^2 \text{ such that} \\ a^{\varepsilon}(\mathbf{u}^{\varepsilon}, \mathbf{v}) = \displaystyle\int_{\Omega} \mathbf{f} \cdot \mathbf{v}\, d\Omega \quad \forall \mathbf{v} \in \left(H_0^1(\Omega)\right)^2 \end{array} \right. \qquad [\text{F}36_{\varepsilon}]$$

where the bilinear form is defined on $\left(H_0^1(\Omega)\right)^2$:

$$a^{\varepsilon}(\mathbf{u}, \mathbf{v}) = \nu \sum_{i=1,2} \int_{\Omega} \nabla u_i^{\varepsilon} \cdot \nabla v_i d\Omega + \frac{1}{\varepsilon} \int_{\Omega} \operatorname{div} \mathbf{u}^{\varepsilon} \operatorname{div} \mathbf{v}\, d\Omega.$$

By the continuity of the mapping div, a^{ε} is continuous on $(H_0^1(\Omega))^2$; observe that we can choose the value $M_{\varepsilon} = \nu + 2\varepsilon^{-1}$ for the continuity modulus of a^{ε}. On the other hand, a^{ε} is coercive on $(H_0^1(\Omega))^2$, since:

$$a^{\varepsilon}(\mathbf{v}, \mathbf{v}) = \nu \sum_{i=1,2} \int_{\Omega} |\nabla v_i|^2 d\Omega + \frac{1}{\varepsilon} \int_{\Omega} (\operatorname{div} \mathbf{v})^2 d\Omega \geq \nu\, |\mathbf{v}|_1^2.$$

We can choose $\alpha_{\varepsilon} = \nu$ as the coercivity constant of a^{ε}. From the Lax–Milgram theorem, the variational formulation [F36$_{\varepsilon}$] therefore admits a unique solution.

h) If we choose $\mathbf{v} = \mathbf{u}^{\varepsilon}$ in [F36$_{\varepsilon}$], we get:

$$\nu |\mathbf{u}^{\varepsilon}|_1^2 + \varepsilon^{-1} \|\text{div}\mathbf{u}^{\varepsilon}\|_{L^2(\Omega)}^2 = \int_{\Omega} \mathbf{f} \cdot \mathbf{u}^{\varepsilon}\, d\Omega. \qquad [\text{E}36_{\varepsilon}]$$

On the left-hand side, there are two positive terms. Therefore, [E36$_{\varepsilon}$] implies that:

$$\nu |\mathbf{u}^{\varepsilon}|_1^2 \leq \int_{\Omega} \mathbf{f} \cdot \mathbf{u}^{\varepsilon}\, d\Omega \leq \|\mathbf{f}\|_{(L^2(\Omega))^2} \|\mathbf{u}^{\varepsilon}\|_{(L^2(\Omega))^2} \leq \sqrt{C_p}\, \|\mathbf{f}\|_{(L^2(\Omega))^2} |\mathbf{u}^{\varepsilon}|_1$$

where $C_p > 0$ is Poincaré's constant. And then $|\mathbf{u}^{\varepsilon}|_1 \leq \frac{\sqrt{C_p}}{\nu} \|\mathbf{f}\|_{(L^2(\Omega))^2}$.

Revisiting [E36$_{\varepsilon}$], we can now see that:

$$\varepsilon^{-1} \|\text{div}\mathbf{u}^{\varepsilon}\|_{L^2(\Omega)}^2 \leq \int_{\Omega} \mathbf{f} \cdot \mathbf{u}^{\varepsilon}\, d\Omega \leq \sqrt{C_p}\, \|\mathbf{f}\|_{(L^2(\Omega))^2} |\mathbf{u}^{\varepsilon}|_1 \leq \frac{C_p}{\nu} \|\mathbf{f}\|_{(L^2(\Omega))^2}^2.$$

Therefore, $\|\text{div}\mathbf{u}^{\varepsilon}\|_{L^2(\Omega)} \leq C\sqrt{\varepsilon}\, \|\mathbf{f}\|_{(L^2(\Omega))^2}$, with $C = \sqrt{\frac{C_p}{\nu}} > 0$.

i) We set $\mathbf{u}^+ = \mathbf{u} - \mathbf{u}^\varepsilon \in (H_0^1(\Omega))^2$, and $p^+ = p + \varepsilon^{-1}\mathbf{div}\mathbf{u}^\varepsilon \in L^2_{zmv}(\Omega)$.

From the "difference" between [F36] and [F36$_\varepsilon$], for every $\mathbf{v} \in \left(H_0^1(\Omega)\right)^2$ we have:

$$\nu \sum_{i=1,2} \int_\Omega \nabla u_i^+ \cdot \nabla v_i d\Omega - \int_\Omega p^+ \operatorname{div} \mathbf{v}\, d\Omega = 0.$$

Furthermore, for every $q \in L^2_{zmv}(\Omega)$ we have:

$$\int_\Omega q\operatorname{div} \mathbf{u}^+ \, d\Omega = - \int_\Omega q\operatorname{div}\mathbf{u}^\varepsilon \, d\Omega.$$

From this, we see that the pair $(\mathbf{u}^+, p^+)$ is a solution of [F36$^+$] with the data $\mathbf{f}^+ = 0$ and $g^+ = -\operatorname{div}\mathbf{u}^\varepsilon \in L^2_{zmv}(\Omega)$. Since we suppose that [F36$^+$] is well posed, we know that:

$$|\mathbf{u}^+|_1 + \|p^+\|_{L^2(\Omega)} \leq C^+ \, \|g^+\|_{L^2(\Omega)}$$

where $C^+ > 0$ is independent of g^+. This leads us to the estimate:

$$|\mathbf{u} - \mathbf{u}^\varepsilon|_1 + \|p + \varepsilon^{-1}\mathbf{div}\mathbf{u}^\varepsilon\|_{L^2(\Omega)} \leq C^+ \sqrt{\frac{C_p}{\nu}} \, \sqrt{\varepsilon} \, \|\mathbf{f}\|_{(L^2(\Omega))^2}$$

and hence the sought after estimate for $|\mathbf{u} - \mathbf{u}^\varepsilon|_1$.

j) Denote by $\mathbf{V}_0^h = (V_0^h)^2$ the approximation space of $(H_0^1(\Omega))^2$. Introduce the discretized variational formulation:

$$\left| \begin{array}{l} \text{Find } \mathbf{u}_h \in \mathbf{V}_0^h \text{ such that} \\ a^\varepsilon(\mathbf{u}_h, \mathbf{v}_h) = \displaystyle\int_\Omega \mathbf{f} \cdot \mathbf{v}_h d\Omega \quad \forall \mathbf{v}_h \in \mathbf{V}_0^h. \end{array} \right. \qquad [\text{F36}_{h,\varepsilon}]$$

In applying the Lax–Milgram theorem to $\mathbf{V}_0^h$, we see that [F36$_{h,\varepsilon}$] admits a unique solution.

k) The error estimate is obtained in noting that $|\mathbf{u} - \mathbf{u}_h|_1 \leq |\mathbf{u} - \mathbf{u}^\varepsilon|_1 + |\mathbf{u}^\varepsilon - \mathbf{u}_h|_1$. The left-hand term is bounded by Question i. As a^ε is symmetric, we know from Céa's lemma ([3.8] bound) that:

$$|\mathbf{u}^\varepsilon - \mathbf{u}_h|_1 \leq \left(\frac{\nu + 2\varepsilon^{-1}}{\nu}\right)^{1/2} \inf_{\mathbf{v}_h \in \mathbf{V}_0^h} |\mathbf{u}^\varepsilon - \mathbf{v}_h|_1.$$

In revisiting the proof of theorem 4.1, and since $(|u_1^\varepsilon|_{2,\Omega})_\varepsilon$ and $(|u_2^\varepsilon|_{2,\Omega})_\varepsilon$ are uniformly bounded with respect to ε, we find that:

$$|\mathbf{u}^\varepsilon - \mathbf{u}_h|_1 \leq C\frac{h}{\sqrt{\varepsilon}}$$

where $C > 0$ does not depend on ε. Putting these two terms together, we get:

$$|\mathbf{u} - \mathbf{u}_h|_1 \leq C\Big(\frac{h}{\sqrt{\varepsilon}} + \sqrt{\varepsilon}\Big) \qquad [I36]$$

where $C > 0$ depends only on the data, $\mathbf{f}$. As in the previous exercise, we see that $\varepsilon = h$ is a optimal: the global convergence then is in the order of $\sqrt{h}$.

1) The solution $\mathbf{u}_h$ can be decomposed as:

$$\mathbf{u}_h = \begin{pmatrix} \sum_{J=1,M} U_1^J w_J \\ \sum_{J'=1,M} U_2^{J'} w_{J'} \end{pmatrix}.$$

After injecting this decomposition into [F36$_{h,\varepsilon}$] and considering $\mathbf{v}_h = \begin{pmatrix} w_I \\ 0 \end{pmatrix}$, we obtain:

$$\sum_{J=1,M} U_1^J\Big(\nu\int_\Omega \nabla w_J \cdot \nabla w_I \, d\Omega + \frac{1}{\varepsilon}\int_\Omega \frac{\partial w_J}{\partial x_1}\frac{\partial w_I}{\partial x_1}\, d\Omega\Big)$$
$$+\frac{1}{\varepsilon}\sum_{J=1,M} U_2^J\Big(\int_\Omega \frac{\partial w_J}{\partial x_2}\frac{\partial w_I}{\partial x_1}\, d\Omega\Big) = \int_\Omega f_1 \, w_I \, d\Omega$$

which can be written in the matrix form:

$$\mathbb{A}_{11}\vec{U}_1 + \mathbb{A}_{12}\vec{U}_2 = \vec{B_1}$$

with $B_I^1 = \displaystyle\int_\Omega f_1 \, w_I \, d\Omega$ and:

$$(\mathbb{A}_{11})_{IJ} = \nu\int_\Omega \nabla w_J \cdot \nabla w_I \, d\Omega + \frac{1}{\varepsilon}\int_\Omega \frac{\partial w_J}{\partial x_1}\frac{\partial w_I}{\partial x_1}\, d\Omega$$

$$(\mathbb{A}_{12})_{IJ} = \frac{1}{\varepsilon}\int_\Omega \frac{\partial w_J}{\partial x_2}\frac{\partial w_I}{\partial x_1}\, d\Omega$$

If we choose $\mathbf{v}_h = \begin{pmatrix} 0 \\ w_I \end{pmatrix}$, then we obtain the equations:

$$\mathbb{A}_{21}\overrightarrow{U}_1 + \mathbb{A}_{22}\overrightarrow{U}_2 = \overrightarrow{B_2}$$

with $B_I^2 = \int_\Omega f_2\, w_I\, d\Omega$ and:

$$(\mathbb{A}_{22})_{IJ} = \nu \int_\Omega \nabla w_J \cdot \nabla w_I\, d\Omega + \frac{1}{\varepsilon}\int_\Omega \frac{\partial w_J}{\partial x_2}\frac{\partial w_I}{\partial x_2}\, d\Omega$$

$$(\mathbb{A}_{21})_{IJ} = \frac{1}{\varepsilon}\int_\Omega \frac{\partial w_J}{\partial x_1}\frac{\partial w_I}{\partial x_2}\, d\Omega$$

We end up with a linear system of order $2M$ that couples together the two components of $\mathbf{u}_h$:

$$\begin{bmatrix} \mathbb{A}_{11} & \mathbb{A}_{12} \\ \mathbb{A}_{21} & \mathbb{A}_{22} \end{bmatrix} \begin{pmatrix} \overrightarrow{U}_1 \\ \overrightarrow{U}_2 \end{pmatrix} = \begin{pmatrix} \overrightarrow{B}_1 \\ \overrightarrow{B}_2 \end{pmatrix}.$$

The matrices $\mathbb{A}_{11}$ and $\mathbb{A}_{22}$ are symmetric. Moreover, note that $(\mathbb{A}_{12})_{IJ} = (\mathbb{A}_{21})_{JI}$ for $1 \leq I, J \leq M$, which means that $\mathbb{A}_{21} = \mathbb{A}_{12}^t$.

m) By restricting p_h to the triangle T_ℓ, we obtain:

$$p_{h|T_\ell} = (\text{div}\, \mathbf{u}_h)_{|T_\ell} = \text{div}(\mathbf{u}_{h|T_\ell}).$$

If we denote by ℓ_i the global numbering of the ith vertex M_i^ℓ of the triangle T_ℓ (see [3.51]), and by τ_i^ℓ the local basis function associated with the vertex M_i^ℓ in T_ℓ, we get:

$$\mathbf{u}_{h|T_\ell} = \begin{pmatrix} \sum_{i=1,3} U_1^{\ell_i}\tau_i^\ell \\ \sum_{i=1,3} U_2^{\ell_i}\tau_i^\ell \end{pmatrix}$$

which leads to:

$$\text{div}(\mathbf{u}_{h|T_\ell}) = \sum_{i=1,3}\left(U_1^{\ell_i}\frac{\partial \tau_i^\ell}{\partial x_1} + U_2^{\ell_i}\frac{\partial \tau_i^\ell}{\partial x_2}\right).$$

However, $\partial_1 \tau_i^\ell$ and $\partial_2 \tau_i^\ell$ are constants since τ_i^ℓ is a first-degree polynomial. Therefore, we can effectively write that $p_h = \sum_{\ell=1,L} p_\ell \mathcal{X}_\ell$, along with the relation:

$$p_\ell = \sum_{i=1,3} \left(U_1^{\ell_i} \frac{\partial \tau_i^\ell}{\partial x_1} + U_2^{\ell_i} \frac{\partial \tau_i^\ell}{\partial x_2} \right), \qquad \text{for } \ell = 1, L.$$

Solution 37. Solving the Maxwell equations

a) For the edges C_x parallel to Ox, we have $\mathbf{t} = (\pm 1, 0)^t$ and $\mathbf{v} \cdot \mathbf{t}_{|C_x} = 0$, giving us $v_{1|C_x} = 0$. For the edges C_y parallel to Oy, we have $\mathbf{t} = (0, \pm 1)^t$ and $\mathbf{v} \cdot \mathbf{t}_{|C_y} = 0$, giving us $v_{2|C_y} = 0$.

b) Let $\mathbf{v} \in \mathbf{V}$. Multiply the first equation of [P37] by curl $\mathbf{v}$, and then integrate over Ω. Let us then add the second equation, multiplied by div $\mathbf{v}$, before then integrating over Ω, to give:

$$\int_\Omega \operatorname{curl} \mathbf{E} \operatorname{curl} \mathbf{v} \, d\Omega + \int_\Omega \operatorname{div} \mathbf{E} \operatorname{div} \mathbf{v} \, d\Omega = \int_\Omega f \operatorname{curl} \mathbf{v} \, d\Omega.$$

Furthermore, we have assumed that $\mathbf{E} \in \mathbf{V}$; $\mathbf{E}$ is therefore a solution of:

$$\left| \begin{array}{l} \text{Find } \mathbf{E} \in \mathbf{V} \text{ such that} \\ \displaystyle\int_\Omega \operatorname{curl} \mathbf{E} \operatorname{curl} \mathbf{v} \, d\Omega + \int_\Omega \operatorname{div} \mathbf{E} \operatorname{div} \mathbf{v} \, d\Omega = \int_\Omega f \operatorname{curl} \mathbf{v} \, d\Omega \quad \forall \mathbf{v} \in \mathbf{V}. \end{array} \right. \qquad \text{[F37]}$$

c) Suppose that q belongs to $H_0^1(\Omega)$ and furthermore that it is a solution of the Dirichlet problem, $\Delta q = \operatorname{div} \mathbf{E}$ in Ω. As Ω is convex, we know that $q \in H^2(\Omega)$ (see section 2.5.2). Thus, we have $\nabla q \in (H^1(\Omega))^2$. As $q = 0$ on $\partial\Omega$, according to the hint contained in the statement of the question, $\nabla q \cdot \mathbf{t}_{|\partial\Omega} = 0$, and consequently $\nabla q \in \mathbf{V}$. Let us choose $\mathbf{v} = \nabla q$ in the variational formulation [F37]. Then, by taking into account that curl $(\nabla q) = 0$, and that div $(\nabla q) = \Delta q = \operatorname{div} \mathbf{E}$ in Ω, we obtain:

$$\int_\Omega (\operatorname{div} \mathbf{E})^2 \, d\Omega = 0$$

proving that div $\mathbf{E} = 0$ on Ω.

d) From the above, since $\operatorname{div}\mathbf{E} = 0$, from [F37] we see that:

$$\int_\Omega \operatorname{curl}\mathbf{E}\operatorname{curl}\mathbf{v}\,d\Omega = \int_\Omega f\operatorname{curl}\mathbf{v}\,d\Omega \ \ \forall \mathbf{v} \in \mathbf{V}.$$

Consider a solution $p \in H^1(\Omega)$ of the Neumann problem:

$$\begin{cases} \Delta p = \operatorname{curl}\mathbf{E} - f & \text{in } \Omega \\ \dfrac{\partial p}{\partial n} = 0 & \text{on } \partial\Omega. \end{cases}$$

The compatibility condition [2.37], that is, that $\int_\Omega (\operatorname{curl}\mathbf{E} - f)\,d\Omega = 0$, is satisfied, on the one hand, as the mean of f vanishes (by hypothesis) and, on the other hand, by integrating by parts:

$$\int_\Omega \operatorname{curl}\mathbf{E}\,d\Omega = \int_\Omega (\partial_1 E_2 - \partial_2 E_1)d\Omega = \int_{\partial\Omega} (E_2 n_1 - E_1 n_2)d\Gamma = -\int_{\partial\Omega} \mathbf{E}\cdot\mathbf{t}\,d\Gamma = 0.$$

As Ω is convex, we know that $p \in H^2(\Omega)$ (see section 2.5.2) and that $\mathbf{curl}\,p \in (H^1(\Omega))^2$ with $\mathbf{curl}\,p \cdot \mathbf{t}_{|\partial\Omega} = \nabla p \cdot \mathbf{n}_{|\partial\Omega} = \frac{\partial p}{\partial n} = 0$. In the previous reduced formulation, choose $\mathbf{v} = \mathbf{curl}\,p$ and, since $-\operatorname{curl}\,(\mathbf{curl}\,p) = \Delta p = \operatorname{curl}\mathbf{E} - f$, we see that:

$$\int_\Omega (\operatorname{curl}\mathbf{E} - f)^2 d\Omega = 0$$

hence proving that $\operatorname{curl}\mathbf{E} = f$ on Ω. In conclusion, the results of this question, along with those from the previous one, show that any solution of the variational formulation [F37] satisfies the Maxwell equations [P37]. Thus, [P37] and [F37] are equivalent.

e) After integrating by parts, we obtain:

$$\begin{aligned}\int_\Omega \operatorname{div}\mathbf{u}\operatorname{div}\mathbf{v}\,d\Omega &= -\int_\Omega \nabla(\operatorname{div}\mathbf{u})\cdot\mathbf{v}\,d\Omega + \int_{\partial\Omega} \operatorname{div}\mathbf{u}\,(\mathbf{v}\cdot\mathbf{n})\,d\Gamma \\ &= -\int_\Omega \left((\partial_1^2 u_1 + \partial_1\partial_2 u_2)v_1 + (\partial_2\partial_1 u_1 + \partial_2^2 u_2)v_2\right)d\Omega + \int_{\partial\Omega} \operatorname{div}\mathbf{u}(\mathbf{v}\cdot\mathbf{n})d\Gamma\end{aligned}$$

If we take the following formula into account:

$$\int_\Omega \operatorname{curl}\mathbf{u}\operatorname{curl}\mathbf{v}\,d\Omega = \int_\Omega \left((\partial_1\partial_2 u_2 - \partial_2^2 u_1)v_1 + (\partial_1\partial_2 u_1 - \partial_1^2 u_2)v_2\right)d\Omega$$

then we get:

$$\begin{aligned}&\int_\Omega (\operatorname{curl}\mathbf{u}\operatorname{curl}\mathbf{v} + \operatorname{div}\mathbf{u}\operatorname{div}\mathbf{v})\, d\Omega \\ &= -\int_\Omega ((\partial_1^2 u_1 + \partial_2^2 u_1)v_1 + (\partial_1^2 u_2 + \partial_2^2 u_2)v_2)\, d\Omega + \int_{\partial\Omega} \operatorname{div}\mathbf{u}(\mathbf{v}\cdot\mathbf{n})d\Gamma \\ &= -\int_\Omega ((\Delta u_1)\, v_1 + (\Delta u_2)\, v_2)\, d\Omega + \int_{\partial\Omega} \operatorname{div}\mathbf{u}(\mathbf{v}\cdot\mathbf{n})\, d\Gamma\end{aligned}$$

Now, after integrating the second-order terms by parts, we arrive at the expression:

$$\begin{aligned}&\int_\Omega \operatorname{curl}\mathbf{u}\operatorname{curl}\mathbf{v}\, d\Omega + \int_\Omega \operatorname{div}\mathbf{u}\operatorname{div}\mathbf{v}\, d\Omega \\ &= \int_\Omega (\nabla u_1\cdot\nabla v_1 + \nabla u_2\cdot\nabla v_2)\, d\Omega - \int_{\partial\Omega}\sum_{i=1,2}\frac{\partial u_i}{\partial n} v_i d\Gamma + \int_{\partial\Omega} \operatorname{div}\mathbf{u}\,(\mathbf{v}\cdot\mathbf{n})\, d\Gamma \\ &= (\mathbf{u},\mathbf{v})_{\mathbf{V}} - \int_{\partial\Omega}\Big(\sum_{i=1}^{2}\frac{\partial u_i}{\partial n}\, v_i\Big)d\Gamma + \int_{\partial\Omega} \operatorname{div}\mathbf{u}\,(\mathbf{v}\cdot\mathbf{n})\, d\Gamma\end{aligned}$$

From Question (a), we know that $v_{1|C_x} = 0$ along the edges C_x parallel to Ox. Moreover, we have that $\mathbf{n} = \begin{pmatrix} 0 \\ \pm 1\end{pmatrix}$, and therefore that $\frac{\partial u_i}{\partial n}_{|C_x} = \pm\partial_2 u_i$, for $i = 1.2$. Finally, by definition, since $0 = \mathbf{u}\cdot\mathbf{t}_{|C_x} = \pm u_1$, we (formally) obtain $\operatorname{div}\mathbf{u}_{|C_x} = (\partial_1 u_1 + \partial_2 u_2)_{|C_x} = \partial_2 u_{2|C_x}$. In calculating the integral over C_x, we find:

$$\int_{C_x} \operatorname{div}\mathbf{u}\,(\mathbf{v.n})\, d\Gamma - \int_{C_x}\Big(\sum_{i=1}^{2}\frac{\partial u_i}{\partial n}\, v_i\Big)d\Gamma = \int_{C_x} (\partial_2 u_2(\pm v_2) - (\pm\partial_2 u_2)v_2)\, d\Gamma = 0.$$

We proceed in the same way for the edges parallel to Oy. Finally, we have proven that:

$$\int_\Omega (\operatorname{curl}\mathbf{u}\operatorname{curl}\mathbf{v} + \operatorname{div}\mathbf{u}\operatorname{div}\mathbf{v})\, d\Omega = (\mathbf{u},\mathbf{v})_{\mathbf{V}}\,.$$

f) To show that the variational formulation:

$$\left|\begin{array}{l}\text{Find } \mathbf{E}\in\mathbf{V} \text{ such that} \\ (\mathbf{E},\mathbf{v})_{\mathbf{V}} = \displaystyle\int_\Omega f \operatorname{curl}\mathbf{v}\, d\Omega \qquad \forall\mathbf{v}\in\mathbf{V}\end{array}\right.$$

is well posed, we will check that the assumptions of the Lax–Milgram theorem are satisfied. Let us introduce the following semi-norm of $(H^1(\Omega))^2$, denoted by $|\cdot|_1$: $\mathbf{v} \mapsto (|v_1|^2_{1,\Omega} + |v_2|^2_{1,\Omega})^{1/2}$.

- $\mathbf{V}$ is a closed vector subspace of $(H^1(\Omega))^2$ by the continuity of the tangential trace; indeed, $\mathbf{V}$ is the preimage of $\{0\}$ (a closed subset of $L^2(\partial\Omega)$) under the continuous mapping:

$$\left| \begin{array}{ll} (H^1(\Omega))^2 & \to L^2(\partial\Omega) \\ \mathbf{v} & \mapsto \mathbf{v}\cdot\mathbf{t}_{|\partial\Omega} \end{array} \right. .$$

$\mathbf{V}$ is therefore a Hilbert space.

- Observe that:

$$(\mathbf{v},\mathbf{v})_{\mathbf{V}} = \sum_{i,j=1}^{2} \int_\Omega (\partial_i v_j)^2 \, d\Omega = |v_1|^2_{1,\Omega} + |v_2|^2_{1,\Omega} = |\mathbf{v}|_1^2 \qquad \forall \mathbf{v} \in \mathbf{V}.$$

And furthermore, after using Weber's inequality for $\mathbf{v} \in \mathbf{V}$, we obtain the following arrangement:

$$\begin{aligned} |\mathbf{v}|_1^2 \leq \|\mathbf{v}\|^2_{(H^1(\Omega))^2} &= \|v_1\|^2_{H^1(\Omega)} + \|v_2\|^2_{H^1(\Omega)} \\ &= \|v_1\|^2_{L^2(\Omega)} + |v_1|^2_{1,\Omega} + \|v_2\|^2_{L^2(\Omega)} + |v_2|^2_{1,\Omega}\,. \\ \text{(Weber's inequality)} &\leq C_W\,(\mathbf{v},\mathbf{v})_{\mathbf{V}} + |\mathbf{v}|_1^2 = (C_W+1)\,|\mathbf{v}|_1^2 \end{aligned}$$

We can therefore endow $\mathbf{V}$ with the $|\cdot|_1$ *norm*, which we assume from now on.

- $a(\mathbf{u},\mathbf{v}) = (\mathbf{u},\mathbf{v})_{\mathbf{V}}$ is bilinear, continuous and coercive from previous discussions.

- $\ell(\mathbf{v}) = \int_\Omega f \,\mathrm{curl}\,\mathbf{v}\, d\Omega$ is a continuous linear form, since:

$$|\ell(\mathbf{v})| \leq \|f\|_{L^2(\Omega)} \, \|\mathrm{curl}\,\mathbf{v}\|_{L^2(\Omega)} \leq \sqrt{2}\, \|f\|_{L^2(\Omega)}\, |\mathbf{v}|_1$$

by bounding the first derivatives in curl $\mathbf{v}$ (see solution 36, Question (a)).

g) To construct an internal approximation of the space $\mathbf{V}$, we must take the essential conditions into account which, according to Question (a), are the same as having $v_1 = 0$ along the edges parallel to Ox, and $v_2 = 0$ along the edges parallel to Oy. Let us introduce the following indexing sets: $\mathcal{I}_1$ is the indexing set for the vertices of the triangulation which do not belong to any edge of $\partial\Omega$ parallel to Ox, and $\mathcal{I}_2$ is the indexing set for the vertices of the triangulation, which do not belong to any edge of $\partial\Omega$ parallel to Oy. We then have:

$$\mathbf{V}_h = \text{vect}(\mathbf{w}_1^i)_{i\in\mathcal{I}_1} \oplus \text{vect}(\mathbf{w}_2^j)_{j\in\mathcal{I}_2} \subset \mathbf{V}.$$

In particular, the *two basis functions* corresponding to each corner of the square are excluded.

h) The discrete variational formulation is subsequently written as:

$$\left| \begin{array}{l} \text{Find } \mathbf{E}_h \in \mathbf{V}_h \text{ such that} \\ (\mathbf{E}_h, \mathbf{v}_h)_{\mathbf{V}} = \displaystyle\int_\Omega f \,\text{curl}\, \mathbf{v}_h \, d\Omega \qquad \forall \mathbf{v}_h \in \mathbf{V}_h. \end{array} \right.$$

i) We write the decomposition of $\mathbf{E}_h$, relative to the basis of $\mathbf{V}_h$, by:

$$\mathbf{E}_h = \sum_{j\in\mathcal{I}_1} E_1^j \mathbf{w}_1^j + \sum_{j\in\mathcal{I}_2} E_2^j \mathbf{w}_2^j.$$

In choosing $\mathbf{v}_h = \mathbf{w}_1^i$ in the discrete formulation, for each $i \in \mathcal{I}_1$, we get:

$$\sum_{j\in\mathcal{I}_1} E_1^j \left(\mathbf{w}_1^j, \mathbf{w}_1^i\right)_{\mathbf{V}} + \sum_{j\in\mathcal{I}_2} E_2^j \left(\mathbf{w}_2^j, \mathbf{w}_1^i\right) = \int_\Omega f \,\text{curl}\, \mathbf{w}_1^i \, d\Omega.$$

That is, by setting $\mathbb{A}_{11}^{ij} = \left(\mathbf{w}_1^j, \mathbf{w}_1^i\right)_{\mathbf{V}}$, $\mathbb{A}_{12}^{ij} = \left(\mathbf{w}_2^j, \mathbf{w}_1^i\right)_{\mathbf{V}}$ and $F_1^i = \displaystyle\int_\Omega f \,\text{curl}\, \mathbf{w}_1^i \, d\Omega$, we have:

$$\sum_{j\in\mathcal{I}_1} \mathbb{A}_{11}^{ij} E_1^j + \sum_{j\in\mathcal{I}_2} \mathbb{A}_{12}^{ij} E_2^j = F_1^i \qquad \forall i \in \mathcal{I}_1.$$

Now, if we choose $\mathbf{v}_h = \mathbf{w}_2^i$ for each $i \in \mathcal{I}_2$, and if we set $\mathbb{A}_{21}^{ij} = \left(\mathbf{w}_1^j, \mathbf{w}_2^i\right)_{\mathbf{V}}$, $\mathbb{A}_{22}^{ij} = \left(\mathbf{w}_2^j, \mathbf{w}_2^i\right)_{\mathbf{V}}$, and $F_2^i = \int_\Omega f \,\text{curl}\, \mathbf{w}_2^i \, d\Omega$, then:

$$\sum_{j\in\mathcal{I}_1} \mathbb{A}_{21}^{ij} E_1^j + \sum_{j\in\mathcal{I}_2} \mathbb{A}_{22}^{ij} E_2^j = F_2^i \qquad \forall i \in \mathcal{I}_2.$$

We combine these equations into the form of a linear system of order card$\mathcal{I}_1$ + card$\mathcal{I}_2$:

$$\begin{pmatrix} \mathbb{A}_{11} & \mathbb{A}_{12} \\ \mathbb{A}_{21} & \mathbb{A}_{22} \end{pmatrix} \begin{pmatrix} \overrightarrow{E}_1 \\ \overrightarrow{E}_2 \end{pmatrix} = \begin{pmatrix} \overrightarrow{F}_1 \\ \overrightarrow{F}_2 \end{pmatrix}. \qquad [S37]$$

A priori, we have $\mathbb{A}_{11} \neq \mathbb{A}_{22}$, since $\mathbb{A}_{11}$ has order card $\mathcal{I}_1$, whereas $\mathbb{A}_{22}$ has order card $\mathcal{I}_2$.

j) By the definition of $(\cdot,\cdot)_{\mathbf{V}}$, we have:

$$\begin{aligned} (\mathbf{w}_1^k, \mathbf{w}_2^l)_{\mathbf{V}} &= \left(\begin{pmatrix} w_k \\ 0 \end{pmatrix}, \begin{pmatrix} 0 \\ w_l \end{pmatrix} \right)_{\mathbf{V}} \\ &= \sum_{i,j=1}^{2} \int_\Omega \frac{\partial}{\partial x_i} \left[\begin{pmatrix} w_k \\ 0 \end{pmatrix} \right]_j \frac{\partial}{\partial x_i} \left[\begin{pmatrix} 0 \\ w_l \end{pmatrix} \right]_j d\Omega = 0 \end{aligned}$$

since, in each of the four products, one term vanishes. Thus, we get $\mathbb{A}_{21} = \mathbb{A}_{12} = 0$. Moreover, as curl $\mathbf{w}_1^i = \partial_1[\mathbf{w}_1^i]_2 - \partial_2[\mathbf{w}_1^i]_1 = -\partial_2 w_i$ for each $i \in \mathcal{I}_1$, we have:

$$F_1^i = \int_\Omega f \operatorname{curl} \mathbf{w}_1^i \, d\Omega = -\int_\Omega f \, \partial_2 w_i \, d\Omega \qquad \forall i \in \mathcal{I}_1.$$

Similarly:

$$F_2^i = \int_\Omega f \, \partial_1 w_i \, d\Omega \qquad \forall i \in \mathcal{I}_2.$$

The task of finding the solution of the linear system [S37] therefore can be reduced to finding the solutions of the two independent linear systems:

$$\begin{cases} \mathbb{A}_{11} \overrightarrow{E}_1 = \overrightarrow{F}_1 \\ \mathbb{A}_{22} \overrightarrow{E}_2 = \overrightarrow{F}_2. \end{cases}$$

Appendix 3

Formulas

– *In* d *dimensions*, where u is a scalar field and $\mathbf{v} = \begin{pmatrix} v_1 \\ \vdots \\ v_{\mathrm{d}} \end{pmatrix}$ is a vector field:

$$\nabla = \begin{pmatrix} \frac{\partial}{\partial x_1} \\ \vdots \\ \frac{\partial}{\partial x_{\mathrm{d}}} \end{pmatrix}, \quad \nabla u = \begin{pmatrix} \frac{\partial u}{\partial x_1} \\ \vdots \\ \frac{\partial u}{\partial x_{\mathrm{d}}} \end{pmatrix}, \quad \nabla \mathbf{v} = \begin{pmatrix} \frac{\partial v_1}{\partial x_1} & \dots & \frac{\partial v_{\mathrm{d}}}{\partial x_1} \\ \vdots & \ddots & \vdots \\ \frac{\partial v_1}{\partial x_{\mathrm{d}}} & \dots & \frac{\partial v_{\mathrm{d}}}{\partial x_{\mathrm{d}}} \end{pmatrix}$$

$$\operatorname{div} \mathbf{v} = \nabla \cdot \mathbf{v} = \sum_{i=1,\mathrm{d}} \frac{\partial v_i}{\partial x_i}$$

$$\Delta u = \operatorname{div} \nabla u = \sum_{i=1,\mathrm{d}} \frac{\partial^2 u}{\partial x_i^2}, \quad \Delta \mathbf{v} = \begin{pmatrix} \Delta v_1 \\ \vdots \\ \Delta v_{\mathrm{d}} \end{pmatrix}.$$

– *In three dimensions*, where u is a scalar field and $\mathbf{v} = \begin{pmatrix} v_1 \\ v_2 \\ v_3 \end{pmatrix}$ is a vector field:

$$\mathbf{curl}\, \mathbf{v} = \nabla \times \mathbf{v} = \begin{pmatrix} \frac{\partial v_3}{\partial x_2} - \frac{\partial v_2}{\partial x_3} \\ \frac{\partial v_1}{\partial x_3} - \frac{\partial v_3}{\partial x_1} \\ \frac{\partial v_2}{\partial x_1} - \frac{\partial v_1}{\partial x_2} \end{pmatrix}, \qquad \begin{array}{l} \mathbf{curl}\, \nabla u = 0, \ \operatorname{div} \mathbf{curl}\, \mathbf{v} = 0 \\ -\Delta\, \mathbf{v} = \mathbf{curl}\, \mathbf{curl}\, \mathbf{v} - \nabla \operatorname{div} \mathbf{v}. \end{array}$$

– *In two dimensions*, where u is a scalar field and $\mathbf{v} = \begin{pmatrix} v_1 \\ v_2 \end{pmatrix}$ is a vector field:

$$\operatorname{curl} \mathbf{v} = \frac{\partial v_2}{\partial x_1} - \frac{\partial v_1}{\partial x_2}, \quad \mathbf{curl}\, u = \begin{pmatrix} \frac{\partial u}{\partial x_2} \\ -\frac{\partial u}{\partial x_1} \end{pmatrix} \qquad \begin{array}{l} \operatorname{curl} \nabla u = 0, \ \operatorname{div} \mathbf{curl}\, u = 0 \\ -\Delta\, u = \operatorname{curl} \mathbf{curl}\, u \end{array}.$$

References

Abraham, R., Marsden, J.E., Ratiu, T. (1988). *Manifolds, Tensor Analysis and Applications*. Springer Verlag, Berlin.

Adams, R.A. and Fournier, J.J.F. (2003). *Sobolev Spaces*. Academic Press, Cambridge.

Apel, T. (1999). *Anisotropic Finite Elements: Local Estimates and Applications*. B.G. Teubner, Leipzig.

Brenner, S.C. and Scott, L.R. (2008). *The Mathematical Theory of Finite Element Methods*. Springer, Berlin.

Brezis, H. (1983). *Analyse fonctionnelle. Théorie et applications*. Masson, Paris.

Brezis, H. (2011). *Functional Analysis, Sobolev Spaces and Partial Differential Equations*. Springer, Berlin.

Ciarlet, P.G. (1989). *Introduction to Numerical Linear Algebra and Optimisation*. Cambridge University Press, Cambridge.

Ciarlet, P.G. (2002). *Basic Error Estimates for Elliptic Problems*. SIAM, Philadelphia.

Cohen, G. and Perret, S. (2017). *Finite Element and Discontinuous Galerkin Methods for Transient Wave Equations*. Springer, Berlin.

Dautray, R. and Lions, J.L. (1990). *Mathematical Analysis and Numerical Methods for Science and Technology*. Springer, Berlin.

Després, B. (2010). *Lois de conservation eulériennes, lagrangiennes et méthodes numériques*. SMAI Springer, Berlin.

Ern, A. and Guermond, J.-L. (2004). *Theory and Practice of Finite Elements*. Springer, Berlin.

Gilbarg, D. and Trudinger, N.S. (1977). *Elliptic Partial Differential Equations of Second Order*. Springer Verlag, Berlin.

Girault, V. and Raviart, P.-A. (1986). *Finite Element Methods for Navier-Stokes Equations*. Springer Verlag, Berlin.

Grisvard, P. (1985). *Elliptic Problems in Nonsmooth Domains*. Pitman, Boston.

Grisvard, P. (1992). *Singularities in Boundary Value Problems*. Masson, Paris.

Lenoir, M. (1987). *Approximation par éléments finis des problèmes elliptiques*. ENSTA, Paris.

Lions, J.-L. and Magenes, E. (1973). *Non-homogeneous Boundary Value Problems and Applications*. Springer, Berlin.

McLean, W. (2000). *Strongly Elliptic Systems and Boundary Integral Equations*. Cambridge University Press, Cambdridge.

Meurant, G. (1999). *Computer Solution of Large Linear Systems*. Elsevier, Amsterdam.

Necas, J. (1967). *Les méthodes directes en théorie des équations elliptiques*. Masson, Paris.

Raviart, P.-A. and Thomas, J.-M. (1983). *Introduction à l'analyse numérique des équations aux dérivées partielles*. Masson, Paris.

Saad, Y. (2003). *Iterative Methods for Sparse Linear Systems*. SIAM, Philadelphia.

Shen, J., Tang, T., Wang, L.-L. (2011). *Spectral Methods. Algorithms, Analysis and Applications*. Springer, Berlin.

Solin, P., Segeth, K., Dolezel, I. (2004). *Higher-Order Finite Element Methods*. Chapman & Hall, London.

Verfürth, R. (2013). *A Posteriori Error Estimation Techniques for Finite Element Methods*. Oxford University Press, Oxford.

Yosida, K. (1980). *Functional Analysis*. Springer, Berlin.

Index

D

E

F

G, H

I

S

T

U, V

Other titles from

ISTE

in

Numerical Methods in Engineering

2022

BOUCLIER Robin, HIRSCHLER Thibault
IGA: Non-conforming Coupling and Shape Optimization of Complex Multipatch Structures (Isogeometric Analysis Tools for Optimization Applications in Structural Mechanics Set – Volume 1)

DERVIEUX Alain, ALAUZET Frédéric, LOSEILLE Adrien, KOOBUS Bruno
Mesh Adaptation for Computational Fluid Dynamics 1: Continuous Riemannian Metrics and Feature-based Adaptation
Mesh Adaptation for Computational Fluid Dynamics 2: Unsteady and Goal-oriented Adaptation

GRANGE Stéphane, SALCIARINI Diana
Deterministic Numerical Modeling of Soil–Structure Interaction

2021

GENTIL Christian, GOUATY Gilles, SOKOLOV Dmitry
Geometric Modeling of Fractal Forms for CAD (Geometric Modeling and Applications Set – Volume 5)

2020

GEORGE Paul Louis, ALAUZET Frédéric, LOSEILLE Adrien, MARÉCHAL Loïc
Meshing, Geometric Modeling and Numerical Simulation 3: Storage, Visualization and In Memory Strategies
(Geometric Modeling and Applications Set – Volume 4)

SIGRIST Jean-François
Numerical Simulation, An Art of Prediction 2: Examples

2019

DA Daicong
Topology Optimization Design of Heterogeneous Materials and Structures

GEORGE Paul Louis, BOROUCHAKI Houman, ALAUZET Frédéric, LAUG Patrick, LOSEILLE Adrien, MARÉCHAL Loïc
Meshing, Geometric Modeling and Numerical Simulation 2: Metrics, Meshes and Mesh Adaptation
(Geometric Modeling and Applications Set – Volume 2)

MARI Jean-Luc, HÉTROY-WHEELER Franck, SUBSOL Gérard
Geometric and Topological Mesh Feature Extraction for 3D Shape Analysis
(Geometric Modeling and Applications Set – Volume 3)

SIGRIST Jean-François
Numerical Simulation, An Art of Prediction 1: Theory

2017

BOROUCHAKI Houman, GEORGE Paul Louis
Meshing, Geometric Modeling and Numerical Simulation 1: Form Functions, Triangulations and Geometric Modeling
(Geometric Modeling and Applications Set – Volume 1)

2016

KERN Michel
Numerical Methods for Inverse Problems

ZHANG Weihong, WAN Min
Milling Simulation: Metal Milling Mechanics, Dynamics and Clamping Principles

2015

ANDRÉ Damien, CHARLES Jean-Luc, IORDANOFF Ivan
3D Discrete Element Workbench for Highly Dynamic Thermo-mechanical Analysis
(Discrete Element Model and Simulation of Continuous Materials Behavior Set – Volume 3)

JEBAHI Mohamed, ANDRÉ Damien, TERREROS Inigo, IORDANOFF Ivan
Discrete Element Method to Model 3D Continuous Materials
(Discrete Element Model and Simulation of Continuous Materials Behavior Set – Volume 1)

JEBAHI Mohamed, DAU Frédéric, CHARLES Jean-Luc, IORDANOFF Ivan
Discrete-continuum Coupling Method to Simulate Highly Dynamic Multi-scale Problems: Simulation of Laser-induced Damage in Silica Glass
(Discrete Element Model and Simulation of Continuous Materials Behavior Set – Volume 2)

SOUZA DE CURSI Eduardo
Variational Methods for Engineers with Matlab®

2014

BECKERS Benoit, BECKERS Pierre
Reconciliation of Geometry and Perception in Radiation Physics

BERGHEAU Jean-Michel
Thermomechanical Industrial Processes: Modeling and Numerical Simulation

BONNEAU Dominique, FATU Aurelian, SOUCHET Dominique
Hydrodynamic Bearings – Volume 1
Mixed Lubrication in Hydrodynamic Bearings – Volume 2
Thermo-hydrodynamic Lubrication in Hydrodynamic Bearings – Volume 3
Internal Combustion Engine Bearings Lubrication in Hydrodynamic Bearings – Volume 4

DESCAMPS Benoît
Computational Design of Lightweight Structures: Form Finding and Optimization

2013

YASTREBOV Vladislav A.
Numerical Methods in Contact Mechanics

2012

DHATT Gouri, LEFRANÇOIS Emmanuel, TOUZOT Gilbert
Finite Element Method

SAGUET Pierre
Numerical Analysis in Electromagnetics

SAANOUNI Khemais
Damage Mechanics in Metal Forming: Advanced Modeling and Numerical Simulation

2011

CHINESTA Francisco, CESCOTTO Serge, CUETO Elias, LORONG Philippe
Natural Element Method for the Simulation of Structures and Processes

DAVIM Paulo J.
Finite Element Method in Manufacturing Processes

POMMIER Sylvie, GRAVOUIL Anthony, MOËS Nicolas, COMBESCURE Alain
Extended Finite Element Method for Crack Propagation

2010

SOUZA DE CURSI Eduardo, SAMPAIO Rubens
Modeling and Convexity

2008

BERGHEAU Jean-Michel, FORTUNIER Roland
Finite Element Simulation of Heat Transfer

EYMARD Robert
Finite Volumes for Complex Applications V: Problems and Perspectives

FREY Pascal, GEORGE Paul Louis
Mesh Generation: Application to finite elements – 2nd edition

GAY Daniel, GAMBELIN Jacques
Modeling and Dimensioning of Structures

MEUNIER Gérard
The Finite Element Method for Electromagnetic Modeling

2005

BENKHALDOUN Fayssal, OUAZAR Driss, RAGHAY Said
Finite Volumes for Complex Applications IV: Problems and Perspectives

Printed and bound by CPI Group (UK) Ltd, Croydon, CR0 4YY
13/09/2026
14961041-0006